BIOLOGY OF THE MAMMAL

Biology of the Mammal

by

P. CATHERINE CLEGG, B.Sc., Ph.D., M.B.Ch.B.
*Sometime Lecturer in Physiology, University of
Sheffield*

and

ARTHUR G. CLEGG, B.Sc., F.I. Biol.
*Sometime Principal Lecturer in Biology, City College
of Education, Sheffield*

FOURTH EDITION

HEINEMANN
LONDON

First Published 1962
Second Edition 1963
Reprinted 1965
Reprinted 1966
Reprinted 1967
Reprinted 1968
Third Edition 1970
Reprinted 1971
Fourth Edition 1975
Reprinted 1978
Reprinted 1979
Reprinted 1982
Second reprint 1982

ISBN 0 433 06122 7

Published by WILLIAM HEINEMANN MEDICAL BOOKS LTD

Printed and bound in Hong Kong by Wing King Tong Co Ltd

CONTENTS

PART I

THE PHYSICOCHEMICAL BACKGROUND OF LIFE

PART II

DIFFERENTIATION AND INTEGRATION

PART III

ORGAN SYSTEMS

To Our Parents

PREFACE

This book was conceived several years ago when the authors were teaching 'A' and 'S' level candidates in Biology and Zoology. It was found then that there was no suitable textbook of mammalian physiology at this level. The books available were of three main kinds: elementary introductions at 'O' level, medical textbooks dealing almost entirely with human physiology and its derangements and works comparing the physiology of a wide range of mainly invertebrate types. This book attempts to satisfy the needs of the student who is beginning a serious study of the physiology of the mammal.

Although this book was written with the needs of 'A' and 'S' level students in mind, we feel that it will continue to be useful to them when they proceed to studies far beyond 'S' level, including first year University courses, Teacher Training and Technical College courses in Biology, Zoology, Physiology, Pharmacy, Medicine and Dentistry.

Whilst the contents of this book are physiological in spirit, much information from other disciplines is incorporated including morphology, histology, biochemistry and medicine. References to wider issues e.g. evolution and adaptation to environment are included in order to give biological perspective. For these reasons we have called this book Biology of the Mammal rather than Physiology of the Mammal.

We wish to thank Professor J. Z. Young for his kindness in allowing us to use figs. 105 and 131 (taken from the Life of Mammals, O.U.P.) and to our publishers for supplying fig. 141. All the remaining figures have been drawn by the authors in an attempt to attain a close unity between the text and illustrations. We are grateful to the Air Ministry for providing information about the regulations for use of oxygen in aircraft. Dr. E. T. B. Francis, Reader in Vertebrate Zoology, University of Sheffield, has always been ready to give up time to discuss anatomical points and to make available specimens for drawing. Above all we are grateful for his friendly advice. It is a pleasure to record our gratitude to our publishers and in particular Mr. Owen R. Evans for his technical advice and encouragement.

January 1962

P.C.C.
A.G.C.

PREFACE TO THE THIRD EDITION

Each year it becomes increasingly possible to explain more and more of life processes in physico-chemical terms. In this new edition we have tried to cover some major advances which have been made since the publication of our last edition. We have thus explored in more detail the process of protein synthesis and its regulation, the basis of cellular differentiation, the mode of action of hormones etc.

Scientific ideas are often best taught and remembered when related to human life. With this in mind we have included sections on human chromosomes and their abnormalities, cancer and chromosomes and the applied physiology of diets and exercise.

Particular new inclusions are listed below.

1. An account of the electron microscope.
2. A discussion of the significance of mitosis.
3. Some human genetics, sex chromosomes, sex-linked inheritance, abnormal chromosomes in man, chromosomes and cancer.
4. An account of the primary, secondary, tertiary and quaternary structure of proteins which includes a new section on the technique of chromatography related to the determination of amino-acid sequence in a polypeptide.
5. Nucleic acids, the genetic code, protein synthesis, the basis of cellular differentiation, factors involved in the regulation of gene activity.
6. The basis of the specificity of enzyme-substrate interaction.
7. The mode of action of hormones.
8. Constituents of the diet, dieting, trace elements, regulation of water balance.
9. The regulation of the secretion of ADH and aldosterone.
10. A new chapter on tissue respiration.
11. A new chapter on energy exchange in the organism and its regulation, the techniques of calorimetry, B.M.R., energy expenditure of various activities and the regulation of food intake and body temperature.

We would like to thank all those who have suggested where more careful explanation was needed in previous editions. We hope that the new additions, the more abundant illustrations and the introduction of colour and photographs will help to ease the student's task and perhaps afford some enjoyment.

We are specially indebted to Mr. Richard Emery of William Heinemann Medical Books Ltd for his assistance in the preparation of this edition.

Brighton, August 1969
A.G.C. & P.C.C.

PREFACE TO THE FOURTH EDITION

New material has been added and changes made, mainly in chapters 1 and 14. In the first chapter the section on human genetics has been expanded and includes a discussion of the means of prevention of genetically determined disease. Chapter 14 is entirely rewritten to take into account current modern concepts in immunology; there is a discussion of the functions of lymphocytes and the thymus gland and there is a historical account of the development of vaccines. In revising these two chapters we have leaned heavily upon script and illustrations from our 'Man Against Disease' and we are grateful to Heinemann Educational books for allowing us to do so.

We have been helped considerably by many teachers of biology who have written to us, often in great detail, about errors and omissions in the last edition and accordingly we have added a section on colour vision in chapter 7 and made various corrections, particularly in the accounts of the genetic code and the optical activity of sugars.

Barrowford, 1975
A.G.C. & P.C.C.

PART I

THE PHYSICOCHEMICAL BACKGROUND
OF LIFE

Chapter One

The Animal Cell

PROTOPLASM

The basic unit of structure of mammals, and of all living things except the most primitive in organization is the cell. In 1835 Dujardin observed emerging from squashed cells, a sticky substance to which he gave the name sarcode. Purkinje (1840) gave the name protoplasm to this substance, which later came to be regarded as 'the physical basis of

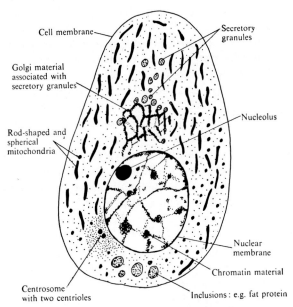

Cell membrane

Secretory granules

Golgi material associated with secretory granules

Rod-shaped and spherical mitochondria

Nucleolus

Centrosome with two centrioles

Nuclear membrane

Chromatin material

Inclusions: e.g. fat protein

Fig. 1. *Diagram of a basic animal cell.*

life', a substance common to all living things. Protoplasm was originally regarded as a homogeneous fluid or jelly, because it appeared structureless under microscopes then available. However in 1833 Robert Brown noticed an area in the cells of orchids which was more opaque than the rest of the protoplasm of the cell and he called this the areola or nucleus of the cell. The term cytoplasm was later introduced to describe the protoplasm of the cell contents other than the nucleus.

The concept of protoplasm as a simple homogeneous substance disappeared when improvements in microscopy and analysis of cell contents showed that protoplasm consists of a complex substance containing particles of a vast range of size, from ions to complex molecules, and containing larger structures including mitochondria, Golgi bodies, centrosome, lysosomes, inclusion bodies of glycogen, fat, pigments, excretory products, and vacuoles of various sizes. Fig. 1 shows in diagrammatic form the contents of a 'basic' animal cell that can be detected by techniques using the light microscope. The cell is seen to be bounded by an apparently structureless cell membrane in which is contained the cytoplasm with its various inclusions, and the nucleus. The nucleus is itself bounded by a nuclear membrane and contains nucleoplasm or nuclear sap, containing one or more nucleoli and chromatin material. The above is a description of a basic type of cell and none but the most undifferentiated type of cell resembles this. In mammals, as in other animals and plants, the vast numbers of cells which comprise the individual are differentiated into populations of special types of cells, each designed to carry out particular functions. Thus the cells of muscle in addition to the above features of the 'basic' cell have special components related to their function of contraction (p. 172).

Using the information of the structure of cells obtained from studies with the light microscope it was still possible to regard protoplasm as a relatively undifferentiated medium in which were suspended particles of various sizes and types. This fluid medium when investigated was found to have many properties which could be explained by assuming that the fluid part of protoplasm is a colloidal solution.

Colloids

Graham is regarded as the founder of colloid chemistry on account of the work he did from 1851–61 on the rates of movement of molecules from a region of high concentration to one of lower concentration. Such a movement is called diffusion. Graham classified the substances he studied as fast diffusers or slow diffusers. He found that substances like salt and sugar diffused rapidly and crystallized easily, so he called this group crystalloids. The slow diffusers he called colloids (kolla in Greek means glue) examples of which were glue, gelatin and starch.

EXPERIMENTAL DEMONSTRATION OF THE DIFFERENT RATES OF DIF-FUSION OF CRYSTALLOIDS AND COLLOIDS. It is difficult to measure the rate of diffusion of a substance in solution because other physical factors like convection distribute the solute through the solvent. Spread by convection is more rapid than by diffusion and therefore the effect of diffusion is masked. The substances whose diffusion rates are to be compared are therefore dissolved in gelatin. Dissolve a little gelatin in warm water. Pour about 3 mls. of the solution into a test tube, add three drops of potassium dichromate. Shake the solution and allow it to cool.

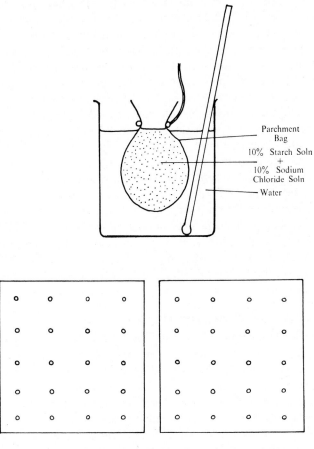

Parchment Bag

10% Starch Soln + 10% Sodium Chloride Soln

Water

Fig. 2. *The experimental investigation of the effect of a membrane on the diffusion of crystalloids and colloids.*

When the gelatin has solidified add 2 mls. of water. Repeat the procedure using congo red instead of the dichromate. In each case note the time taken for the colour to diffuse from the gelatin into the water. The crystalloid dichromate diffuses more rapidly than the colloid congo red.

EXPERIMENTAL INVESTIGATION OF THE EFFECT OF A MEMBRANE ON THE DIFFUSION OF CRYSTALLOIDS AND COLLOIDS. (See fig. 2.) Soak a piece of parchment about a foot square in water until it becomes pliable, then fold it into the shape of a flask. Make up 25 mls. of 10% starch solution and 25 mls. of 10% sodium chloride solution. Half fill a beaker with distilled water and suspend the parchment flask above the water. Do not put the 10% solutions into the parchment flask yet. Take two clean tiles and using clean glass rods make 20 spots of iodine solution on one tile and 20 spots of silver nitrate on the other. Now pour the two solutions of starch and salt into the parchment flask and lower it into the water. Take a little of the fluid out of the beaker every 20 seconds dropping it onto the spots which you have already prepared on the tiles. If the starch can penetrate the parchment the iodine spots will turn blue, if the chloride ion from the salt penetrates the parchment the silver nitrate solution will turn milky as it is converted to a white insoluble precipitate of silver chloride. (It is easier to see this precipitate if the silver chloride is spotted out onto a black tile.)

Results. The iodine spots remain yellow whilst the silver nitrate goes milky after about the twelfth spot. This is true even if the fluid in the beaker is tested after an hour.

Conclusion. Parchment prevents the diffusion of the colloid starch but allows the crystalloid salt to pass through it. (The passage of the sodium ion can be demonstrated.)

The separation of crystalloid and colloid by means of a membrane is called dialysis.

A brief outline of the properties of colloids

1. Colloids are substances that diffuse very slowly in solution.

2. They will not pass through the pores in a parchment membrane. Both these facts are the result of the size of the particles dispersed in the liquid. Colloids are peculiar kinds of solution in which there are always two phases, a continuous one and a dispersed one. In the case of starch solution the disperse phase is starch and the continuous one is water.

3. The characteristic feature of a colloidal solution is the molecular size of the dispersed material. In true solutions the disperse phase is ionic or molecular in size but in a colloid the dispersed particle is multimolecular (or micellar) in size. The particles are not usually big enough to be observed with a hand lens. Using the light of the visible spectrum the form of things less than $0.3\ \mu$ cannot be seen clearly. The size of the colloidal particles has to be measured in $m\mu$ ($1\ m\mu = 10^{-7}$ cm.) and is usually between 1 and 100 $m\mu$. Almost any substance can be prepared in colloidal form, provided that aggregates of the correct size are dispersed in a continuous phase. Colloidal solutions of silver and gold were known to Graham. In chemical analysis colloidal

solutions are often formed inadvertently when solutions are precipitated under certain physical conditions, e.g. barium sulphate precipitated in the cold. It is then impossible to remove all the barium sulphate by filtering, as particles less than 500 mμ pass through the filter paper and several of the particles are of 100 mμ size. One is therefore advised to precipitate barium sulphate in a warm solution to facilitate its subsequent removal by filtration. The point to grasp about the size of these particles is that they are too small to precipitate and yet too big to dissolve.

4. Colloids are classified depending on the nature of the two phases.

A solid dispersed in a liquid	.	.	.	.	sol
liquid	liquid	.	.	.	.	emulsion
gas	liquid	.	.	.	.	foam
solid	gas	.	.	.	.	smoke
liquid	gas	.	.	.	.	fog

All the above mentioned colloids are collectively called sols and may be represented diagrammatically as in fig. 3.

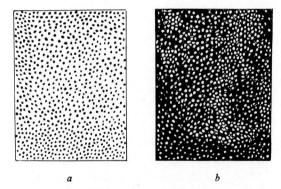

a   b

Fig. 3. *Diagram illustrating the difference in structure o, a sol and gel.* a. shows a sol e.g. starch powder or gelatin suspended in water. The black dots represent the discontinuous phase, in the above example particles of starch or gelatin, and the white areas represent the continuous phase, in the above example water. b. Shows a gel e.g. the result of heating starch and water. The starch, originally in the discontinuous phase is now in the continuous phase and the water is restricted to the discontinuous phase.

There is a whole group of colloids known as gels typified by a solution of gelatin in water or by a jelly. Here the colloid appears as a semi-solid. Its structure is that shown in fig. 3, where the gelatin is the continuous phase and the water the disperse phase. It is often possible to transform sols into gels and vice versa. When starch is placed in cold water it forms a sol but when the sol is heated the whole solution turns into a gel. The sol, white of egg, turns irreversibly to a gel on the addition of alcohol or on heating. If some living Amoebae, small protozoa one-hundredth of an inch long, are observed through a microscope the effects of this process of sol–gel transformation can be watched. At the posterior end gel is turned to sol and then the outer gel contracts forcing the sol forwards. At the front end the sol turns to gel once again. So Amoeba moves forward. (See fig. 4.)

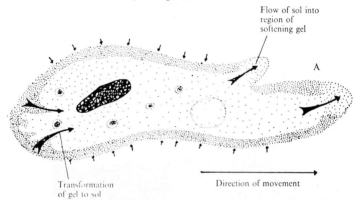

Fig. 4. *Diagram illustrating movement in an amoeba.* The outer densely shaded region represents protoplasm in the gel form, the inner lighter region protoplasm in the sol form. The contracting outer coat of gel protoplasm forces the fluid sol forward into the softening gel of the moving end A. At the hind end there is a transformation of protoplasm from the gel to the sol form, so replacing the sol which has been displaced.

5. A very important feature of colloidal sols is that they bear electrical charges on their surfaces. The charge may be negative as in colloidal clay, silver, arsenic sulphide and many acid dyes, or positive as in ferric hydroxide, haemoglobin, aluminium hydroxide and basic dyes. The important thing is that all these charges are the same, in a given colloid and therefore repel each other. This keeps the particles spatially separate. It is well known that colloidal particles 'flocculate' or clog together in the presence of electrolytes. The electrolytes supply charged ions which remove the effect of the charges on the colloidal particles, which are then precipitated. This can easily be demonstrated using suspensions of silt from a river bed. Put equal amounts of silt into three gas jars and after adding equal amounts of distilled water to all three shake them thoroughly. Leave one as a control, to the second add a few drops of alkali and to the third a few drops of acid. You should try this for yourself. It is noticeable in the streams of the Derbyshire dales that the water

is very clear, and in the slower reaches the bottom of the streams is very muddy. This is because the clay particles picked up by the headwaters are flocculated by the calcium ions from the limestone and precipitated as soon as the speed of the river falls. Colloidal sols are very unstable then and if a permanent sol is required, as in making up creams containing solids in pharmacy where it is essential that the valuable ingredient does not settle out onto the bottom of the bottle, or as when mixing ice into the cream in the icecream trade, it is necessary to protect the sols. The dispersed particles are stabilized by surrounding them with gelatin. This stabilization of sols by organic materials can be a nuisance in qualitative analysis in chemistry. If all organic material is not removed at the beginning, the sulphur in colloidal form is often stabilized after passing in hydrogen sulphide in group two. The addition of ammonium chloride in the next group does not precipitate the colloid if it is protected by an organic sol. Thus substances may escape detection.

6. Dyes are often easily adsorbed onto the surface of colloidal particles. This can be demonstrated by pouring a solution of eosin in water onto sand and clay in separate vessels. The clay has particles of colloidal dimensions but the sand has larger particles. The clay removes the colour from the eosin solution by adsorbing the dye onto the surface of the colloidal particles. The sand has no such effect. Water may be taken up by colloids. This is well illustrated by the way that seeds take up their own weight of water to hydrate their colloids when germination starts.

7. The suspended particles of a colloidal sol show Brownian movement. If Indian ink is observed through the microscope the small black particles are seen to be continuously vibrating. They are so small that each time a molecule of water hits an ink particle the ink particle moves just as a football would move if tennis balls were constantly being hurled against it. The water molecules are too small to be seen but we can see their effect on the sol particles, if they are big enough to see. This constant jiggling motion is called Brownian movement. When seen under the ultra microscope the effects of the small suspended particles can be seen. The ultra microscope enables the particles to show their presence by each one producing an area of halation as light is reflected from them. This is why dust shows up in a sunbeam.

8. Finally it must be emphasized that colloids have a tremendous surface area for their mass, because of their very fine state of division. Now that we know a few facts about these peculiar solutions called colloids we can look at protoplasm once more and discuss the evidence that many of its properties are understandable if we consider that it is a colloid.

Evidence for the colloidal nature of protoplasm

(a) Life can only exist within a very restricted temperature range. Although there are exceptions to be found in bacteria that live in the very hot temperatures of hotwater springs and geysers at 65°C, most living things live at temperatures between 4°C and 45°C. Temperatures above this range usually coagulate the protoplasm irreversibly. This action can be compared to the action of heat on the colloidal solution of egg albumen in the white of a bird's egg. Protoplasm is also coagulated by chemicals like formaldehyde and alcohol. This is called fixation and is the first process to be carried out in the preparation of material to be sectioned for histological examination. The object is to coagulate the protoplasm in the position that most nearly corresponds to its structure in the living state.

(b) Protoplasm is pH sensitive. The pH values encountered in living things are between 4 and 9 although in vitro a much greater range is found from negative numbers to +14. The protoplasm of animal cells is usually found to be remarkably constant at just slightly acid conditions (pH 6·9). Even slight changes in pH markedly affect the chemical reactions occurring in the protoplasm. (See Bohr effect.) We have seen that colloids are precipitated by the addition of acids and this sensitivity to the presence of any charged ions from either acids or alkalies or salts is almost certainly related to the effect that these charged ions have on the surface charges of the colloids.

(c) We have seen that colloids may be present in the form of sols or gels and that these can be converted one into the other in some cases, as when Amoeba and some white blood cells move. The viscosity of protoplasm is commonly about the same as olive oil. Measurement of the viscosity of protoplasm is very difficult as the sol may turn to a gel as soon as instruments are pushed into it. Evidence for it being a liquid rather than a solid is to be found in the following facts.

  i. Liquids injected into protoplasm assume the form of droplets.
  ii. Tiny granules in protoplasm show Brownian movement.
  iii. Protoplasm is seen to move in certain cells, e.g. streaming in cells of the Canadian pond-weed.

Determinations of the viscosity by measurements on Brownian movement or rate of movement of particles in protoplasm under the force of gravity show the viscosity to be between two and ten times that of water.

(d) When viewed by the ultra microscope protoplasm is shown to be heterogeneous even when it appears homogeneous under the ordinary microscope.

(e) Protoplasm displays many properties in common with colloidal solutions that have long fibrous molecules in them. If particles of nickel are pulled through a cell by a magnetic force they move through the protoplasm until the force is removed. When this occurs the nickel particles not only stop but even spring back a little as if they were encountering elastic fibres. Secondly protoplasm is thixotropic like the non drip paint one can buy. When it is in movement its viscosity is smaller than when it is static. This is explained by assuming that the long molecules become aligned during movement and then take up more irregular patterns when at rest.

(f) We have seen that colloids have charges on the surfaces of the particles. Some colloids (lyophilic) attract water to their surfaces. It is well known that when tissues are dehydrated they rapidly absorb water. This may be to satisfy the demands of the lyophilic colloids in the protoplasm. Electrical phenomena are well known in living things, nervous conduction depends on it and most biological membranes are thought to bear electrical charges. This could be explained on the assumption that there are charged colloidal particles in the protoplasm, although it may also be due to ions of various elements present in the protoplasm.

From a consideration of all this data taken together we see that many of the properties of protoplasm may be better understood when it is assumed that protoplasm is colloidal. When the diversity of all the chemistry of a single cell is considered we should bear in mind the tremendous area of the interface between the various phases in the colloid. This great area may be the factor that makes the wealth of chemical activity possible.

The composition of protoplasm from chemical analysis data. Now that the colloidal nature of protoplasm is established we may look at the various components present to see if they fit in with the general picture we have already constructed. Although various authorities give slightly different figures the general picture is as follows

substance	% by weight	relative number of molecules
water	80–90	18,000
protein	7–10	1
lipids	1–2	10
carbohydrate	1–1·5	20

By taking the molecular weight of the constituents into account we see that the molecular ratios are as shown above. Thus we have the picture of protoplasm as a watery fluid containing protein molecules each surrounded by ten lipid molecules and twenty carbohydrate ones.

This fits in well with our previous picture of a lyophilic colloidal sol containing long molecules (protein) suspended in it.

THE PROBLEMS OF THE MICROSCOPIC EXAMINATION OF PROTOPLASM. Since 1660 when Leeuwenhoek the Dutch draper started observing his little animalcules through small but high-powered hand lenses there have been great advances in microscopy. Robert Koch made a big contribution by designing the Abbé condenser. Refinement of detailed construction has gone on and on, and we have been able to see smaller and smaller things in greater and greater detail. There comes a limit however at which we can no longer see smaller things clear enough and further magnification only serves to magnify the blurring. The power to separate two adjacent objects is called the power of resolution. Resolution then is the power to see detail. The resolving is dependent upon the amount of light received by the instrument and the wavelength of the illumination. In order to get a lot of light we use a lens of very short focus in a microscope so that we can get very near to the object. The lenses usually found in the nosepiece of a microscope have focal lengths of $\frac{2}{3}$ inch, $\frac{1}{6}$ inch or $\frac{1}{12}$ inch. The shorter the wavelength of light used the greater is the resolution of the instrument. Many microscopes have in them a filter holder in which is placed a blue filter, which improves resolution slightly since blue light has the shortest wavelength of all the visible spectrum. Using blue light we can only see objects of diameter greater than $0.2\,\mu$ (approximately). Using ultraviolet light we can resolve objects of $0.1\,\mu$ diameter. Unfortunately we cannot see ultraviolet light nor will it pass through glass, therefore if we must use this source of illumination we have to resort to quartz lenses and photographic records. The electron microscope uses a beam of electrons of very short wavelength and is theoretically capable of resolving objects of diameter about $0.03°A$

$1°A = 1/10,000,000$ mm.

$1\,\mu = 1/1,000$ mm.

$1\,\mu = 10,000°A$

The electron beam can resolve objects down to $0.03/10,000\,\mu$ = $0.000003\,\mu$ or $0.000,000,003$ mm. diameter. Now this is smaller than atomic dimensions and therefore theoretically we ought to be able to investigate atoms using the electron microscope. So far we have only been able to see things bigger than $20°A$. However this is quite small enough for looking at biological molecules. $20°A$ is about the size of a molecule containing 300 atoms and we know that many biological molecules are bigger than this.

The electron microscope, in the arrangement of its working parts is similar in principle to the light microscope, figs. 5 and 6. The illumination source which generates the beam of electrons is produced as in a cathode-ray tube by a hot cathode. The beam of electrons is accelerated

Fig. 5. *Photograph of an electron microscope. (By courtesy of Dr. S. Bradbury.)*

towards an anode which is perforated by a hole so allowing a narrow beam of electrons to pass through it. Because of the vacuum in the electron microscope there is little chance of collision of electrons and the electrons travel in straight paths towards the section of tissue to be examined. These parallel beams of electrons can be focussed onto the object by means of a magnetic field generated by electromagnets.

These electromagnets serve the same function as the condenser of the light microscope. When the electron beam passes through the section of the material there is some scattering of electrons—mainly due to collision with electrons of the atoms of the specimen material. But the materials of which cells are composed—nucleic acids, proteins and

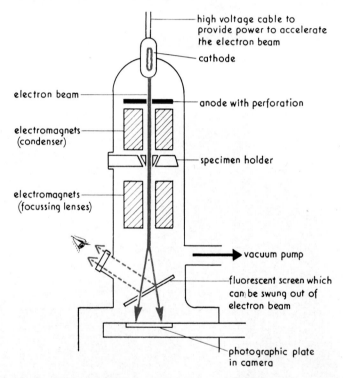

high voltage cable to provide power to accelerate the electron beam

cathode

electron beam

anode with perforation

electromagnets (condenser)

specimen holder

electromagnets (focussing lenses)

vacuum pump

fluorescent screen which can be swung out of electron beam

photographic plate in camera

Fig. 6. *Diagram showing the component parts of the electron miscroscope.*

carbohydrates—have very similar electron densities and thus deflect the electron beam to a similar extent. Thus there is little contrast between the components of the cell. Contrast can however be increased by the impregnation of the tissue section with heavy atoms of high electron density e.g. manganese or osmium. Since the more physically dense areas of the cell take up more of the electron dense material this procedure will provide contrast in the image; the electron beam will be deflected most in these regions of the tissue section. As the electron beam emerges from the specimen material it is focussed by another series of electromagnets, which form the equivalent of the lens system of the light microscope. The image can be viewed on a screen which fluoresces in the electron beam or it can be photographed.

From the point of view of resolution the electron microscope has far surpassed the optical instrument. Another advantage is that due to its construction it is never out of focus for the depth of focus of its optical system is greater than the thickness of the objects it can investigate (30 mμ). The advantage of always having the object in focus is obvious but there is the disadvantage that it is not possible to focus up and down and investigate the structure in three dimensions. Using the electron microscope everything appears to be on the same level. Electrons are very easily deflected by anything in their path and it is this property which is used in forming the image, since the dense parts of the object deflect electrons away most and therefore appear palest. If the electrons had to travel through air they would continually be being deflected and therefore the whole microscope has to have all the air removed from it by a vacuum pump. The electrons cannot be seen and the image is projected onto a fluorescent screen, or photographs are taken. Because the radiation is so intense only dead tissue can be examined. The sections have to be very thin for the electrons to get through at all and the usual thickness is 30 mμ. Here of course is another problem—special microtomes had to be designed to cut these very thin sections.

The ultra microscopic structure of protoplasm. As our knowledge of physiological processes occurring in cells has grown it has become apparent that a great number of different chemical reactions are proceeding simultaneously inside the same cell. Obviously the reactants and products of all these reactions do not mingle freely or there would be chemical chaos. We are just beginning to find out things about the finer structure of protoplasm. In 1945 workers using the electron microscope showed that there was a lace-like three dimensional reticulum in the cytoplasm of cells grown in tissue culture. The investigations were made on smears of whole cells and at high magnification they found that this three dimensional network had vesicles attached to it. (See fig. 7.) It was fortunate that the first work was done on whole cells or else it might have been very many years before anyone would have understood the three dimensional structure of the reticulum within the cytoplasm, for all the subsequent work has been done on sections of such thinness that they cut through the vesicles so that they are only seen in profile. The sections are about 30 mμ thick whereas the vesicles measure 100 mμ. The appearance of the vesicles in section obviously depends upon the level of the section. Thus different workers claim the vesicles to be of different shapes but now that many workers have had time to look at many thousands of cells some measure of agreement is emerging on the structure of what is called the endoplasmic reticulum. The cells of different tissues have different endoplasmic reticula and there is variety in the size, number, distribution and texture of the vesicles.

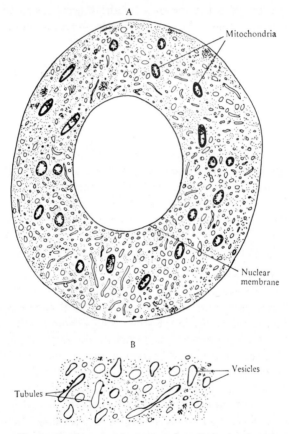

Fig. 7. **A.** *Diagram of a section of a cell showing the endo-plasmic reticulum* as it occurs in its simplest form (e.g. in spermatocytes), consisting of vesicles and short tubules arranged in strings and connected with one another in a random fashion to form the reticulum. In a section of the cell these vesicles and tubules are mainly seen in section and their interconnections are not apparent. B. Shows a portion of the section of the endoplasmic reticulum at greater magnification.

What is of interest to us here is that there is a very fine network in the cytoplasm which could be used to isolate the different chemical reactions.

It is no longer possible to regard protoplasm as a homogeneous structureless medium in which are suspended various bodies. Protoplasm has an organized structure of its own, a complex system of membranes on which may be localized biochemical events essential for the life of the cell. Thus the granules present on the endoplasmic reticulum are very rich in ribose nucleic acid (p. 121) and are active in protein synthesis. These granules are called microsomes, or ribosomes (figs. 8, 9). Another possible function of the endoplasmic reticulum is that by invaginations from the surface membrane the reticulum is connected to the external environment of the cell and forms a vast surface area for exchange of substances with the fluid bathing the cells. In muscle cells the endoplasmic reticulum is probably concerned in the inward conduction of the electrical impulse of excitation from the surface membrane of the contractile elements (p. 614).

The cell is not a blob of structureless protoplasm in which are suspended the various bodies which are necessary for the continued life of the cell. All the components of the cell form an integrated unit, each component having its own special role to play in the life of the cell. We can now proceed to consider in more detail the structure and function of the various cell components, which are listed on page 18.

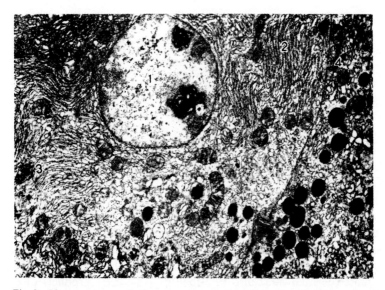

Fig. 8. *Electron micrograph of a pancreatic exocrine cell showing a low-power view of general cell structure.* 1. nucleus. 2. endoplasmic reticulum bearing ribosomes. 3. mitochondrion. 4. plasma membrane. (*By courtesy of Dr. S. Bradbury.*)

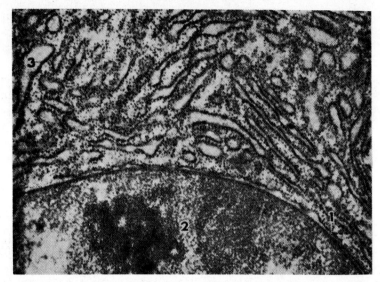

Fig. 9. *Electron micrograph of a pancreatic exocrine cell.* High power view showing part of the nucleus, nuclear membrane and endoplasmic reticulum bearing ribosomes (microsomes). 1. Perinuclear cisterna. 2. nucleus. 3. membrane of endoplasmic reticulum bearing ribosomes. (*By courtesy of Dr. S. Bradbury.*)

Plan of the contents of the cell

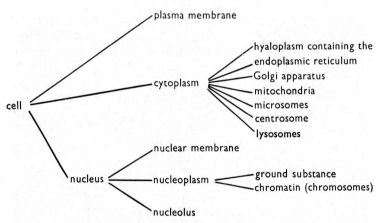

The cell membrane (Plasma membrane). An active cell is constantly using food as it maintains itself, thus carbohydrate, fat and protein must be supplied to it. Oxygen is necessary for the efficient respiration of most cells and must also be supplied. The waste products of the metabolic activity must obviously be removed. There is consequently a constant coming and going of a variety of chemical substances in the region of the cell boundaries. There is a constant flux of substances from the tissue fluid to the cell protoplasm.

The blood stream supplies the tissue fluid with food and oxygen and takes away the waste products. The blood has to receive food from the gut wall and oxygen from the lungs. In the tissues of these organs there is a great deal of activity involved in transferring these substances. Similarly in the excretory organs the cells and their plasma membranes have many molecules passing through them. The plasma membranes of all cells are obviously important in regulating the exchange of the many substances between the environment outside the cell and the protoplasm within the cell. It is essential, in the mammal especially, that the composition of the protoplasm is kept as constant as possible whilst at the same time there is this tremendous exchange of material. In short the function of the plasma membrane is to regulate the dynamic flux of the materials in the protoplasm. We must look a little more closely at the membrane and the forces which operate there.

1. DIFFUSION. The molecules in a gas are constantly vibrating and moving about at random in space. When a gas is compressed the molecules have less individual space in which to move and there are more actual molecules of gas in a given volume. The gas has been concentrated. If a concentrated volume of gas is brought into contact with a volume of less concentrated gas the two concentrations will soon become equal as molecules of the gas move from the region of high concentration to the region of lower concentration. Such a movement is called diffusion.

If a membrane is put between the 'weak' and concentrated gases then the two concentrations may retain their original values or they may reach a mean value depending upon whether the membrane is impermeable or permeable to the molecules. The same principles apply to membranes separating solutions, for in solutions the molecules are vibrating just as they are in gases although they are more restricted in their movements in solutions. Thus if two solutions of sugar in water are separated by an impermeable membrane, the molecules of sugar hit the membrane on each side as they vibrate. If there were holes in the membrane just big enough for the sugar molecules to pass through then the sugar molecules might pass from one side to the other and vice versa if they managed to hit the membrane in the region of the pore. The number of sugar molecules hitting the pore depends upon the concentration of molecules on either side of the membrane. Thus if the sugar solution on one side of the membrane had a concentration of $4n$ molecules and on the other side of $2n$ molecules then more molecules would pass from the $4n$ concentration than into it. Thus a dynamic equilibrium is set up when there is a concentration of $3n$ molecules on each side. Then $3n$ molecules pass from right to left and $3n$ from left to right as the rates of diffusion are equal in both directions. If on one side of the membrane there are molecules bigger than the size of the pores, e.g. protein molecules, the membrane would prevent them diffusing across it. Thus membranes can prevent the diffusion of some molecules whilst allowing the diffusion of others, i.e. they allow differential diffusion.

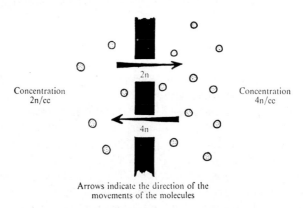

Concentration
2n/cc

2n

4n

Concentration
4n/cc

Arrows indicate the direction of the
movements of the molecules

Fig. 10.   *Diagram illustrating rate of diffusion.*

Rate of diffusion. Many factors are important in governing the rate of diffusion of substances through membranes. They include the following.

(*a*) The concentration of molecules.

(*b*) The size of the pores in the membrane. We have already seen that if the pores are too small diffusion is prevented. Recent work on the structure of membranes indicates that the rate of diffusion across cell membranes cannot be explained unless special points in the membrane are present where diffusion can occur very quickly. Most non-electrolytes diffuse into red blood cells by diffusion at normal rates, but certain substances enter very rapidly. Glycerol enters at 10^4 times the calculated rate. This fast diffusion is thought to occur through the 'polar pores'. (See fig. 11.)

(*c*) The solubility of the diffusing molecules in the membrane. Protoplasmic membranes contain a matrix of long protein molecules with lipid molecules dispersed in the interstices. (See fig. 11.) Molecules which are too big to pass through the pores may pass through if they can dissolve in the lipid molecules.

(d) If the diffusing molecule can enter into combination with large molecules on one side of the membrane then the effective concentration on that side of the membrane would be reduced.

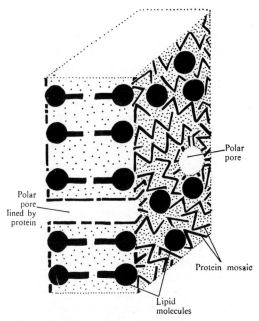

Fig. 11. *Diagram illustrating the structure of the plasma membrane.* The membrane consists of two layers of protein with lipid molecules dispersed between the protein molecules. Two polar pores are shown in the diagram.

(e) The area of the membrane is obviously important. The lungs of a mammal have a very large surface area because of the many small pockets in the lung membranes called alveoli. This allows a large area to be presented for the diffusion of oxygen into the body and carbon dioxide out.

(f) The thickness of the membrane is important as the speed of movement from the outside of the cell to the protoplasm inside would be reduced more by a thick barrier than a thin one. There are perhaps three problems in diffusing molecules across a membrane; getting into the membrane, passing through it and getting out of the membrane.

There is one type of diffusion operating across cell membranes called facilitated diffusion. This process brings about the movement of substances down a concentration gradient but the rate of transfer is greater than can be accounted for by the forces of simple diffusion. It has been postulated that such a type of diffusion may involve the combination of the substance with a 'carrier molecule' on the surface of the membrane which moves across the membrane to release the substance on the inner face of the membrane.

2. ELECTROCHEMICAL GRADIENTS AND DIFFUSION. It is apparent that the rate of movement of substances across membranes by diffusion depends upon properties of the membrane and upon the concentration gradient. The more permeable the membrane and the steeper the concentration gradient the faster will be the rate of transfer of substance across it. The membranes of cells in addition to having permeability properties influencing diffusion also have electrical properties which influence the diffusion of particles bearing an electrical charge e.g. ions. The development and maintenance of the electrical charge on cell membranes, the inside of the cell being electrochemically negative to the outside of the membrane, is described in detail on p. 284. Here it is sufficient to note that the outside of the cell membrane is positively charged compared to the interior of the cell. This positive charge on the outside of the cell membrane will tend to repel positively charged particles which are tending to diffuse out of the cell. The electrochemical negativity of the interior of the cell will tend to attract positively charged particles into the interior of the cell. Thus the electrical potential across the cell membrane may tend to oppose or support the movement of charged particles moving across the membrane along a diffusion gradient.

3. ACTIVE TRANSPORT. Some substances are moved across cell membranes even in the absence of a concentration gradient and often against a concentration gradient. Thus in the kidney, glucose is absorbed from the glomerular filtrate by the cells of the proximal tubule against a concentration gradient, all the glucose being absorbed from the filtrate and transferred to the blood. Active transport is involved in the reabsorption of other substances from the glomerular filtrate, and in the absorption of many of the products of digestion from the small intestine. Active transport requires the provision of energy by the cell in order to move these substances against their concentration and thus diffusion gradients. Because active transport requires the provision of energy, anything which interferes with the vitality of the cell will interfere with the transport. Thus, if one cools the kidney or reduces its

oxygen supply, or interferes with metabolism by the administration of metabolic poisons such as cyanide, then active transport stops and the glucose and other materials of the glomerular filtrate are no longer reabsorbed but are excreted in the urine.

A special example of active transport common to all cells is the transport of ions across the cell membrane. Normally the interior of the cell contains much fewer sodium ions and more potassium ions than the fluid bathing the cell. Although the cell membrane is not fully permeable to these ions there is a net tendency for sodium and potassium ions to move across the cell membrane along their concentration gradients and for an equilibrium to become established between the sodium and potassium of the two fluids. This tendency is opposed by an active transport system which extrudes sodium ions from the interior of the cell and takes in potassium from the exterior of the cell.

4. PHAGOCYTOSIS AND PINOCYTOSIS. Amoeba obtains its food by engulfing food particles into its cytoplasm. Amoeba approaches a food particle and extends projections of its protoplasm to engulf the particle which becomes contained in a vacuole in the cytoplasm. Here it undergoes digestion, the products of digestion being absorbed into the cytoplasm. This process has been called phagocytosis. For many years this behaviour was regarded as a special form of feeding found in some protozoa, and in certain special defence cells of mammals (macrophages, polymorphonuclear leukocytes). However a similar process has been found in other types of cells isolated in 'tissue culture' so that they can be observed in the living state under the microscope. This process has been called pinocytosis and it involves invaginations of the surface membrane of the cells which deepen and separate away from the membrane, passing into the cytoplasm as small vacuoles (fig. 12). In this way substances near to the surface of the cell can be taken directly into the body of the cell; they can enter without having to traverse the surface membrane directly. This process of pinocytosis also occurs in Amoeba in addition to the process of phagocytosis. If an Amoeba is placed in a solution containing mineral salts or proteins then a number of invaginations of the surface membrane appear, these deepen to form channels in the cytoplasm at the base of which vacuoles form. Digestion and absorption of the contents occurs as in the larger food vacuoles. There is a limit to the amount of pinocytosis that can occur and after a while pinocytosis channels disappear and the Amoeba cannot be persuaded to undergo further pinocytosis or phagocytosis until a period of time has elapsed.

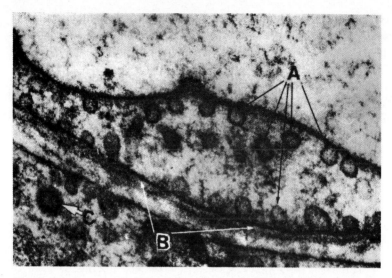

Fig. 12. *Electron micrograph showing the structure of a capillary endothelial cell with pinocytic vesicles.* A—pinocytic vesicles. B—basement membrane of endothelial cell. (*By courtesy of Dr. S. Bradbury.*)

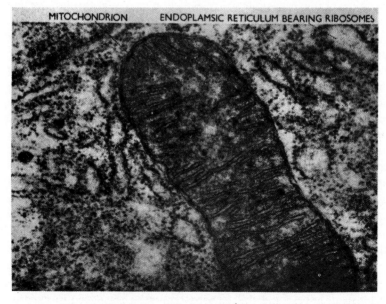

Fig. 13. *Electron micrograph of part of a pancreatic exocrine cell showing mitochondrion.* (*By courtesy of Dr. S. Bradbury.*)

Summary. The movement of substances across cell membranes is determined by the number, size and character of the particles involved, various physical properties of the membrane, active transport mechanisms and pinocytosis.

Mitochondria. These are spherical or rod shaped bodies in the cytoplasm of cells and are visible under the light microscope. There may be up to several hundred mitochondria in a cell, those cells performing secretory or mechanical work having large numbers. A mitochondrion has a double walled membrane consisting of protein and lipid. The inner wall of the membrane is infolded into the centre of the body of the mitochondrion (fig. 13). The contents of the mitochondrion include the enzymes of Krebs' cycle. In the wall are localized the respiratory chain enzymes which transfer hydrogen from respiratory substrates to oxygen. The mitochondrion is thus the site of the transformation of energy in the cell, the energy of respiratory substrates being transformed into metabolically useful forms of energy in the form of energy-rich phosphate compounds.

Golgi apparatus. This is a system of interlocking fibrils, or sometimes in some cells just scattered threads, usually localized around the centriole. The apparatus consists of lipids and proteins. Under the electron microscope the Golgi body is seen to be composed of a series of membrane bound channels similar in appearance to the structure of the endoplasmic reticulum with which it is often continuous. The function of the Golgi body has been something of a mystery. Functionally the Golgi body has a close relationship with the secretory activity of cells, and an association with the formation of the spindle in cell division. In view of its similarity in structure to and its relationship with the endoplasmic reticulum it has been suggested that it forms a site for the continuous formation of new membrane for the endoplasmic reticulum.

Centrosome. This is a clear area of the cytoplasm attached to the outer side of the nuclear membrane and contains the two centrioles which are important as the apices of the spindle in cell division. They are found in animals but not in the higher plants. Under the light microscope the centrosome is only visible when the cell begins to undergo division.

Chromophile substance. This is found in small masses in the cytoplasm and stains with basic dyes in a manner similar to the chromatin of the nucleus. Cells with a lot of this substance are the basophiles of the anterior pituitary gland, and many nerve cells. In the latter the chromophile substance forms Nissl granules. This material is a ribose nucleoprotein.

Ribosomes are very minute granules in the cytoplasm (figs. 9, 13). They are so small (about 100 millimicrons diameter) that they are invisible in the light microscope. They have been studied under the electron microscope and several types have been found. They are believed to be attached to, or part of, the endoplasmic reticulum. They contain RNA and lipid. Sixty per cent of the entire RNA of the cell is contained in them. Although they are so very tiny they are present in enormous numbers and actually make up one fifth of the total cell mass by weight and this suggests that they are an important part of the cell. They are involved in protein synthesis and it has been suggested that they are, or contain, the plasmagenes. Others have reported that they are important in cell oxidations and also that they have thromboplastic activity. (See p. 215.) The list of their probable function grows as research proceeds. Sometimes they are called microsomes.

Lysosomes. These bodies occur in many if not in all animal cells and although they are often above the lower limits of resolution by the light microscope they were not discovered by optical methods. Their essential structure is that of a small envelope containing digestive enzymes of many kinds including phosphatases, cathepsin (acting upon proteins), ribonuclease and desoxyribonuclease (acting upon the corresponding nucleic acids).

The lysosomes are present in large numbers in those cells with a prominent digestive role in the life of the organism e.g. macrophages and some leukocytes. It is suggested that after phagocytosis of bacteria and parts of degenerating cells the lysosomes approach the 'food vacuole' and fuse with its wall, emptying the digestive enzymes into the vacuole. If lysosomes were ruptured inside the living cell they would cause breakdown of the structural elements of the cell. They may well be responsible for the degeneration of tissues after death, a process called autolysis ('self-dissolving').

The nucleus

In stained cells the most conspicuous inclusion is the nucleus. In actively dividing cells it is generally spherical although in some cells it may have a more complex shape, e.g. the polymorph nuclear leucocyte has a lobate nucleus. When the cell is not dividing the nucleus may have the same optical properties as the cytoplasm but as the cell starts to divide the nucleus becomes more refractable. The dyes used to stain the nucleus, e.g. haematoxylin, crystal violet, methyl green and basic fuchsin, do so because of their power to stain the hereditary material of the nucleus which is called chromatin. The stained nucleus shows that the chromatin is arranged in a fine network structure, called the reticulum in tne non dividing cell, whilst the rest of the nucleus consists of a colourless nuclear sap. In this sap is the dark rounded nucleolus which is rich in ribosenucleic acid. Some parts of the chromatin called prochromosomes stain deeper than the remainder which in general stain poorly during the time in which the cell is not dividing, and this is perhaps because the nucleic acids which make the reticulum are too diffuse to absorb much stain. The chromatin is highly hydrated at this stage and this is perhaps a reason why the reticulum does not accept stain. Because the chromatin only stains well after the cell has started to divide the early workers were misled into thinking that the chromatin disappeared during the resting phase. However the existence of the prochromosomes in the resting phase is now well established and by adjustment of the salt concentrations of the medium containing cells of the grasshopper it was shown in 1949 that even in the resting phase the hereditary material can be made to become clearly visible in living cells. Thus it seems that at least in some species the chromosomes retain their integrity in the resting phase of the cell.

Mitosis is the name given to the nuclear division which occurs in the normal body cell and although division is a continuous process it is usually considered to consist of several phases: prophase, metaphase, anaphase, telophase followed by the resting phase (interphase).

1. PROPHASE. The nucleus takes up water and swells and the chro-matin begins to be more distinct as longitudinally split threads called chromosomes appear which become increasingly easy to stain as they are progressively dehydrated during prophase. Each strand in the chromosome is called a chromatid and the two chromatids are coiled round each other to form the chromosome. In good preparations of protozoa the chromatids have been seen to consist of two halves each of which is called a chromonema strand, but these are difficult to distinguish in higher animals and the important thing to grasp is that the functional unit in mitosis is the chromatid. As the chromosomes dehydrate they become shorter and they spiralize into a great number of coils, but as the end of propase approaches although they continue to shorten the number of coils gets less, each coil being of greater diameter so that the chromosomes appear thicker. This reduction in the number of coils is called despiralization. In animal cells the short thick chromosomes migrate to the nuclear membrane until the latter disappears. The nucleoli usually get smaller and finally disappear in late metaphase. In animal cells the centriole divides into two within the centrosome, as it lies against the nuclear membrane, and then the two halves migrate to the poles of the nucleus along the membrane, before it disappears.

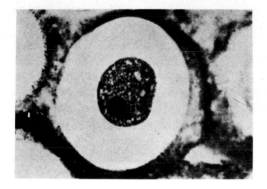

(a) Interphase.

(b) Prophase.

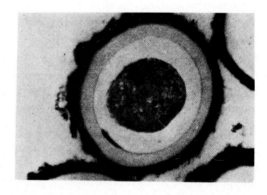

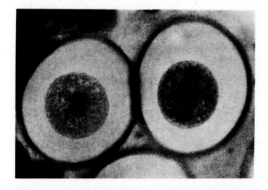

(c) Metaphase: both cells are in the stage of metaphase but the chromosomes are seen from different views. The cell on the left is viewed from the pole of the cell while the cell on the right shows a side view of the metaphase plate.

Fig. 14. *Photomicrographs of the zygote of Ascaris (round worm) showing various stages of mitotic cell division.*

(d) Anaphase.

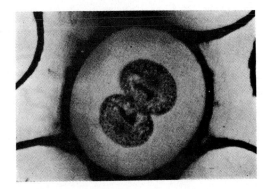

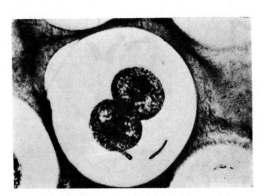

(e) Telophase.

Fig. 14. (*a–e by courtesy of Dr. S. Bradbury.*)

2. METAPHASE starts when the nuclear membrane has disappeared and the spindle has begun to form between the two halves of the centriole. The chromosomes come to lie on the equator midway between the poles and they are each attached to the spindle by a spindle attachment, which can be seen under the microscope as a marked constriction of the chromosome. The spindle attachment is often called the centromere constriction.

3. ANAPHASE is said to have begun when the centromere becomes functionally double, the two halves repelling each other and starting to move polewards, dragging the chromatids apart. Later in anaphase in some organisms the spindle may elongate and help in separating the sister chromatids. Since the two chromatids have identical hereditary properties the function of anaphase can be said to consist in the separation of two groups of like genetical constitution.

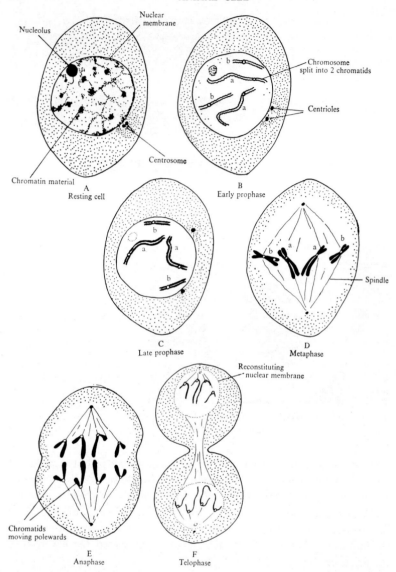

Fig. 15.   Diagram to illustrate the process of mitosis. A. shows a *resting cell* in which no chromosomes are apparent inside the nucleus. B. *Prophase*. The chromosomes are now distinguishable in the nucleus each is split into its two chromatids. Two pairs of homologous chromosomes are represented, a and b. C. *Late prophase* in which the chromosomes have become shorter and thicker. D. *Metaphase*. The chromosomes are arranged on the spindle. Shortening and thickening has proceeded. E. *Anaphase*. The centromeres of the chromosomes have divided and the chromosomes have split into two quite separate chromatids, one passing to one pole of the spindle and the other passing to the opposite pole. F. *Late Telophase*. The chromatids have approached near the poles of the spindle and are beginning to stain less easily, and the nuclear membrane is being reconstituted.   The cytoplasm of the cell is in the process of dividing into approximately equal halves.

4. TELOPHASE covers the period in which the two groups of chromatids each become surrounded by their own nuclear membrane. This period of regrouping may only be transitory in actively dividing tissue, but if a resting phase is to follow the chromatids loosen their coiling and once again lose the power, for the most part, to take up stain easily. The nucleolus and the prochromosomes reappear and interphase has started. During the late telophase the cytoplasm of the parent cell has divided into two equal parts each of which invests one of the daughter nuclei. The cytoplasmic division strictly speaking is not part of mitosis and the latter term should be restricted to the activity of the nucleus only.

5. DURING INTERPHASE each chromatid makes another absolutely identical one so that when the chromatin takes up stain at the next prophase it is seen to be a chromosomal strand which is longitudinally split into two identical chromatids.

This means that from the fertilised egg repeated cell division gives rise to populations of cells—later of tissues and organs—which have an identical chromosome complement. Each individual of a species has the same chromosome number and appearance; in man all normal individuals have forty-six chromosomes/cell nucleus, all cats have thirty-six chromosomes/cell nucleus, all mice have forty chromosomes/cell nucleus and so on. Although all men have forty-six chromosomes/cell the hereditary information carried by these chromosomes varies somewhat from one man to another—some men are short, some tall, some negroes, some white and so on.

Thus all the cells of one individual have similar chromosomes and similar hereditary material—as copies of the chromosomes of the fertilised egg. All individuals of one species have the same chromosomes but the hereditary material varies somewhat from one member to another. Although all cells of an individual have the same hereditary material, cells may vary widely in their form and function—compare a red blood cell with a muscle cell. All body cells have the necessary 'information' for performing the activities of a red blood cell or a muscle cell but this information is not put to use—we say that the genes which carry this information are suppressed. In a later section we shall look at what is known of the mechanisms which can 'switch the genes on and off'.

Mitotic cell division occurs not only in the development of the individual from the fertilized egg but also to replace dead cells in the tissues of the adult. The cells of some body tissues are replaced at a much faster rate than others. Skin is being continually replaced by mitotic cell division of the basal later of the skin (germinative layer).

The epithelial layer covering the intestinal villi are continually replaced by mitotic division of cells lying at the bases of the crypts of Lieberkuhn. There is an enormous turnover of red blood cells, each cell being replaced after about 120 days; worn-out red blood cells are replaced by mitotic cell division of 'stem cells' in the red bone marrow. These are the kinds of tissues to examine to observe cells engaged in mitotic division. In some tissues e.g. central nervous system cell division is a much rarer event and mitotic cells are very rarely seen when nervous tissues are examined under the microscope.

Estimation of the mititic activity in a tissue (i.e. the number of cells undergoing mitotic division in unit time) can be facilitated by administering the drug colchicine to an animal (see also page 45). This drug halts cell division at the stage of metaphase by interfering with spindle formation. After a suitable interval e.g. several hours—tissue is taken from the animal for preparation for examination under the microscope. All cells which started mitotic division since the drug was given will be arrested at the stage of metaphase and they can be counted. The speed at which a cell passes through the various stages of mitotic division varies between species and cell types; a mouse spleen cell passes through mitosis in about one hour.

HOMOLOGOUS CHROMOSOMES

We have seen that the chromosomes are the carriers of the hereditary factors or genes as they are called. In some cells, for example in the salivary gland of the fruit fly Drosophila, the chromosomes are very large and are called giant chromosomes. In these giant chromosomes transverse bands are seen and swellings all along the length of the chromosome rather like beads on a string. Each bead represents a gene and the position of this gene is called its locus. All chromosomes have genes located along their length but they are not as easy to observe directly as in the giant chromosomes. On a chromosome there may be a locus for eye colour another one for size, another for texture of the hair and so on.

On the metaphase plate at mitosis it is seen that there are two of each kind of chromosome and that each chromosome is split into two chromatids. The pairs of similar chromosomes are called pairs of homologous chromosomes. Of each pair of homologous chromosomes in the fertilized egg one member of the pair arises from the father, the other from the mother. Because of the nature of mitotic cell division, each cell of the organism comes to contain a copy of these chromosomes. Homologous chromosomes are identical in their appearance under the microscope and identical in the linear arrangement of factors—genes—which determine various aspects of body form and function. Thus at a particular point along their length each member of a pair of homologous chromosomes contain a gene which determines a particular feature, say eye colour. In other words there is a *locus* on each homologous chromosome which carries a gene determining eye colour. In this they are identical. However the two loci of a pair of homologous chromosomes may differ in that they determine different kinds of eye colour. At one locus there may be a gene which determines blue eyes and at the same locus on the other chromosome of the pair there may be a gene which determines brown eyes. In this example the effect of only one of these genes becomes manifested in the individual i.e. the individual has brown eyes. The gene which determines brown eyes is said to be dominant to that for blue eyes, and the latter is said to be recessive.

If at a given locus both chromosomes bear identical genes e.g. dominant genes determining brown colouration of the eye, then the animal is said to be homozygous at this locus. If, however, one chromosome carries a recessive factor at the locus and the other carries a dominant factor then the animal is said to be heterozygous at this locus. Homologous chromosomes are thus very similar but they are not identical. In contrast the two chromatids which make up one member of a pair of homologous chromosomes are absolutely identical and this is necessarily so for one chromatid manufactures its sister chromatid during the interphase of mitotic division.

We have seen that in the case of eye colour the gene which determines brown colour produces its effect whether it is present on one or both chromosomes of a homologous pair i.e. it is a dominant gene. If one gene *is* dominant then its alternative form (allele) must be recessive. A recessive gene only produces its effect when it is present on both members of a homologous pair of chromosomes. Some genes however are neither dominant or recessive and the appearance of an individual heterozygous for such a gene is intermediate in appearance between the homozygotes.

To clarify the question of dominant and recessive genes we will look at an example of an inherited dominant condition—that of anonychia in man. In anonychia some or all of the nails of toes and fingers are absent or are present in a very rudimentary condition. Anonychia is due to a mutant (changed) gene on one homologous pair of chromosomes. At a particular point—locus—on each member of this homologous pair of chromosomes there may be either the normal gene or its alternative (allelomorphic) form which determines anonychia.

There are three possible combinations of the genes on the homologus pair of chromosomes and this is shown diagrammatically below.

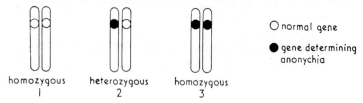

homozygous   heterozygous   homozygous
     1   2   3

○ normal gene

● gene determining anonychia

Fig. 16. *Possible combinations of genes, determining normal and abnormal nail growth, on a pair of homologous chromosomes.*

Because the gene for anonychia is dominant then individuals 2 and 3 show the disorder of nail growth. The condition of anonychia is very rare. Thus almost all of the individuals with the condition are hetero-zygous (2) for the gene; this arises because of the remoteness of the chance of the marriage of two individuals with the disorder and the production of an individual homozygous for the condition. Figure 17 shows the result of the marriage of an individual heterozygous for the gene, with a normal mate. Only one of the twenty three pairs

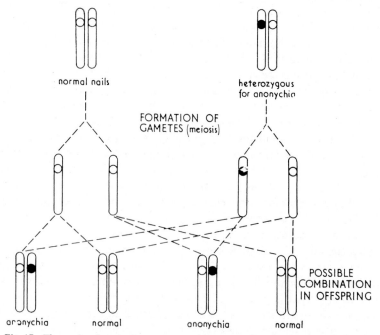

normal nails

heterozygous for anonychia

FORMATION OF GAMETES (meiosis)

POSSIBLE COMBINATION IN OFFSPRING

anonychia   normal   anonychia   normal

Fig. 17. *The results of mating between a normal person and one heterozygous for anony-chia.* Only the chromosomes determining nail growth are shown

of chromosomes is shown i.e. the pair of chromosomes carrying the genes which determine nail structure.

Meiosis and the conservation of the chromosome number. Since the characteristic features of an animal are determined in part at least by the hereditary material which is carried from one generation to another in the genes on the chromosomes, it is essential that the chromosome number remains constant if the characteristics of the species are to be maintained reasonably constant. In sexual reproduction two cells called the gametes fuse in a process called fertilization, to form a zygote which then develops into the embryo. Each gamete brings with it a set of chromosomes. It is obvious that if the gamete had the full set of chromosomes, i.e. the diploid number, then the zygote would have double this number and it would then have a different chromosome complement from its parents. In order to avoid this eventuality, in mammals during the formation of the eggs and sperms the number of chromosomes is reduced to half the diploid number, the so called haploid number of chromosomes. The reduction of the number of chromosomes occurs in the reduction division of meiosis which is a special kind of nuclear division, which is described below. The reduction division does not occur in the process of gamete formation in all organisms but occurs at other stages of the life cycle. In many green algae for example the whole body of the plant is haploid and the diploid number is attained at fertilization. During the germination of the zygote the reduction division occurs so that the embryo has only the haploid number. All the body cells of mammals, and all the vertebrates, are diploid and this confers obvious advantages on the cell. In a haploid cell if there is any spontaneous change (a mutation) in any of the genes then the effect of the change will be shown in the characteristics (phenotype) of the organism. This is so despite the fact that most mutations are recessive ones. The recessive genes in an haploid organism find expression because only one chromosome of an homologous pair is found in such an organism; the normal dominant gene, present on the homologous partner to the chromosome bearing the recessive gene thus cannot suppress the expression of this recessive gene. In the diploid organism the genes are always in pairs, one on each homologous chromosome of the pair of chromosomes. and there is always the possibility that a recessive gene is accompanied by a dominant one at the same locus. so that in a diploid organism there may be many recessive genes being carried in the heterozygous condition and these do not find expression in the phenotype. Therefore diploid organisms will tend to show fewer variants than haploid ones; they will not suffer from deleterious recessive mutations until the recessive gene is present in the homozygous condition.

The stages of Meiosis (figs. 18, 19, 21, 22)

PROPHASE. The most significant difference from the prophase of mitosis is that in meiosis the chromosomes are not double strands when they first become stainable, i.e. they are not at this stage split into their constituent chromatids. The homologous chromosomes come to lie very close together in pairs whereas in mitosis the two members of the homologous pair were scattered at random through the cell. A pair of

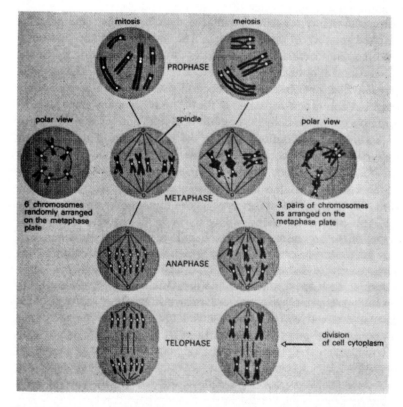

Fig. 18. *Diagrammatic representation of mitotic and meitotic cell division.* All cells are seen from the side except for those labelled polar view which are seen from the ends of the cells. For the sake of simplicity the nuclear membranes are not included.

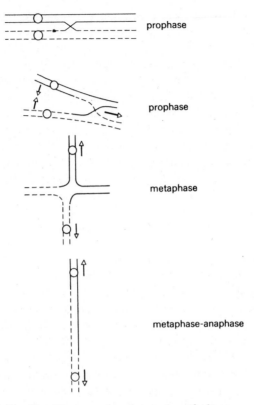

prophase

prophase

metaphase

metaphase-anaphase

Fig. 19. *Diagrammatic representation of chiasma formation and separation of homologous chromosomes.*

homologous chromosomes lying side by side in meiosis is called a bivalent.

The pairs of homologous chromosomes at the metaphase plate in meiosis have an unusual shape because they have become bound very closely together during the stage of prophase. This binding together of homologous chromosomes is a necessary mechanical process to ensure that pairs of homologous chromosomes arrange themselves on the metaphase plate, and not single chromosomes. If the homologous chromosomes did not undergo this special physical union then they would arrange themselves on the metaphase plate as independent units, and when they started to move to the poles in anaphase they would do so quite randomly, resulting in gametes having an odd number of chromosomes. This kind of failure can result in serious disorders in the offspring.

Figs. 19, 21 illustrate the way in which the members of a pair of homologous chromosomes become physically united and how later they separate from one another. In prophase the homologous chromosomes come together, each split into two chromatids. At one or more points of contact between the chromosomes, breakage occurs in the chromatids of the two chromosomes. These breaks quickly unite again but in such a way that the chromatid of one chromosome becomes united to the chromatid of the other chromosome. These physical bonds between the chromosomes are called chiasmata. Eventually the chromosomes have to separate from one another and figs. 18-19, 20-21 show how this occurs. As the chromosomes separate from one another the chiasmata move towards the end of the chromosome, and by the time the chromosomes have become attached to the metaphase plate the chiasmata have reached or nearly reached the ends of the chromosomes. We say that the chiasmata have terminalized.

The formation of chiasmata thus has a mechanical function in bringing homologous chromosomes to the metaphase plate in pairs. The process also has genetic effects. Each chromosome of a homologous pair that moves to the poles during anaphase is no longer identical to the corresponding chromosome in the rest of the body's cells, or to the chromosomes provided by the mother or father of which the latter are a replica. Because of the exchange of material between the chromosomes, the single chromosomes that move in anaphase contain a mixture of the genetic material originally supplied by the mother and father of the individual. Fig. 22 illustrates the formation of new combinations of genes that results from the formation of a chiasma. Only six genes are indicated on the chromosomes—a, b, c, provided by one parent and A, B, C, by the other parent. After the formation of the chiasma there has been a recombination of these genes.

IN METAPHASE the spindle appears and the nuclear membrane disappears, the centromeres of the bivalents now repel each other and the members of the pair of homologous chromosomes move apart. Note that the centromere of each chromosome is not divided and the whole chromosome moves to the pole, whereas in mitosis the centromeres are split and the chromatids are the units which migrate polewards.

ANAPHASE sees the continued movement of the chromosomes to the poles whilst the chromatids, splaying apart, are attached to each other only at the centromere. The splaying of the chromatids allows the chiasmata to separate and the chromosomes to leave their homologues.

TELOPHASE AND INTERPHASE do not usually occur but instead a mitotic

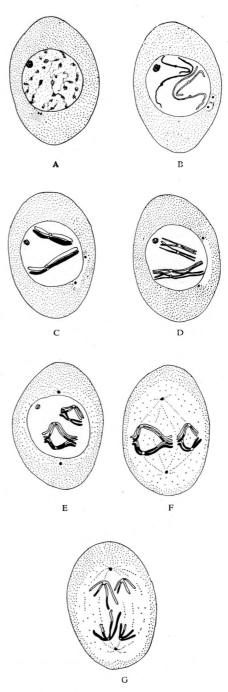

Fig. 20. *Diagram of the reduction division of meiosis.* A. The end of interphase. Chromosomes are not clearly visible. B. Early prophase (leptotene stage). Long thin chromosomes are present and are not split into chromatids. C. Later prophase (zygotene stage). The chromosomes have paired with their homologues to form bifids. They are shorter and thicker. D. Late prophase (diplotene stage). The chromosomes have each split into two chromatids. Chiasmata are seen where chromatids have broken and exchanged material. E. Diakinesis. The visible points of contact between the chromosomes are moving to the ends of the chromosomes, but the true chiasmata where chromatid interchange has occurred remain at the places of the original chiasmata. F. Metaphase. The nuclear membrane has gone and the centrosomes are beginning to move apart. G. Anaphase. Whole chromosomes move towards the poles their constituent chromatids splaying apart.

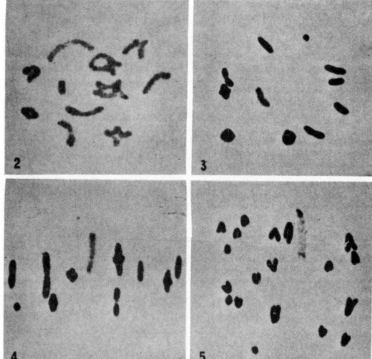

Fig. 21. *Photomicrograph of cells from a grasshopper testis showing some stages o meiotic cell division.*

1. Polar view of a metaphase plate of a cell engaged in *mitotic* division for comparison with 2–5 which are of stages of *meiosis*. 20 chromosomes are arranged on the metaphase plate and each chromosome is split into two chromatids. Homologous chromosomes are *not* paired.

2. Meiosis. The stage of diplotene-diakenesis. All chromosomes except the sex chromosome (in this species the male is XO, the female XX i.e. there is no Y chromosome) are in bivalents i.e. pairs of homologous chromosomes. Seven autosomal bivalents have one chiasma and two have two chiasmata.

3. Meiosis. Metaphase 1. Polar view. The X chromosome is in the centre. The chromosomes are condensed and chiasmata are not readily seen. Compare with 1.

4. Meiosis. Metaphase 1. Side view of metaphase plate. All bivalents show one chiasma.

5. Meiosis. Anaphase 1. Side view. Homologous chromosomes are now separating.

(*By courtesy of Dr. K. R. Lewis*)

type of division ensues so that the chromatids of each chromosome become separate and lie in different cells. The result of the reduction division followed by the mitotic (equational) division is to form four cells in each of which the homologous pair of chromosomes of the parent cell (which consisted of four chromatids) is represented by one chromatid only.

CROSSING OVER AND ITS SIGNIFICANCE

In the prophase of meiosis crossing over occurs and points of contact called chiasmata are formed between the two homologous chromosomes which lie side by side. During the process of shortening, the chromatids may break and then rejoin on to the wrong piece of chromatid forming a chiasma. When this occurs new groupings of genes are produced as shown in fig. 22. If crossing over had not occurred the two homologous

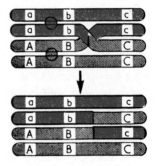

Fig. 22. *Diagram illustrating the formation of new combinations of genes resulting from chiasma formation.*

chromosomes would separate at the metaphase of the reduction division and then in the following equational division the chromatids of each chromosome would separate. The four cells produced would have genes *abc, abc, ABC, ABC*; but after crossing over the four cells produced have the genes *abc, abC, ABc, ABC*. Two new combinations have occurred—*abC* and *ABc*. It is known that genes have an effect on each other—the so-called position effect. It is obvious that crossing over has given variety in the gene composition of the gametes. It is well known that sexual reproduction gives rise to a variety of offspring whereas vegetative reproduction produces exact replicas of the parent (the expert rose breeder propagates his new variety vegetatively and not from seed). The source of variety in sexual reproduction in animals is to be found in (1) crossing over in gamete formation, (2) the fact that in the two gametes which fuse together in fertilization there may be different genes. This variety in genetic material, together with the interaction of the genes with the environment, provide the raw material on which the processes of natural selection can work.

Sex chromosomes

We have stressed the fact that each member of a pair of homologous chromosomes are identical in appearance. This is not true for a pair of chromosomes which are especially associated with sex. In the human male this pair of chromosomes consists of a larger chromosome —called the X chromosome—and a much smaller member—called the Y chromosome. In the female the sex chromosomes consist of an identical pair of X chromosomes similar to the X chromosome of the male. The remaining chromosome pairs—i.e. non-sex chromosomes are called autosomes. In all vertebrates the sex chromosomes are an unequal pair in the male and an equal pair in the female, which are difficult to distinguish from the autosomes. This is not necessarily the case in other classes of animals and in butterflies for example the male is XX and the female XY.

The behaviour of sex chromosomes during the meiotic division in the maturation of germ cells is shown in fig. 23. It is clear that the male produces two kinds of spermatozoa, one type carrying X chromosomes and another type carrying Y chromosomes. The female produces only one class of ova all containing an X chromosome. The sex of an individual is thus determined by the father—at the moment of fertilisation of the egg. If the egg is fertilised by a sperm bearing a Y chromosome the sex is male (XY) whereas a female (XX) results from fertilisation of the egg by a sperm bearing an X chromosome.

The Y chromosome is very important in the determination of the sex of an individual. In the absence of a Y chromosome the sexual development is female. In the presence of a Y chromosome the sexual development is male even if there is an abnormal number of X chromosomes; thus an abnormal sex chromosome constitution of XXXY does not prevent a reasonably normal male development (see also p. 46).

Sex linked inheritance

In addition to determining the pattern of sexual development the sex chromosomes, just like the autosomes, carry various genes which affect body structures and functions which have no direct relationship to sexual development. These genes are rather special in that their inheritance is inevitably associated with that of sex. Whether or not an individual shows a particular trait may thus depend on the sex of the individual.

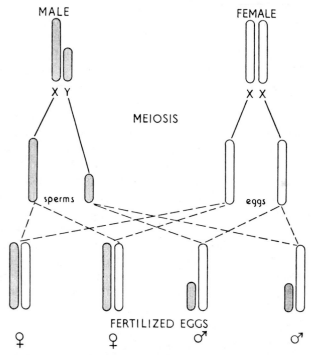

Fig. 23. *The behaviour of X and Y chromosomes during meiotic division.*

Sex linked inheritance describes inheritance via the X chromosome. The genes responsible are carried only on the X chromosome and there are no corresponding loci on the Y chromosome. There are other possibilities e.g. genes exclusive to the Y chromosome or genes represented on both X and Y chromosomes but they are not of practical significance—at least in man. We can best illustrate the effect of inheritance of a character by way of the X chromosome by looking at a gene which determines an abnormal condition. If the male carries the abnormal gene then the condition will be apparent in the individual i.e. it shows in the phenotype. There is no such thing as a recessive or dominant gene here for all genes on the X chromosome which do not have normal alleles on the Y chromosome will find expression. The situation is very different for the female who has a pair of homologous X chromosomes; in the female a sex linked gene can be dominant, recessive or intermediate, as in autosomal genes.

An example of sex linked inheritance, haemophilia, is discussed in detail on page 216. Another sex linked recessive condition is an inherited deficiency of an enzyme found in red blood cells. This enzyme, glucose-6-phosphate dehydrogenase, is necessary for the normal metabolism of red blood cells; this and other enzymes are able to protect the cell from damage by many oxidising agents. As red cells age the enzyme content declines and the cells eventually are broken down. The inheritance of G-6-PD deficiency is sex-linked and may affect 10 per cent of negroes; it also occurs in Caucasians and in these the deficiency when it occurs may be more severe. It has been estimated that many millions of people are affected by the condition. The effect of G-6-PD deficiency is shown in a shorter survival of red blood cells in the circulation. Under normal conditions an individual with deficient enzyme may compensate for this shortened life of red blood cells by an increased rate of production of red blood cells from the bone marrow. But if oxidant drugs are given e.g. sulphonamides for bacterial infections or anti-malarial drugs then large numbers of red blood cells are damaged and rapidly destroyed, producing an acute anaemia. These drugs are particularly dangerous in men carrying the factor (i.e. with no corresponding normal allele on the Y chromosome) and in women who are homozygous for the factor. Most women are heterozygous for the factor and although they show some detectable deficiency of the enzyme in red blood cells they are not as prone to develop the anaemia.

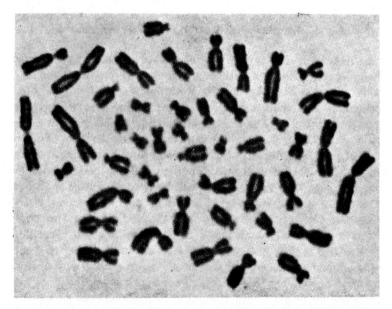

Fig. 24. *Photomicrograph of a cell growing in culture with growth arrested at metaphase.* Each chromosome at this stage is made up of two chromatids joined at the centromere. (*From 'The chromosome disorders', G. H. Valentine, Heinemann Medical.*)

Abnormal chromosomes in man

We have stressed the importance of maintaining normal chromosome structure and number. We can now look at some examples of the effects of variations in chromosome number and structure in man. Detailed examination of the chromosomes of man has only recently become possible and it was not until 1956 that the correct number of chromosomes was established—23 homologous pairs, including the sex chromosomes. In the study of human chromosomes, cells are obtained from blood (white blood cells), skin or bone marrow and are cultured in the laboratory. A drug called colchicine is added to the culture medium which prevents the formation of the spindle in the process of mitotic cell division. This effect arrests cell division

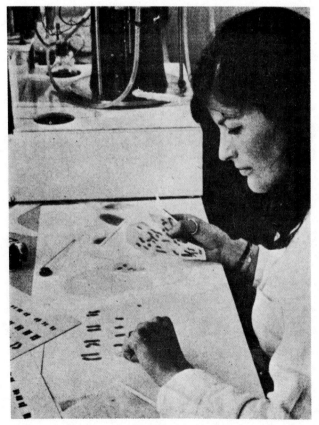

Fig. 25. *Preparation of a Karyotype.* A photograph of a cell at mitotic metaphase is being cut up, matching pairs of chromosomes being pasted on a sheet. (By courtesy of the Wellcome Research Laboratories.)

at the stage of metaphase when the chromosomes are conveniently distributed for observation. It also increases the number of cells showing chromosomes since all cells which entered mitotic division since the drug was added to the culture medium will be arrested at the stage of metaphase. The cells are then suspended in a hypotonic solution—water enters the cells and spreads out the chromosomes. The cells are then spread on slides, dried and suitably stained. Fig. 24 shows a photograph of such a preparation. Such photographs can be cut up and the individual chromosome pairs pasted on a sheet (fig. 25). Fig. 26 is a picture of such a preparation showing the chromosomes, which have divided into daughter chromatids, still attached by the centromeres. The individual chromosomes are identified by such features as length and position of the centromere. Such a preparation is called a karyotype.

One cause of abnormal chromosome number is a process called non-disjunction of chromosomes. This term describes the situation during the reduction division in the formation of germ cells when individual

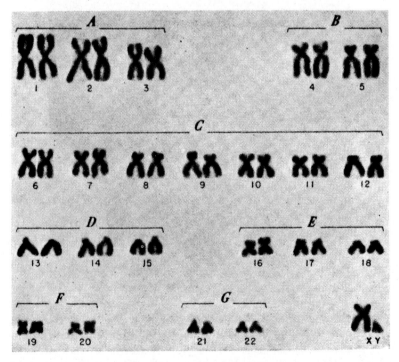

Fig. 26. *Normal karyotype from a male, X Y. (From 'The Chromosome Disorders', G. H. Valentine, Heinemann Medical.)*

members of a pair of homologous chromosomes fail to separate during the stage of anaphase so that one gamete has two members of a homologous pair and the other gamete has neither (fig. 27). We will look

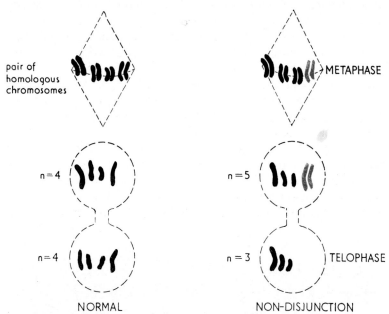

Fig. 27. *Diagram illustrating the process of non-disjunction of chromosomes.*

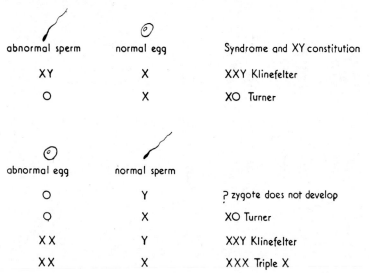

Fig. 28. *Diagram showing the origin of sex chromosome anomalies.* O refers to the absence of an X or Y chromosome.

now at what happens when non-disjunction occurs in relation to the sex chromosomes. Non-disjunction which occurs during the development of eggs may lead to an egg containing two X chromosomes or to one containing no X chromosomes. In the formation of spermatozoa non-disjunction may lead to sperms containing both X and Y chromosomes or to sperms containing no sex chromosomes. When these abnormal gametes unite with normal gametes a variety of possible abnormal chromosome complements results (see fig. 28).

Individuals who show Klinefelter's syndrome (XXY) are usually outwardly normal males except that the testes may be small and there may be some breast development. The testes do not produce sperm and the individuals are thus infertile. Fig. 29 shows a karyotype of a subject with Klinefelter's syndrome. A karyotype with 3 X chromosomes is sometimes found when otherwise normal women are subjected to chromosome studies. The extra X chromosome does not usually interfere with normal development and fertility. There is however a incidence of mental deficiency with XXX which is higher than in the

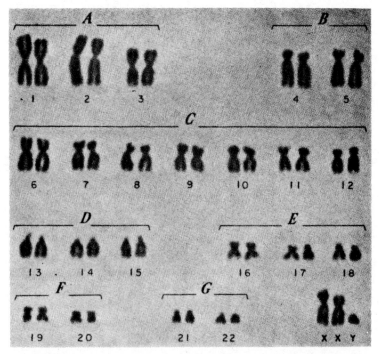

Fig. 29. *Karyotype of an individual with Klinefelter's syndrome (XXY). (From 'The Chromosome Disorders', By G. H. Valentine, Heinemann Medical.)*

normal population, and this also applies to Klinefelter's syndrome. Because of the normal fertility of women with triple X it is possible for the chromosomal abnormality to be passed on to subsequent generations since her ova will presumably be of two kinds, X and XX. An XX ovum when fertilised by a Y carrying sperm will result in a male with Klinefelter's syndrome; if the ovum is fertilised by an X carrying sperm this will result in another triple-X female.

Individuals with a karyotype containing XO show the features which are called Turner's syndrome. These individuals (figs. 30, 31) are outwardly female but the external genital organs remain small and the uterus is infantile. These deficiencies arise because the ovaries do not develop with a resulting deficiency of female sex hormones. But abnormalities may also occur outside the reproductive system and there may be 'webbing' of the neck, mental deficiency, dwarfism, and an abnormal development of the aorta.

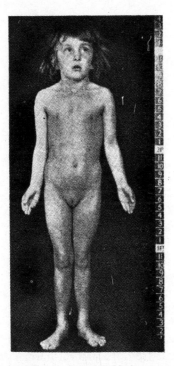

Fig. 30. *Turner's syndrome showing webbed neck, deformity of right fifth finger.* The photograph was taken before treatment. Treatment consists of administration of female sex hormones (mainly oestrogen) to replace the secretions of the absent ovaries. (*By courtesy of Dr. Bonham Carter.*)

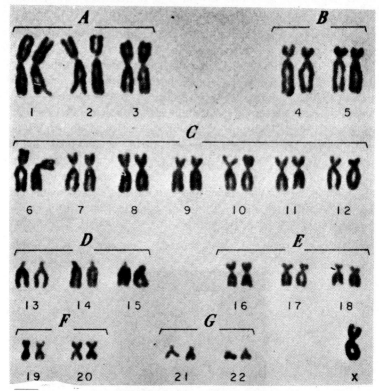

Fig. 31. *Karyotype of X O Turner's syndrome.* (*From 'The Chromosome Disorders', G. H. Valentine, Heinemann Medical.*)

Much more complex abnormalities of the sex chromosomes have recently been described. Some cases of Klinefelter's syndrome have been found to show XXXY or even XXXXY composition. The organism can thus tolerate gross changes in sex chromosome composition—far greater than would be tolerated for autosomes.

Abnormal or ambiguous sexual development—so called intersexuality—has causes other than variations in XY constitution. The development of normal genital organs is the result of a complex sequence of processes in embryological and post-natal development. The organs of males and females are moulded from similar basic rudiments and defects in development can arise at many stages. The cause of derangements from the path of normal development may or may not be genetic in origin. Non-genetic causes may include the treatment of pregnant women with progesterone or androgens (e.g. for recurrent or threatened miscarriage of pregnancy) which can divert the normal sexual development of a female foetus onto a more male course. A similar result can arise from abnormal activity of the adrenal cortex of a female foetus, the gland producing large amounts of masculinising hormones (androgens) which stimulate the growth of the 'more male' components of the female genitalia (e.g. overgrowth of the clitoris—the homologue of the male penis). This disorder of the adrenal cortex is genetically determined.

If an infant is born with any ambiguity of external genital organs then the situation calls for urgent investigations so that the child can from the start be brought up in the more appropriate sex. This is important not only for the peace of mind of the parents but also because the kind of early upbringing (i.e. as a boy or as a girl) has an important influence of the mental sexual identity of the individual in later life. The first investigations which are carried out will include culture of cells (e.g. blood or skin cells) to determine the karotype. Clues as to the nature of the sex chromosome composition can also be obtained by looking at

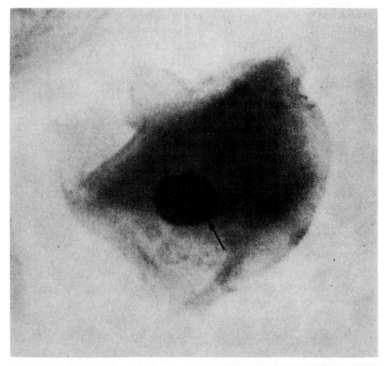

Fig. 32. *Photomicrograph of a squamous epithelial cell from inside the cheek showing a Barr body, indicated by the arrow.* (*By courtesy of Dr. S. Bradbury.*)

certain differentiated cells. One readily accessible cell type can be obtained by scraping the inside of the cheek with a blunt knife. The cells obtained are smeared onto slides, fixed, stained and examined under the microscope. A large proportion of the cells from females show a darkly staining body—the sex chromatin body or Barr body—applied to the inner surface of the nuclear membrane (fig. 32). These Barr bodies are of special importance because the X chromosomes of cultured cells are rather difficult to distinguish from some other chromosomes. Studies of Barr bodies can give important confirmatory evidence to studies of sex chromosomes. In XXX females a proportion of cells show two Barr bodies and in XXXX or XXXXY individuals some cells show three Barr bodies. It seems that Barr bodies are X chromosomes; the maximim number of Barr bodies is one less than the number of X chromosomes. These bodies are presumably inactive X chromosomes (Lyon hypothesis). Males (XY) thus have no Barr bodies.

Another sex difference if shown by polymorphonuclear leukocytes which, in females, have a small pedunculated nodule on the nucleus. This is called a drumstick (fig. 33).

These kinds of studies are now being used in international athletics to exclude women with a 'male' chromosome composition because they may have an unfair advantage in terms of muscular power over normal

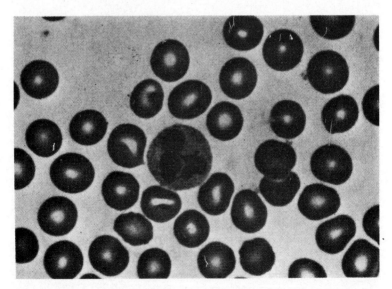

Fig. 33. *Photomicrograph of a polymorphonuclear leukocyte* (*surrounded by erythrocytes*) *showing a tiny 'drum-stick' projecting from the lobed nucleus.* (*By courtesy of Dr. S. Bradbury.*)

women. The greater muscle size and power of males is mainly due to the effects of male sex hormones (androgens) on muscles. The scientific basis of these 'sex tests' seem somewhat questionable for they exclude from competition XY individuals who are externally normal females because they possess a gene which prevents body tissues, both genital and non genital, from responding normally to androgens (the testicular-feminisation syndrome). But the tests do not exclude normal females who have deliberately taken androgens to improve their muscle power. Even complex studies of the excretion of hormones in the urine would not reveal that the athletes had taken androgens provided that the hormone treatment was discontinued some time before the tests.

Abnormalities of autosome number

The frequency of sex chromosome abnormalities is fairly high (see table below). But even more frequent is an abnormality of the autosomes which causes the condition called mongolism (or Down's syndrome).

Frequency of disorders due to non-disjunction of chromosomes.

Syndrome	Frequency
XXX	1/1000 females
Turner's syndrome	1/3000
Klinefelter's syndrome	1/500
Down's syndrome	1/500
	(if mother's age is 35-39 the incidence rises to 1/300).

The mongol shows both severe mental retardation and abnormal physical development (fig. 34). Disordered growth of the skull produces a small skull which is flattened from before backwards. The nose is short and flat and the mouth cavity is small—hence the protruding tongue. There may be squint, abnormal ears, cataract, abnormal teeth etc. The abnormal physical development is not restricted to the skeleton and there may be abnormalities of the heart e.g. defects in the interventricular septum.

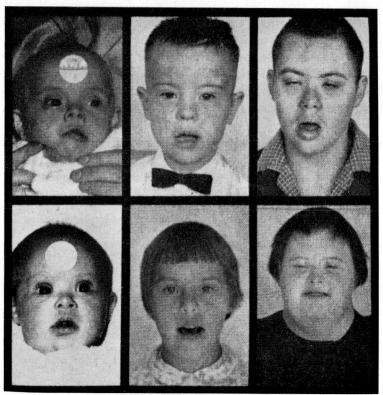

Fig. 34. *Mongolism or Down's syndrome.* The top right-hand male is about eighteen years old, the bottom right female is middle aged. The discs on the foreheads of the babies are for identification in a survey. (*From 'The Chromosome Disorders', G. H. Valentine, Heinemann Medical.*)

The chromosome abnormality in mongolism may in many cases be a non-disjunction affecting the autosomes—thus there are three chromosomes of pair 21 instead of the normal two i.e. tristomy of chromosome 21 (fig. 35).

Usually mongols are infertile. Very rarely mongol women have produced children; half of the children are normal, the other half are mongol. The ova produced by a mongol woman would be of two kinds, one containing two chromosomes 21 and the other with one chromosome 21. The chromosomal non-disjunction which produces a mongol must also result in some germ cells which carry no chromosome 21 yet there is no known syndrome in which individuals carry only one chromosome 21 (the chromosome derived from the father). Presumably some normal sperms fertilise ova which have no chromosome 21 but the abnormal chromosome composition does not permit

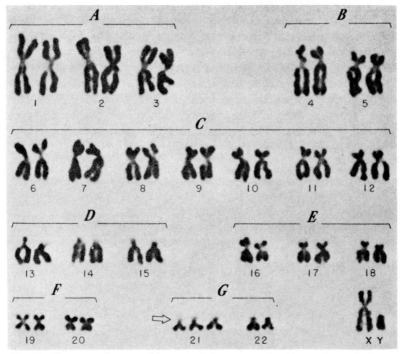

Fig. 35. *Karyotype of a mongol (trisomy of chromosome 21). (From 'The Chromosome Disorders', G. H. Valentine, Heinemann Medical.)*

normal development of the embryo. It would seem then that the loss of a chromosome has more severe effects than has the adding of a chromosome. Additional autosomes produce more effects on the organism than do additional sex chromosomes—compare the defects of the mongol (trisomy of chromosome 21) with those of individuals with an extra X chromosome (triple X, Klinefelter XXY). Chromosome 21 is only a small chromosome yet the presence of an extra 21 produces widespread defects in development.

Other causes of abnormal chromosome composition

A whole range of abnormalities of human development can arise from breakage of chromosomes. A segment of a chromosome can be lost from the nucleus of a germ cell—a so-called deletion. The broken segment may become reattached to its parent chromosome only after it has rotated through 360°, producing what is called an inversion. A broken segment may undergo non-disjunction so that one gamete has a segment of a chromosome duplicated. Breakage may involve more than one chromosome and members of two different homologous pairs may in fact exchange pieces of chromosomes; this is called translocation. Like non-disjunction, these major changes in the chromosomes are mutations, that is they produce a change in the genetic nature of the organism that results in a new variation. These major rearrangements of the genetic information will understandably have profound effects on development when compared with the effect of mutation of a single gene. Often the change is incompatible with normal development and survival. Sometimes however, this kind of chromosome rearrangement

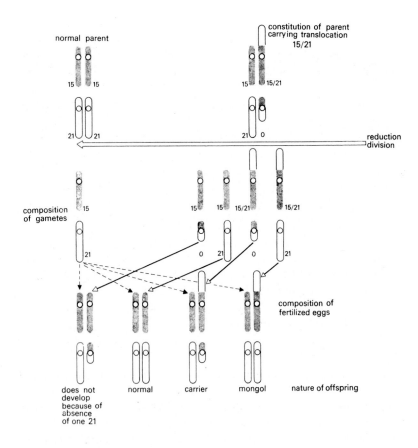

Fig. 36. *The inheritance of translocation mongolism.*

seems to have been important in evolution, and this is particularly true of inversions in which the total genetic information is unchanged, being merely redistributed among the chromosomes. Translocations also may be compatible with survival particularly if they are balanced translocations, i.e. with an extra segment on one chromosome being balanced by a corresponding deficiency in a chromosome of another pair. Fig. 36 shows the cause of translocation mongolism, the exchange of chromosome pieces between chromosomes 15 and 21. The left-hand side of the figure shows the normal behaviour of the homologous pairs of chromosomes 15 and 21 during the reduction division in the formation of gametes; all gametes contain one chromosome 15 and one chromosome 21. On the right-hand side of the figure is shown the effects of exchange of chromosome material between one chromosome 15 and one chromosome 21. When these chromosomes separate in the anaphase of the reduction division, four combinations are possible, as shown in the figure. When gametes carrying these combinations unite with normal gametes, the outcome varies. Some unions can result in normal offspring, some cannot develop, some produce a mongol and yet others can produce an apparently normal individual who is a 'carrier' for this type of mongolism. A translocation mongol has forty-six chromosomes, as does a normal person, but there is an extra chromosome 21 attached to chromosome 15. A 'carrier' of translocation mongolism has only forty-five chromosomes. The carrier has developed normally because most of the missing chromosome 21 has in fact been translocated onto chromosome 15. The only missing portion of this translocated chromosome is a small part close to the centromere (see Fig. 36) and it is thought that this part of a chromosome carries little or no genetic information.

Chromosomes and cancer

Cancer is a disease which can arise in virtually any tissue of the body. It is characterised by the appearance of cells which do not obey the laws which govern the growth of a particular tissue. Cancer cells continue to multiply in the absence of any need for the repair or enlargement of an organ. As the cancer grows so the normal cells of the tissue may become obliterated at the expense of the cancer. Many cancers spread far beyond the tissue in which they originate. Cancer cells which invade veins or lymph vessels may be carried to distant sites where they colonise other tissues and organs (the process of metastasis). Some cancer cells bear little to no resemblance to the cells of the tissue in which they arise and they consist of very undifferentiated cells. These primitive cells—anaplastic cells—are usually the most malignant.

In view of what we know of the overwhelming importance of the chromosomes in determining the activities of cells it will come as no surprise that there is a widely held theory which explains the origins of cancer cells in terms of abnormal chromosome composition. Certainly cancer cells do have abnormal chromosome composition and it seems very plausible that the abnormal chromosomes form the determining factor in the development of cancer cells.

The alterations in the chromosomes of cancer cells are both numerical and structural. Changes in chromosome number may involve the presence of extra chromosomes or loss of chromosomes. Some kinds of cancer seem to show a progressive evolution of chromosome structure as the cancer ages. One reported case of congenital acute leukaemia (cancer of the white-blood forming cells in the bone marrow which over-populates the blood with white cells) showed an evolution which involved secessive acquisition and duplication of extra chromosomes. It has been suggested that this evolution of the chromosome composition of cancer cells is due to changes in the metabolism of the cells. The presence of increased amounts of chromatin in a cell determines the production of increased amounts of particular enzymes (some tissues of Mongols—trisomy of chromosome 21—show increased amounts of some enzymes) and this produces a need for increases in the amounts of other enzymes to maintain metabolic equilibrium. Thus 'natural selection' may determine the evolution of cancer cells just as it does the evolution of bacteria, viruses and higher organisms. It would seem that many cancer cells die because of gross changes in their metabolism caused by abnormal chromosomes. Cells are 'selected' which show chromosome structure consistent with metabolic equilibrium.

Changes in chromosome structure may characterise particular kinds of cancer. Almost all individuals with chronic myeloid leukaemia show what is called the Philadelphia chromosome which is produced by the partial deletion of about 2/5 of a G chromosome (fig. 37). Studies have shown that this abnormal chromosome is present before the leukaemia shows itself—indeed the chromosome may be present at birth. This kind of specific abnormal chromosome is called a marker and they have been described for several kinds of cancer.

Further support for a genetic cause for cancer comes when we consider something of the causes of cancer. We know that certain external factors e.g. X-rays and virus infections can cause abnormal chromosome composition which may later be associated with the appearance of cancer. There are also internal factors which predispose to cancer; cancer frequency is high in families with chromosomal abnormalities and cancer of the gonad is more frequent in individuals who show intersexuality (with abnormal XY constitution).

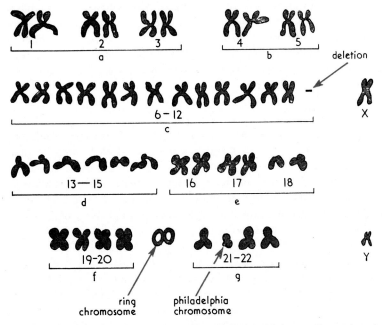

Fig. 37. *Drawing of a karyotype of a male with chronic myeloid leukaemia showing loss of a C chromosome, the presence of a ring chromosome and a Philadelphia chromosome.*

Summary

There exists a variety of chromosome abnormalities that can be detected by cytological methods. These mutations thus produce *visible* changes not only in various bodily features but also in the structure of of the chromosomes. About one in a hundred of all live-born infants have some such chromosome abnormality. Probably even more embryos start off with chromosome abnormalities, but these are not compatible with normal development (particularly if there is a *loss* of chromosome material) and the embryos dies in utero.

Common effects of chromosome abnormalities are mental retardation and sterility. At the moment there is no treatment for such conditions; the only possible measure is *prevention* by giving appropriate advice to fertile individuals who carry chromosome abnormalities, e.g. carriers of translocation 15/21.

Diseases due to dominant and recessive genes

These diseases are at present estimated to seriously affect about one in a hundred of all live-born individuals at some time in their lives. Most of the diseases are due to dominant genes, but some are due to recessive genes and a few are sex-linked. Most of the diseases due to dominant genes are sufficiently mild to be transmitted through several generations, but only about a third of them may be severe enough to prevent the affected individual from having children. By contrast, most of the serious recessive conditions prevent the affected individual from producing offspring.

Recessive conditions

We have already looked at the inheritance of recessive conditions. One important feature of recessive conditions is that recessive genes can be passed on through many generations by means of apparently normal individuals who are heterozygous for the condition. Only when two heterozygotes happen to marry one another is there a chance of the affected individual, the homozygote, appearing. We will refer back to this when we discuss the possible effects of the marriage of related individuals.

When an abnormal recessive gene is present in the homozygous condition, then the effect may be very serious and the individual may not survive long enough to marry and have children. Examples of such conditions are cystic fibrosis, infantile amaurotic idiocy and phenylketonuria. Cystic fibrosis (fibrocystic disease of the pancreas) is the most common of all diseases in this country which are due to recessive mutant genes that produce large effects. The disease affects about one child in 2000. The basic abnormality is that the mucus in the respiratory tract and digestive glands is abnormally sticky and the sweat glands produce a sweat which contains abnormally large amounts of salt. The main effects of this disease are caused by the sticky mucus. In the lungs the mucus tends to cause obstruction of the small air passages and this tends to lead to repeated chest infections. In the pancreas the ducts of the gland tend to become blocked so that there is a deficiency in the release of digestive enzymes; the digestion of fats and protein is thus incomplete and this affects the nutrition of the individual. In the past no cases of cystic fibrosis survived to adult life. Nowadays patients are being treated with pancreatic enzymes (obtained from animal glands) by mouth so as to improve the state of the digestion of fats and proteins. In addition, the introduction of a large range of antibiotics and other drugs that kill or damage bacteria means that the respiratory infections can be more effectively treated. Because of the frequency of this disease (1:2000), the prolonged nature of the treatment and the often frequent admissions to hospital for the management of respiratory infections, it imposes a fair-sized burden on medical facilities. The carrier rate in the population is abour 1:20–1:25.

Phenylketonuria affects one child in 40,000 in this country and in Northern Europe. The basic defect is an inability of the body to convert the amino acid phenylalanine into another amino acid called tyrosine. This defect results in widespread changes in the body. About 80 per cent of affected persons are blue-eyed blondes who may have eczema. Very many have marked mental deficiency of a degree that produces what we call idiots or imbeciles. The remainder are usually mentally defective to a degree. The disease can be diagnosed by demonstrating the products of phenylalanine which appear in considerable amounts in the urine of affected individuals. The urine test is simple, and once this simple screening test was discovered it was found that one in a hundred of patients in hospitals for the severely mentally subnormal suffered from phenylketonuria. In Great Britain all infants are examined for this disease. The appropriate test reagents are impregnated on a strip of paper and a drop of urine or a wet nappy will turn the paper a characteristic colour if the urine contains abnormal amounts of the products of phenylalanine. Widespread testing is carried out because there is the possibility that if affected infants are reared from an early age on a diet free from phenylalanine, then relatively normal mental development may be possible. This practice is based upon the principle that the mental defect is not an essential part of the disease but results from poisoning by the by-products of phenylalanine. This is one of the rare instances in which one can prevent mental deficiency.

Diseases due to dominant genes

The inheritance of dominant genes is characteristic. Whenever the gene is present its effects are manifested in the individual and the inheritance is readily shown by examining the family tree. Many disorders have been shown to be carried by dominant genes; these include blindness (due to degeneration of the retina), porphyria, intestinal polyposis, dwarfism due to achondroplasia, hereditary spherocytosis (acholuric jaundice), Huntington's chorea and so on. We will look briefly at one example—familial polyposis of the large bowel. In this disease there are many small wart-like growths called polyps on the inner surface of the large bowel. These polyps appear in childhood, although they may cause no symptoms until the late teens or even later. The danger is that one or more of these polyps may transform into malignant tumours which spread and ultimately kill the patient. The risk is so great that once the condition is recognized the patient is advised to have all the affected bowel removed before cancer develops. The large bowel is removed and the end of the small intestine is made to open onto the surface of the abdominal wall (ileostomy), or the small bowel is joined onto the stump of the rectum if this is relatively unaffected by the disease. In spite of this apparently radical kind of treatment, patients are able to lead relatively normal lives. Children of individuals with the disease (who are almost always heterozygous for the abnormal gene) have a one in two chance of developing the condition. The children can be told whether or not they carry the gene by examining the bowel for the presence of polyps (this examination is carried out by inserting an illuminated tube, a sigmoidoscope, up the rectum and into the colon). If they have no polyps then they do not carry the gene and their own children will be unaffected. If polyps are present then their own children will in turn have a one in two chance of developing polyposis and of needing surgical treatment. This disease is rather rare and one would expect this because it kills, certainly it used to kill prematurely, by causing cancer of the bowel.

Diseases with a more complex inheritance

In the examples we have been discussing, the diseases are believed to be maintained in the population by means of recurring mutations. Hence it would be impossible to eradicate many of the diseases by the control of marriages or of the reproduction of affected individuals. Because of these spontaneous mutations, the diseases can crop up in families in which there has never been a previous case of the disease. This fact is of considerable importance to those concerned with public health because various factors in our environment are known to increase the rate of mutations, i.e. they are mutagenic. The control of these factors and the search for as yet undiscovered mutagenic agents is a vital responsibility of those concerned with our health. This activity will be discussed further in a later section.

In some inherited diseases there are other factors which operate to maintain the frequency of a particular gene at a high level from one generation to another. This is because the abnormal gene can paradoxically increase the fitness of the individual to withstand particular circumstances. Thus we have the unusual situation in which a gene damages and reduces fitness under some circumstances, and yet may have the reverse effect and increase fitness under other circumstances. The frequency of the gene in a population will depend on the balance of these forces. We call this a balanced polymorphic system.

One such condition which has been widely studied is sickle-cell anaemia. This disease is so-called because of the characteristic sickled shape of the red blood cells under certain conditions. These cells contain an abnormal form of haemoglobin, called haemoglobin-S. The disease results in a chronic anaemia which is punctuated by 'crises' in which the anaemia rapidly worsens due to a massive break-down of the abnormal cells in the circulating blood. The seriousness of the disease depends on whether an individual is heterozygous or homozygous for the gene. In the heterozygous condition, the gene results in nearly 50 per cent of the haemoglobin of the red cells being in the abnormal S-form. This fraction rises to much higher levels in the homozygous state, when the disease is often fatal. The gene responsible for this disease is very unevenly distributed in the world. In some areas it is *very* rare but in parts of Asia and Africa the frequency of homozygotes may be as high as 10 per cent. The reason for the accumulation of the gene in these areas is that the heterozygote for this trait benefits by having an increased resistance to malignant malaria. Such is the complexity of nature.

Factors governing the inheritance and distribution of other genes may be even more complex. There are many malformations of infants and diseases of later life, the inheritance of which may be complex and the expression of which is determined by the interaction of the genetic make-up with one or more factors in the environment. Many of the diseases in this class can be looked upon as the result of the action of certain environmental factors on an individual already susceptible to the disease by virtue of his genetic constitution. The genetic factor may be single recessive or dominant genes, or it may be the product of the action of many genes at a number of loci on the chromosomes. Conditions in which there is a genetic predisposition are many and varied, and include schizophrenia, manic-depressive psychoses, mental deficiency, pernicious anaemia, diabetes mellitus, some types of epilepsy, diseases of blood vessels, susceptibility to various infections, still-births, asthma and duodenal ulceration. The accurate assessment of the part played by genetic factors in these various conditions is exceedingly difficult. Studies of family trees and comparisons of health of identical twins are very important tools in trying to unweave the nexus of heredity and environment.

Mutations, mutagens and the effect of modern life on mutations

The mechanisms of heredity ensure that the offspring are similar to their parents. In the words of Dobzhansky heredity 'is a conservative force; evolutionary innovation demands that heredity be occasionally thwarted'. This thwarting occurs when a mutation takes place. We have seen a variety of examples of mutations in man, ranging from changes in the effects of single genes to relatively major changes in the number or composition of the chromosomes themselves. Major changes of the genetic material are often incompatible with normal development and survival. These major changes include duplication of chromosomes, losses of chromosomese or repatterning of chromosom segments by translocation. Most mutations, however, cause no change in the structure of the chromosomes that can be seen even with the most powerful of electron microscopes. The changes are in the individual genes, i.e. in the molecular structure of the minute sub-units of the hereditary material of the chromosomes.

All the mutations that we have considered in relation to human nature have been harmful. Indeed all mutations in all organisms are usually harmful. This state of affairs is inevitable if we consider for a moment the significance of the hereditary material. The complex array of genes in the nucleus of the fertilized egg carries the information that not only guides the development of this tiny cell into the complete organism but also enables it to survive and reproduce its kind in its particular environment, be it arctic tundra, arrid desert or the depths of the ocean. Any random change in the complex array of interacting genes is almost bound to result in an organism that is less fitted to survive in its own particular niche in the world. However, environments change and man himself has altered his own environment on a vast scale, and some genes that were unfavourable in the old circumstances may become favourable in the new; of course, the converse may be equally true. We have seen such an example in the case of the gene that determines the presence of an abnormal type of haemoglobin in the red cells (sickle-cell anaemia). In certain areas of the world this gene is relatively common because it confers some resistance to malignant malaria. In countries where malaria does not occur, only the damaging effect of the gene expresses itself and only rare individuals carry the gene. The decline in the frequency of this gene in non-malarious areas is due to loss of the gene by the death of individuals with sickle-cell anaemia (homozygotes) and to a somewhat increased mortality of the heterozygotes for this gene compared to normal individuals.

If natural selection carries out this role of 'weeding out' harmful genes, we may ask ourselves why it is that mankind suffers from so much hereditary ill health. The answer to this question can be given under three headings: the occurrence of harmful mutations, the less than perfect efficiency of the process of natural selection, and the effects of applying the advances of modern medical science in the treatment of individuals who suffer from hereditary illness.

The various genetically determined illnesses that occur in human populations do not all arise because of the inheritance of harmful genes from the parents; many occur in individuals who have no family history of such illness, and of these many are due to mutations that have occurred early in the development of the individual, possibly during the development of one of the germ cells from which he arose. This means that even when a genetically determined illness kills before the sufferer is able to reproduce, new cases of the disease may continually appear in the population because of recurring new mutations.

Even when a harmful gene does not kill the individual before he has a chance of producing offspring, natural selection may still operate to reduce this possibility. The individual may suffer from chronic ill health or may be otherwise less acceptable as a marriage partner, particularly if the disorder markedly alters the appearance of the individual. Achondroplastic dwarfs usually survive and enjoy fairly normal health, but they produce few children; this is due not only to the small pelvis of the female dwarf which makes for difficulty in childbirth, but also because many dwarfs remain unmarried. Thus some mutant genes are less efficiently passed on from one generation to another than is the corresponding normal gene. However, natural selection is much less efficient in removing harmful genes of a recessive character. Many individuals in a population may carry harmful recessive genes in a heterozygous condition (i.e. paired with a normal gene) without producing any effect on their general health or reproductive capacity. Natural selection can only operate when a gene manifests itself in the individual. These harmful recessive genes that are carried 'hidden' in the heterozygous state form what is called the 'concealed genetic load' of a population. This concealed load of recessive genes is immense compared with the visible genetic load that becomes revealed when two individuals heterozygous for the same recessive gene marry and produce children who are homozygous for the gene (but only one in four of their children will reveal the genetic defect). The recessive nature of many genetically determined diseases thus reduces the efficiency of the processes of natural selection in eradicating the gene from a population. The activities of man himself further hinder the process of natural selection by enabling affected individuals to survive and reproduce, individuals who would, in the absence of the benefits of modern medicine, die in their infancy or their youth.

Man not only hinders the forces of natural selection, he actually promotes the appearance of genetically determined diseases, either by contaminating his environment with substances that increase the rate of mutations or by means of social habits, such as the marriage of blood relations (consanguineous marriages), that encourage the revealing of the 'hidden genetic load'.

The effects of agents in the environment on mutation rate

Mutation takes place regularly in all species of organisms. Under 'natural conditions' individual genes mutate very rarely, perhaps once in 100 000–1 000 000 germ cells. However, since the number of genes in most organisms is very high the total number of mutations may be quite considerable. If we make a reasonable assumption that the 23 chromosomes in a human germ cell carry 10 000 genes, then one gamete in ten may carry a mutated gene. This then is the estimate of the 'natural' rate of mutation. During his evolution, particularly in that short period following the industrial revolution, man has so altered his environment that this 'natural' rate of mutation *can* be considerably increased. In the main this is due to the introduction of radio-active materials and chemicals that can stimulate mutations—the so-called mutagens.

Radiation

Much is known and much has been written about the effects of ionizing radiations in living organisms. Radiation is known to produce mutations in all organisms so far studied, and there is no reason at all for thinking that man might be an exception to this. For very obvious reasons it is difficult to estimate the effects of radiation in man, and it is easier to obtain information about bacteria or animals such as fruit flies that reproduce rapidly.

There are various natural sources of radiation that affect man and other organisms—cosmic rays, radiation from natural radio-active elements in the crust of the earth and in the air, and radiation from radio-isotopes that are normally present in the body. The dose of radiation to the soft tissues of the body from all natural sources of radiation averages about 0·1 rad per year (a rad is a unit of absorbed dose of radiation: one rad $= 100$ erg/G; a submultiple is the millirad, Mrad $= 10^{3-}$ rad) although the dose varies according to latitude, altitude and variations in the composition of soil and rocks. 0·1 rad/year is, then, an average dose of radiation from natural sources. Man's various activities in medicine (X-rays, radiotherapy, etc.) occupations (radiographers, radiologists, workers in 'atomic-energy' institutions) and in war (nuclear explosions) can considerably increase this 'natural' dose of radiation for certain individuals. The following table shows the dose of natural radiation that reaches the gonads (ovaries or testes) compared with the dose received during various X-ray examinations.

Examination	Gonad dose per examination (m rad)
X-ray hips	53–3600
X-ray femur	50–1650
X-ray pelvis	20–3580
X-ray kidneys (retrograde pyelography)	200–3800
Pelvimetry (measurement of pelvis)	76–2500
X-ray chest	0·01–450

(Annual dose from natural sources 100)

Data from 'Ionizing Radiation and Health' by Bo. Lindell and R. Lowry Dobson, *Public Health Papers 6*, World Health Organization. The wide variation in the measured values results from studies by different workers often using different techniques, exposing different numbers of films and in some cases shielding of the gonads of males with lead.

The average yearly dose from diagnostic X-ray examinations in technologically well developed countries is about the same as that from natural radiation, although for individuals the dose may be much higher and concentrated over a very much shorter period of time. When radiation is used to *treat* disease rather than to diagnose disease, larger doses of radiation may be used. This is particularly true when radiation is used to treat malignant tumours. Because of the serious nature of the illness for which the radiation is used, it is unlikely that long-term genetic effects will result. In many countries, however, some fairly harmless conditions such as acne, plantar warts and other skin problems may be treated by radiotherapy. Obviously this treatment should only be carried out with a full realization of the potential dangers and using special precautions to protect the gonads. Exposure to radiation in industries is subject to stringent regulations, but in some countries it is still permitted to use X-ray fluoroscopy in the fitting of shoes and this can involve considerable hazard both for the customer and the shop assistant.

Chemical mutagens

Much less is known about the genetic effects of chemicals that can stimulate the appearance of mutations. Two things, however, are certain: mankind is being exposed to an increasing variety of chemicals particularly in connection with medical treatment, food processing (preservatives, colouring and flavouring agents, etc.) pesticides and so on; and a large number of chemical compounds can be shown to have genetic effects. Research in this field is still very fragmentary; many different types of organisms may be used in the studies—bacteria, moulds, fruit flies, mice, etc.—and different effects of drugs searched for (visible effects on chromosomes, abnormalities in development of the newborn, appearance of inherited single-gene mutations). It is very difficult to apply this information to man, for not all species respond in the same way to the same chemical.

The most powerful of all chemical mutagens which act on all test organisms belong to a group of compounds called the alkylating agents. These chemicals can produce effects in cells that are super-ficially similar to those caused by ionizing radiations—hence their name, radiomimetic drugs. The discovery of the radiomimetic action of these drugs arose out of wartime research on the chemical weapon called mustard gas which was first used in the 1914–18 war. Quite early in the investigations of the action of alkylating agents it was found that, like radiation, the chemicals could stop cell division. Soldiers dying of mustard-gas poisoning showed virtual atrophy of the bone marrow and the presence of widespread ulceration of the small bowel; both of these tissues have a very rapid turnover of cells and it is these tissues that are affected first by any agent which interferes with cell division. The discovery in 1943 by Charlotte Auerbach that mustard gas could also produce true mutations was a finding of great importance.

Mustard gas is the parent compound of a whole range of alkylating agents. They are called alkylating agents because their prime action is to replace a hydrogen atom in a molecule with which they react by an alkyl group, $R.CH_2$—. All alkylating agents react by the formation of an intermediate compound bearing a positively charged carbonium ion, $-CH_2{}^+$. This positively charged carbonium ion can react with a variety of negatively charged centres in biological molecules; when a dose of alkylating agent is administered to an animal, most of the compound is 'mopped up' by biological molecules such as tissue proteins (including enzymes), phosphoric acid, and so on. The most important action of alkylating agents is, however, on the chemical substance of the genetic material itself, that substance called deoxyribonucleic acid

(DNA). The DNA of a chromosome is a very long and complex molecule. It is a duplex structure consisting of a pair of intertwining strands. Prior to cell division these two strands separate, each going to a different daughter cell where each strand manufactures its partner to re-establish the 'double helix'. Some alkylating agents bear two reactive groups in their molecule, i.e. $CH_2^+ - R - CH_2^+$, and these bifunctional molecules can 'cross-link' two biological molecules. They can thus cross-link the two strands of the double helix of DNA and prevent the normal separation of the strands that precedes cell division. One can visualize (Fig. 38) that it would take only one relatively small

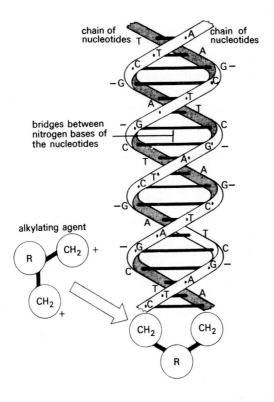

Fig. 38. *Mode of action of a molecule of an alkylating agent on the hereditary material (DNA) of the chromosome. The chemical can cross-link the two strands of DNA and thus disturb cell division. (From Clegg & Clegg 'Man Against Disease', Heinemann Educational Books.)*

molecule of an alkylating agent to produce this effect in a very long molecule of DNA. Such is the sensitivity of the genetic material to alkylating agents. This sort of action can disturb cell division and cause breakage of chromosomes. Alkylating agents can also affect single genes, but we will not pursue the detailed biochemistry of this kind of effect.

In addition to alkylating agents, a large number of other chemicals have been found to produce visible changes in the chromosomes or invisible effects on single genes that can be detected only in the offspring of affected individuals. Many of these studies are carried out on microorganisms, plants and fruit flies, and their relevance for human genetics is unknown. Certainly some substances do produce chromosome changes in man, substances such as LSD or methyl mercury, but again the significance of these effects is unknown. Recent work at the Karolinska Institute in Stockholm, for example, has shown that people who have eaten fish contaminated by methyl mercury (mercury levels ranging from 1–7 mg/kg fish, i.e. 1–7 parts per million) showed evidence of abnormally frequent chromosome breakages in cells. Methyl mercury is one of those persistent poisons that contaminate our environment. The mercury arises from industrial wastes that are discharged into waterways or into the sea. The inorganic mercury in the effluent is converted into organic mercury (methyl mercury) by the microorganisms and minute plants and animals of fresh and sea water. As these minute organisms are eaten by larger animals the organic mercury is passed along the food chain, where it can become concentrated in the bodies of the larger fish that are eaten by man. The presence of methyl mercury (1–2 parts per million) in tinned tuna fish resulted in the condemning of vast stocks of this food in the U.S.A.

These examples serve to illustrate the great care that has to be taken before new chemicals are added to foodstuffs. The last example shows how our garbage may return to us in an unsuspected and dangerous guise. The tipping of industrial waste (not merely radio-active wastes) into the vast oceans may *seem* to be a harmless practice, but in reality it may have potential far-reaching consequences for man's health in the future.

Consanguineous marriages and other patterns of behaviour

There are various aspects of human reproductive behaviour that can affect the number of children affected by genetic disease. Some habits (in particular marriage of blood relations) can increase the risk of bearing diseased children while others (e.g. efficient contraceptive techniques used in older women) can have the opposite effect.

The marriage of blood relatives increases the risk that characters that are determined by recessive genes or by polygenes will appear in the offspring. In terms that we have previously defined, consanguineous marriages tend to reveal the hidden genetic load of a population. Results of recent studies show that consanguineous marriages do not produce a greater number of miscarriages or stillbirths, but from birth onwards the death rate in children of such marriages is higher than that of a control group. There is also an increase in the number of children born with major congenital abnormalities. In one Japanese city it was found that in the offspring of first cousins there was a death rate of 116 per 1000 during the first eight years of life, compared to 55 per 1000 in controls. The number of major congenital abnormalities in children of first-cousin marriages was about double that of controls. Studies in America, where there is a lower total death rate, showed even greater differences. In one city the offspring of first-cousin marriages had a death rate of 81 per 1000 by the age of ten years compared with 24 per 1000 in controls.

In many communities first-cousin marriages are now uncommon (not more than 3 per 1000 marriages) and is still decreasing; but in some communities these marriages are still relatively common. One factor which probably affects the frequency of consanguineous marriages is the size of the family. A reduction in the size of families might result in a decrease in the frequency of consanguineous marriages; with fewer children in the various branches of a family tree and with increasing mobility of families, particularly in developed countries, there are fewer cousins, with fewer opportunities to meet and marry.

In developed countries there is an increasing tendency for women to produce their children while they are young, and then perhaps to return to work when they are in their thirties or forties. This pattern, which depends, of course, on efficient means of family planning, has various advantages for a community, not the least of which is that it can reduce the frequency of several inherited diseases which appear more frequently in the offspring of older mothers. Mongolism and disease due to Rhesus incompatibility of mother and foetus are well-known examples of this effect of maternal age. In mongolism the frequency of affected infants rises rather slowly until the mother reaches the age of thirty-five years (page 53) but then it becomes very rapid as the mother gets near the end of her child-bearing period. The overall frequency of mongolism is about 1·5–2·5:1000, although there are geographical variations which *may* be due to differences in maternal age in different populations. In the age group 35–39 there is a 2–4 fold increase in the frequency of mongolism and this rises to a 5–10 fold increase for the age group 40–44; after the age of forty-five there is a 10–20 fold increase in the frequency of mongolism.

Prevention and treatment of genetically determined disease

We have already mentioned many aspects of the control of these diseases and the future possibilities of nullifying the effects of some abnormal genes. The following is a list of the important preventative measures.

1. Control of exposure to mutagenic agents—X-rays and other ionizing radiations and chemicals.

2. Scientific study of the many substances used as drugs, food additives and preservatives to detect their ability to promote mutations, particularly in mammals. When suitable test procedures have been developed, legislation should follow to prescribe tests for mutagenic properties, particularly for newly introduced compounds.

3. Some radiomimetic chemicals are used in the treatment of various diseases but mainly for malignancies. These chemicals tend to be strongly mutagenic and this raises the question of whether patients under treatment should procreate. At the moment this is a small problem, at least in terms of the numbers involved.

4. Health education of the public about the known risks of consanguineous marriages and the effect of maternal age on genetic disease in the offspring. This education would be of little value in the absence of parallel developments of efficient and acceptable family planning services.

5. Genetic counselling. The implementation of many of the above measures awaits the development of government awareness and decisions, of scientific research and the establishment of appropriate health services. Genetic counselling is, however, a service that is available now for those individuals who need advice on the known risks of producing children that are affected by a particular genetic abnormality. These services will be discussed more fully in a later section.

6. Eugenics, which is the science of improving the human stock. Genetic counselling is one aspect of eugenics although the benefits of counselling are felt mainly by the particular family. Eugenic measures are designed to improve the health of the population as a whole. At present the main aim of these measures is to reduce the total burden of genetic disease, and this aspect is called negative

eugenics. Attempts to actually improve health—to improve such features as resistance to disease, intelligence or other attributes—are called positive eugenics.

Negative eugenic measures can never hope to eradicate completely the inherited burdens that handicap many individuals. Spontaneous mutations and the coming together of carriers of recessive genes will always bring about new cases of genetic disease. Even if every male haemophiliac were compulsorily sterilized, new cases of the disease would always appear, either because of spontaneous mutations in the germ cells of normal individuals or because a woman carrying the abnormal gene 'hidden' on one of her X chromosomes contributes the abnormal gene to one or more of her sons. The aims of negative eugenics are thus to try to *reduce* the frequency of hereditary diseases.

One obvious measure of negative eugenics is to discourage individuals from having children if they suffer from serious inherited diseases that carry a high risk of transmission to their offspring. This aim can sometimes be achieved by means of education and counselling. Some might argue that these individuals should be compulsorily sterilized or that pregnancies of such potentially dangerous unions should be compulsorily terminated. These decisions are vexed by moral and political issues and by a popular view of the inalienable right of every human being to reproduce his or her kind, whatever the consequences for the offspring or for the economic or medical burden to society as a whole. Compulsory measures have, however, been advocated, and sometimes used, for cases where genetic counselling and family planning advice (including *voluntary* sterilization) are likely to be ineffective. This is the eugenic problem of those individuals who suffer from severe degrees of mental retardation. There is clear evidence of a genetic influence in many mentally-subnormal individuals. These individuals moreover tend to have larger than average families and nowadays these children have an increased chance of survival because of improvements in medical and social care.

Positive eugenics does not really come within the scope of our present discussion but we will consider it briefly for the sake of completeness. Positive eugenics involves a planned programme to make the whole population (not just particular families) less prone to disease, more intelligent and so on. This sort of programme involves even more violation of the liberty of the individual than does negative eugenics. It might mean considerable restriction on the choice of a mate or even the use of artificial insemination of women using semen from a donor of optimum genetic constitution. Artificial insemination *is* carried out today, usually because a husband is sterile or because he is unable, for some reason, to impregnate his wife with semen. In the former case a donor's seminal fluid, and in the latter the husband's own seminal fluid is introduced into the neck of the womb of his wife. In the future it is possible that some forms of sterility in women may be treated by implanting into the womb of the barren woman an egg obtained from a normal woman. This procedure, like that of artificial insemination, could have important implications for the techniques of positive eugenics. However, in a world that is coping quite inadequately with the techniques of family planning to limit the astronomical growth of populations, these notions of 'positive eugenics' are mere pipe dreams. It is worth remembering, however, that in recent history positive eugenics of a sort was practised in Nazi Germany, involving not only the brutal elimination or sterilization of unwanted 'non-Aryan' (usually Jewish) stock, but also the pairing of chosen 'brood mothers' with approved males who showed the physical features of presumed excellence.

The treatment of inherited disease

Many of the diseases caused by genetic defects are virtually untreatable. It is, however, becoming increasingly true that a number of these diseases can be prevented or treated. We have looked at the treatment of that inherited metabolic disorder called phenylketonuria (page 61). Another inherited metabolic disorder, galactosaemia, can also be treated. If the diagnosis of galactosaemia is not made then early death is common, and even if the disease is not fatal there is a severe disability. Treatment consists in removing the sugar lactose from the diet; if this is done then physical and mental development are normal. Another example of a completely successful treatment of an inherited disease is the surgical correction of congenital pyloric stenosis. In this disease the pylorus, that part of the stomach that leads into the first part of the small intestine, is elongated and thickened. As it is difficult for food to pass through this thickened part of the stomach, the disorder shows itself by vomiting of feeds, beginning a few weeks after birth. This disease is inherited but in a rather complex manner. Before the advent of surgical correction of the defect, mortality was high, but in spite of this the disease maintained itself at a fairly high frequency, affecting as many as one in 150 in the population. The mortality of the disease is now very low following the introduction of a simple surgical procedure —Ramstedt's operation—in which the hypertrophied muscles of the pylorus are split, leaving the underlying mucous membrane intact.

Even if the genetically determined disorder cannot be completely corrected then considerable improvement in the condition is often possible. Congenital hare-lip and cleft palate and many orthopaedic disorders such as congenital dislocation of the hip or club foot can be considerably improved by means of appropriate treatment.

There is one special group of diseases in which there is an inherited susceptibility to certain drugs. The affected individual may be completely or relatively healthy until the particular drug is given, and then there may be diastrous consequences. General anaesthesia today commonly involves the use of a drug which relaxes muscles so that only small amounts of anaesthetic are needed, just sufficient to maintain unconsciousness. A commonly used muscle relaxant is one called 'scoline' (suxamethonium). This drug, like other muscle relaxants, paralyses all muscles, including the muscles used in breathing: this means that patients have to be artificially ventilated whilst they are under the influence of the drug. Most people metabolize and destroy this drug fairly rapidly, within ten minutes or so after receiving a relaxing dose, after which they begin to breathe using their own muscles. Some individuals, however, have a genetically determined abnormality of the enzyme in the body, called pseudocholinesterase, which destroys this relaxant drug. The action of the drug may be very prolonged in these individuals and sometimes normal breathing never returns and the patient dies. It is probable that the number of people who are sensitive to this drug is as high as 1:100.

We have already discussed the defect of enzymes in the red blood cells caused by a sex-linked recessive gene (page 43). Homozygotes for the gene may be extremely sensitive to some commonly used drugs, such as sulphonamides or antimalarials, and the use of these drugs results in a rapid destruction of many of the circulating red cells.

It is unlikely that the list of genetically determined sensitivities to drugs will grow rapidly and become a subject of great practical importance. Once these sensitivities are recognized then public health measures can be aimed at controlling their effects.

Genetic counselling
Genetic clinics

We can now go on to look at the main service available today for the prevention of genetically determined disease; this service is genetic counselling. From the preceding discussions it is obvious that the number of people who could benefit from genetic counselling is not enormous, but on the other hand the number is not negligible. Genetic counselling can not only prevent inherited disease, but it can also help the early diagnosis of these diseases; if parents who are likely to produce children who are at risk refuse to take contraceptive advice, then their offspring can be intelligently examined early in life with a view to determining the presence of inherited disease. Early diagnosis can be important because it enables treatment to be started early in life. In clinics that specialise in genetic counselling it has been found that the bulk of enquiries come from couples who have already had a child suffering from some disorder and who want some reassurance that there is little risk of another child being affected in a similar way, or who at least want to know the risks involved. The remaining 10 per cent or so of enquiries are from people with some abnormality in themselves or in their family history that they fear they will pass on to their children. This, then, is one aspect of genetic counselling, the advising of couples who come to the clinics for information about the risks of passing on particular disorders to their children. A second and potentially much more important aspect of genetic counselling is education of the public by way of health studies in schools, lectures to adults, and the spread of information by way of the mass media—newspapers and television. This reaching of large audiences is important in areas of the world where certain harmful genes are common in the population (e.g. sickle-cell disease or thalassaemia—see below) or where habits such as consanguineous marriages encourage the appearance of genetic ill health.

Our increasing recognition and understanding of genetic disease makes it more and more possible to give useful advice. One important advance is the ability to recognise carriers of harmful genes, that is those individuals who are outwardly healthy but who are able to transmit genetic disease to their offspring. In the study of mongolism we have seen how the advances in the study of human chromosomes has made it possible to distinguish between parents that are unlikely to have a second mongol child from those in which one partner carries an abnormal chromosome so that the chances of having a second mongol child are very high. Advances in biochemistry also allow us to make more accurate predictions. The commonest form of a disease called muscular dystrophy is due to a sex-linked recessive gene. In this disease the muscles gradually weaken and atrophy, and there is an increasing disability and deformity as the disease progresses. This progressive disease eventually results in death, often as a result of a chest infection in the weakened and inactive patient. Since the disease is determined by a sex-linked recessive gene (i.e. carried on the X chromosome) it is boys who are mainly affected (see the discussion of the inheritance of haemophilia, another disease due to a sex-linked recessive gene). Females who are heterozygous for the gene suffer no disability but they are capable of passing the disease on to their sons. When a family produces a boy suffering from muscular dystrophy, it is always a problem as to how to advise sisters who may carry the gene. In the past, sisters of boys with the disease could be told only that there was a one in two chance of them being a carrier and that the risk of any of their sons being affected was one in four. However, recent progress in biochemistry has shown that those who suffer from the disease have abnormally large amounts of a certain enzyme in the blood. This enzyme, creatine phosphokinase, arises from the degenerating muscles affected by the disease; even individuals who only 'carry' the gene (i.e. heterozygotes) show raised levels of creatine phosphokinase. It is now possible to tell most of the sisters of boys suffering from the disease whether or not they carry the abnormal gene. If they are found to carry the gene then the risk of producing a son affected by muscular dystrophy can be put precisely at a one in two chance.

It must be emphasised that a high-risk figure for affected offspring need not necessarily deter a couple from producing children, provided that treatment is available that will permit an affected child to lead a relatively normal life. A decision has to be taken with full regard for the background—medical, financial, social, psychological—of a particular family. The mentally stable woman without other children and perhaps with finances adequate to ensure domestic help may be able to cope well with the perhaps prolonged treatment of an affected child. If the woman has other children, affected or not, and is lacking psychological and/or financial stability then she may well be incapable of supporting the affected child through its treatment.

Genetic advice in populations with a high frequency of particular harmful genes

A few special clinics are quite unsuitable for countries where there is a high frequency of particular harmful genes. In tropical Africa there are some areas where the sickle-cell gene is carried by a quarter or more of the entire population; this size of problem requires planning at government level, although in this particular case genetic advice of any sort is rarely, if ever, available. Some governments, however, have responded to this kind of large-scale problem. In Italy there is network for the detection of the hereditary disease called thalassaemia and this consists of an institute in Rome and sixteen provincial centres which have facilities for the detection of the disease and for premarital counselling. Thalassaemia is an important inherited cause of anaemia due to a deficiency in the synthesis of the red pigment (haemoglobin) of the red blood cells. Individuals who receive the abnormal gene from both parents (i.e. are homozygous for the gene) either die before birth or suffer from a severe form of anaemia (Cooley's anaemia) and die, usually at an early age. Individuals who receive the gene from only one parent (i.e. are heterozygous for the gene) suffer from a mild anaemia called thalassaemia minor. There is no specific cure for thalassaemia although much can be done to treat the disease. Repeated blood transfusions may be needed for severe anaemia, and surgical removal of the spleen may help some cases; in areas where thalassaemia is common this sort of treatment could obviously impose a severe burden on medical resources. In Italy tests for the detection of carriers (i.e. those suffering from thalassaemia minor) are freely available and are applied especially to school children. The population is made aware by public propaganda of the dangers involved if carriers marry one another. If finances and services were available, the same measures could be applied to areas of the world where sickle-cell anaemia is common; this could lead to a dramatic fall in the number of cases even within a single generation.

Conclusions

The third Report of the World Health Organisation Expert Committee on Human Genetics ('Genetic Counselling', World Health Organisation, Technical Report Series, No. 416, 1969) made, among others, the following recommendations: 'Genetic counselling centres should be established in sufficient numbers in regions where infectious disease and nutritional disorders are being brought under control, and the relative importance of hereditary disorders is increasing, and in areas where genetic disorders have always constituted a serious public health problem.

In some parts of the world the high frequency of certain lethal or sub-lethal genes, such as those responsible for sickle-cell and thalassaemia, will require a special genetic counselling service for carriers, as well as suitable medical facilities for the care of afflicted individuals. Since genetic counselling centres and the specialised medical and laboratory services they require are an integral part of medical care, they should be covered by health and social insurance schemes.

It will not be possible to implement these recommendations unless trained personnel are available'.

Chapter Two   Chemicals of Life

Introduction. The progress of science is often dependent upon exact observation of the material world. Before any speculation is worth while we must be certain that our data is sound. Until we know something of the structure of the animal body we cannot hope to understand how it works. Therefore we must study carefully the physical matrix upon which is centred all those activities which we call 'life'. The common factor in every living cell is a complex physical mixture of chemical compounds which we call protoplasm. The many thousands of chemical compounds which are found in protoplasm fall into three main groups so that we shall look at three groups of compounds in the faith that a knowledge of these fundamental units of protoplasm will lead us to a sounder understanding of the complex activity which occurs in protoplasm.

The study of the chemicals of the animal body is called biochemistry. Let us apply what knowledge we have of chemistry to a study of these three major groups of compounds:

1. Carbohydrates.
2. Lipids.
3. Proteins.

CARBOHYDRATES

Carbohydrates are complicated compounds consisting of the elements carbon, hydrogen and oxygen. There are always twice as many hydrogen atoms as oxygen atoms in the carbohydrate molecule. Glucose has a formula $C_6H_{12}O_6$, sucrose $C_{12}H_{22}O_{11}$.

The general formula for all carbohydrates can be stated as $C_x(H_2O)_y$ where x and y are any numbers. You will not be surprised that there are no carbohydrates with a formula CH_2O or $C_2(H_2O)_2$. Living things are never quite so simple.

Carbohydrates are commonly classified into groups depending upon the length of the molecule. One of the characteristic features of the chemistry of the element carbon (i.e. organic chemistry) is that very often one finds that the molecules are very complex. This is partly due to the fact that carbon atoms have got the capacity to make compounds by connecting themselves together into long strings or into cyclic figures thus

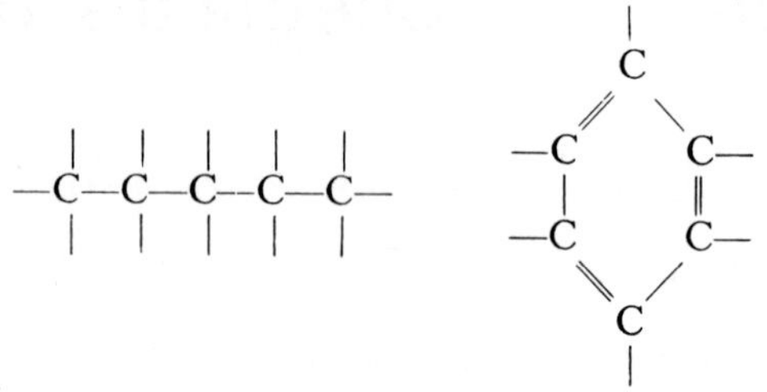

Other atoms may be attached to this carbon skeleton.

Relatively short chains of six or less carbon atoms form the framework for the MONOSACCHARIDES, e.g. glucose, $C_6H_{12}O_6$. Longer ones of twelve carbon atoms are called DISACCHARIDES, e.g. cane sugar (sucrose) $C_{12}H_{22}O_{11}$ whilst even longer ones with from 18 to 100 carbon atoms are called POLYSACCHARIDES.

We shall study the compounds in the order shown in the following classification.

1. *Monosaccharides.*
   Pentoses $C_5H_{10}O_5$ (ribose, xylose and arabinose)
   Hexoses $C_6H_{12}O_6$ (glucose, fructose, galactose)

2. *Disaccharides.* $C_{12}H_{22}O_{11}$ (lactose, sucrose, maltose)

3. *Polysaccharides.* $(C_6H_{10}O_5)_n$ (starch, cellulose, glycogen, dextrin).

Monosaccharides

These are sugars which cannot be broken down into sugars of smaller size.

All monosaccharides produce an orange red precipitate of cuprous oxide when they are boiled with Fehling's solution. This reaction is spoken of as a chemical reduction because a cupric compound is converted into a cuprous compound, that is the conversion of a more to a less electropositive compound. Any decrease in the electropositive nature of an element is spoken of as a reduction. The reaction can be thought of simply in the following way.

Fehling's solution *A* (cupric sulphate, blue copper sulphate, $CuSO_4$ $5H_2O$) is mixed with Fehling's solution *B* (which is a mixture of sodium hydroxide and potassium ammonium tartrate). The following reaction occurs:

$$CuSO_4 + 2NaOH \rightarrow Cu(OH)_2 + Na_2SO_4$$

    cupric   sodium   cupric   sodium
  sulphate + hydroxide $\rightarrow$ hydroxide + sulphate

If copper sulphate and sodium hydroxide were mixed in the absence of the tartrate the cupric hydroxide would be precipitated; the action of the tartrate is to keep the cupric hydroxide in suspension so that when the Fehling's solution is added to sugar and the mixture heated the following reaction occurs.

$$2Cu(OH)_2 + \underset{\text{from sugar}}{2H} \rightarrow Cu_2O + 3H_2O$$

cupric   → cuprous
hydroxide   oxide

We may think of the monosaccharide as supplying hydrogen to reduce the cupric hydroxide to the cuprous oxide.

Any sugar which produces an orange red precipitate when boiled with Fehling's solution is called a reducing sugar.

ISOMERISM

Monosaccharides are good examples of substances showing stereo-isomerism. Very often several compounds contain exactly the same atoms and the same numbers of each atom but nevertheless the compounds have different properties. Such compounds are called *isomers*.

It is the manner in which the atoms are linked together in the molecule which determines the properties of the compound. Thus dimethyl ether and ethyl alcohol both contain two carbon atoms, six hydrogen atoms and one oxygen atom but the structural formulae show that these two compounds have the same atoms combined in different ways.

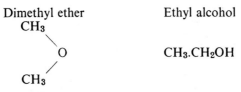

Dimethyl ether   Ethyl alcohol

CH₃

        CH₃.CH₂OH

CH₃

Ethers and alcohols have very different properties. Thus ethers do not give hydrogen chloride when treated with phosphorus pentachloride but ethyl alcohol does because it contains an OH group. In chemical language, all compounds having the same empirical formula (C_2H_6O in the above case) but different structural formulae are said to be isomeric.

Dimethyl ether and ethyl alcohol are *structural* isomers and to convert one to another would involve considerable re-arrangement of the atoms in the molecule; several bonds would have to be broken and the bonds rejoined to different atoms. Some isomeric substances differ only in the way that the constituent atoms are arranged in three dimensions. These are called *sterioisomers* and to convert one isomer to another the broken bonds would be rejoined to the same atoms but arranged to produce a different spacial arrangement of the atoms.

The commonest cause of stereoisomerism is the presence of an asymmetric carbon atom in the molecule (see below). Imagine the carbon atom as being at the centre of a triangular based pyramid and the atoms (P, Q, R, S) to which the carbon atom is linked as being at the corners of the pyramid (see fig. 39). The left-hand figure is a mirror

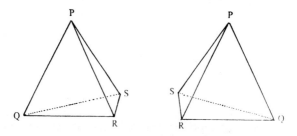

Fig. 39.   *Diagram to illustrate stereoisomeric molecules.*   The two pyramids are mirror images of one another.

image of the right-hand figure and the two are stereoisomers. Carbon atoms which are attached to four different groups, like the ones in the figure, are called asymmetric carbon atoms.

The simplest sugars can be regarded as derivatives of glyceraldehyde. This is a stereoisomer and the structure of the two isomers is shown below.

$$C^1HO \qquad\qquad C^1HO$$
$$HC^2OH \qquad\qquad OHC^2H$$
$$C^3H_2OH \qquad\qquad C^3H_2OH$$
$$\text{D form} \qquad\qquad\qquad \text{L form}$$

The isomers result from the presence of the asymmetric carbon atom 2. By convention the left-hand structure is called the D form and the right-hand structure is the L form. Naturally occurring sugars exist in the D form and L forms are usually laboratory artefacts.

Optical activity.   Stereoisomers differ in their optical activity.   One isomer will turn a beam of polarized light to the right (dextrorotatory, signified by the sign $+$) and the other isomer will turn it to the left (laevorotatory, signified by the sign $-$).   What is polarized light? This can be best understood by means of an analogy of a man flicking a long rope so that waves pass along the rope. Provided that the man only moves his hand up and down in the vertical plane the waves in the rope will only move in the vertical plane as they pass along the rope. Polarized light is like this, it vibrates only in one plane. Now let us think of non-polarized light. Here, vibrations occur in several planes simultaneously as if several ropes were being made to writhe, some in the vertical plane, some in the horizontal plane and others in oblique planes. If these several ropes were all passed through a vertical slit in a piece of wood which was firmly screwed to the floor, so that the slit was about half way along the rope, what would happen when the ropes were made to move in the several planes we have mentioned? The vibrations would only move beyond the slit in the rope which was vibrating in the vertical plane. (See fig. 40). Thus the vertical slit has

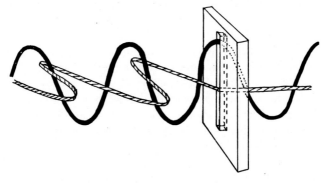

Fig. 40.   *Figure to illustrate the principle involved in the polarization of light.*

acted as a polarizing filter. Certain substances, e.g. calcite, will polarize a beam of light which is passed through them.

The investigation of optical activity—the Polarimeter.   The study of the optical activity of a compound is made with an instrument called a polarimeter.   In the polarimeter (fig. 41) light is provided by a mercury arc or a sodium flame, and the beam of light passes through

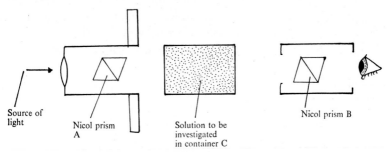

Fig. 41.   *The polarimeter.*

a Nicol prism A which is made of calcite or tourmaline. This Nicol prism polarizes the light. The eyepiece with prism B inside it is rotated so that no light can pass through it. Prism B is now horizontal whilst A is vertical. If a solution which has no optical activity is poured into container C, light will still fail to pass through the prism B. If a few crystals of sugar are now placed into container C light will quickly be seen when we look into the eyepiece. The sugar has twisted the beam of polarized light so that it now passes through the prism B. The prism B is now turned until the light disappears again. If the eyepiece has to be turned in a clockwise direction the sugar is dextrorotatory whereas if it has to be turned anticlockwise the sugar is laevorotatory.

We have now seen that monosaccharides have asymmetrical carbon atoms in them and that this gives them the property of optical activity. There are other causes of optical activity but these need not concern us here.

The interesting fact is that living organisms have the power to discriminate between D and L isomers. It is only the D isomers of monosaccharides which are useful biologically. If yeast is grown in a solution of d- and l-glucose it will ferment the d-glucose but it will not use the l-glucose at all. Such a lowly organism as yeast can thus detect the d- and l-isomers as efficiently as the most up-to-date polarimeter.

We have so far not discussed why molecular asymmetry gives rise to optical activity nor have we indicated which of the mirror-image arrangements (D and L) of, say, glyceraldehyde is dextrorotatory ($+$) and which is laevorotatory ($-$). There has been some progress in the understanding of these problems but it is beyond the scope of this book to discuss them. It must, however, be made clear that when we use the letters D and L we refer to the stereochemical structure of compounds and when we use the symbols $+$ and $-$ we refer to the optical activity of the compounds. These two descriptions of a compound may correspond; thus the 'natural' form of glucose is $D(+)$ but the natural form of fructose is $D(-)$. Historically these differences between structural and optical activity have given rise to much confusion in terminology. Because fructose is laevorotatory it was known to many chemists as l-fructose, and today the sugar is still sometimes called laevulose.

Classification of the monosaccharides is based upon the number of carbon atoms in the molecule.

(a) 5 CARBON MOLECULES—PENTOSES $C_5H_{10}O_5$

The most important pentose we shall be concerned with is ribose. Ribose and a derivative desoxyribose, $C_5H_{10}O_4$, are important units in the structure cf very complex molecules called nucleic acids. Nucleic acids are important constituents of the nuclei of cells. Ribose is also built into a complex molecule called adenosine triphosphate; this is an enzyme which is essential for the energy producing reactions of the body. (See p. 421.)

(b) 6 CARBON MOLECULES—HEXOSES

The most important examples for us are glucose and fructose. The general formula for a hexose monosaccharide is $C_6H_{12}O_6$. But this formula does not tell us much about the properties of hexoses and therefore the structural formula is more often used.

FORMULAE OF HEXOSES. The hexoses may be separated into two groups depending on whether they contain a —C═O (ketone) or

$$a —C \overset{\diagup H}{\underset{\diagdown O}{}}$$

(aldehyde) group. The former group of hexoses are called ketoses and the latter aldoses. Simpler ketones like acetone (CH_3COCH_3), will not reduce Fehling's solution whereas all aldehydes will. It may be confusing to the student of chemistry to learn that a ketone, fructose, will reduce Fehling's solution. The presence of the CH_2OH group next to the C═O group modifies the action of the ketone and allows it to reduce Fehling's solution. The formula of fructose may be written thus.

$$CH_2OH—\overset{\overset{H}{|}}{C}—\overset{\overset{H}{|}}{\underset{\underset{OH}{|}}{C}}—\overset{\overset{OH}{|}}{\underset{\underset{H}{|}}{C}}—\overset{}{\underset{\underset{O}{\|}}{C}}—CH_2OH$$

This formula explains most but not all the properties of fructose and therefore sometimes the formula is written thus

Fructofuranose

The crystalline form of fructose is the pyranose form the furan form is only known in solution.

Such a five-membered ring in a 6-carbon atomed sugar is called a furan form. Fructose is therefore often referred to as fructofuranose.

Now we must turn our attention to a very important aldose sugar called glucose. As its name suggests this is an aldehyde and of course will reduce Fehling's solution. Its formula may be written thus.

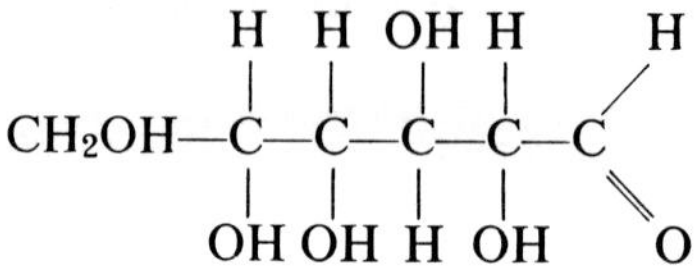

But as in fructose this straight chain formula is not useful for explaining a few of the molecule's properties and therefore a cyclic formula is sometimes used. It will be seen that this is a six sided figure—it is called the pyranose form and hence glucose is often referred to as glucopyranose,

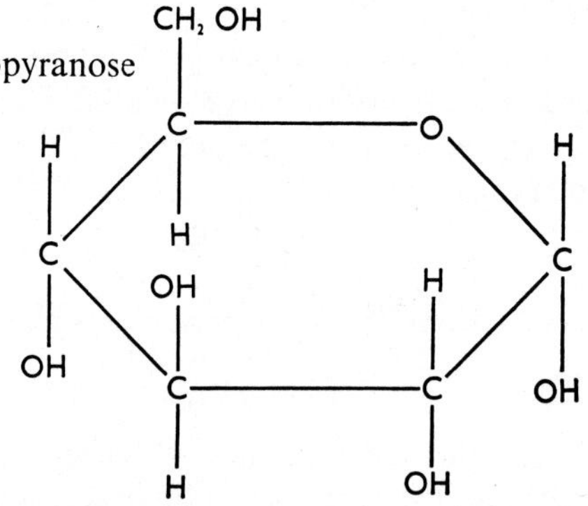

The relationship between the ring and straight line forms can be assumed to take place by the addition and then the removal of one molecule of water.

One further complication must be mentioned concerning the structure of glucose. This will be seen shortly to be of extreme biological importance. It will be noticed that the symbol α appears before the word glucopyranose in the figure. There are two forms of glucose (beside the D and L forms) depending on whether the H—OH groups attached to carbon atom number 1 are arranged like this

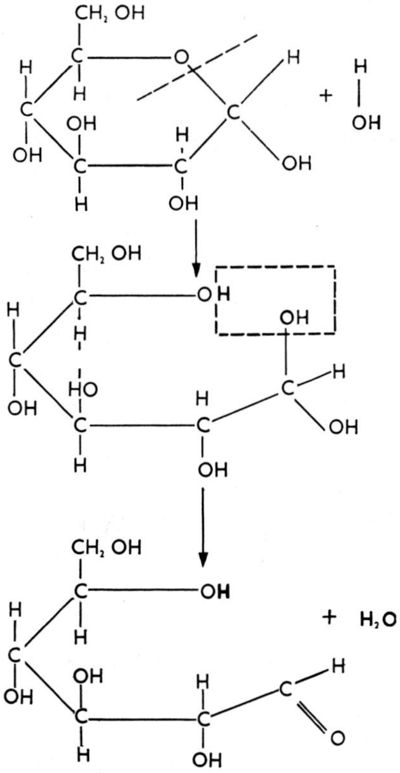

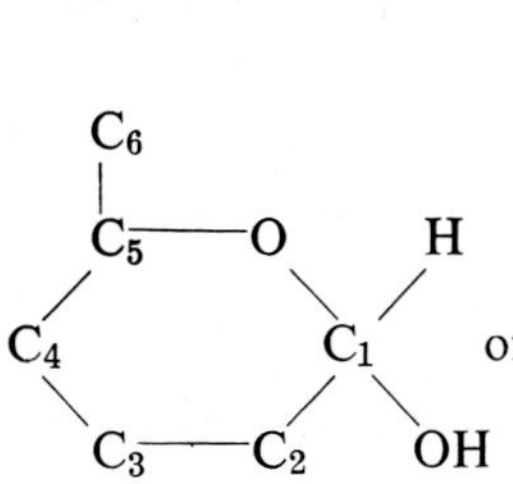

α glucose   β glucose

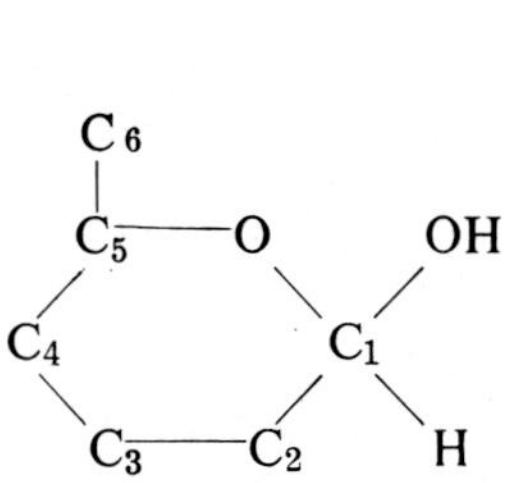

The former is called α glucose and the latter β glucose. It may seem a very trifling matter whether the H is above the OH or vice versa but it is very important. We shall find continually as we proceed in this account of life processes that minute differences in chemical structure have effects in the whole organism out of all proportion to their size. We shall see that human beings have no power to digest materials which are made of long chains of β glucose molecules but can easily digest foods consisting of long chains of α glucose units. Mammals have to call bacteria to their aid before cellulose can be digested. We shall learn more about this when we study digestion. (p. 397.)

In *summary* we now can say this about monosaccharides. They may be pentoses like ribose or hexoses like glucose and fructose. All monosaccharides are reducing sugars even although some like fructose are ketones. We have many isomers of hexoses; the optically active D and L forms; the ketoses and aldoses: the pyranose and furanose forms and the α and β forms. They all have a sweet taste, are crystalline substances, soluble in water and have known molecular weights.

Role of monosaccharides in the body. Much time has been spent describing the structure of monosaccharides because they are of extreme importance in the body. They are of importance for several reasons.

(*a*) They have the ability to move across living membranes, where other carbohydrates meet an impenetrable barrier. It is only as monosaccharides that carbohydrates can enter our bodies through the wall of the small intestine. Therefore we are dependent upon them for our supply of carbohydrates, and since carbohydrates are one of our main sources of energy we are almost dependent upon monosaccharides as a source of energy.

(*b*) Monosaccharides are an important stage in the long chain of reactions which ultimately releases the actual energy from all carbohydrates. All carbohydrates have to be converted in some way to fructofuranose sugar before they can proceed to the energy releasing chemical reactions. They form the biochemical entrance to the hall of energy release. (p. 430.)

(*c*) Because of their small molecular size and their high solubility in water when compared to other carbohydrates, monosaccharides are the form in which carbohydrates are carried in the blood stream.

Disaccharides

This is a class of carbohydrates containing sweet tasting sugars like cane sugar, common on every household table. Each molecule consists of two monosaccharide units which are joined together. Suppose two units of monosaccharide could combine together water would be eliminated: such a synthesis is called a condensation reaction. e.g.

$$C_6H_{12}O_6 + C_6H_{12}O_6 \rightarrow C_{12}H_{22}O_{11} + H_2O$$
$$\text{mono-s} + \text{mono-s} \rightarrow \text{disaccharide} + \text{water}$$

The type of disaccharide produced depends upon the kind of monosaccharide involved. Disaccharides are not synthesized by the chemist but obtained as natural products.

When disaccharides are made to react with water, either by using an enzyme like invertase or by boiling with a dilute acid, the constituent monosaccharides are released. Such a breakdown involving the uptake of water is called a hydrolytic reaction.

We shall only consider two disaccharides, maltose and sucrose.

Maltose. $C_{12}H_{22}O_{11}$

1. When hydrolyzed by the enzyme maltase (found naturally in the small intestine of many mammals) or by boiling with dilute hydrochloric acid, maltose is hydrolyzed and breaks down completely into glucose.

$$\text{Maltose} + \text{Water} \rightarrow \text{Glucose} + \text{Glucose}$$
$$C_{12}H_{22}O_{11} + H_2O \rightarrow C_6H_{12}O_6 + C_6H_{12}O6$$

We may thus represent maltose as being the condensation product of two glucose molecules.

2. Maltose will reduce Fehling's solution so that we must conclude that the aldehyde groups are not all involved in the linkage of the two glucopyranose molecules but are left free to reduce the cupric hydroxide.

Sucrose (Cane sugar) $C_{12}H_{22}O_{11}$

1. When hydrolyzed by the enzyme invertase or by boiling with dilute mineral acids a mixture of glucose and fructose is produced.

$$\text{Sucrose} + \text{Water} \rightarrow \text{glucose} + \text{fructose.}$$
$$C_{12}H_{22}O_{11} + H_2O \rightarrow C_6H_{12}O_6 + C_6H_{12}O_6$$

We may represent the structure of sucrose graphically as follows

Glucopyranose unit   Fructofuranose unit

2. Sucrose will not reduce Fehling's solution unless it is first hydrolyzed into its component monosaccharides. It is therefore classified as a non-reducing sugar.

Demonstration of the presence of sucrose. Dissolve a little sucrose in water. Add a few drops of dilute hydrochloric acid and boil. The sucrose will now be hydrolyzed to monosaccharide. Make sure the solution is neutral or slightly alkaline by adding caustic soda solution. Add equal quantities of Fehling's solution A and B. Warm the solution gently. An orange red precipitate of cuprous oxide is formed. (N.B. If the HCl is not neutralized the copper oxide dissolves as soon as it is formed and no red ppt is seen.)

This is not a specific test for sucrose as the same result may be obtained using any non-reducing sugar. If it were necessary to identify the particular sugar present more complex tests would be needed, e.g. the formation of osazones using phenyl hydrazine. Osazones have crystals of characteristic shapes depending upon the kind of sugar used in their formation. This can be verified by trying the following experiment.

Preparation of osazones using glucose, maltose and lactose. Place 10 mls. of 1% glucose into a boiling tube. Add enough phenyl hydrazine hydrochloride as will cover a new penny. Add enough sodium acetate to cover two pence then add 1 ml. of glacial acetic acid. Mix and dissolve by gently warming. Filter into a clean tube and place the tube in a beaker of rapidly boiling water for 30 minutes. Turn off the gas and allow the tube to cool in the beaker of water. In several hours a yellow crystalline osazone will have crystallized out. Transfer a few crystals onto a microscope slide and examine under the microscope. You will see fine yellow needles of glucosazone arranged in fans, small groups or crosses.

The same procedure is followed to make maltosazone or lactosazone but the solutions should be boiled for 60 minutes instead of 30 as the osazones do not form as easily as does glucosazone.

Maltosazone crystallizes in broad plates whilst lactosazone crystallizes in clusters resembling hedgehogs.

To find out whether a non-reducing sugar is present in a mixture of unknown composition. The investigation is in two parts

1. First it is necessary to find out whether a reducing sugar is present. If a reducing sugar is present we cannot easily find out whether a non-reducing sugar is there at the same time by using Fehling's solution. The reason for this will be clear if we look back at the 'demonstration of the presence of sucrose'. A reducing sugar like glucose would give a red precipitate with Fehling's solution. It would be necessary to remove all traces of red ppt and of glucose before we could boil with HCl and add Fehling's to test for the non-reducing sugar.

2. If a reducing sugar is *not* present we can add dilute HCl, boil and add Fehling's A and B. If an orange red ppt is formed then a non-reducing sugar is present.

Summary of investigation

A. Add equal amounts of Fehling's A and B. Warm.

Red ppt . . . . Reducing sugar present

No red ppt . . . Reducing sugar absent, go on to B.

B. Add dilute HCl to some fresh unknown solution. Boil. Add Fehling's A and B. Warm.

Red ppt . . . . A non-reducing sugar present.

No red ppt . . . No sugars of any kind present.

Polysaccharides

These are long chain carbohydrate molecules consisting of many condensed monosaccharide units. Unlike the disaccharides they do not taste sweet. We shall mention only four polysaccharides: starch, cellulose, glycogen, inulin.

Starch. It is the principal reserve carbohydrate in plants and is stored in plant tissues as granules of different sizes and shapes in different species of plants. The type of starch grain is peculiar to each species of plant. The food plant of the diet of an animal may be identified by looking at the structure of the starch grains.

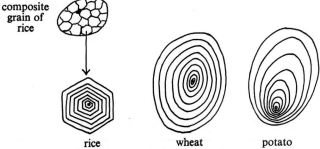

composite grain of rice

rice   wheat   potato

Fig. 42. *Diagram of various kinds of starch grains.*

STRUCTURE OF THE STARCH MOLECULE. The general formula is $(C_6H_{10}O_5)_n$ where n is a number from 300–1,000. The monosaccharide units are entirely glucopyranose units and the linkages are mainly between α glucose units. The link is between C atom 1 and C atom 4 and called a 1 : 4α link. There may be a few β linkages in the chain as well. We may represent the starch molecule graphically as follows:

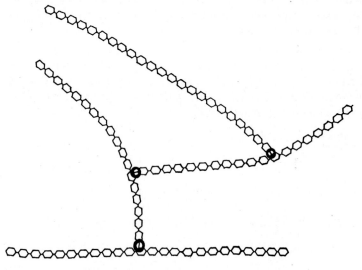

1 : 4 α Link

By allowing starch to soak in water for several hours it is possible to separate it into two portions called amylose and amylopectin. Amylose is a helically coiled molecule which is made up of an unbranched chain of about 300 glucopyranose molecules. It is soluble in water and on hydrolysis it will break down fully to maltose.

Amylopectin is not as soluble in water and consists of a maze of interlocking chains as shown in the diagram. Each chain has from 20–24

Fig. 43. *Representation of an amylopectin molecule.* The circles represent 1 : 6 linkages.

monosaccharide units in it. There are an unknown total number of chains in the molecule. The points at which circles are drawn in the diagram have peculiar linkages which are not of the 1:4α type; they are linkages between carbon atom 1 and 6 and thus they cannot be broken by means of 1: 4α glucoside enzymes, but only by a 1: 6α glucoside enzyme.

When a 1 : 4α glucoside enzyme, such as the amylase found in plants, attacks starch it attacks the straight chains of amylose and then attacks amylopectin until it reaches the 1 : 6α links which it cannot destroy. Thus the amylopectin loses the ends of its chains as shown in the diagram. The 'pruned' molecule of amylopectin is called dextrin, and gives a purplish colour with iodine.

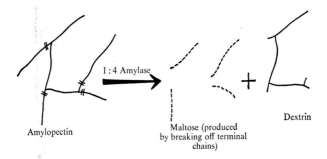

Amylopectin   1 : 4 Amylase   Maltose (produced by breaking off terminal chains)   Dextrin

If allowed to act for long periods of time the 1 : 4 amylase in some way, which at present is not understood, can straddle the 1 : 6 links and may degrade about 80–90% of amylopectin. Thus after a very long time the main product of breakdown is 1 : 4 maltose; but in addition there is a small percentage of disaccharide with a 1 : 6 link (e.g. iso-maltose). The breakdown of these 1 : 6 links needs an enzyme called an amylo-1 : 6- glucosidase. These enzymes capable of attacking 1 : 6 links have been isolated from muscle, beans and potatoes; they are sometimes called R enzymes.

The digestive enzyme ptyalin and pancreatic amylase of mammals can attack the amylopectin in a similar way to amylase in plants but they attack the amylopectin without needing long periods of time (see later experiments with starch and saliva).

TEST FOR STARCH. Place enough starch to cover the tip of a small spatula into a boiling tube. Add 10 mls. of water and boil. The opaque white solution produced is a peculiar solution called a colloidal gel. (More is said about these solutions in Chapter I.) Cool the starch solution and then add a few drops of iodine dissolved in potassium iodide solution. (Iodine is insoluble in water.) A deep blue colour is seen.

This colour is characteristic and indicates the presence of a complex starch iodide molecule. This molecule is very unstable and when the solution is warmed it breaks up. This is why you can only test for starch in cool solution.

Try the effect of warming a solution of starch iodide. What happens when the solution cools again?

Glycogen, animal starch. $(C_6H_{10}O_5)_n$. Glycogen consists of hundreds of monosaccharide units condensed together. It resembles amylopectin in that it has a branching molecule. Each short chain is about eighteen units long. The linkages (except in the branch-chain links) are of the α type and the number of chains of saccharides involved is not known exactly, but it is a very large number.

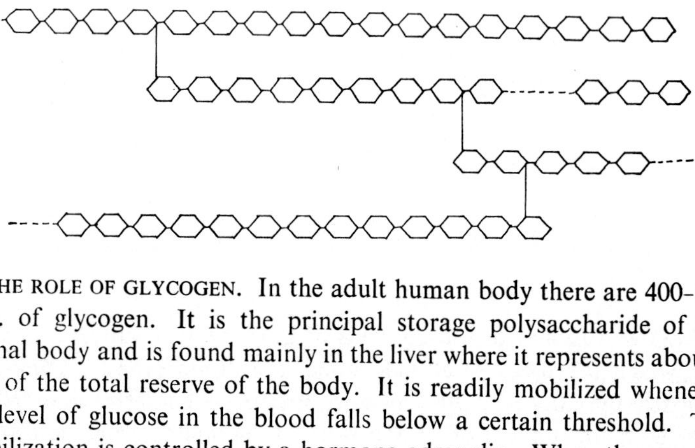

THE ROLE OF GLYCOGEN. In the adult human body there are 400–500 gms. of glycogen. It is the principal storage polysaccharide of the animal body and is found mainly in the liver where it represents about a half of the total reserve of the body. It is readily mobilized whenever the level of glucose in the blood falls below a certain threshold. The mobilization is controlled by a hormone adrenalin. When the amount of sugar in the blood is above the normal level as it will be for instance after a meal, the excess sugar is converted to glycogen in the liver. This reaction is controlled by insulin. (See p. 262.) Glycogen is also found in the brain and in muscle. Starvation will remove liver glycogen but the glycogen in brain and muscle is not rapidly removed. In brain and muscle glycogen acts as a store of energy and is converted into glucose when needed.

Cellulose. This substance is not found in mammals but since it is the principal constituent of the cell walls of plants it is a common substance in the diet of many mammals. Its function in the plant is one of structure and rarely one of food storage. Like amylose it is made of glucopyranose units linked together in long chains. It differs from amylose in two respects. Firstly the linkages are predominantly β linkages and secondly the molecule is not coiled in cellulose. The long chains of saccharide units are built into bundles called micelles. The micelles are part of a micro-fibril, many of which go to make one fibre of cellulose. (See fig. 44.)

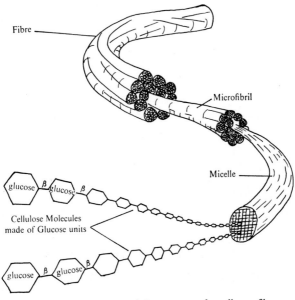

Fibre

Microfibril

Micelle

glucose — β — glucose — β

Cellulose Molecules made of Glucose units

glucose — β — glucose — β

Fig. 44. *Representation of the structure of a cellu ose fibre.*

Because mammals, even herbivores, have no enzyme to split the β glucose links they cannot digest cellulose unaided. Herbivores commonly harbour bacteria inside their guts, and these bacteria split the β links for the mammal (see p. 397). The bacteria in return get protection inside the gut. Such a cooperative association for mutual benefit is called symbiosis.

Inulin. This is a polysaccharide found commonly in the group of plants known as the Compositae (Sunflower, Dahlia, Artichoke and Dandelion family). It is not found in animals. Its best known source is perhaps the swollen underground stem tuber of the Jerusalem artichoke. It is a food storage substance. The basic repeating unit here is fructose the ketohexose monosaccharide.

Other saccharide containing compounds

There are many substances which contain saccharides as just a part of their molecule, e.g. gums, mucilages and many potent drugs like digitalis the heart stimulant which comes from the foxglove.

LIPIDS

Under this heading we shall consider four groups of substances:
1. Fats.
2. Phospholipids.
3. Waxes.
4. Sterols.

Fats

Fats are neutral, water insoluble organic compounds containing carbon, hydrogen and oxygen. They have only a small proportion of oxygen relative to hydrogen (cf. carbohydrate with the ratio of H to O of 2 : 1). Chemically they are classified as esters of glycerine and fatty acids.

In inorganic chemistry a very fundamental equation is:—

$$\text{acid} + \text{base} = \text{salt} + \text{water}$$

$$HCl + Na\,OH = NaCl + H_2O$$

$$\begin{array}{ccc} \text{Hydrochloric} + \text{Sodium} & = & \text{Sodium} + \text{Water} \\ \text{acid} & & \text{hydroxide} \;\; \text{chloride} \end{array}$$

In organic chemistry the homologous equation is

$$\text{acid} + \text{alcohol} = \text{ester} + \text{water}$$

$$CH_3COO\,H + C_2H_5\,OH = CH_3COOC_2H_5 + H_2O$$

$$\begin{array}{ccc} \text{Acetic acid} + & \text{Ethyl} & = & \text{Ethyl} + \text{Water} \\ & \text{alcohol} & & \text{acetate} \end{array}$$

In the formation of fats the alcohol must be glycerol. It is the —OH group which gives it the properties of an alcohol.

$$
\begin{array}{lllll}
R_1\ COO\,H & CH_2\,OH & & CH_2\ COOR_1 & \\
& | & & | & \\
R_2\ COO\,H & + & CH\,OH & = & CH\ COOR_2 & + 3H_2O \\
& | & & | & \\
R_3\ COO\,H & CH_2\,OH & & CH_2\ COOR_3 & \\
\text{Fatty} & + & \text{Glycerol} & = & \text{Fat} & + \text{Water} \\
\text{acids} & & & &
\end{array}
$$

Note that there can be three different kinds of fatty acid associated together with glycerol. The fatty acid involved has an even number of carbon atoms.

e.g. Stearic acid $C_{18}H_{36}O_2(C_{17}H_{35}COOH)$, a saturated acid.
Oleic acid $C_{18}H_{34}O_2(C_{17}H_{33}COOH)$, an unsaturated acid.
Palmitic acid $C_{16}H_{32}O_2(C_{15}H_{31}COOH)$, a saturated acid.

Note the low percentage of oxygen in the above fatty acids. The initial letters of the above fatty acids can be arranged to form the word SO(A)P. It is perhaps useful to remember this fact because when fats are boiled with caustic soda, sodium soaps are produced.

The solidity of fats depends upon two factors, the temperature and their saturation (see below). Esters of glycerol which are solid at room temperature are called fats, whereas if they are liquid at room temperature they are called oils. It should be noted that we are referring here to oils like olive oil or whale oil and not to the mineral oils used in motor cars. The latter are not esters of glycerol.

SATURATION. Stearic acid has the formula

i.e. $C_{17}H_{35}$————————————COOH

It can be seen that the available valency bonds of the carbon atoms are fastened on to hydrogen atoms. Such an acid is described as being fully saturated. In oleic acid there are two hydrogen atoms less, therefore there are two carbon atoms which are not saturated. This is an unsaturated acid.

i.e. Oleic acid $C_{17}H_{33}$————————————COOH

The interesting fact is that most oils contain a high proportion of unsaturated acids whereas the more solid fats contain a high proportion of saturated fatty acids.

e.g. Linseed oil from flax contains 70% unsaturated fatty acids.
Solid fat from cocoa seed (cocoa butter) contains 75% saturated fatty acids.

Margarine is made by blowing hydrogen in the presence of a catalyst into oils (which may be obtained for example from groundnuts). The unsaturated liquid oil takes up hydrogen and becomes a saturated solid fat which is given the name margarine.

EMULSIONS. When olive oil is poured steadily into water it forms a layer on the top because it is insoluble in water and lighter than water. If the oil and water are shaken vigorously a creamy liquid is produced. The oily layer has disappeared, but the oil has not dissolved in the water. If the mixture is allowed to stand for a few minutes the oil settles out into a surface layer once again. The creamy liquid is formed when the fat is broken up into fine droplets and dispersed throughout the water. Such a mixture of two liquids is called an emulsion. Emulsions are colloidal solutions. In this example, water is called the continuous phase of the solution and olive oil the disperse phase (see p. 7). We shall see later that emulsification of fats plays an important part in their digestion. A few drops of 10% caustic soda shaken with the oil and water makes a very stable emulsion as the caustic soda forms a soap with free fatty acid which is present in commercial olive oil. The soap stabilizes the emulsion.

TEST FOR FATS

1. Fats turn red when the red stain called Sudan III is added to them. If olive oil is added to water and shaken up to form an emulsion and then a few drops of Sudan III added the whole solution goes pinky red. On standing the oil separates out and the red colour is only present in the oily layer.

2. They are stained black with osmic acid. This is because the fats reduce the osmium tetroxide to the black metal osmium.

3. They are neutral substances, soluble in ether or chloroform and make a translucent area when applied to a piece of paper.

ROLE OF FATS. (See also Chapter VIII.)

1. As a source of energy.
2. Specific functions.
3. They have some vitamins dissolved in them.

1. One gramme of fat when oxidized gives rise to 9,000 calories of heat. (A calorie is the amount of heat required to raise the temperature of 1 gm. of water by 1°C.) This is often referred to as 9 Calories, i.e. when the capital letter is used Calorie means 1,000 calories. Carbohydrates liberate only about 4 kilo calories per gramme.

1 gm. protein	5·6 Calories
1 gm. carbohydrate	4·2 Calories
1 gm. fat	9·3 Calories.

2. When fully oxidized, fat gives rise to more water than any other class of food, because of the high ratio of hydrogen to carbon in the molecule.

1 gm. protein	0·41 gm. water
1 gm. carbohydrate	0·55 gm. water
1 gm. fat	1·07 gm. water.

The production of water during the oxidation of fat is very important for some land animals which lay shelled eggs. Here the supply of water which can be given to the egg is strictly limited but this is continually supplemented by the breakdown of fat which is stored in large quantities in the eggs of reptiles and birds. The hump of a camel's back is a fat store and thus an indirect source of water.

3. In the mammal, fat is stored in various tissues. It occurs under the skin and serves as a heat insulator, reducing heat loss from the body. It is significant that it is only in the warm blooded animals, i.e. birds and mammals that we find significant amounts of subcutaneous fat. Fat is also stored around various organs of the body. For example the kidney, in mammals, is embedded in a mass of fat which has a protective function.

Phospholipids

Phospholipids are fats which contain nitrogen and phosphorus in addition to carbon, hydrogen and oxygen. An important phospholipid in the body is lecithin which has the following formula

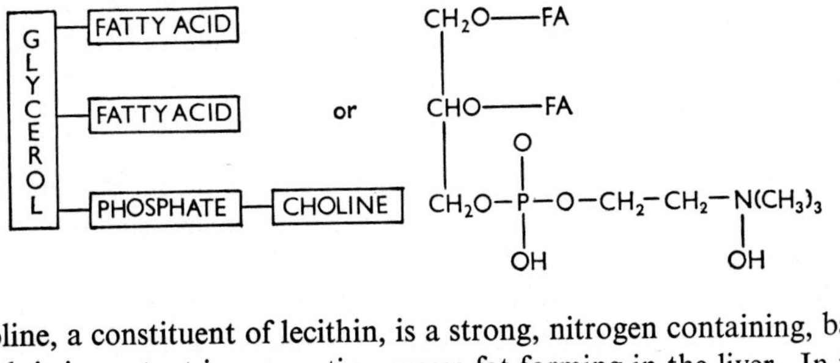

Choline, a constituent of lecithin, is a strong, nitrogen containing, base which is important in preventing excess fat forming in the liver. In the absence of choline a condition called fatty liver, which is the first stage of liver degeneration, occurs. When choline is acetylated it forms acetyl-choline, a substance which is very important in transmission of nerve impulses (Chapter VII).

Phospholipids combine the water loving (hydrophilic) properties of the phosphate group with the water repellant properties of the fatty part of the molecule. They are therefore said to be bipolar substances and at an interface where water is present the phospholipid molecules align themselves with the phosphate radicles buried in the water and the fatty tails sticking out of the water. These bipolar molecules may thus help in stabilizing interfaces and this property may explain part of their role in the structure of the plasma membrane (see fig. 11). Phospholipids are constituents of all animal and plant cells but are present in abundance in nervous tissue, heart, kidney and egg yolk.

The amount of phospholipids in nervous tissue (both in grey and white matter) is not reduced by starvation and must have some functional significance. The presence of phospholipid in the structure of the neurone may be useful as a handy source of choline for the synthesis of acetyl choline.

Phospholipids are soluble in water to some extent and are found in high concentration in the blood when fat stores are being mobilized and when fatty acids are being absorbed from the gut. (When neutral fat is absorbed via the gut it is not turned into phospholipids.) Fatty acids are absorbed by forming compounds with bile salts, then the bile salts are freed as the fatty acids are converted to phospholipids in the cells of the gut wall (see fig. 394).

The role of phospholipids may be summarized as

   i. Structural—an important component of plasma membranes.
  ii. Transport of fat.
 iii. Some are a source of choline, others are concerned with the clotting of blood (cephalin is a thromboplastin).

It is known that cobra venom contains an enzyme which splits off a fatty acid from lecithin giving a substance which causes the breakdown of red blood cells.

Waxes

Waxes are esters of fatty acids with complex alcohols other than glycerol. The alcohols are of high molecular weight; cholesterol is a common alcohol found in animal waxes.

Wool wax from the sheep is used commercially to produce lanolin which is used in pharmacy as a base for many ointments. Lanolin can absorb 80% of its weight of water and is also readily absorbed by the skin. It is therefore often called a good skin 'food'. Waxes are not used in animal nutrition as they are very resistant to hydrolysis and are not attacked by lipases.

Sterols

These are complex monohydroxy alcohols found in animals and plants. They are either found free as alcohols or combined with fatty acids to form esters. Because they are esters of fatty acids they are often classified under lipids.

Chemically the sterols are derivatives of phenanthrene and are related to the salts of the bile, produced by the liver. They have a basic ring structure like this.

CHOLESTEROL $C_{27}H_{45}OH$. Cholesterol is an important sterol which is abundant in the grey matter of nervous tissue and in the adrenal glands and ovaries. It is an alcohol present in the body both in the free state and in combination with fatty acids as esters. It is present in all tissues of the body, in the cells and the blood.

Parts of the diet are rich in cholesterol, e.g. egg yolk, butter, liver, but the body can synthesize its own supply of cholesterol from proteins or carbohydrates. Excess of cholesterol is excreted in the bile.

The functions of cholesterol include,

1. Cholesterol is probably a constituent of all cell membranes.
2. The sex hormones of the ovary, e.g. oestradiol and progesterone and the hormones of the adrenal cortex, e.g. cortisone, all have a ring structure very similar to that of cholesterol. The large amounts of cholesterol which can be extracted from the adrenal glands or the ovaries probably act as stores of raw materials from which these hormones are synthesized.
3. Vitamin D is chemically related to cholesterol. De-hydrocholesterol is a substance, present in the skin, which can be easily converted to vitamin D by the action of sunlight (ultraviolet light). (See p. 418.)

PROTEINS

Proteins are extremely complex large molecules containing the elements carbon, hydrogen, oxygen and nitrogen, and several other elements often, but not always, including phosphorus and sulphur.

The molecular weight is in the range 20,000 to several million. The complexity of the problem of working out the precise structure makes a satisfactory classification at present impossible. Classification at present is based upon certain physical properties. Thus proteins may be divided into

1. *Albumins*, e.g. egg albumin (egg white) or serum albumen. These are soluble in water and coagulate on heating.
2. *Globulins* are insoluble in water and coagulated by heat, e.g. globulins found in blood serum, with which the antibodies are associated.

There are many such classifications.

On hydrolysis proteins can produce up to about twenty types of units called amino acids. The nature of a particular protein depends upon the number, kind and arrangement of the constituent amino acids. There is an astronomical number of possible combinations of amino acids and almost as many proteins. Each species of animal builds up the amino acids to form proteins in a particular and characteristic way.

Amino acids. An amino acid is a fatty acid in which one hydrogen atom attached to the α carbon atom has been replaced by an amino group.

$$
\begin{array}{cc}
\overset{\displaystyle H}{\underset{\displaystyle |}{}} & \overset{\displaystyle H}{\underset{\displaystyle |}{}} \\
R.C-H \rightarrow & R.C^x-NH_2 \\
| & | \\
COOH & COOH \\
\text{Fatty acid} & \text{Amino acid}
\end{array}
\qquad C^x = \alpha \text{ C atom}
$$

The simplest known amino acid is glycine (amino acetic acid)

$$
\begin{array}{c}
NH_2 \\
| \\
H-C-COOH \\
| \\
H
\end{array}
$$

The next simplest is Alanine

$$
\begin{array}{c}
H \quad NH_2 \\
| \quad\; | \\
H-C-C-COOH \\
| \quad\; | \\
H \quad H
\end{array}
$$

All the naturally occurring amino acids are colourless crystalline solids and most of them are soluble in water. All, except glycine, are optically active, there being D and L isomers.

The amino group behaves in a basic manner and the carboxyl group in an acidic manner; this is the basis of the amphoteric nature of amino acids, i.e. they have acidic and basic properties.

Thus glycine (aminoacetic acid) can give two kinds of salts:

1. Sodium amino acetate
2. Glycine hydrochloride

In solution amino acids produce an ion which has a positive and negative ion of equal strength. Such an ion is called a dipolar ion or zwitterion.

$$\underset{NH_2}{R.CH-COOH} \rightleftharpoons \underset{NH_3^+}{R.CH.COO^-}$$

This ion behaves like an acid with caustic soda

$$\underset{NH_3}{H.CH.COO} + NaOH \rightarrow Na^+ \left[\underset{NH_2}{H.CH.COO}\right]^- + H_2O$$

sodium amino acetate

$$\left[\begin{array}{cccc} Cf. & HCl & + & NaOH & \rightarrow & Na^+[Cl]^- + H_2O \\ & acid & & base & & salt \qquad water \end{array}\right]$$

or as a base with hydrochloric acid.

$$\underset{NH_3^+}{H-CH.COO^-} + HCl \rightarrow \left[\underset{NH_3}{H.CH.COOH}\right]^+ Cl$$

glycine hydrochloride

Here the amino acid has combined with a hydrogen ion to produce a positive ion comparable to the ammonium ion, $(NH_4^+.)$

ESSENTIAL AND NON-ESSENTIAL AMINO ACIDS. The mammal is able to synthesize some amino acids using other amino acids which it already possesses. These amino acids which can be made by the mammal are not an essential part of the diet and are called non-essential amino acids. The amino acids which can only be obtained ready made from plants or from other animals and which the individual mammal cannot synthesize are essential constituents of the diet. They are the essential amino acids. (See p. 349.) Animal protein supplies larger amounts of essential amino acids per weight of protein than does plant protein. Animal proteins are thus called first class proteins.

ROLE OF AMINO ACIDS. Amino acids are the form in which protein is absorbed into the body. They are incorporated into the tissues of the body and built up into the various constituent proteins.

Synthesis of protein. Because amino acids are amphoteric in nature they can combine with each other, the amino group of one amino acid combining with the carboxyl group of the other amino acid. The product of combination of two amino acids is called a dipeptide. Water is eliminated in the condensation process as indicated below.

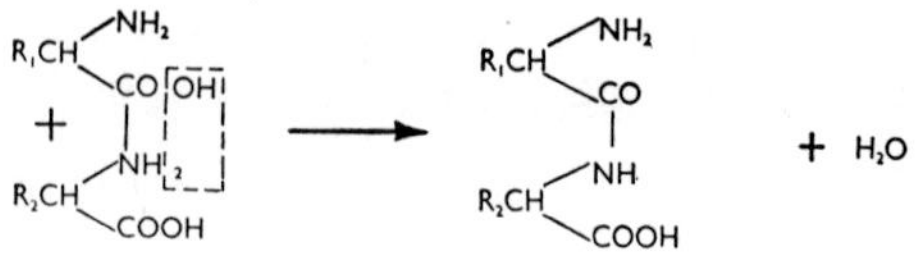

The 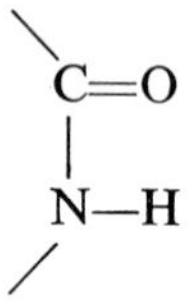
$$\begin{matrix} \diagdown \\ C{=}O \\ | \\ N{-}H \\ \diagup \end{matrix}$$
link is called the peptide link.

In space the dipeptide has the following structure.

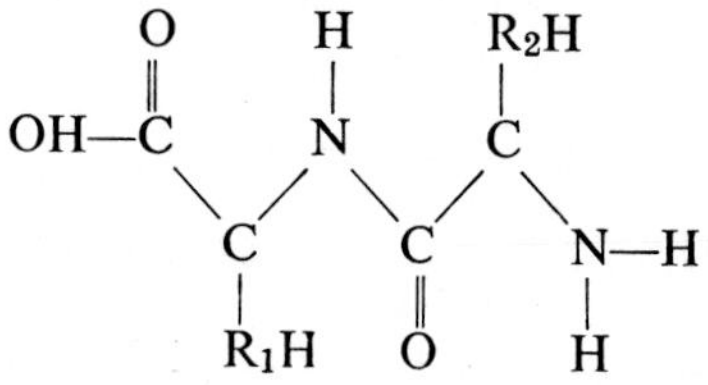

We shall see that this zig-zag structure is apparent in X-ray analysis of protein. The dipeptide still has a free amino group at one end and a free carboxyl group at the other end, and at these points more amino acids can be attached.

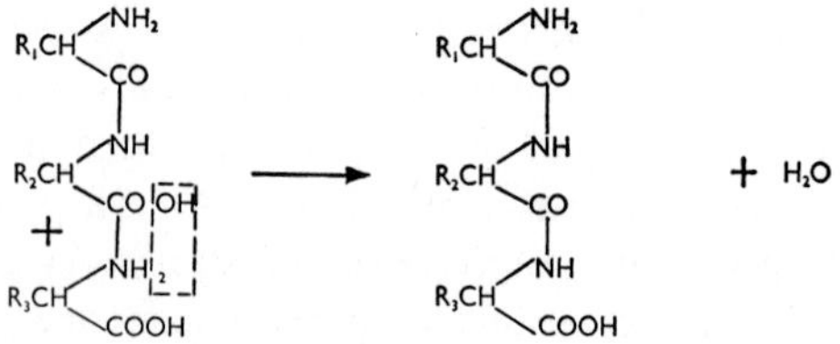

This process can go on indefinitely and eventually very long molecules are built up. The amino acids are built up into dipeptides, tripeptides and polypeptides. When the number of amino acid units reaches about three hundred we speak of proteins.

It is obvious that the backbone of the protein molecule is consistently repetitive in structure. The only variable feature is in the structure of the side chains (R_1, R_2, R_3, etc.). Each amino acid has a different side chain and these are shown below.

The twenty primary amino-acids and their side chains (R-groups).

Glycine (Gly)	—H
Alanine (Ala)	—CH_3

Valine (Val)　　　　　$-CH\begin{smallmatrix} CH_3 \\ \\ CH_3 \end{smallmatrix}$

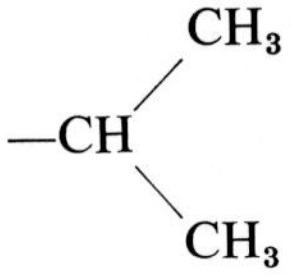

Leucine (Leu)　　　　$-CH_2-CH\begin{smallmatrix} CH_3 \\ \\ CH_3 \end{smallmatrix}$

Isoleucine (Ileu)　　　$-CH\begin{smallmatrix} CH_2-CH_3 \\ \\ CH_3 \end{smallmatrix}$

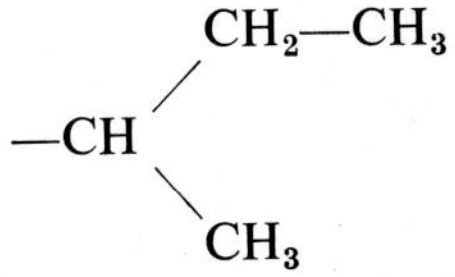

Phenylalanine (Phe)
$$-CH_2-C \overset{\displaystyle CH=CH}{\underset{\displaystyle CH-CH}{\Big\langle}} CH$$

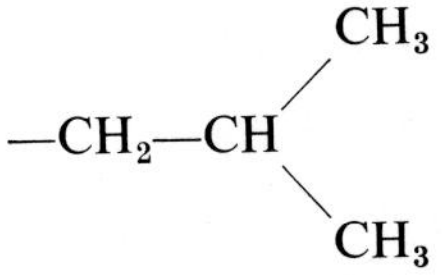

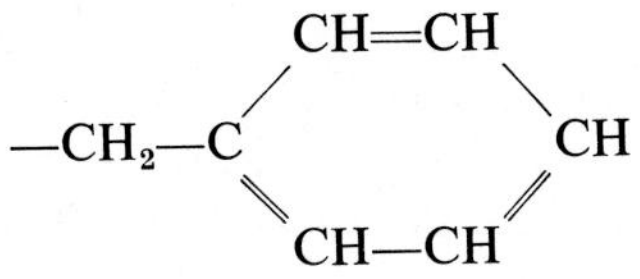

Proline (Pro)
$$O=C \overset{\diagup}{\diagdown} \begin{matrix} CH-CH_2 \\ | \quad\quad CH_2 \\ N-CH_2 \end{matrix}$$

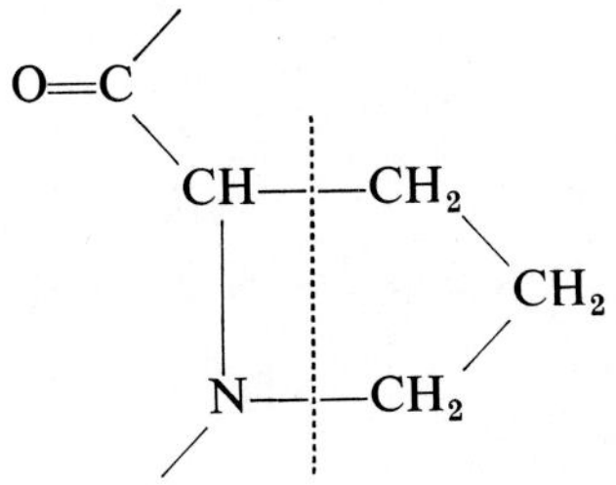

Tryptophan (Try)
$$-CH_2-C-C\overset{\displaystyle CH}{\diagdown}CH$$
$$CH \quad C \quad CH$$
$$NH \quad CH$$

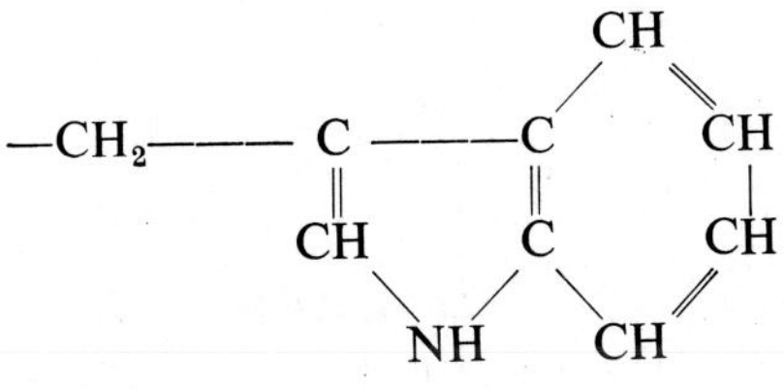

Serine (Ser)　　　　　$-CH_2-OH$

Threonine (Thr)

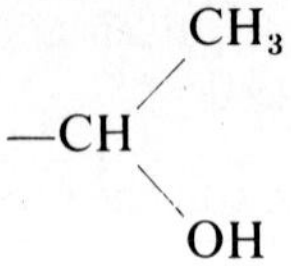

$$-CH \overset{\displaystyle CH_3}{\underset{\displaystyle OH}{\big<}}$$

Cysteine (CysH) $-CH_2-SH$

Methionine (Met) $-CH_2-CH_2-S-CH_3$

Aspartic acid (Asp)

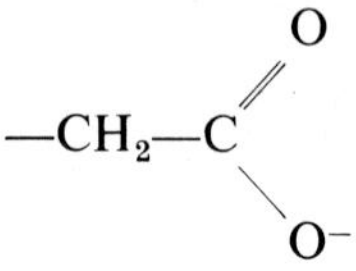

$$-CH_2-C \overset{\displaystyle O}{\underset{\displaystyle O^-}{\big<}}$$

Glutamic acid (Glu)

$$-CH_2-CH_2-C \overset{\displaystyle O}{\underset{\displaystyle O^-}{\big<}}$$

Asparagine (Asp NH_2)

$$-CH_2-C \overset{\displaystyle O}{\underset{\displaystyle NH_2}{\big<}}$$

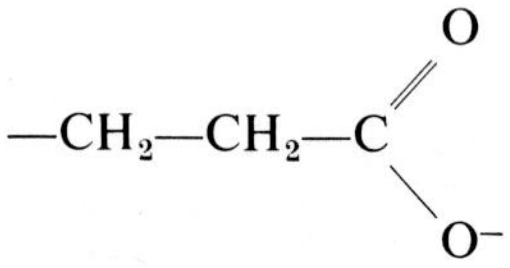

Glutamine (Glu NH_2)

$$-CH_2-CH_2-C \overset{\displaystyle O}{\underset{\displaystyle NH_2}{\big<}}$$

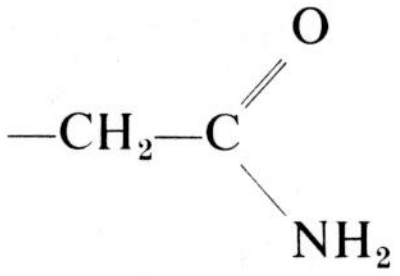

Tyrosine (Tyr)

$$-CH_2-C \overset{\displaystyle CH=CH}{\underset{\displaystyle CH-CH}{\big<}} C-OH$$

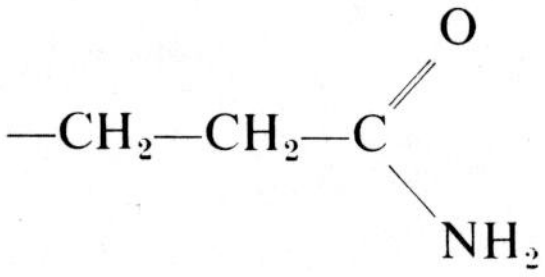

Histidine (His)

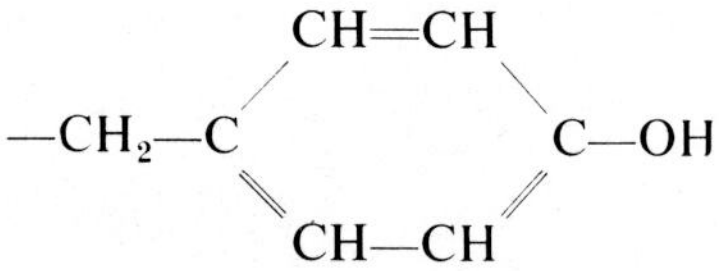

$$-CH_2- \; C \!=\!=\! CH$$
$$\qquad NH \qquad N$$
$$\qquad\quad \searrow \quad \swarrow$$
$$\qquad\qquad CH$$

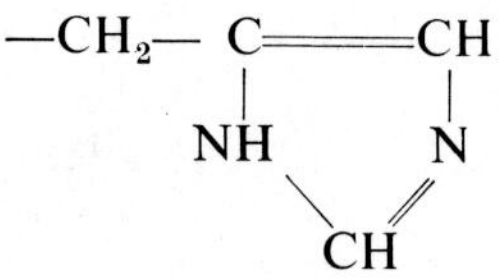

Lysine (Lys) $-CH_2-CH_2-CH_2-CH_2-\overset{+}{N}H_3$

Arginine (Arg)

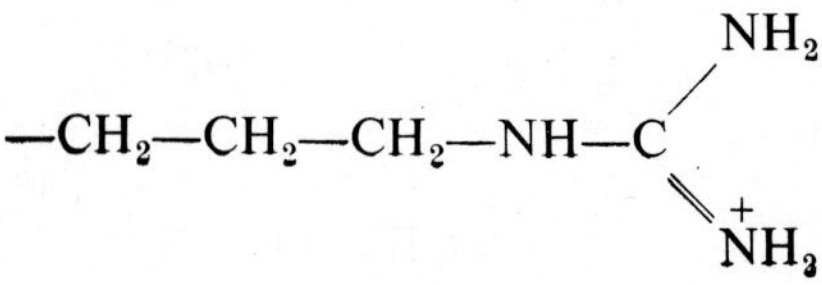

$$-CH_2-CH_2-CH_2-NH-C \overset{\displaystyle NH_2}{\underset{\displaystyle \overset{+}{N}H_3}{\big<}}$$

Because of the various side chains along the molecule different parts of the molecule may show different properties. Some side chains may be acidic, some basic, depending upon whether there is an exposed carboxyl (acidic) or amino (basic) group. Carboxyl and amino groups are hydrophilic, or water loving, and may be surrounded by a blanket of water molecules. Other side chains may possess hydrophobic or water repellant groupings. The side group of the amino acid cysteine plays a special role in some molecules. Bridges can be formed between cysteine side chains in different parts of the protein chain so producing loops in the molecule or joining two chains together (fig. 46). Union between two cysteine residues is achieved by the following reaction:

$$....-CH_2-SH + HS-CH_2....$$

$$\text{cysteine} \qquad \text{cysteine}$$

$$+$$

$$O$$

$$\Updownarrow$$

$$....-CH_2-S-\!\!-S-CH_2-....$$

$$+$$

$$H_2O$$

The primary structure of the protein chain

The primary structure of the protein chain is thus determined by the pattern of side chains along the molecule. Each molecule of a particular protein e.g. beef insulin, has an identical sequence of amino-acid residues along the peptide chain and this sequence is unique and is not shown by any other beef protein. The sequence of amino acids in some peptide chains is shown in figs. 45 & 46. Of the twenty different amino acids some are used more frequently than others in the construction of protein molecules. Thus amino acids such as alanine,

Fig. 45. *Diagram illustrating the primary structure of the hormone vasopressin (ADH). S—S bonds are shown as lines between folds of the chain.*

glycine and serine occur more frequently than amino acids such as tryptophan or histidine. This parallels the construction of words where some letters such as the vowels occur more frequently than letters such as z or x.

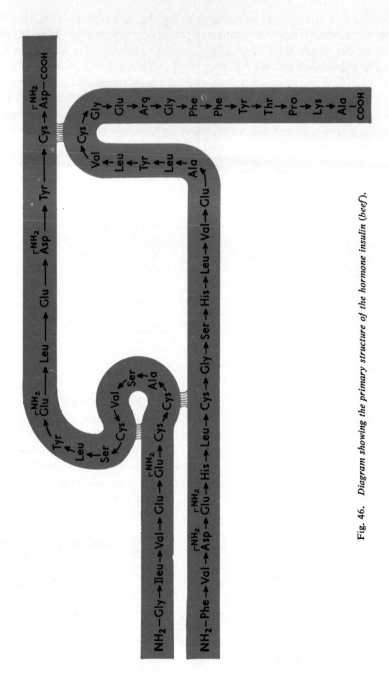

Fig. 46. *Diagram showing the primary structure of the hormone insulin (beef).*

DETERMINATION OF THE STRUCTURE OF THE
PROTEIN CHAIN

Amino acid composition

If we possess a pure sample of a particular protein then the amino acid composition of the protein can be determined by the technique of chromatography. This term chromatography was first used in an early application of the technique to the separation of a mixture of plant pigments. However, the technique is used to separate substances whether or not they are coloured. The principle of the technique is shown in figure 47. In this example a drop of a solution containing a mixture of amino acids to be separated is applied near one edge of a sheet of filter paper. When the spot is dry the filter paper is hung over the edge of a trough which contains an appropriate solvent. The solvent may be methyl alcohol, propyl alcohol, butyl alcohol etc. or a mixture of solvents which are chosen depending upon the kind of substances to be separated from one another. The initial determination of the nature of the solvent is often a matter of trial and error—of finding which solvent 'works best'. The solvent flows slowly down the filter paper under the influence of gravity and capillarity. As the solvent flows past the spot on the paper the components dissolve and move down the filter paper. At various points along the path of the solvent different components of the mixture are deposited on the filter paper. The position at which this occurs on the filter paper depends on the solubility of the component in the solvent and for a particular kind of supporting material (e.g. filter paper) the distance a particular substance moves is related to its solubility in the solvent.

The filter paper is now dried and examined for the appearance of spots of the various components of the mixture which was originally applied to the paper. These spots are circled in pencil. The distance from the starting point is now measured and the spot cut out with scissors. These pieces of filter paper containing the components of the original mixture of the amino acids can now be treated with solvents to dissolve out the amino acids. Thus using the technique of chromatography we can obtain a series of solutions of individual amino acids from the original mixture of amino acids. These are now available in pure form for qualitative or quantitative studies.

The spots on the filter paper may not be very clearly visible in daylight and it may be necessary to examine the paper under ultra violet light or to 'develop' the paper by spraying it with a developer such as ninhydrin which stains the various amino acids different colours. In the example we have discussed a wider separation of the various spots can be obtained by turning the paper at right angles running another solvent across the filter paper. This is called two-dimensional chromatography (figure 47).

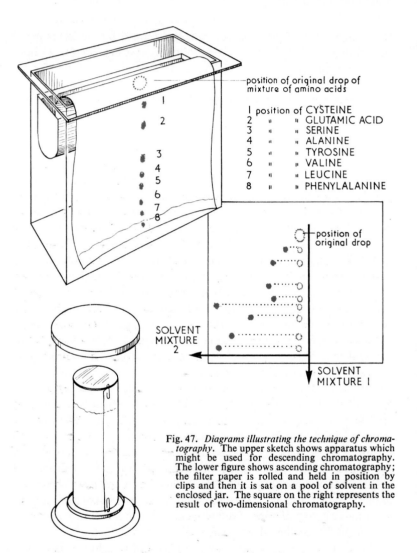

Fig. 47. *Diagrams illustrating the technique of chroma-tography.* The upper sketch shows apparatus which might be used for descending chromatography. The lower figure shows ascending chromatography; the filter paper is rolled and held in position by clips and then it is sat on a pool of solvent in the enclosed jar. The square on the right represents the result of two-dimensional chromatography.

We have described a very simple form of the technique—descending uni-dimensional chromatography on filter paper. But in some systems the movement of fluid may be upwards—depending upon the force of capillarity (figure 47). In addition the filter paper may be replaced by various substances ranging from columns of alumina, silica and starch etc. to thin suspensions of materials applied to glass plates (so called thin-layer chromatography or T.L.C.). But even with the simplest form of apparatus useful work can be carried out—often with amounts of substance in the range of micrograms (0·001 mg). In some studies it may be useful to apply small spots of known substances to the chromatogram so that any component of a mixture of unknown substances which moves to the same position on the chromatogram as a known substance becomes immediately identifiable.

If we apply the technique of chromatography to a pure protein which has been broken down into its constituent amino acid by acid hydrolysis, we can determine the kind and amount of the various amino acids in the protein. But in order to determine the primary structure of the protein molecule we need to know the sequence of amino acids along the protein chain.

The determination of amino acid sequence

This was first achieved by Sanger of Cambridge in 1954 when he described the primary structure of the molecule of insulin. Insulin is a relatively small molecule as proteins go, having a molecular weight of only 5700. But even so it took years of effort to unravel its primary structure. We will look briefly at the method used by Sanger in this great achievement.

Using prolonged hydrolysis with acid it is possible to breakdown all the peptide bonds in a protein chain, so obtaining the constituent amino acids. Using briefer periods of acid hydrolysis or hydrolysis with mammalian digestive enzymes which attach the protein chain only at certain points (page 390) it is possible to break down the protein molecule into fairly large fragments. These large pieces of the molecule are called oligopeptides. The oligopeptides can now be separated by the process of chromatography. We now have specimens of various sections of the protein molecule. Some of this material can now be completely hydrolysed into constituent amino acids which are further identified by chromatography. This provides us with information about the amino acid composition of different segments of the protein chain— but of course gives us no information about the order of the amino acids.

However, it is possible to obtain information about amino acid sequence due to the fact that an oligopeptide has an amino group at one end (called the N-terminal) and a carboxyl group at the other (called the C-terminal). The substance dinitrofluorobenzene is capable of reacting with the terminal amino group of an oligopeptide so that a dinitrophenol group is substituted for a hydrogen atom. The combination of the dinitrophenol group with the peptide is very stable and will survive even the complete hydrolysis of the oligopeptide into its amino acids. The amino acids can now be separated by chromatography. The one bearing the dinitrophenol group can be identified by its ·yellow colour—and it is this amino acid which occupied the N-terminal position in the original oligopeptide.

We can look now at how the structure of a simple peptide could be determined using this technique. The peptide contains 6 different amino acids R_1–R_6. So much could be determined simply by complete acid hydrolysis and chromatographic separation and identification of the amino acids. The next stage is to produce partial hydrolysis, in this case producing three dipeptides, R_1–R_2, R_3–R_4, R_5–R_6. These are now separated by chromatography and each pure dipeptide exposed to dinitrofluorobenzene which reacts with the N-terminal end of the dipeptides. Each dipeptide is now hydrolysed into its constituent amino acids which are separated by chromatography. Each dipeptide yields 1 amino acid coloured yellow because of its dinitrophenol group and one uncoloured amino acid. We know that the yellow amino acid must have been at the N-terminal end of the dipeptide. The six amino acids can now be arranged in the order they must have occupied in the original molecule of the polypeptide.

Of course the problem is greater when the original molecule under investigation is much longer than the sextapeptide we have discussed, but the principle is the same. A large number of different oligopeptides has to be prepared from the original molecule, their N-terminals determined, their amino acid composition determined, and finally the oligopeptides with their known N-terminals must be laid out so that from their overlaps the complete amino acid sequence can be determined. This is illustrated below.

1. The protein molecule when completely hydrolysed produces the following amino acids:—

   A.B.C.D.E.F.G.H.

2. Oligopeptides were produced and labelled with dinitrophenol (N-terminal underlined).

   A—C, C—E—F—B, E—F—B—D,
   G—H—A, H—A—C.

3. Solution of the problem. The areas of overlap of the various oligopeptides are placed one above the other.

```
G—H—A
   H—A—C
      A—C
         C—E—F—B
            E—F—B—D
```

G—H—A—C—E—F—B—D

In summary we can say that the primary structure of a protein molecule is a long chain made up of constituent amino acids linked together by peptide bonds. Each particular protein shows a characteristic number and sequence of amino acids. In a later section we shall see how this sequence is genetically determined.

Secondary structure of proteins

We have described the primary structure of the protein molecule. In this form the molecule is visualised as a long flexible structure. This flexibility arises because the backbone of the protein molecule is non-rigid. This is because the atoms of the $C\backslash C/N\backslash C/C\backslash N$ backbone are linked by single bonds and these permit rotation around the axis of the bond. Further the single bonds on either side of the peptide group lie at an angle of about 110°.

Rotation around bonds set at an angle results in a fully flexible chain.

Some protein molecules, particularly those of low molecular weight do seem to exist in flexible extended forms which, in solution, undergo a randomly folding changing form. A study of the rates of diffusion (under the influence of random thermal movement) and sedimentation (under the influence of an applied gravity force) indicates that these proteins do in fact exist in a long flexible form which undergoes random flexing and folding.

However diffusion and sedimentation studies of other proteins suggest that in these the peptide chains are folded up in some way as to produce a compact molecule. The term secondary structure of the protein molecule describes the nature of the folding of the polypeptide chain.

The methods which have been used to study the secondary structure of proteins are highly complex depending as they do on computer analysis of x-ray diffraction studies of crystals of protein and the construction of molecular models of peptide chains to see how they might fold up. We will not attempt to give details of these studies. We can however say that crystals are made up of regularly repeating molecular or atomic sub units, the component electrons of which scatter an x-ray beam directed onto the crystal. The angles at which the x-ray are scattered (diffracted) depends on the distance between the atoms or molecules within the crystal. Knowing the wavelength of the x-ray beam and the angles and intensities of scattered rays from the crystal it is possible to build up a picture of the electron dense parts of the crystal and so determine its composition in three dimensions.

This kind of study led Pauling and Corey (1951) to suggest that the peptide chain might fold up into a long spiral or helix (figure 48).

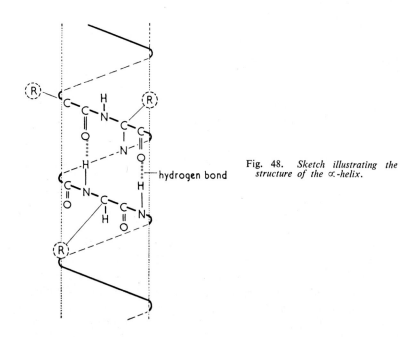

Fig. 48. *Sketch illustrating the structure of the α-helix.*

This configuration of the peptide chain is stabilised by a type of electrical attraction between turns of the spiral called hydrogen bonding. These hydrogen bonds are relatively weak, having bond energies (the energy needed to pull the atoms apart) which are only 1/10th those of covalent protein molecules and of nucleic acids (page 123). Hydrogen bond formation is always the result of the drawing away of electrons from a hydrogen atom. The hydrogen atom is now left with a net positive charge and becomes attracted to the negatively charged nitrogen or oxygen atom of a nearby molecule. In the example shown below hydrogen bonding is shown between a carbonyl and a hydroxyl group. The carbonyl group is electrically asymmetrical, the positively charged carbon at one end and the negatively charged oxygen at the other. Similarly the hydroxyl group is asymmetrical with its terminal exposed hydrogen nucleus which is negatively charged. There is thus an attraction between the carbonyl and hydroxyl groups which brings the two groups together.

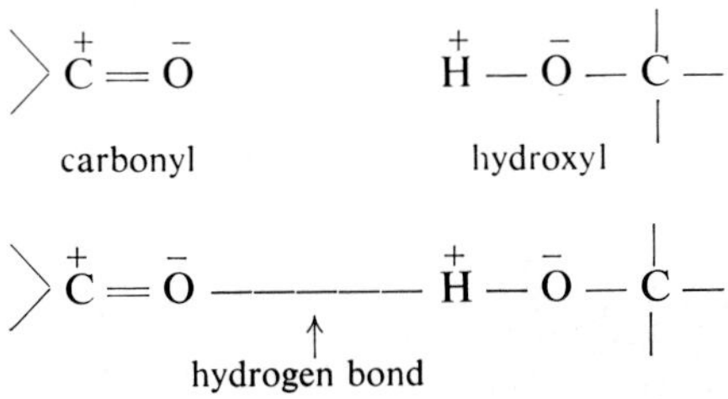

$$\overset{+}{C} = \overset{-}{O} \qquad\qquad \overset{+}{H} - \overset{-}{O} - \overset{|}{\underset{|}{C}} -$$

carbonyl   hydroxyl

$$\overset{+}{C} = \overset{-}{O} ------- \overset{+}{H} - \overset{-}{O} - \overset{|}{\underset{|}{C}} -$$

hydrogen bond

In the helix of the protein molecule hydrogen bonding occurs between the —NH group of a peptide linkage on one turn of the helix with a —CO on the adjacent turn.

The hydrogen bonds between the various turns of the helix makes the peptide chain less flexible than in its primary structure of an extended chain. But in some parts of the molecule of a protein there may be no organisation into a helix and the greater flexibility in these regions permits the backbone to be folded to assume a more compact structure. This shape may be stabilised by other forces such as union between two cysteine residues as described on page 107.

The helix formed by natural proteins is always a right handed spiral— the so called $\propto$ helix. This is somehow related to the fact that natural proteins are formed only from the L-isomers of amino acids.

The $\propto$ helix is not the only type of secondary structure shown by by proteins. In collagen there is a triple-stranded coiling. In the keratin of horn and feather a number of $\propto$ helices may coil together to form super-helices.

Tertiary structures of proteins

Tertiary structure of a protein describes the way in which the helix is folded to form a compact structure. As we have seen, this folding occurs in parts of the molecule in which the peptide chain does not

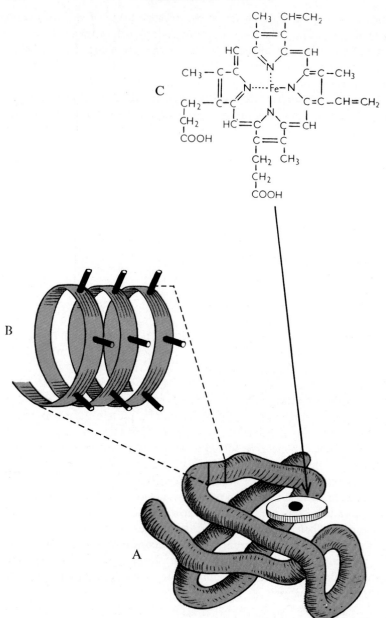

Fig. 49. A is a sketch of a model of myoglobin, made from X-ray diffraction studies, showing its tertiary structure (modified after Kendrew, 1961). In the straight portions of the 'sausage' the single long polypeptide backbone is folded into an α helix. B shows a schematised magnified part of a straight portion of the 'sausage' with the α helix studded with side chains of amino acid residues. The spaces between the folds of the sausage are in reality filled with these side chains. The sketch A also shows the orientation of the haem group which is conjugated to the protein and which confers on myoglobin the property of oxygen transport. The dark centre of the disc represents the iron atom at the centre of the haem group.
C shows the chemical formula of the haem group.

form a helix, i.e. in regions of greatest flexibility. Few proteins have been studied sufficiently to describe their tertiary structure. Figure 49 shows a simplified sketch of the tertiary structure of myoglobin. Myoglobin is a conjugated protein, that is a protein in which there is linkage with a non-protein or prosthetic group. In the case of myoglobin the prosthetic group is a pigmented 'haem' group similar to that in haemoglobin. The sketch shows only the backbone of the molecule and in the complete molecule the spaces between the loops would be filled by the various side chains.

Quaternary structure of proteins

This describes the situation where protein molecules may aggregate to form longer structures, e.g. fibres or form complexes with non-protein molecules such as those of nucleic acids or lipids. This interaction of the folded ∝ helices thus represents a further level of complexity of organisation of proteins.

Summary

We have seen that the smallest unit of protein structure is the amino acid. These amino acids are condensed together in a specific sequence to form a long flexible polypeptide chain. This is the primary structure of the protein molecule. Some proteins, particularly those of low molecular weight, exist in this form. In other proteins the flexible polypeptide chain becomes folded into an ∝ helix at least over considerable stretches of the polypeptide chain. This is what we call secondary structure.

In this form the chain may undergo a process of folding to produce a more compact molecule, and this is tertiary structure. These compact molecules may then aggregate with one another or with other non-protein molecules producing units of greater complexity. This is quaternary structure.

Denaturation of protein. A colloidal solution of protein can be coagulated or precipitated by heat, the ions of heavy metals (e.g. zinc, mercury) and by strong acids and strong bases. This coagulation is an irreversible reaction and is called denaturation and is associated with the loss of all biological properties. We shall see elsewhere that there is a good deal of evidence to show that the properties of enzymes and protoplasm can be largely explained on the assumption that their activities are centred on protein molecules. Denaturation of the proteins destroys these activities and removes those properties of protoplasm which we regard as 'life'. We can think of denaturation as a breaking of the bonds which hold the proteins in their tightly spiralized forms. As soon as these bonds break, as they would for instance under the strain of increased vibration caused by heat, the molecules despiralize. In the despiralization the spatial relationships of one part of the molecule relative to another part is altered. When the spatial relationships are altered the many complex chemical chain reactions break down because certain catalysts which are necessary are now too far away to be able to enter the reactions. Thus denaturation causes death.

Role of proteins. They are essential constituents of tissues in all living things. Muscle for example contains 20% protein (dry weight) and blood serum contains 10%. The protoplasmic membranes of cells are made up of a latticework of protein molecules. We shall see that proteins are the basis of the structure of protoplasm and enzymes. Their functions are both structural, as in the membranes of cells, and functional in the role of enzymatic catalysts. Protein is not stored in large quantities in the animal (except in eggs) and excess protein is broken down to provide energy (5·6 K. cal/gm.), the amino residues being excreted in urea.

Test for proteins. The tests commonly used for proteins are really tests for specific amino acids or specific links like peptide links. Amino acids like tyrosine occur so frequently in proteins that it is fairly safe to assume that if a protein is present tyrosine will also be. Millon's test is a test for a phenolic group which is present in tyrosine.

1. MILLON'S TEST FOR PROTEIN. The material to be tested is broken up into small pieces and then placed in a test tube with a few drops of Millon's reagent with 1 ml. of water. Boil. If protein is present a brick red colour is produced in the material tested.

2. BIURET REACTION, FOR PROTEINS IN SOLUTION. To the solution of protein add a few drops of 1% copper sulphate. Then carefully add a few drops of 40% sodium hydroxide. A mauve colour develops slowly in the solution if protein is present, due to the presence of two peptide links. Peptones give a rose pink colour rather than a mauve colour.

It is important to avoid using too much copper sulphate in this test as the blue colour obscures the mauve or pink.

NUCLEOPROTEINS

These are associated with inheritance, synthesis of proteins (growth) and disease (viruses) and they therefore merit our close attention.

Structure. Nucleoproteins consist of nucleic acid plus a protein molecule. The protein molecule we have learned is composed of a selection of about 22 amino acids polymerized together into a long chain molecule. The nucleic acid consists of units called nucleotides which are polymerized together (see fig. 50). There are usually only four kinds of nucleotide in a given nucleic acid, but nucleic acids are very long chain molecules consisting of hundreds of nucleotides, each of the four nucleotides being repeated many times. Let the nucleotides be called A, B, C and D. Then the structure of a possible nucleic acid might be

ABCDABCDABCDABCDABCD   –   –   ABCDABCD

and another might be

AADACBAADACBAADACB   –   –   AAADCBAAADCB.

Thus using just this four letter code we could produce many different strings of letters. The letters used, their relationship to other letters (i.e. the order in which they occur) and the number of times they occur are important in the kind of 'word' they produce. So is it with nucleotides and nucleic acids; although only four kinds of nucleotide are usually found in a nucleic acid. There are many different kinds of nucleic acid

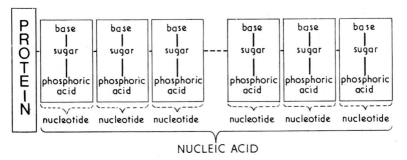

Fig. 50. *Structure of a nucleoprotein.*

NUCLEOTIDES. On referring to the above figure it will be seen that each nucleotide consists of a nitrogen containing base (a purine or a pyrimidine), a sugar (ribose) and phosphoric acid. The chemical structure of a nucleotide is shown below (fig. 51).

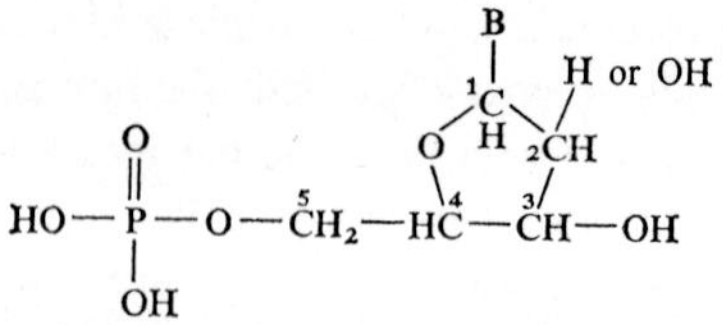

Fig. 51. *The unit of nucleic acid structure—a nucleotide, consisting of phosphoric acid, a sugar—desoxyribose in DNA (—H at 2) or ribose in RNA (—OH at 2), and a base B.*

B represents a variety of bases, the commonest ones being shown below. (Fig. 52)

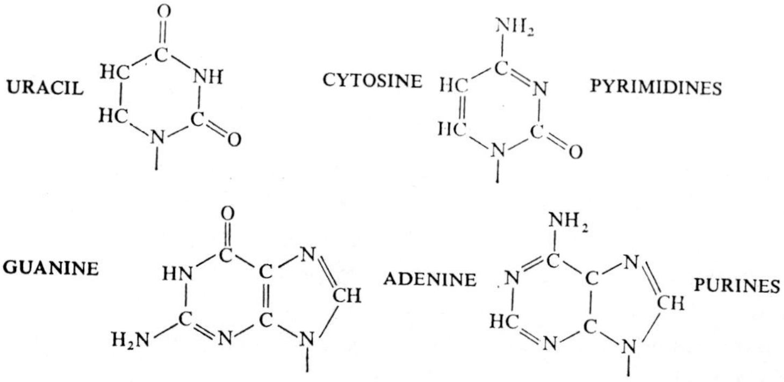

Fig. 52. *Formulae of nucleotide bases, the pyrimidines unracil and cytosine and the purines, guanine and adenine.* All four are found in all types of nucleic acid although in RNA the pyrimidine base is uracil while in DNA it is 5-methyl-uracil or thymine (formula not shown).

Figure 53 shows how individual nucleotides are linked together in the nucleic acid molecule, with phosphate groups alternating with sugar units, so forming the backbone of the chain. The bases project from the side of the chain as indicated in the figure. Just as proteins are formed of units called amino acids linked to form a long chain so do nucleic acids consist of a long chain of nucleotide units.

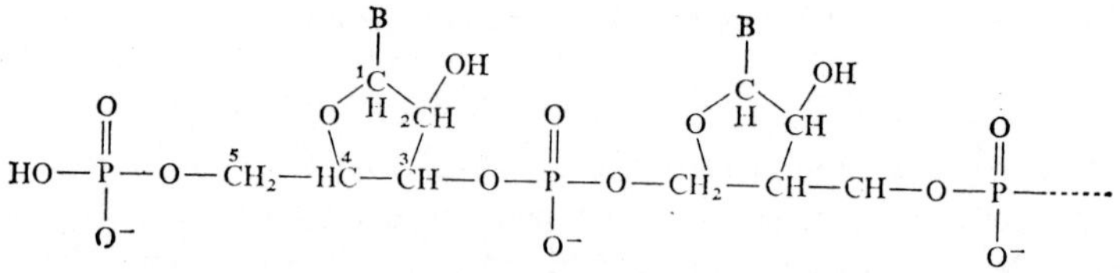

Fig. 53. *Part of the backbone of a nucleic acid chain showing union of two nucleotides.*

Types of nucleic acids

There are two classes of nucleic acid, ribonucleic acid (RNA) and deoxyribonucleic acid (DNA). These differ markedly in structure, stability, distribution and biological functions.

Both RNA and DNA have in common the basic structure as described above, but in RNA the sugar of the nucleotide is ribose (—OH at C_2 in figure 51) instead of the deoxyribose of DNA (—H at C_2 in figure 51). The two nucleic acids also contain different kinds of nitrogenous base. The pyrimidine base uracil is rarely found in DNA. Instead it is replaced by 5-methyl-uracil called thymine. This base is rarely, if ever, found in RNA. The proportions of the different bases also are different in these two types of nucleic acid. In almost all kinds of DNA the purine and pyrimidine bases are found in equal proportions—and this is of considerable significance for the three-dimensional structure of DNA which we shall be describing in a later section. The chains of nucleotides in DNA tend to be much longer than those of RNA; the molecular weight of purified RNA's is in the range of $20,000–2 \times 10^6$ whereas for DNA's the molecular weight is of the order $100,000–120 \times 10^6$. The longer nucleotide chains of DNA are organized into complexes each consisting of two chains forming a double helix. The structure of these double helices will be dealt with in detail below. In nature at least most types of RNA seem to exist in a single stranded form.

We can now briefly introduce the question of distribution and functions of these two kinds of nucleic acid although we shall be examining these in more detail in a later section. DNA is a stable substance found mainly in the nucleus of cells and in particular in the chromosomes, those vehicles of heredity. DNA is in fact the carrier of genetic information. Since nucleic acids differ very little in their base composition (compare the number of types of bases along the nucleotide chain with the number of kinds of side chains in proteins) then we must assume that it is the sequence of bases along the chain of nucleotides rather than variety of base which determines the uniqueness of a particular nucleic acid. We can regard genes as stretches of a nucleotide with a specific base sequence. This genetic material (DNA) in the nucleus of the cell controls the activities of the cell by regulating the synthesis of proteins and especially the synthesis of enzymes. These enzymes, the 'biological catalysts' of chemical activity are not, however, manufactured in the nucleus of the cell but are synthesized in the cell cytoplasm—in particular on the ribosomes. Since the site of protein synthesis is in the cytoplasm and the genetic material is in the nucleus it seems clear that there must be some link between these two sites, a link which involves the transfer of information from the genes to direct the activities of the ribosomes in protein synthesis. This link is formed by a kind of RNA called messenger RNA. This kind of RNA which carries information for protein synthesis is manufactured in the nucleus which it leaves to reach the cell cytoplasm.

Other kinds of RNA are found in the cytoplasm. According to their function they are named transfer RNA and ribosome RNA. We will examine the functions of these RNA's in a later section.

We can now summarize the distinguishing features of DNA and RNA.

Ribonucleic acid (RNA)	*Deoxyribonucleic acid (DNA)*
Found in cytoplasm of cells (also nucleus).	Found mainly in the nucleus—associated with chromosomes.
Concerned with the synthesis of proteins. Concentration in the cell varies with the rate of protein synthesis.	Concerned with heredity. Controls the synthetic activities of the cell. The amount present does not vary with the rate of protein synthesis.
Molecular weight: 20,000 to 2,000,000.	Molecular weight: 100,000 to 120,000,000.
Exists mainly as single-strand form.	Exists as double strands.
Sugar component ribose.	Sugar component deoxyribose.
Contains the pyrimidine bases cytosine and uracil and purine bases adenine and guanine.	Contains the pyrimidine bases cytosine and thymine and purine bases adenine and guanine.

The secondary structure of DNA. The Watson–Crick model of a double helix

The basic primary structure of DNA (see above) has been known for some time. This is a fully flexible randomly folded chain of nucleotides. Solutions of DNA however may show marked changes in various properties when subjected to alteration of pH, salt concentration etc. which suggests that the thread like molecules normally have a secondary structure—a well defined structure in three dimensions. This parallels the situation in proteins where destruction of the highly organized secondary and tertiary structure by 'denaturation' with heat, heavy metals and the like results in marked changes in the physical and biological properties of the protein.

Although a variety of suggestions had been put forward for the secondary structure of DNA it was not until 1953 that fundamental progress was made. Basing their proposal on a model of the structure of DNA they had made, which incorporated the then known facts about the chemistry of DNA (and the X-ray crystallography of prepared threads of DNA) Watson and Crick suggested that DNA existed as a two stranded helix with a structure as shown in figure 54. In this model a regular structure is obtained from two nucleic acid chains twined into a double helix in which the adenine bases are always paired with thymine by hydrogen bonding (see page 115) and guanine bases with cytosine. Each strand of the helix consists of a backbone of alternating deoxyribose sugar and phosphate units. The various bases are attached to this backbone and project into the core of the

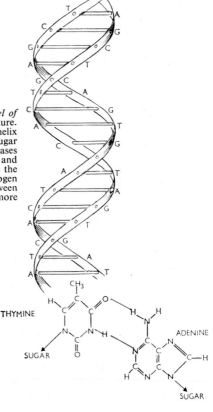

Fig. 54. *Diagram of Watson–Crick model of DNA.* Showing the double helical structure. The backbone of each member of the helix consists of alternating phosphate and sugar units to which are attached the various bases —thymine (T) adenine (A), guanine (G) and cytosine (C). Complementary bases on the two strands are shown linked by hydrogen bridges. The hydrogen bridges between thymine and adenine are shown in more detail below.

helix, stacked above one another 'like a pile of plates'. The pairing of specific bases (cytosine with guanine and thymine with adenine) accounts for the known facts about the base composition of DNA (page 121). Because of specific base pairing the nucleic acid chains which coil round each other in the double helix are therefore identical but head in opposite directions.

A G C T A G C T A G C T
T C G A T C G A T C G A

This simple yet brilliant suggestion has had widespread repercussions in biology and indeed has fathered a new discipline—that of molecular biology. For this achievement Watson and Crick were awarded a Nobel Prize in Medicine. The model—which has been amply substantiated by later studies in biochemistry, genetics and X-ray crystallography—has very important implications for genetics and evolution. In their original letter to the magazine 'Nature' they pointed out that 'the specific pairing we postulate immediately suggests a possible copying mechanism for the genetic material'. Thus the two identical strands of the helix of DNA could separate, each then forming a template on which a complementary sequence of nucleotides could be constructed. By this means each daughter cell produced by a dividing cell could receive an identical complement of genetic material. Indeed each cell of the body can, by this means, carry a virtually identical copy of the original genetic material of the fertilized egg. The hereditary material—DNA—contains in an unbelievably compressed form the information needed to construct, maintain and repair an entire organism. If the molecules of DNA of a single living cell was uncoiled they would stretch about three feet. The information in this stretch of DNA is not determined by the kind of bases present for the DNA of a mouse contains the same kind of bases as that from an elephant. Instead the information is determined by the sequences of bases along the chain of nucleotides.

Protein synthesis

The information necessary for the synthesis of particular proteins lies in the DNA of the chromosomes. This information is transcribed in the form of a molecule of a kind of RNA called messenger RNA. This RNA is a single stranded molecule very similar to DNA except that one of the four bases is different—thymine of DNA is replaced by uracil, and the sugar in the backbone of the molecule is ribose rather than deoxyribose. Messenger RNA is a transcription of a stretch of DNA. The bases of DNA make specific base pairing with bases on messenger RNA. Thus the sequence of bases on the DNA determines the base sequence of messenger RNA. The bases of DNA do not pair with identical bases on RNA; adenine in DNA pairs with uracil of RNA and so on—see below:—

DNA	*RNA*
Adenine	Uracil
Guanine	Cytosine
Cytosine	Guanine
Thymine	Adenine

It appears that only one of the DNA strands is copied in RNA synthesis —and since the two strands are different (although complementary) the same strand is copied on each occasion of RNA synthesis.

Evidence that DNA controls protein synthesis

1. TRANSFORMING PRINCIPLES. Interesting results have come from work on Pneumo-cocci, the bacteria that cause pneumonia. The many bacteria in this group are identified by the shape of the capsule which forms a protective covering to the bacterium. The bacterium capsule is

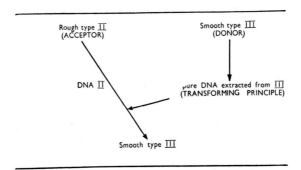

formed from polysaccharides and is made by the enzymes of the bacterium. Sometimes bacteria are obtained which have no ability to make a capsule, they are called 'rough' bacteria. Pneumococci are classified into types I, II, III, IV, and so on depending on the nature of their capsule.

The results of the experiment were as shown above. DNA from a normal smooth type III with a capsule, when added to a culture of rough type II pneumococci caused these bacteria to grow into smooth type III bacteria. It is concluded that the DNA had altered the enzymes of rough type II so that they could make the type III capsule. The type III capsulated offspring bred true. Thus the DNA has affected the control of protein synthesis and also heredity.

2. BACTERIOPHAGES. These are very small viruses the individual particles of which are shaped like a drumstick. They are peculiar in that they contain only DNA and no RNA at all. They are viruses which attack bacteria.

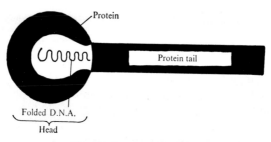

Fig. 55. *Structure of a phage.*

The 'phage' sticks to the bacterium by its 'tail' and the DNA is injected into the bacterium leaving the phage as an empty protein sac. The bacterium now stops making bacterial protein and produces phage protein instead. The bacterium formerly produced RNA and its own DNA but it now makes no RNA only phage DNA. DNA has an obvious role in the synthesis of nucleic acids.

3. BACTERIUM STAPHYLOCOCCUS AUREUS. The shells of bacteria can be cracked by the use of supersonic vibrations and the nucleic acids can be removed by suitable solvents. Before the nucleic acid is removed the bacteria have the ability to synthesize protein but when the nucleic acid is removed the bacteria lose the power to make protein. When suitable nucleic acids are added the ability to make protein is restored.

The genetic code for messenger RNA

The sequences of bases on DNA is transcribed into a corresponding sequence on messenger RNA in the form of a code made up from ACAG representing the bases uracil, cytosine, adenine and guanine. This sequence is later translated into a sequence of amino acids in a protein chain. Some evidence for the RNA code was obtained in 1961 when a synthetic molecule of RNA made entirely of uridine nucleotides (containing uracil) was added to a cell free extract of ribosomes from colon bacilli. The protein which was made consisted entirely of the amino acid phenylalanine. Thus a sequence of uracil on mRNA caused a sequence of phenylalanine in the protein.

Obviously one single base on RNA cannot code for one amino acid for there are only four bases but 20 amino acids. A code of two bases to represent each amino acid cannot work either for there are only 16 possible combinations of pairs of bases from four different bases. With three bases (nucleotides) in the code there are 64 possible combinations of U.C.A.G. (i.e. 4^3). Thus there are 64 combinations of 3 bases to code for 20 amino acids. Such a code is said to be degenerate i.e. there is more than one 'code word' for most of the amino acids. All amino acids except tryptophan and methionine have more than one code word and some e.g. leucine, have six possible codes. The term degenerate does not mean imperfect for there is no code word which specifies more than one amino acid.

Crick and others have produced considerable evidence to support this triplet theory. They refer to a triplet of bases as a codon. Their work was done mainly on a type of bacteriophage (T_4) and they found that a particular region of DNA determines whether or not this phage can attack the bacterium E.coli (strain K). Various mutants of this phage have been produced by adding or removing single bases from the

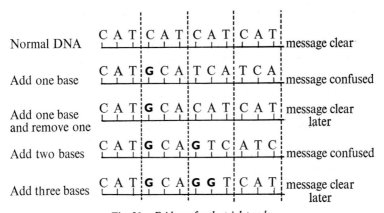

Fig. 56. *Evidence for the triplet code.*

sequence of bases on the phage DNA. None of these mutants with one altered base can attack strain K but if three bases are added the ability to infect K is restored. It is also restored if the addition of one base is accompanied by removing a neighbouring one. If two bases are added close together the ability to infect is lost. These results are explicable if the coding is in triplets. The addition of one or two bases makes the code read nonsense, whereas if one base is added and one removed the code reads sense. Figure 56 makes this clear.

There are four major characteristics of the triplet code. It is degenerate, 3 of the 64 triplets do not code for any known amino acid, and no punctuation is necessary between one triplet and the next until the end of the polypeptide chain is reached. Finally the code seems to be universal for all kinds of organisms.

The following schema representing the messenger RNA code is taken from Biochemistry by Lehninger published by Worth, New York, in 1970.

	U	C	A	G
U	UUU Phe UUC Phe UUA Leu UUG Leu	UCU UCC UCA ⎬Ser UCG	UAU ⎫Tyr UAC ⎭ UAA Ochre UAG Amber	UGU ⎫Cys UGC ⎭ UGA Umber UGG Trp
C	CUU Leu CUC Leu CUA Lev CUG Leu	CCU CCC CCA ⎬Pro CCG	CAU ⎫His CAC ⎭ CAA Gln CAG Gln	CGU CGC CGA ⎬Arg CGG
A	AUU Ile AUC Ile AUA Ile AUG Met	ACU ACC ACA ⎬Thr ACG	AAU ⎫Asn AAC ⎭ AAA ⎫Lys AAG ⎭	AGU ⎫Ser AGC ⎭ AGA ⎫Arg AGG ⎭
G	GUU Val GUC Val GUA Val GUG Val	GCU GCC GCA ⎬Ala GCG	GAU ⎫Asp GAC ⎭ GAA Asp GAG Glu	GGU GGC GGA ⎬Gly GGG

Many attempts have been made to find out which amino acids are coded for by the triplets UAG, UGA and UAA. These triplets have been called amber, umber and ochre. They are now not thought to code for any amino acid but to be signals for ending the synthesis of a chain of amino acids. UAA (ochre) is the normal signal for this but UAG and UGA may also be used.

One triplet leads directly to the next in a linear sequence on mRNA and there is no code for commas between the triplets. It is therefore very important that the linear code of mRNA is correctly set at the beginning of the read out and then moved sequentially from one triplet to the next as the code is translated into a protein chain. At the end of the chain UAA gives the signal. The beginning of a new chain is probably signalled by the triplet AUG which transfers N formyl methionine into the beginning of the chain. There is still some ignorance about this as AUG can also be used for coding methionine into the middle of a protein chain.

The genetic code seems to be identical in all species but it may be modulated slightly in different species, for example in the bacterium E.coli arginine is incorporated by triplets AGG and CGG only weakly but the same triplets in mammals binds nucleotides containing arginine rather well. Perhaps these two triplets are not used much by E.coli but are more important in mammals.

Proteins and genes

The sequence of amino acids in a protein determines the nature of the protein. The protein haemoglobin has 287 amino acids and it is known that only one amino acid in this sequence needs to be altered to cause the disease known as sickle cell anaemia.

	Sequence of bases in normal nucleic acid	Sequence of bases in abnormal nucleic acid
DNA	—TAC—	—AAC—
RNA	—AUG—	—UAG—
	Glutamic acid replaced by valine	
	normal Hb	Sickle cells

A single mutation causes this substitution of one amino acid for another at one point of the chain. If a mutation can be imagined as an alteration in the sequence of bases in DNA it can be seen how this could cause sickle cell anaemia. A change in a gene may be just a change in the sequence of bases in DNA and the length in which a mutation occurs is thereby reduced to the distance between successive bases on the DNA chain. The gene, or what the modern geneticist calls the cistron, can be defined as the smallest piece of DNA which can synthesize a particular protein. This length has been variously estimated as something from 400–8,000 nucleotides. It is obviously possible to have many different changes of sequence of bases within the strip of 8,000 nucleotides, i.e. mutations can occur in many ways within a cistron (gene).

The translation process: the assembly of amino acids to form specific proteins

The materials needed for the construction of molecules of protein in the cell cytoplasm include the following:—

1. Instruction from the cell nucleus in the form of a nucleotide chain of messenger RNA. Along the chain the instruction for the synthesis of individual amino acids is in the form of sequences of three nucleotides—a triplet. Other triplet sequences along the nucleotide chain determine at which point translation of the message is to start and finish—the capital letter and full stop codons.

2. A supply of the necessary individual amino acids and, in addition, enzymes and a supply of energy (ATP) to prepare the amino acids for incorporation into a growing protein chain.

3. A special type of RNA which gathers up appropriate amino acids for assembly on the ribosome.

4. The ribosomes on which the polypeptide chains are constructed.

Activating enzymes

The first stage in the assembly of amino acids consists in the reaction of individual amino acids with ATP to form an adenyl-amino compound in which the energy of ATP is stored. This reaction is somewhat unusual in that when ATP activates a compound it does so by transferring the terminal energy rich phosphate to the compound. Here, however, the adenine monophosphate combines with the amino acid and pyrophosphate is released—

$$\text{A.T.P.} + \underset{\substack{\diagdown \diagup \\ R_1CH \\ \text{amino acid}}}{HOOC \qquad NH_2} \rightarrow \underset{\substack{\diagdown \diagup \\ R_1CH \\ \text{pyrophosphate}}}{AMP-OOC \qquad NH_2} + PP$$

This reaction requires the presence of activating enzymes of which there are at least twenty different kinds. These enzymes catalyse this first important step in protein synthesis whereby the amino acid is charged with sufficient energy to unite with other amino acids by peptide links. These adenyl-amino acids are very reactive substances and if they are mixed together in a solution they will spontaneously react to form polypeptides (see over).

$$
\begin{array}{c}
\overset{\displaystyle NH_2}{\underset{\displaystyle COOH}{R_1CH}} \\[2em]
\overset{\displaystyle NH_2}{\underset{\displaystyle COOH}{R_2CH}} \quad \rightarrow \\[2em]
\overset{\displaystyle NH_2}{\underset{\displaystyle COOH}{R_3CH}}
\end{array}
\qquad
\begin{array}{c}
\overset{\displaystyle HN_2}{\underset{\displaystyle CO}{R_1CH}} \\
NH \\
\overset{\displaystyle }{\underset{\displaystyle CO}{R_2CH}} \\
NH \\
\overset{\displaystyle }{\underset{\displaystyle COOH}{R_3CH}}
\end{array}
\qquad
\begin{array}{c}
+\ H_2O \\[4em]
+\ H_2O
\end{array}
$$

This random union of activated amino acids does not occur in living cells because each amino-acyl adenylate molecule remains bound to the activating enzyme which initiated its formation. The activating enzyme now performs a second function. It brings a molecule of transfer RNA into contact with the amino-acyl adenylate (figure 57).

Transfer RNA

The structure of transfer RNA is such that in one region of the molecule it can combine with an amino acid and in another region it can combine with a special part of the molecule of messenger RNA which is the template on which protein is constructed. Compared to other nucleic acids it is a fairly small molecule containing 65–70 nucleotides. The molecule is a single chain which is bent back upon itself to form a double helix with a loop at one end (Fig. 58). In this terminal loop is the site for the attachment of amino acids and if this loop is removed from the molecule amino acids cannot become attached. However, since all types of transfer RNA's seem to have the same terminal base sequence in this loop (cytidylyl-cytidylyl-adenylic acid) there must be some other part of the molecule which recognizes individual amino acids. There are large numbers of different kinds of transfer RNA which are specific for each amino acid, sometimes for more than one. At the point on the molecule where the strand folds back on itself there are some unpaired bases which probably form the recognition site for the RNA template.

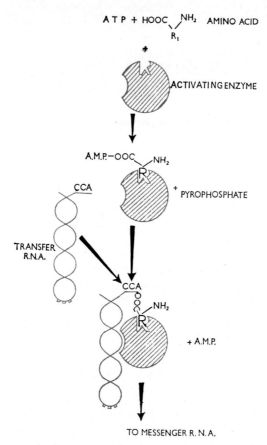

ATP + HOOC NH₂ AMINO ACID

ACTIVATING ENZYME

A.M.P.—OOC NH₂
R
CCA + PYROPHOSPHATE

TRANSFER R.N.A.

CCA
NH₂
R
+ A.M.P.

TO MESSENGER R. N. A.

Fig. 57. *Stages in protein synthesis.* 1. The activation of amino acids and attachment to transfer RNA. (*From 'Introduction to Mechanisms of Hormone Action', P. C. Clegg, Heinemann Medical Books Ltd.*)

The Translation Process

The next stage in the synthesis of protein is the attachment of transfer RNA's with their attached amino acids to the messenger RNA template. Messenger RNA does not appear to operate as a template in protein synthesis until it has been bound to ribosome. Each molecule of messenger RNA becomes related to a group of ribosomes (to form a polysome) during protein synthesis. It is thought that the ribosomes may actually move along the messenger molecule, reading and translating each codon as it progresses along the molecule.

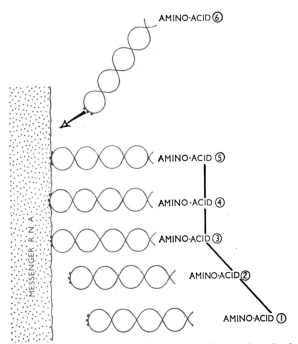

AMINO-ACID ⑥

AMINO-ACID ⑤

AMINO-ACID ④

AMINO-ACID ③

AMINO-ACID②

AMINO-ACID ①

MESSENGER R.N.A.

Fig. 58. *Stages in protein synthesis.* 2. The assembly of the protein molecule on the surface of messenger RNA (ribosome not shown). (*From 'Introduction ot Mechanisms of Hormone Action'*, P. C. Clegg, Heinemann Medical Books Ltd.)

The exact way in which the complex of transfer RNA-activated amino acid is transferred from the activating enzyme to the molecule of messenger RNA has not been elucidated. It is thought that the transfer RNA molecules with their attached amino acids are assembled on the messenger RNA in order by means of the pairing of bases on the transfer RNA molecule with complementary bases on the messenger RNA. When this happens peptide bond formation occurs between adjacent amino acids and the protein peels off the ribosome. After releasing their amino acid the molecules of transfer RNA leave the ribosome (fig. 58). The incorporation of amino acids into the growing peptide chain is thought to occur at the site of attachment of messenger RNA to the ribosome. This site changes of course as the ribosome moves along the messenger RNA molecule, translating the code of messenger RNA into amino acid sequence as it does so.

These concepts are illustrated in fig. 59 showing the movement of a ribosome along the molecule of RNA; as it progressively decodes the message the peptide chain grows. Appropriate amino acids are plugged into the peptide chain by means of transfer RNA.

Until very recently (1968) the structure of ribosomes and the way in which they decode the message of RNA and translate it into protein synthesis has remained a mystery. We know that all active ribosomes are composed of two units, one being about twice the size of the other. The two units are identified by the rate at which they sediment out of solution—this rate being expressed in Svedberg units. A complete

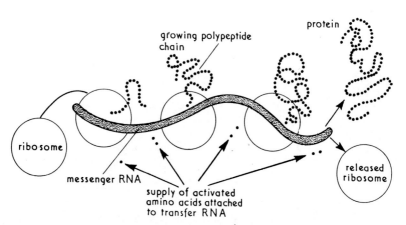

Fig. 59. *Diagram illustrating the movement of a ribosome along a molecule of messenger RNA.* As it moves along the molecule, decoding its message, and supplied by activated amino acids attached to transfer RNA, the protein chain grows.

bacterial ribosome sediments at 70s and its two sub units at 50s and 30s. Breaking up the two sub units further yields two kinds of RNA molecule and at least 80 different kinds of protein molecules. Obviously the ribosome has a complex structure. It is no longer looked upon as a passive framework on which the polypeptide chain grows, instead it is regarded as having an active part to play in protein synthesis.

It seems that only when engaged in protein synthesis do the two sub units come together to form a complete ribosome. After completion of protein synthesis the two parts separate to join a pool of 50s and 30s units in the cell cytoplasm. The first step in the synthesis of a new protein molecule appears to be the binding of a 30s sub unit with the capital letter codon of the messenger RNA (page 127). Now the capital letter transfer RNA is added and only now does the 50s particle join to form a complete ribosome. Protein synthesis now begins.

To look now at theories which explain the growth of the peptide chain on the ribosome. It seems clear that the ribosome must bear active sites or receptors which can receive the molecules of transfer RNA carrying amino acids. During the growth of the peptide chain there must be a constant flux of transfer RNA's, some approaching and binding to the ribosome while others leave the ribosome after they have deposited these amino acids. One theory is that the ribosome has two sites for handling transfer RNA's. One site handles newly arrived transfer RNA with amino acid whilst the other site binds transfer RNA carrying the growing peptide chain. These two sites are called the 'amino acid' site and the 'peptide' site respectively. The process of protein synthesis with such a two site mechanism is illustrated in Fig. 60.

In 1 the ribosome is attached to the messenger RNA. The molecule of transfer RNA carrying the growing polypeptide chain is plugged

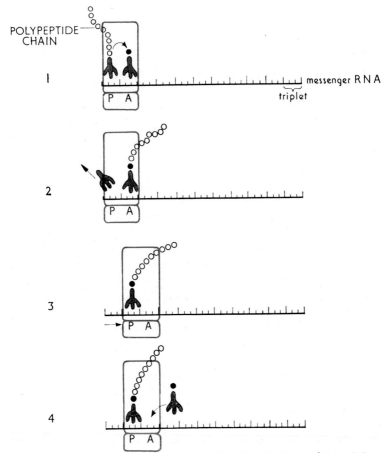

Fig. 60. *Scheme to illustrate the two receptor hypothesis for ribosomal transcription.*

into the peptide side P. The amino acid site (A) is occupied by another molecule of transfer RNA bearing an amino acid which is the next one to be incorporated into the protein. The growing polypeptide chain now shifts (or is translocated) from P to A. The peptide site now bears transfer RNA with no amino acid. This leaves the ribosome. The amino acid site is now occupied by transfer RNA bearing the growing polypeptide chain. In order that a new molecule of transfer RNA bearing the next amino acid can be accepted the peptide chain has to be shifted back to site P. This is achieved by the ribosome moving along the molecule of RNA by a distance of one codon (3). A new molecule of transfer RNA–amino acid can now be accepted (4). There are various other refinements of theory to account for the tricky process of translocation but we will leave theories here.

The Basis of Cellular Differentiation

Every cell of the body contains the same kind and amount of DNA. However, in different types of cells different types of proteins, different enzymes are produced. In a particular differentiated cell the enzymes which are produced thus form only a small fraction of the number of enzymes which the cell is potentially capable of producing. The nucleus contains mechanisms which determine which genes are active in the production of appropriate messenger RNA. This is the basis of cellular differentiation. Thus the cells of the pancreas produce the specific enzymes of pancreatic secretion, the activities of the erythrocyte are directed almost exclusively to the synthesis of the protein haemoglobin, and muscle cells produce large amounts of the proteins actin and myosin. All body cells contain the appropriate genetic information for the manufacture of all these proteins but in most body cells this information is not put to use, in other words the particular genes concerned are repressed. That *all* body cells—even when highly differentiated—do contain all the information for *all* metabolic processes was dramatically illustrated by recent experiments with toad larvae. The intestine was removed from larval toads and the highly differentiated epithelial cells were isolated. The cell nucleus was then removed from an epithelial cell and injected into a toad egg which had been treated with ultraviolet light to inactivate its own nucleus. This egg, containing a nucleus from a fully differentiated and specialized cell of a toad larva, continued to develop normally into a larva and eventually into a normal adult toad.

If all cells produced all the proteins for which genetic information is available then chaos would reign in the differentiated organism. All cells would produce plasma proteins, haemoglobin, contractile proteins, all hormones and so on. Some hint of this state of affairs can be seen in the case of some tumour tissue in which the normal regulation of gene activity is disturbed and genes previously inactive begin to express themselves in terms of protein synthesis. Thus in some tumours of breast or lung tissue genes which are active only in adrenal tissue become derepressed and flood the organism with hormones which drastically disturb the body's regulating mechanisms.

Adaptation

Even in a particular type of differentiated cell the activity of the genetic apparatus is variable. Genetic repression which can be modulated permits a more flexible organization of function in time. This is the basis of adaptation of the organism to fluxes in the internal or external environment. A pre-requisite of this type of adaptation is that genes previously repressed can be derepressed when a situation arises which call for the expression of a particular function.

Thus some species of mammals and birds have restricted periods during the year during which they breed. The timing of this activity is often related to seasonal fluctuations in environmental conditions and food supplies. It is obvious that in periods of the year in which breeding does not occur it would be inappropriate and uneconomical to maintain the structure of the reproductive organs at the level which is attained during the breeding season. Thus in female mammals during the anoestrous period a variety of retrogressive changes occurs in the reproductive organs. Sexual behaviour is lost, the uterus shrinks in size, there is a thinning of the vaginal epithelium—indeed the vaginal aperture may become sealed, the ovary shrinks, egg production stops and sex hormone production declines. The re-awakening of many of these structures as the breeding season approaches must be associated with the 'switching on' of very many genes. In the uterus for example, there is an intensive synthesis of the contractile proteins. The entire energetics of the uterine muscle cell is awakened and energy-rich phosphate compounds accumulate in increasing amounts. The activities of the cell membrane also changes which expresses itself in a rise in the trans-membrane electrical potential.

Some Possible Factors which are Concerned in Regulation of Gene Activity

1. *Nucleohistones.* In the nuclei of most body cells the DNA of the chromosomes is associated with two other types of protein, protamines and histones. These are basic proteins containing a high proportion of the amino acids arginine and lysine. The structure of nuclear DNA

is that of a double helix about 2mμ in diameter, but the finest chromosomal threads that can be seen with the light microscope are much thicker
than these, being 100–200 mμ diameter. It is not clear how the double
helices of DNA are organized to form these thicker chromosomal
threads but it may involve an aggregation of many DNA double helices
or a complex folding of a single long DNA double helix.

The nucleoprotamines and histones appear to be wound round the
double helix of DNA lying in the grooves of the helix (Fig. 61).

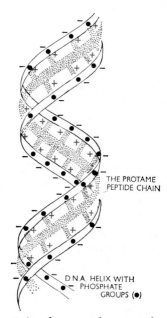

THE PROTAME
PEPTIDE CHAIN

D.N.A. HELIX WITH
PHOSPHATE
GROUPS (●)

Fig. 61. *Representation of a suggested structure of nucleoprotamines.*

The function of these basic proteins is not clear. It is possible that
they may carry some genetic information but it seems likely that they
fulfil other functions. Because they are basic proteins they neutralize
the net negative charge of the DNA molecule and this may facilitate
the folding or aggregation of DNA molecules, for example to permit
close packing of DNA into the head of the sperm. They may also
perform other structural functions by acting as bridges between DNA
molecules in the structure of the chromosomes.

Another possible role of histones which is of special relevance in
this discussion is that thay may play a part in the differentiation of cells
by inactivating part of the genetic material. Thus it may be that during

the development of the cell histones could be manufactured and become bound to particular regions of DNA, filling in the grooves of the double helix and making it inactive in the synthesis in messenger RNA. In the differentiated cell large areas of DNA could be structurally and functionally 'covered' by histones. This view is supported by some experimental evidence. In some tissues the act of derepression of the genetic material is associated with a loss of histone from the chromatin. Allfrey and Mirsky removed part of the histone from a suspension of thymus cell nuclei by treatment with the proteolytic enzyme trypsin. They found that this was followed by a marked increase in the synthesis of messenger RNA (page 121).

Thus it seems likely that histones are in some way concerned in the modulation of genetic activity of the cell, and thus in the processes of cellular differentiation and adaptation.

2. *Hormones.* It seems now certain that in many tissues modulation of gene activity is effected by certain hormones. The various lines of evidence which support this view include the following:—

(*i*) An early increase in the rate of nuclear RNA synthesis following administration of the hormone.

(*ii*) The appearance of increased amounts of cytoplasmic RNA in particular that with the characteristics of messenger RNA after treatment with the hormone.

(*iii*) Inhibition of the effects of the hormone by previous treatment with the agents actinomycin D and puromycin. Actinomycin D is a substance which base pairs specifically with guanine as it is found in the double helix of DNA and the complex formed prevents DNA from acting as a template for RNA synthesis. Thus actinomycin blocks the synthesis of messenger RNA.

(*iv*) The appearance of new types of cytoplasmic protein following the administration of the hormones.

3. *Enzymic induction and repression.* Some idea of the nature and complexity of gene regulation has been derived from studies of micro-organisms, and these may have some bearing on the nature of differentiation and adaptation of cells in higher organisms.

In micro-organisms the manufacture of particular enzymes may require in addition to possession of an appropriate gene (the so-called structural gene) the presence of a particular substrate in the environment to act as an inducer. The presence of a particular substrate in the medium in which micro-organisms are cultured will often lead to the appearance of various enzymes concerned in the stages of metabolism of the substrate. This phenomenon is called enzymic induction. The reverse situation also occurs in which the presence of a particular substrate can inhibit the synthesis of a particular enzyme.

From genetic and biochemical studies in micro organisms Jacob and Monod were able to show that the synthesis of a particular enzyme by a micro organism is determined not only by a gene which specifies the amino acid sequence of the enzyme in terms of messenger RNA (the structural gene) but also by a hierarchy of other genes. One such gene is the regulator gene which may control the activity of a group of distant structural genes by means of a repressor substance. The repressor substance produced by the regulator gene does not seem to act directly on the structural genes. Instead it acts by way of another gene which is located adjacent to the group of structural genes controlled by the regulator (Fig. 62). To this gene is given the name of operator. The intracellular repressor produced by the regulator acts

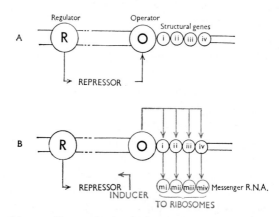

Fig. 62. *Scheme to illustrate Jacob and Monod's theory of genetic regulation.*

A. Shows four regulator genes (i–iv) and the adjacent operator gene (O). These genes are inactive, i.e. repressed, because of the activity of a regulator gene (R) which produces a repressor substance inhibiting the activity of the operator gene.

B. Shows the effect of adding an inducer substance to the medium in which the bacteria are living. The inducer substance combines with and inactivates repressor substance. The operator gene is now released from inhibition and it activates the structural genes i–iv. These now produce specific messenger RNA, mi–iv, which passes to the ribosomes to instruct the synthesis of four distinct enzymes.

Note that although there pressor substance is shown arising from the regulator gene this protein, like all cell proteins, is coded for by a segment of DNA which produces a specific messenger RNA instructing the synthesis of the repressor molecules on the ribosomes.

(*From 'Introduction to Mechanisms of Hormone Action'. P. C. Clegg, Heinemann Medical Books Ltd.*)

on the operator to switch off the activity of the whole groups of structural genes.

Jacob and Monod suggested that the repressor substances produced by regulator genes are protein in nature and contain two active sites, one capable of binding with the operator gene and one specific to the inducer substance. They suggested that on introduction of the inducer substance into the medium this combines with the repressor protein and this results in such a modification of the structure of the repressor molecule that it can no longer combine with the operator gene. The activity of the operator gene is now released from inhibition and it then activates the structural genes.

When Jacob and Monod put forward this scheme of regulation of metabolic activity in bacteria there was no direct evidence of the nature or indeed even of the presence of the repressor substance. In 1966 studies by American teams of scientists provided this direct evidence. The repressor has been isolated for the metabolic system which first led Jacob and Monod to put forward their concept—that is the repressor substance which controls the genes determining the enzymes which metabolize lactose in the bacterium E. coli. The repressor does indeed bind itself to a section of DNA. In 1968 an account was published of work which identified the region of the genetic material of E. coli where the synthesis of messenger RNA for lactose splitting enzymes begins. When a stretch of DNA is copied in terms of messenger RNA an enzyme called RNA polymerase appears to become attached at a particular point on the DNA chain and works along like a zip fastener, attaching itself to successive stretches of DNA and producing messenger RNA as it moves along. It seems that the operator gene lies between the point of attachment of the RNA polymerase and the structural genes which code for messenger RNA. This discovery points to a straightforward way in which the repressor substance can prevent the copying of the structural genes. When the repressor substance attaches itself to the operator it acts like a thread caught in a zip fastener to prevent the progress of RNA polymerase along the segment of DNA.

Inducible and repressor enzymes have now been described in animal cells. In general the response of animal cells to changes in substrate composition are slower and smaller than those of bacteria. The system described in bacteria may provide some pointers to the way in which hormones can alter gene activity, e.g. by inactivation of an intracellular repressor or inducer.

The very brilliance of this concept by Jacob and Monod has for a few years somewhat blinded us to the possibility that other mechanisms may also be involved in the changes in protein synthesis which some hormones produce. For example there is evidence that growth hormone can effect not only messenger RNA formation but also the transcription of this message by the ribosomes.

Rather different mechanisms which regulate gene activity are indicated by the work of Professor Harris of Oxford. In 1965 he described experiments in which he introduced the nucleus of a cell of one species of animal into the cell of a different species. The donor cells were bird erythrocytes (red blood cells) which are so highly specialized that virtually all of the genes are switched off. These cells produce no DNA and only very small amounts of RNA. Harris introduced the nuclei of bird erythrocytes into human cells which were active in the synthesis of both DNA and RNA. He found that after introduction into the human cell the bird erythrocyte nucleus increased several times in size, its highly condensed chromatin became opened up and it became active in RNA synthesis. He argued that these results were not compatible with Jacob and Monod's repressor hypothesis. The results indeed suggested an *activation* of the DNA of bird erythrocyte nuclei by a cytoplasmic factor(s) from the recipient cell. It is suggested that the activator causes nuclear swelling and opening up of the dense inactive chromatin so permitting the DNA to be copied by RNA polymerase. The amounts of RNA produced by the erythrocyte nuclei were indeed proportional to the degree of nuclear enlargement. These results have been confirmed in studies of the effect of incubation of erythrocyte nuclei with mouse liver cell cytoplasm.

It would appear that there may be a variety of mechanisms involved in the regulation of the activity of DNA.

Chapter Three

Enzymes

Enzymes and the diversity of reactions in a living cell. We have studied the structure of protoplasm and have seen how in different cells differentiation of function has occurred. Tissues and organs are constructed specially to carry out certain functions for the whole body. These millions of specialized cells all working in harmony allow the mammal to live a very complex life. If we stop to reflect on the complexity of the reactions which are occurring simultaneously in a single cell of an organ like the liver we will be amazed that such a variety of chemical activity can take place in such a small space. The liver cell may be

(a) turning glycogen into sugar or sugar into glycogen

(b) synthesizing urea from the amino-groups of excess amino acids in the body

(c) secreting bile

(d) intercepting the flow of amino acids coming from the gut and synthesizing the proteins of the liver and the plasma proteins

(e) generating energy for all these complex activities by its own respiration. It is releasing energy from carbohydrates and building up molecules of adenosine triphosphate (ATP) which are used to supply energy into the other reactions in the cell.

Such a list of activities is indeed impressive expecially when one considers the size of the cell concerned—about $5\,\mu$. As one begins to realize that the reactions involved in the releasing of energy from glucose are extremely complex, amounting to dozens of reactions linked together like one great assembly line, the complexity of reaction occurring within one cell assumes an almost astronomical dimension. It is beyond our comprehension to think of all this vast maelstrom of diverse activity. And all this diverse reactivity is occurring in an ordered manner. It is only necessary for one step to be broken in this meshwork of chemical intricacy for the whole cell and often the whole organism to suffer (see insulin and its effect on metabolism). Whilst all this activity goes on the organs are also maintaining themselves, constantly replacing worn parts and often expanding the whole works if the organism is growing.

We must try to answer the question 'how does all this chemistry occur simultaneously in such restricted spaces?' Imagine trying to make copper sulphate, sodium hydroxide and nitric acid in one flask at the same time. You would probably end up with sodium sulphate and copper nitrate and the alkali would certainly neutralize the acid. How is this sort of thing prevented in the living cell? The answer is not fully known of course—the investigation of cellular chemistry is in its infancy, but a general idea can be obtained. Different reactions are localized in different centres in the protoplasm. We have seen how the microsomes are centres of protein synthesis in the cell and dependent upon the mitochondria for their energy supply. Perhaps the function of the endoplasmic reticulum may be to separate various actions from one another. We know that there are hundreds of possible reactions located in the protoplasm of a cell and that these reactions are spatially separate. (We know this for some reactions and it is a reasonable guess that it must be so for most reactions.)

We know that in this chemical activity, each reaction is controlled by a specific catalyst which can speed up its own particular reaction. These specific catalysts are called enzymes, and there are almost as many enzymes in the body as there are different chemical reactions. These enzymes are protein containing and this explains how they can be specific to each reaction. We have spoken in Chapter II of the variety of protein it is possible to form from twenty different amino acids. Since these enzymes are responsible for controlling the rate at which the chemistry of our cells proceeds we must look at them more closely.

The first thing to be clear about is that the reactions are going on all the time. They may be going on so slowly that by our crude methods of investigation we cannot measure their rate. The enzymes can alter the rate of a reaction but they have no power to start one. The factors which allow a chemical reaction to proceed are indeed very complex and far outside the scope of this book. A textbook of physical chemistry would be necessary to probe into the subject fully.

WHAT MAKES A REACTION POSSIBLE?

We have said that enzymes are biological catalysts which increase the rate at which reactions proceed in the body. They will only do this provided that the reaction has the appropriate energy conditions to allow it to proceed. These conditions are

1. There must be a decrease in the free energy of the system or energy must be supplied from outside the reacting system.
2. The molecules concerned must be in a state of activation.

Decrease of free energy

Free energy simply means energy that is available for work. In any chemical compound only a part of the total energy of the molecule is available for work whilst the remainder is locked up in the molecule and is not available. The unavailable energy is called the entropy of the molecule.

F	=	H	—	TS
Total free energy		Total energy of the system		Unavailable energy (entropy)

It will be noted that the entropy S is multiplied by the temperature T since the amount of unavailable energy depends on the temperature. Any reaction which is accompanied by an escape of free energy can occur without external assistance. When objects fall downwards they lose some of their potential energy and release some free energy. Objects only move upwards if energy is given to them from some source as when they are lifted by hand to a higher position. They then increase their free energy. Chemical reactions are classified as exothermic or endothermic depending on whether they give out heat or take heat in. The heat given out in an exothermic reaction is not necessarily equal to the loss in free energy. It may be less than the free energy by an amount equal to the entropy. It is not possible to measure the free energy of a system directly but it can be calculated using the following formula:

$$\text{Change in free energy} = -RT \log_e K$$

(R is the gas constant, T is the temperature in degrees absolute and K is the equilibrium constant—see later).

If the change in free energy is zero then the reaction is in equilibrium. If it is positive this means that free energy is going into the system as when an object is lifted and the reaction is impossible unless energy is supplied from outside, i.e. by the hand which is doing the lifting. The only time that a reaction proceeds spontaneously unaided is when there is a negative value for the change in free energy, i.e. when free energy is being released from the system.

Activation of the molecules

Activity in chemical language means readiness to react. Some elements are very ready to react whilst others like the inert gas neon are not so. To activate a molecule is to make it ready to react if it is not already prepared. So whilst a reaction is theoretically possible because it will release free energy the reaction may not proceed until the molecules are activated. There are two main ways of activating reactions:

1. BY HEATING the reactants thus giving their molecules greater speeds so that they presumably collide with more force or greater frequency.

2. BY USING ENZYMES. Many, perhaps all, enzymes act by forming an intermediate compound with the substrate. Thus in a reaction A → B the stages are probably A plus enzyme → enzyme substrate complex → B plus enzyme. In the absence of the enzyme substrate complex there is an energy barrier between A and B. The enzyme has the effect of lowering this energy barrier. This idea is illustrated in fig. 63.

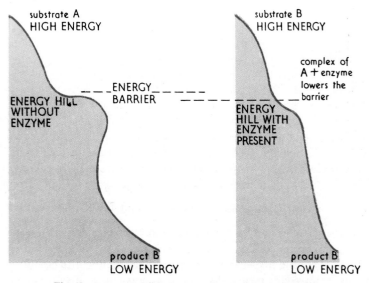

Fig. 63. *Diagram to illustrate the removal of an energy barrier by means of an enzyme.*

We can compare molecule A to a car on the top of a hill with its brakes on. The enzyme releases the brakes (activates molecule A) allowing the car to run down the hill decreasing the amount of free energy in the car. We know that it is the protein part of the enzyme which is responsible for activating the substrate.

FACTORS AFFECTING THE RATE AT WHICH A POSSIBLE REACTION CAN PROCEED

When there is a decrease in free energy and the molecules are in a reactive state the reaction may proceed, but it may do so very slowly, so slowly in fact that it is of no real use to the organism. In living cells the rate at which a reaction proceeds is controlled by three main factors; temperature; concentration of substances; catalysis.

Temperature and the rate of reaction

If the temperature is raised by ten degrees centigrade the reaction rate is doubled. In a reversible reaction the forward and back reaction are equally accelerated. In living systems the rate of reaction is not always doubled by a rise of 10°C, since above a certain temperature the protein of the body becomes denatured and death ensues. (See inhibition of enzyme activity by heat.)

Concentration of substances and rate of reaction

Increase in the number of molecules in a fixed volume will increase the number of collisions between molecules and this will give the chance of a higher rate of reaction between these molecules if the reaction is possible. This idea is expressed quantitatively as the law of mass action which states that the velocity of a reaction is proportional to the product of the active masses of the reactants. (Active mass means the concentration expressed in gram molecules per litre.) Note that the velocity is proportional, not equal to the product of the active masses. Thus in a reaction $A + B = C + D$ the velocity of the forward reaction is $[A] \times [B]$ multiplied by a constant k_1 where the [ ] means concentration in gram molecules per litre. The velocity of the back reaction is $[C] \times [D]$ multiplied by a constant k_2. The equilibrium constant for the whole equation is

$$k = \frac{k_1[A][B]}{k_2[C][D]}$$

If we start with only A and B and they react to give C and D, at first the equilibrium point will be far over to the right. As the concentration of C and D increases the rate of the back reaction will increase provided always that the right energetic conditions prevail. In the body there are many reactions which only go one way because the right energy conditions do not prevail for the reverse reaction. Thus if all the protein in the stomach is converted to amino acids these amino acids will never recombine in the gut to form protein. They will be resynthesized to protein later on in the cells of the liver where different energy conditions prevail.

Catalysis and the rate of reaction

If dry hydrogen and dry chlorine are mixed there is no reaction but if a drop of water is introduced the two gases react rapidly to form hydrochloric acid. The water has speeded up a reaction which although possible energetically was proceeding so slowly that we could not measure its rate. Similarly if a solution of hydrogen peroxide in water is kept in a cool place in a dark bottle it remains as a solution but if a drop of mineral acid is added the peroxide breaks down to water and oxygen.

$$2H_2O_2 \rightarrow 2H_2O + O_2$$
$$HCl$$

The same effect is seen if hydrogen peroxide solution is put onto a cut in the skin. The peroxide froths as oxygen is released. There is something coming from the cut cells which increases the rate of breakdown of the peroxide. This something is a chemical called catalase. The term catalyst refers to any chemical substance which has the power to accelerate or retard a chemical reaction. Catalysts are used extensively in industry to speed up reactions which would otherwise proceed so slowly as to be economically unprofitable. A good example is the use of a catalyst, which is mainly iron, in the Haber synthesis of ammonia from nitrogen and hydrogen.

$$3H_2 + N_2 \rightleftharpoons 2NH_3 + \text{heat energy}$$

It is interesting to note that in this reaction the production on ammonia is favoured by low temperatures. The temperature has to be sufficiently high to give a reasonable rate of reaction. Low temperature gives a good equilibrium point, as far as the production of ammonia goes, but this point would be attained too slowly. It is important not to confuse factors which affect the rate and those which affect the point of equilibrium. High pressure will tend to favour the production of ammonia because four volumes are being reduced to two volumes. Pressure is often very important in fixing the equilibrium point in chemical reactions but since high pressures do not develop in the body we shall not concern ourselves here with pressure effects.

The characteristic features of catalysts

1. They alter the rate of chemical reactions which are already possible energetically. They have no power to initiate reactions.

2. Since they alter the forward and back reaction equally they have no influence upon the position of equilibrium in a reversible reaction.

3. They have an effect out of all proportion to the quantity of catalyst present. The catalytic activity of colloidal platinum in decomposing hydrogen peroxide can be detected when only one gram atom is present in 7,000,000 litres of solution. In some reactions however larger quantities of catalyst are required, e.g. in the Freidel-Craft reaction a lot of

aluminium chloride is involved. This is because the catalyst is involved in the production of an intermediate compound.

4. Catalysts are not altered chemically during the reaction. This explains why a little has a great effect since it can be used over and over again. The catalyst may be altered physically however. If manganese dioxide is used to accelerate the breakdown of potassium chlorate to give oxygen, the manganese dioxide may be changed from a compact solid to a fine powder.

5. Catalysts may be inhibited or poisoned and then they cease to accelerate the reaction. Therefore gases are usually purified in industry before they are introduced to the catalyst chamber. One gram molecule of hydrocyanic acid gas in 20,000,000 litres has an easily detected effect on the decomposition of hydrogen peroxide by platinum.

6. There is no universal catalyst which will work for all reactions. Industry is always searching for new catalysts to make particular reactions work faster.

Chemical catalysts and enzymes compared

1. Enzymes like catalysts alter the rate of possible reactions. We have already mentioned the effect of catalase on the breakdown of hydrogen peroxide.

2. Like catalysts they increase forward and back reactions equally if both reactions are possible energetically. The reverse reaction is not always possible as we have already shown with reference to proteolytic enzymes in the gut, which hydrolyze protein to amino-acids. The reverse reaction takes place in the various tissues of the body where the cells supply energy in a suitable form. An entirely different set of enzymes are involved in the forward and back reaction.

3. Like catalysts a little goes a long way. A single molecule of catalase is capable of decomposing 5,000,000 molecules of hydrogen peroxide per minute at $0°C$.

4. Unlike most catalysts enzymes seem to be used up during reactions. Like all other parts of the body they are constantly being replaced. They may seem to be used up because they are easily prevented from acting by a variety of influences. If an enzyme is altered physically this is often enough to stop it working because there is a close structural relationship between enzyme and substrate.

5. Inhibition. Enzymes are much more easily prevented from acting than are chemical catalysts.

6. Specificity. Enzymes are much more restricted to particular reactions than are chemical catalysts.

The causes of inhibition fall into four main classes
  (*a*) Protein precipitants
  (*b*) Competitive inhibition
  (*c*) High temperatures
  (*d*) Unsuitable pH values.

(*a*) PROTEIN PRECIPITANTS. Heavy metals, ultra violet rays, mechanical shaking and a variety of chemicals will denature proteins and also prevent enzymes from working.

(*b*) COMPETITIVE INHIBITION. Catalysis in the body takes place on surfaces of colloidal particles and is called surface catalysis. Enzymes can only link on to the substrate at fixed points in the molecule and if these points are blocked by other molecules then inhibition occurs, e.g. the enzyme succinic acid dehydrogenase takes hydrogen away from succinic acid to form fumaric acid, and in order to do this a loose compound is formed between the enzyme and the succinic acid. The compound then breaks up to give fumaric acid

$$
\begin{array}{ccc}
\text{COOH} & & \text{COOH} \\
| & & | \\
\text{CH}_2 & -\text{H}_2 & \text{CH} \\
| & \xrightarrow{} & \| \\
\text{CH}_2 & \text{enzyme} & \text{CH} \\
| & & | \\
\text{COOH} & & \text{COOH} \\
\text{Succinic acid} & & \text{Fumaric acid.}
\end{array}
$$

There is another acid called malonic acid which has a structure similar to succinic acid, so similar in fact that it fits into the enzyme molecule as if it were succinic acid. The complex so formed however does not break up and the enzyme is thereby blocked and can be of no further use to the organism

$$
\begin{array}{l}
\text{COOH} \\
| \\
\text{CH}_2 \qquad \text{Malonic acid.} \\
| \\
\text{COOH}
\end{array}
$$

The effect is just like when one pushes the wrong yale key into a lock. The keys may be sufficiently alike to fit into the lock but if you get the wrong one in, the key becomes jammed and the lock is no further use.

A similar case is found when the poison Prussic acid is introduced into the body. The enzyme cytochrome oxidase accepts the Prussic acid at the place on its molecule where it normally takes up oxygen. At this place on the cytochrome oxidase there is a single atom of iron. The iron cannot therefore accept oxygen. There is only a little cytochrome oxidase in the body, although it occurs in every cell, and it relies for its success upon a rapid handing on of the oxygen to a hydrogen molecule, from a dehydrogenase enzyme, to form water. If it cannot do this the dehydrogenases become blocked by hydrogen and the oxidation of sugars, which depends on the removal of hydrogen, stops. Thus cyanide (Prussic acid) blocks cytochrome oxidase and the body very quickly is deprived of energy.

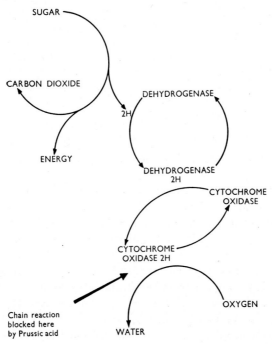

(c) HIGH TEMPERATURES. Enzymes are largely protein and some like pepsin are wholly so. This being the case high temperatures cause denaturation of the protein and consequent cessation of enzymic activity. Enzymes are characterized by having an optimum temperature for their activity. (See fig. 64.)

At X the increase in rate due to the rise in temperature is balanced by the denaturation of the enzyme by heat. To the right of X the enzyme breaks down rapidly and so the rate of reaction slows. It should be noted here that the optimum temperature is not easily determined since the length of time the temperature has to affect the enzyme is very important. If the rate could be measured after a few seconds, before the protein of the enzyme was denatured then the optimum temperature would be very high. If the rate was measured under the same conditions after three hours the optimum would seem to be much lower since much of the enzyme would have been denatured by its longer exposure to high temperature. In this respect enzymes differ markedly from other catalysts which often operate at temperatures of several hundred degrees Centigrade.

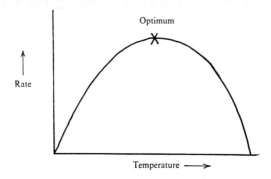

Fig. 64. *The optimum temperature for enzymic activity.*

(*d*) UNSUITABLE pH. Enzymes have an optimum rate within a very restricted pH range. The optimum varies for different enzymes. Thus the protein digesting enzymes of the stomach (pepsin) work at an acid pH of between 1·5 and 2·5. For a particular enzyme there is an exact optimum pH.

The enzymes in the ileum however, work at an alkaline pH, e.g. trypsin and chymotrypsin work best in alkaline conditions. It is interesting to note that proteins other than enzymes show pH optima for properties like solubility, viscosity, conductivity, etc. The optimum value for these properties occurs at what is called the isoelectric point. The pH optimum is an expression of the amphoteric nature of protein.

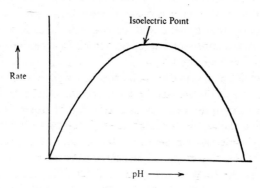

Fig. 65. *The optimum pH for enzymic activity.*

Specificity of enzymes. We have already noted that there is no such thing as a universal catalyst, but the same catalyst will often accelerate many different reactions. Platinum will act as a catalyst in the breakdown of hydrogen peroxide and in the synthesis of ammonia. Enzymes are much more specific on the whole than catalysts. The degree of the specificity is associated with the closeness with which they fit onto the substrate. There are four main degrees of specificity in enzymes.

(a) Low specificity

(b) Stereochemical specificity

(c) Group specificity

(d) Absolute specificity.

(a) LOW SPECIFICITY. A very few enzymes seem to be specific to the linkage group in a compound and do not mind the nature of the radicles which are connected by this linkage. Good examples are the lipases which are specific to the ester link.

(b) STEREOCHEMICAL SPECIFICITY. In Chapter 2 it was stressed that the D isomers of monosaccharides are much more common than the L isomers in the body. In the amino acids it is the L isomer which is the common one. Enzymes show a predilection for these common isomers. Arginase will act only on L arginine (an amino acid) producing L ornithine and urea. Succinic acid dehydrogenase oxidizes succinic to fumaric acid but it never makes maleic acid, although maleic acid is an isomer of fumaric acid. They both have the formula COOH—CH= CH—COOH. This stereo specificity is very common in enzymes.

(c) GROUP SPECIFICITY. The hydrolytic enzymes in the human gut will attack starch but not cellulose. The main difference between these two polysaccharides is that starch (amylose) has α links whilst cellulose has β links between the pyranose units. There is no enzyme capable of breaking β links in man. There are α hydrolytic enzymes but these cannot deal with the β links. So cellulose remains indigestible to man. The human polysaccharases require that the hexose units be linked by α links, they are specific to the hexose radicle plus the α link. Such an enzyme is called an α glucosidase. Maltase from the human gut is an α glucosidase and needs a substrate like maltose with the ⟨⟩⁼ structure, i.e. it is group specific. However maltase from germinating barley is absolutely specific to the disaccharide maltose and no other substance will do whether it contains the ⟨⟩⁼ group or not. (See p. 89.)

(d) ABSOLUTE SPECIFICITY. This is the commonest kind of specificity, found when the activity is confined to one particular substance. If a particular compound S...T is to be attacked by the enzyme then the nature of S and T and the nature of the linkage group must be just right.

Succinic acid dehydrogenase will only oxidize succinic acid and no other substance. It will be noted that an enzyme which is absolutely specific is also stereo and group specific of course.

Summary of enzymes and catalysts compared

Property	Catalyst	Enzyme
Alter the rate of reaction . . .	yes	yes
Do not alter the equilibrium point .	yes	yes
Do not initiate reactions . . .	yes	yes
Effect out of all proportion to quantity .	yes	yes
Not altered chemically . . .	yes	usually
Easily inhibited	slightly	easily
Inactivated by high temperature . .	no	yes
Inactivated by protein precipitants .	no	yes
Restricted range of pH . . .	no	yes
Specific	slightly	very

Apoenzymes and coenzymes. The protein part of an enzyme is called the apoenzyme and is usually the part of the enzyme which is responsible for activating the substrate. The demands which the apoenzyme makes on the substrate results in specificity.

Some enzymes consist wholly of protein, but most have a non protein part called the co-enzyme or prosthetic group. The coenzyme plays a part in the reaction after the apoenzyme has activated the substrate. The coenzyme may be very simple consisting of a single inorganic atom of iron in cytochrome oxidase, or it may be very complex. One such complex coenzyme is DPN (diphospho pyridine nucleotide) which is the coenzyme of dehydrogenase enzymes. Although the dehydrogenases are absolutely specific they may have common coenzymes. Vitamins like riboflavin and pyridoxine are known to be coenzymes.

DEFINITION OF AN ENZYME. In summary we may think of enzymes as being protein containing compounds of high molecular weight, which are thermolabile, pH sensitive and easily inhibited. They are the specific catalysts of biological reactions.

AN OUTLINE CLASSIFICATION OF ENZYMES. Enzymes are classified according to their function in five main groups.

1. Hydrolases. This group includes the digestive hydrolytic enzymes which act outside the cell, e.g. amylases, saccharases, proteinases, lipases etc. Also included are enzymes that operate inside the cell and involve the use of water, e.g. deaminases, arginase, carbonic anhydrase.

2. Adding enzymes, e.g. decarboxylase which adds or removes carbon dioxide.

3. Transferring enzymes. These catalyze the transfer of a radical from one molecule to another, e.g. transphosphorylase which transfers phosphate groups and transaminases which transfer amino groups.

4. Oxidizing enzymes. These may catalyze the removal of hydrogen, e.g. dehydrogenases or the addition of oxygen, e.g. oxidazes.

5. Isomerases catalyze the formation of one isomer from another. They catalyze the regrouping of atoms in the molecule with a consequent redistribution of energy, often making the molecule unstable in the process, e.g. phosphohexose isomerase.

The basis of the specificity of enzyme action—the nature of enzyme substrate interaction

We have seen that many enzymes are highly specific with regard to the substrate with which they will react. For these enzymes even minor modifications in the chemical structure of the substrate will prevent the action of the enzyme.

One of the first explanations of the nature of enzyme specificity was put forward by Fischer in 1894 in his 'lock and key' theory of enzyme action. Fig. 66 illustrates this theory in which the enzyme acts as a 'key'. Only a substrate molecule having the appropriate complementary structure will fit the lock. This concept readily explains why minor chemical modifications of the substrate prevent the enzyme from exerting its action—the lock and key no longer fit perfectly.

This theory has received support from modern X-ray studies of enzyme-substrate complexes. Thus when the enzyme lysosome reacts with its substrate the substrate molecule fits smoothly into a groove on the surface of the enzyme molecule. In this position the chemical groups of the enzyme which exert the catalytic effect are in an appropriate position to rupture a chemical bond in the substrate molecule.

About ten years ago this 'lock and key' theory was presented in a modified form by Koshland. He felt that Fischer's theory did not adequately explain the action of all enzymes and that something more than just binding of the two parts occurs when enzyme and substrate interact. His explanation is called the 'induced fit' hypothesis. In this hypothesis the enzyme is not a rigid lock; rather it is a mobile lock which is triggered into action when the substrate molecule interacts with it. The union of enzyme and substrate is thought to put stress on the enzyme molecule; in response to this stress there is a movement of parts of the enzyme molecule bringing the catalytic groups of the molecule into contact with the chemical bonds in the substrate which are to be broken. Fig. 66 shows in a pictorial way how the 'cutting tool' of the enzyme molecule (the catalytic groups) is brought into action on to the appropriate substrate bonds. In this theory we have two levels at which specificity can operate. In the first the substrate molecule must have an appropriate key structure to fit onto the enzyme. In the second the substrate must have an appropriate structure so that once it has bound to the enzyme it exerts appropriate stresses on the enzyme molecule to bring the catalytic groups into their appropriate position. Because of these refinements the theory goes further to explain how the smallest change in the structure of a substrate can prevent the action of an enzyme.

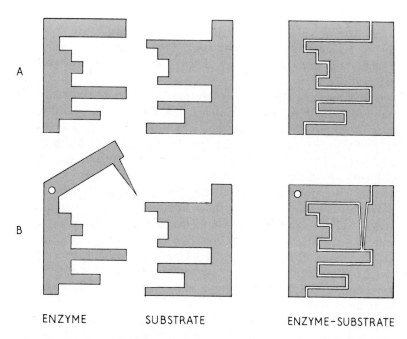

A

B

ENZYME   SUBSTRATE   ENZYME–SUBSTRATE

Fig. 66. *Diagram illustrating the 'lock and key' theory* (A) *and the 'induced fit' hypothesis* (B).

X-ray analysis of the structure of the protein-splitting enzyme carboxypeptidase (page 391) have supported Koshland's theory. Carboxypeptidase is a protein composed of 307 amino acids with an atom of zinc which is vital for the activity of the enzyme. This atom of zinc is at one end of a groove on the surface of the molecule. Near the zinc is a pocket which leads from the surface of the molecule into the interior. When a polypeptide chain (the substrate) unites with the enzyme it lies in the groove on the surface of the enzyme with the peptide bond which is to be split lying near to the atom of zinc and the side chain of the C-terminal amino acid (see page 112) fitting into the pocket. Thus here we have the structure of the lock and key with the side chain of the C-terminal amino acid fitting into the interior of the enzyme molecule. Thus this enzyme can only break peptide bonds at the end of a polypeptide chain since the pocket in the enzyme molecule can accept only one amino acid side chain.

Studies of the union of carboxypeptidase and substrate also show that when the groove and the pocket of the enzyme are occupied by the substrate there are marked structural changes in the enzyme. Further even small chemical modifications of the polypeptide, although permitting union of substrate and enzyme prevented the induced changes in the structure of the enzyme molecule.

We have then an explanation of the nature of enzyme specificity and action. But not all enzymes are put under strain when they react with their substrate. X-ray studies of the enzyme carbonic anhydrase show that there is no strain on the molecule when it unites with the substrate. Other explanations of enzyme action have been put forward for these cases, but they are too specialized to be dealt with here.

PART II

DIFFERENTIATION AND INTEGRATION

Chapter Four

Tissues and Organs— Differentiation

Differentiation and complexity. It seems that increase in complexity above a certain level is dependent upon the body having many cells. Protozoa do not have their bodies divided into cells and they are invariably small creatures, the largest of which is only just visible to the naked eye, yet within their tiny acellular bodies they have attained a remarkable degree of complexity. The ciliated protozoa like Paramoecium are the most complex protozoa and their bodies consist of protoplasm which is specialized into distinct portions called organelles; but they remain acellular. There are organelles called neuronemes for conduction of messages and organelles called myonemes which perform muscular functions, whilst the complex contractile vacuoles regulate the water content of the body. (See fig. 67 of Paramoecium.)

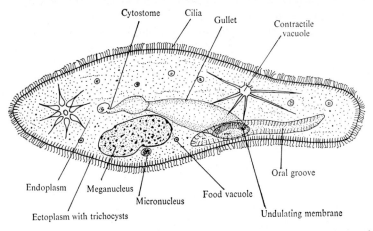

Cytostome   Cilia   Gullet   Contractile vacuole

Endoplasm   Meganucleus   Micronucleus   Food vacuole   Oral groove

Ectoplasm with trichocysts   Undulating membrane

Fig. 67. *Diagram showing the structure of a Paramoecium with the complexity of structure to be found in an acellular organism.*

Mammals attain a much higher degree of complexity than protozoa and consist of many cells which are specialized to perform particular functions on behalf of the whole animal. There are nerve cells for conduction of messages on behalf of the whole body and muscle cells which can contract for the benefit of the whole body, as when they move the limbs in locomotion. It is obvious that where there are enough cells in

161

the body for groups of cells to have a special function then the structure of the cell can be specially adapted to perform this one particular function very efficiently. If the body consists of only one cell then that cell has to perform all the functions itself and its structure has to enable it to be a 'Jack of all trades'. We say that the mammal shows a high degree of differentiation into specialized cells.

Multicellularity allows the specialization of protoplasm for the more efficient performance of all the necessary functions of the body and is a necessary grade of organization for the most complex animals.

The multicellular grade of organization and size. As a general rule once a growing cell attains a certain size it either stops growing or divides into two. This is true of all grades of organization but in protozoa the two daughter cells separate to form new individuals whilst in multi-cellular animals the two daughter cells remain as two cells in the same animal body. We do not fully understand why there should be division when a cell reaches a certain optimum size but two reasons may be suggested:

    i. there is an optimum surface/volume ratio
    ii. above a certain size the nucleus can no longer exert control over the cytoplasm.

    i. SURFACE/VOLUME AND THE CELL SIZE. Supposing that a typical cell

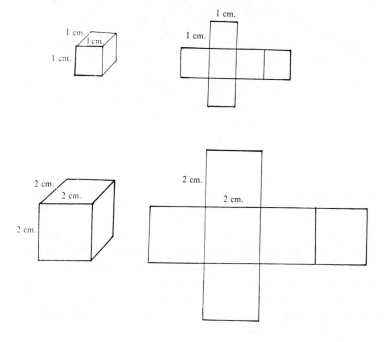

is cubic then as the volume increases the increase in its surface area is less than its volume increase, so that if its linear dimension increases by 8 units its surface area will have increased by 458 square units whilst its volume will have increased by 728 cubic units. (See fig. 68.) This can be expressed by saying that as a cell gets bigger its surface/volume ratio gets less. This creates some difficult problems, for the cell is dependent

Length L	Volume $L^3 = L \times L^2$	Surface area $6 \times L^2$	$\dfrac{\text{Surface area}}{\text{volume}}$	$\dfrac{6L^2}{L^3} = \dfrac{6}{L}$
1	1	6	6/1	6
2	8	24	6/2	3
3	27	54	6/3	2
4	64	96	6/4	$1\frac{1}{2}$
5	125	150	6/5	$1\frac{1}{5}$
6	216	216	6/6	1
7	343	294	6/7	$\frac{6}{7}$
8	512	384	6/8	$\frac{3}{4}$
9	729	464	6/9	$\frac{2}{3}$

Fig. 68. *To show that as a cell gets bigger its surface/volume ratio gets less.*

upon the physical process of diffusion for its food and oxygen and for the removal of its metabolic waste. We have seen in chapter 1 that the rate of diffusion is dependent upon the surface area of the membrane. The demand for food and oxygen however depends directly upon the bulk of the protoplasm involved, that is on its volume. Therefore

supply α surface area
demand α volume.

Now supply cannot equal demand when the volume is great compared to the surface area, that is when the surface/volume is small, that is when the cell is large. In a cell of unit length each cubic unit of protoplasm is supplied by six units of area, but when a cell is nine units long each cubic unit of protoplasm is supplied by only one third of a unit of area. Thus the cell is now bigger but its relative area has decreased eighteen times. Under these conditions it is impossible to get nutrients into the protoplasm fast enough. Therefore active animal cells must be smaller than a certain maximum or must be of such a shape that their surface/volume ratio is big. Diffusion has also only an effectively fast rate over distances less than one millimetre, in the body. For these reasons it is not surprising that a cell divides when it gets bigger than a certain size. It is not proven that the surface/volume ratio is the cause of division but it is a tempting idea. It may be significant that very large active cells like nerve cells have a surface/volume ratio not very different from smaller more spherical cells.

ii. EFFECTIVE RANGE OF INFLUENCE OF THE NUCLEUS, AND CELL SIZE. When a cell increases in size, parts of the cytoplasm will come to lie further away from the nucleus and if the nucleus is only effective over a certain range then some cytoplasm would come to lie outside the sphere of influence of the nucleus. This situation may be avoided in three ways:

(a) The cell may remain small and stop growing.

(b) The cell may divide into two when it gets to a certain size, the nuclear material dividing equally into two and each part growing to full size.

(c) The cell may enlarge without dividing into cells but the nucleus may divide up so that a multi-nuclear cell is produced. This is the common situation in the group of fungi called phycomycetes (e.g. the pin moulds) and in some algae, e.g. Vaucheria.

Whatever the ultimate cause of cell division and cell size may be, we can conclude that the multicellular state of organization is linked with the increase of size and complexity of animals above a certain level.

The integration of highly differentiated cells into tissues. In highly differentiated animals like mammals, cells are specialized to perform a restricted range of the total activities of which protoplasm is capable. The muscle cells are contractile units but do not conduct electrical impulses as well as the nerve cells do. Conversely nerve cells conduct well but are unable to contract. These specialized cells do not only perform one task however, they all retain certain properties of protoplasm. A nerve cell generates its own energy by respiration just as a muscle cell or a protozoan does. Thus in a highly differentiated organism all the cells do certain things, like energy generation, for themselves but they delegate other functions to specialized cells, just as in a society each man digests his own food but delegates other functions like preparing the food and growing the food to specialized workers. In a community all the various specialists must cooperate, the farmers grow the food, and shopkeepers sell it, the tailors make clothes whilst the doctor tends to the sick. The doctors need food and clothes the shopkeeper needs the farmer to produce the food and the tailor to make his clothes and the doctor to tend to him when he is sick. If each man were a hermit he would grow his own food, tend his own illnesses and so on, but his clothes would be of poor quality and probably he would often be hungry and he may even die young. Community life is more efficient provided that all the specialists help each other. In a highly civilized community if cooperation ceases there is widespread trouble. It is so also in the body. Increased differentiation of cells leads to greater efficiency only in so far as there is cooperation between the specialized cells. This linking together of highly differentiated cells is called integration.

In our civilization people of similar interests are found grouped together, farmers on the land, business men in the city, fishermen round the coasts and teachers in schools. This helps them to do their jobs efficiently. In the body, cells which are similarly differentiated for one function in particular, are collected together into communities called tissues. Thus muscle cells are grouped into muscular tissue and nerve cells into nervous tissue and so on. The tissues are classified into four main groups.

1. Epithelial tissues. The cells are generally arranged in sheets for covering the surfaces of the body. Glands are also derived from this tissue.
2. Muscular tissue. There are cells of three main varieties but all are specialized for contraction.
3. Nervous tissue is a collection of cells specializing in conduction of impulses.
4. Connective tissue consists of several types of cell plus an intercellular substance. This tissue includes blood cells, cartilage and bone cells and the cells of 'connective tissue proper' and their products. The intercellular substance is very important.

EPITHELIAL TISSUE

An epithelium consists of a layer of cells arranged to form a complete sheet of tissue. The function of these sheets of tissue is basically to form a delimiting membrane. These delimiting membranes form both the covering layer of the outside, i.e. the skin, and the inner lining layer of hollow internal organs; thus epithelium lines the entire alimentary canal from the mouth to the anus, lines the entire respiratory tract from the nose down to, but probably not including, the alveoli, and forms the lining membranes to many other hollow structures including the pancreatic duct, the bile duct, ureter, bladder, urethra etc.

From their position they are subjected to a variety of mechanical stresses; on the outside of the body epithelia receive considerable mechanical stress in the form of repeated friction, pressure, blows etc. and inside the gut the lining is continually subjected to the friction of food particles passing along it. An epithelium may be adapted in several ways to withstand these forces.

1. Firstly and characteristically epithelial cells may be joined together firmly by

(a) an intercellular cement substance, which although small in amount, fulfils the function of sticking the cells together in a continuous sheet. These sheets of epithelia lie on a material called a basement membrane, which may or may not be a product of the epithelial cells themselves. (See fig. 69.)

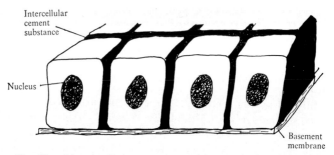

Fig. 69. *Diagram illustrating how epithelial cells are held together.*

(b) Epithelial cells may be interconnected by bridges of protoplasm connecting one cell to another. This is characteristically seen in the epidermis in the skin.

2. Secondly the epithelium may counteract stresses by increasing in thickness. Thus in the skin the epithelium is many cells thick. An epidermis so constructed is called a compound or stratified epithelium as opposed to the simple epithelium which consists of a sheet only one cell thick. (See fig. 70.)

In the skin the epithelium is further adapted to withstand the constant trauma (or damage) by the production of a hard horny substance called keratin. Thus in situations such as the soles of the feet or the palms of the hands which are subjected to much trauma the epidermal epithelium is thicker and harder than in other situations, e.g. the skin of the abdomen, where much less trauma is received.

A   B

Fig. 70. *Simple epithelium (A) and compound epithelium (B).*

3. A third way of maintaining the continuity of the epithelium in the face of repeated physical trauma is seen in the gut. In the mouth, pharynx and oesophagus the main function is to transport the food down to the stomach and intestines. To withstand the friction of the food particles the lining epithelium is compound as in the skin on the outside of the body. But in the lower parts of the alimentary canal, e.g. the duodenum, small intestine etc. the epithelial lining of the tube must be thin to permit the absorption of the products of digestion. A thick stratified epithelium in these circumstances would render the absorption of nutrients almost impossible. But how can a thin delicate epithelium consisting of only one cell layer withstand the constant friction of food particles? The answer is found in a very rapid rate of division of certain cells in the intestinal epithelium, so that cells damaged and lost can be quickly replaced. The site of these cells will be discussed later; here it is sufficient to state that in examining a prepared microscope slide of a section of gut epithelium one can usually find a proportion of cells engaged in cell division whereas this is a fairly uncommon occurrence in examining other adult epithelia or any other tissue, excluding the germ cells and bone marrow.

Summary. Thus we have seen that epithelia can maintain themselves in the face of mechanical trauma in various ways, including the presence of intercellular cement or a basement membrane, by producing stratified epithelia, sometimes adding the horny keratin, or by having a high replacement rate for damaged cells.

Origin of epithelia. True epithelia arise in the embryo from two distinct layers of tissue the ectoderm and the endoderm. The diagram (fig. 71) is of a transverse section through the trunk of the embryo showing the three main layers of tissue from which all the tissues and organs of the body are derived. The outer layer called the ectoderm gives rise to the skin and its structures, the nervous system, and to part of the fore gut and the hind gut. The endoderm gives rise to the major part of the gut (excluding those parts derived from the ectoderm), including various glands, e.g. liver and pancreas. It is from the ectoderm and endoderm that all true epithelia arise. The mesoderm situated between the ectoderm and endoderm give rise to the connective tissues (bone, cartilage etc.) and muscle. The ventral mesoderm is split into outer and inner layers, between which is the coelomic cavity. Parts of these outer and inner layers in the adult form thin sheets of flattened cells which surround the internal organs—thus the pleura surround the lungs, the pericardium surrounds the heart and the peritoneum coats the organs in the abdomen. These sheets of cells are 'epithelial' in structure and function and are given the name mesothelia (indicating their origin from mesoderm) or endothelia (indicating their internal position). The blood vessels are lined by flattened cells which function as epithelia. These cells arise from mesoderm and are called endothelia.

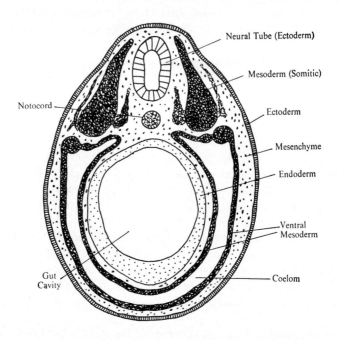

Fig. 71.   *T.S. trunk of embryo.* The three germ layers, endoderm, ectoderm and mesoderm are illustrated. The packing tissue between these three layers consists of a loose undifferentiated tissue called mesenchyme derived by migration of cells from the mesoderm. From it arise various connective tissues including blood vessels, bone, cartilage, tendon, ligament etc.

Simple epithelia.   As we have seen, epithelia are divided into simple and compound (or stratified) types according to whether they consist of one cell or many cells in thickness.

Simple epithelia may be further subdivided; they may be classified according to their shape or to their function, or by both criteria. Thus according to the shape of the cells, simple epithelia may be classified in the following way.

CUBICAL EPITHELIA.   The cells are shaped like a cube, being as tall as they are broad; the nucleus is spherical and is situated centrally.

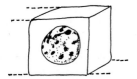

Fig. 72. *Cubical epithelium.*

This may be regarded as a basic shape which may be modified in two main ways.

| The cells may be elongated to produce a *columnar type* | The cells may be flattened to produce a 'pavement' or *squamous type* |

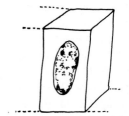

Fig. 73. *Columnar epithelium.*

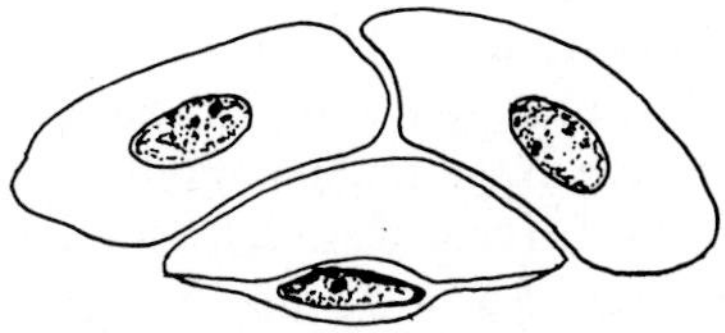

Fig. 74. *Squamous epithelium.*

We can perhaps best understand the significance of these modifications in the shape of cells by referring to the functions of the cells. In the change from a cubical to a columnar type of epithelium it is obvious that for a given area of epithelium there will be a considerable increase in the amount of protoplasm, both nuclear and cytoplasmic, per unit area of epithelium.

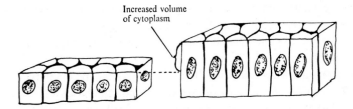

Increased volume of cytoplasm

Fig. 75. *Diagram to show that columnar epithelium has a greater volume per unit area than has cubical epithelium.*

This increased amount of protoplasm will be significant where the function of the epithelium includes not only a delimiting or mechanical function but also a secretory function. Thus columnar epithelia are often found to have glandular or secretory functions, and the increased amount of protoplasm is employed in the secretion or absorption of chemical substances. In the alimentary tract the epithelium is columnar in shape and has a glandular function. Also in certain parts of the kidney there are columnar cells engaged in secretion or reabsorption. The active secreting cells of the thyroid gland are also columnar in shape and it is significant that when they stop secreting the shape alters from a columnar type to a low cubical type. Here is a perfect example of the relationship between shape of cell and function.

Cubical epithelia are also found typically in glandular organs but they are present in those parts of the organ not engaged in secretion and are thus found in ducts which transmit the secretions from one part of the body to another, e.g. salivary ducts, pancreatic duct.

SQUAMOUS EPITHELIUM. The squamous cell is a flattened cell (often called a squame) with a small amount of protoplasm in the cell compared to its area. The nucleus is in the shape of a flattened disc which often causes a bulge in the shape of the cell. (See fig. 74.) From the discussion of the amount of protoplasm in the cell and the cell's function it is obvious that these cells are not secretory. A typical example of a squamous cell is found in the Bowman's capsule of the kidney. The Bowman's capsules, of which there are thousands in each kidney, are the filtration units. Each consists of a hollow ball of flat squamous cells infolded upon itself to form a double walled hollow hemisphere. Projecting into this hemisphere is a knot of blood capillaries, the glomerulus; and under the force of the blood pressure, fluid is filtered out through the capillary wall, through the thin layer of squamous cells which form the wall of Bowman's capsule, into the cavity of the capsule itself from whence it drains into the uriniferous tubule (see fig. 76). As the fluid trickles along the uriniferous tubule various changes occur in its composition until it is discharged into the ureter as urine. The squamous cells which form Bowman's capsule thus act as a passive filter and it is obvious that the shape of these cells is adapted closely to their function.

Squames are found in other situations; thus the cells lining the blood vessels, forming the smooth surface over which the blood moves, form a flattened layer of squamous epithelium. Indeed the capillaries consist of little more than a tube of squamous epithelium (endothelium). Here again the shape of the cells is related to function in that it is through the capillary wall that all the essential substances for the cells must diffuse out, and through which the waste products of metabolism must pass from the tissue fluids into the blood stream; a thin membrane here is of distinct advantage.

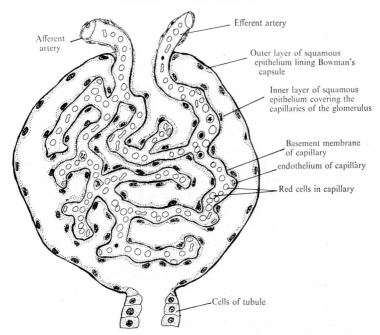

Fig. 76. *Bowman's capsule*, showing the squamous epithelium of the capsule

The edges of the squamous cells which line the blood vessels are irregularly notched so that they fit into one another like the components of a mosaic; these cells are called tessellated squamous cells.

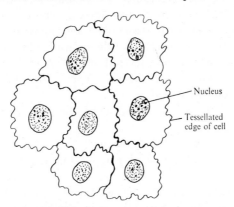

Fig. 77. *Tessellated squames*.

The superficial cells of a compound or stratified epithelium are dead and flattened and superficially resemble squamous cells. If a little saliva is examined under the microscope one finds many flat squamous like epithelial cells, which arise from the dead superficial layers of the stratified epithelium which lines the mouth cavity. These are not true squamous epithelial cells but they are useful to examine since squamous epithelia are found in inaccessible parts of the body.

MUSCLE TISSUE

Muscle tissue is a collection of cells in which the power of contractility is specially developed. The cells are bound together by connective tissue into muscle organs which are the objects which the layman calls muscles. There are three distinct types of muscle cell each specializing in a different type of contraction. The physiological differences in the three types of cell are reflected in their structure and in the structure of the tissues which contain them. We shall consider the three types separately.

Striated muscle. The muscle seen in dissection is an organ made up of several tissues. Around the outside is an envelope of connective tissue called the perimysium which is continuous with the tendons which attach the muscle to the skeleton. Characteristically the force generated in striated muscle is conducted away by the connective tissue so that it can act at a fairly localized point on the skeleton. We shall see later that the other two types of muscle operate without being attached to the skeleton. We may therefore think of striated muscle as skeletal muscle. A transverse section through a muscle of the leg would look like fig. 78.

It is seen that the muscle is made of several bundles of muscle fibres. Both the muscle fibres and the bundles of muscle fibres, have their own connective tissue sheaths, called the endomysium and perimysium respectively. The bundles of muscle fibres are collected into groups by a further layer of connective tissue called the epimysium.

In order to study the structure of the individual fibres it is necessary to take a small piece of fresh skeletal muscle about as thick as a piece of cotton and look at it under the microscope. To do this the following procedure should be adopted.

(a) Warm gently in conc. nitric acid to break down the connective tissue envelopes.

(b) Wash in water to remove the acid.

(c) Place in a watch glass and tease out the strand of muscle with a needle.

(d) Add a few drops of iron haematoxylin stain and leave for about ten minutes until you think the muscle has taken up the stain.

(e) Transfer the muscle to water to wash away the excess stain.

(f) Place on a slide in a drop of dilute glycerol or water and tease out thoroughly so as to separate the cells.

(g) Place the coverslip on and examine under high power.

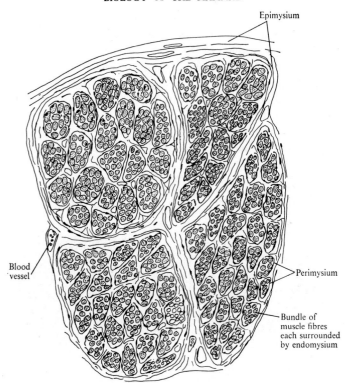

Fig. 78. *Diagram of a transverse section of a part of a muscle to show the connective tissue components.*

The actual cells should now be seen. They are long cylindrical cells, so long in fact that the ends are seldom seen, as the cell may be one centimetre long ormore. The cells consist of:—

i. very long fine myofibrils which are banded, hence this type of muscle is called striated muscle.

ii. protoplasm in which the fibrils are embedded. This protoplasm is called sarcoplasm.

iii. many nuclei. Since there are many nuclei in each cell the cells are said to be syncitial.

iv. a sarcolemma, which is the limiting layer on the outside of the cell. It is not clear whether this is secreted by the cell or whether it is made by the surrounding endomysium. The sarcolemma is closely attached to the sarcoplasm and follows the changing shape of the cell. By its close attachment to both the sarcoplasm and the connective tissue it helps to transmit the force of contraction to the skeleton.

These long thin cells are called muscle fibres and care should be taken not to confuse the term fibre (cell), with the term fibril (myofibril) which is only a constituent of the cell.

When the fibres are stained cross-striations are seen on the myofibrils. This striated appearance is due to the fact that certain parts of the fibrils take up the stain better than others. Thus a highly stained disc (an anisotropic or 'a' disc) is followed by a disc which does not stain (an isotropic or 'i' disc). The 'a' discs and the 'i' discs of adjacent myofibrils are alongside each other, so that the cell as a whole appears banded.

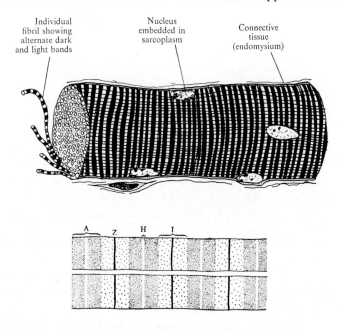

Individual fibril showing alternate dark and light bands

Nucleus embedded in sarcoplasm

Connective tissue (endomysium)

A   Z   H   I

Fig. 79. *Two myofibrils drawn from an electron microphotograph.*

Recent theories suggest that the anisotropic disc contains the protein myosin whilst the isotropic disc contains the protein actin. The actin molecules are thought to slide in and out of the myosin molecules as the fibril alters in length. (See p. 616.)

The amount of collagen, elastin and reticular fibres present in the connective tissue sheaths of muscle is very variable and depends upon the function of the particular muscle. There are a lot of elastic fibres in the muscles of the tongue for instance where the demand made upon the muscle is that it should be mobile and elastic rather than possessed of great tensile strength.

The characteristic feature of the activity of striated muscle is that its fibres shorten rapidly and powerfully. It is the type of muscle responsible for movements of the skeleton. We are all aware that our skeletal muscles become fatigued after exercise, this is the great disadvantage of striated muscle and an important point of comparison with smooth muscle which can work incessantly. Because most of the movements made using striated muscle may be voluntary this type of muscle is often referred to as voluntary muscle. There are exceptions to this, e.g. the diaphragm consists of striated muscle but contracts without any conscious or voluntary action on the part of the mammal. Striated muscle originates from the mesodermal somites and is supplied by somatic motor nerves. The muscles associated with the gills in fishes are striated but are derived embryologically from mesenchyme rather than the somites. Consequently these branchial or visceral skeletal muscles are innervated by the visceral efferent nervous system instead of the somatic nervous system. Usually of course the visceral efferent nervous system supplies smooth muscle. (See Chapter VII.)

Some important muscles in mammals are homologous to the visceral branchial muscles, e.g. the trapezius muscle, which in man is inserted on to the spine of the shoulder blade and originates from the neural spines of the vertebrae, is derived from muscles which in sharks serve to lift the gill arches. Because during evolution of the mammalian stock from primitive vertebrates the gill muscles have been converted to other purposes, we find some striated muscles in mammals with surprisingly peculiar nerve supplies. Therefore it is better perhaps to classify muscles as we have done here on their histological appearance rather than call them voluntary and involuntary.

Before muscles can be useful to the animal they have to be connected to the skeleton by tendons and supplied with blood. They must be related to the outside world by nerves. The somatic efferent nerves end in striated muscle in a rather characteristic fashion at a motor end plate. (See fig. 80.)

Smooth muscle. This type of muscle is not seen as distinct anatomical muscles in dissection but is rather spread out through the organs of the body and is perhaps most in evidence as an important component of the gut. (See figs. 81, 82.)

Seen under high magnification under the microscope the cells are spindle shaped about 200 μ long and 6 μ wide. The dimensions of the cells vary slightly in the different organs as the size is determined to some extent by the demands made on the organ. The cells are pointed at each end and bulge in the middle where the single nucleus is situated. The cells show feint longitudinal striations due to the long protein molecules in the sarcoplasm which run along the length of the cell. The cells are therefore birefringent but not cross-striated.

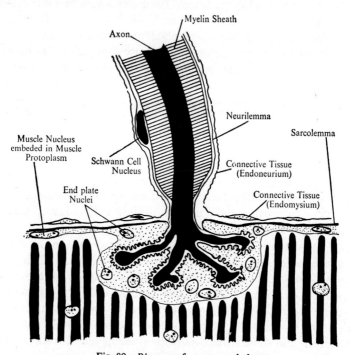

Fig. 80. *Diagram of a motor end plate.*

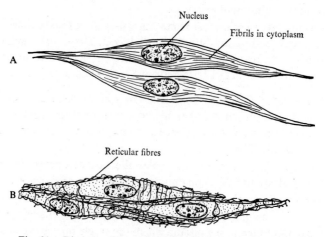

Fig. 81. Diagram of smooth muscle fibres (A) and showing the reticular fibres (B).

There is no distinct sarcolemma but each cell is surrounded by a network of reticular fibres and sometimes by collagen or elastic fibres. Elastic fibres are commonly associated with smooth muscle and some authorities prefer to call this tissue 'myoelastic tissue'. In the walls of the larger arteries many elastic fibres are found, so that when the arteries are stretched by the force of blood entering them the muscles are helped by the elastic tissue to contract and force the blood onwards.

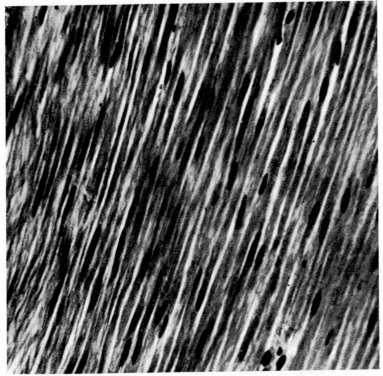

Fig. 82. *Photomicrograph of a section of smooth muscle. (By courtesy of Dr. S. Bradbury.)*

The fibres (cells) do not always all run in the same direction, for instance in the small intestine there is a band of circular smooth muscle surrounded by a band of longitudinal muscle. The nerve supply is from the visceral efferent system and there are no motor end plates but the fine endings of the nerves wrap round the cells instead.

The characteristic feature of the activity of this tissue is its capacity for sustained rhythmic contraction. It cannot contract very rapidly nor very powerfully but it does not suffer from fatigue. It is found in the body where continuous pulsation is required without any great feat of strength, e.g. in the gut from the oesophagus to the rectum, in the walls of the ducts of glands and the urinogenital tract, surrounding the glands where it squeezes out the secretions and in the walls of the arteries and veins (to a lesser extent) and in the skin.

Cardiac muscle (figs. 83, 84). This tissue is the muscle of the heart and may perhaps be considered as a type of striated muscle as the discs A, I, Z, etc. can be seen. There are several differences from normal striated muscle. These are:

(a) the fibres of cardiac muscle branch and anastomose with others running parallel with them to form a network. This does not occur in striated muscle.

(b) there is more sarcoplasm between the myofibrils.

(c) there are conspicuous thick striations crossing the fibres transversely at regular well spaced intervals. These are called intercalated discs and whilst their function is at present unknown they do serve as a useful diagnostic feature of cardiac muscle.

(d) although the cells appear syncitial there is some evidence from tissue culture that the cells may be uninucleate. In culture the surrounding protoplasm of adjacent nuclei have been seen to pulsate independently.

(e) the sarcolemma is more delicate than in striated muscle.

Cardiac muscle has properties intermediate between those of smooth and striated muscle. The contraction is fairly rapid and powerful but

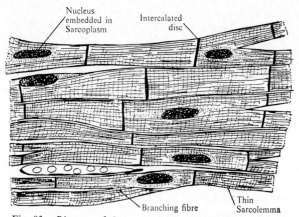

Fig. 83. *Diagram of the structure of cardiac muscle showing the branching fibres and intercalated discs.*

the muscle does not easily fatigue. This is an essential feature for heart muscle of course. Perhaps the most characteristic feature of the physiology is the long refractory period of cardiac muscle. This means that after the cell has contracted there is a relatively long period of time in which the cell cannot contract again. Therefore after a single contraction there must follow a period of relaxation. This feature allows the possibility of the regular beating of the heart (see p. 230).

Fig. 84. *Photomicrograph of an L.S. of cardiac muscle.*
1. blood capillary. 2. intercalated disc. 3. striated myofibril.
(*By courtesy of Dr. S. Bradbury.*)

There is specialization within the tissue. Near the septum which separates the ventricles of the heart there are some specialized cardiac fibres called Purkinje fibres. The anatomical peculiarities need not detain us here but it should be noted that these modified fibres are used to coordinate contractions in various parts of the heart as they are capable of conduction as well as contraction (see p. 228).

NERVOUS TISSUE

In all tissues certain fundamental properties of living matter are singled out for special development and in nervous tissue we find the cells specially designed for conducting information, i.e. the power of irritability and conduction is emphasized. Information is carried through the protoplasm in the form of electrical charges called impulses.

The basic type of cell in nervous tissue is called a neurone (or nerve cell) and the neurones are supported by several types of unspecialized cells called neuroglia. The nerves which are seen in dissection are technically called organs for they contain connective tissues and blood vessels as well as nervous tissue.

A TYPICAL NEURONE is seen in fig. 85, and it is obviously well designed for conduction, for the protoplasm is pulled out into long wire like processes. If these processes conduct impulses towards the cell body of the neurone, which contains the single nucleus, they are called dendrites,

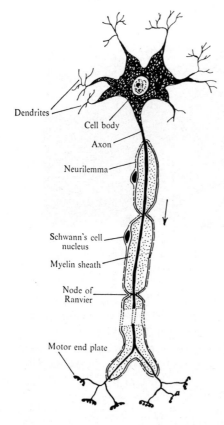

Dendrites

Cell body

Axon

Neurilemma

Schwann's cell nucleus

Myelin sheath

Node of Ranvier

Motor end plate

Fig. 85. *Diagram of a typical motor neurone.* The cell body of a motor neurone is situated within the ventral grey matter of the spinal cord. From the cell body the axon of the neurone passes out of the spinal cord in a spinal nerve and terminates in striated muscle tissue as motor end plates. The arrow on the figure indicates the direction of conduction of the impulse.

whereas the process which conducts the impulses away from the cell body is called the axon. It is not possible by merely looking at the cell to say which process is a dendron and which an axon. The cytoplasm of the cell body contains Nissl granules which are made of nucleic acid and stain well with basic dyes like methylene blue. The axon is covered by a sheath of fatty material called myelin and the neurone is therefore referred to as a myelinated or medullated fibre. Certain fibres have no myelin or very little and they are called non-myelinated or non-medullated fibres. In dissection the nerves which are made of myelinated fibres appear white whereas those made of non-myelinated fibres appear grey. Myelin is thought to be concerned with the electrical insulation of the neurones and it certainly makes the neurones capable of more rapid conduction. Outside the myelin sheath are plastered the Schwann cells and outside these is the neurilemma, which is a tough, thin membrane of scleroprotein. The sheath of Henle fits around the whole lot of these covering layers and is continuous with the connective tissue endoneurium which binds all the neurones together. Like the endoneurium the sheath of Henle is made of non-elastic white collagenous fibres. At various places along the axon the myelin sheath is absent and the neurilemma dips down to the axon. This point is called a node of Ranvier,

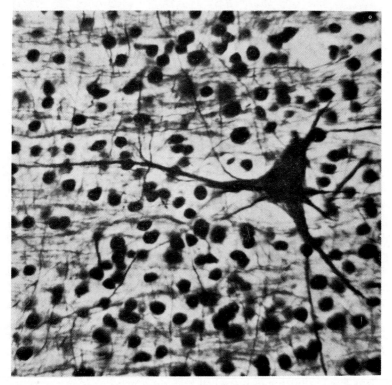

Fig. 86. *H.P. photomicrograph of a neurone in a section of the cerebral cortex (visua area). (By courtesy of Dr. S. Bradbury.)*

and the portion of the axon between the two nodes has one Schwann cell only. When the neurone is first formed the nodes are close together but later they grow apart as new protoplasm which is made in the cell body of the neurone, flows down the axon causing it to extend.

Fig. 87.   *H.P. photomicrograph of a neurone (motor neurone) in the ventral horn of the spinal cord. (By courtesy of Dr. S. Bradbury.)*

DAMAGE TO NEURONES. When a neurone is damaged so that the axon is cut, the distal part of the axon dies as it is separated from its nucleus. When this part of the axon disintegrates the Schwann cells grow out from the intact part of the axon to form a hollow tube into which protoplasm from the cut end of the axon can flow. Regeneration of damaged neurones is very slow as it is limited by the speed of the flow of protoplasm along the axon. It is important to note that nerve cells can only grow so long as their cell bodies remain intact and once the cell body is damaged the neurone cannot be replaced; for nerve cells cannot divide to replace their damaged neighbours. In the section on epithelia we have described how epithelial cells divide in order to replace their damaged neighbours; nerve cells cannot do this for they have not got the power to divide. Cell division in the nervous system is complete before the young mammal is born so that the total number of nerve cells is fixed at birth and if the cell body of a neurone is damaged this cell is lost forever and cannot be replaced.

CONNECTIVE TISSUE

This tissue is very variable in structure, sometimes being very diffuse and spreading through the body as an integral but not very obvious component of many organs as in the case of areolar connective tissue, but often being a more obvious consolidated tissue as in bone. The common feature of all connective tissue is that there are always two elements present: first, the cells, which are mesodermal in origin and secondly, the product of these cells which is called the intercellular substance or matrix. In different kinds of connective tissue there are different kinds of mesodermal cells, bone cells in bone and fibroblast cells in fibrous tissue. Each different kind of cell naturally makes a different kind of matrix and so we have as many different kinds of connective tissue as we have different kinds of specialized cells. The relationship between the cell, the matrix and the tissue is made clear by the following examples.

Tissue	Mesodermal cell	Matrix
bone	osteoblast	bone matrix
cartilage	chondroblast	cartilage matrix
elastic fibrous tissue (ligamentous)	fibroblast	elastin fibre
non-elastic fibrous tissue, collagenous	fibroblast	collagen fibre

Areolar connective tissue

A white sticky substance penetrates and envelops all the organs of the body and packs much of the space between the organs. Since this tissue has many potential spaces in it, which become filled with air in dissection, it is called areolar tissue or loose connective tissue. The structure of areolar tissue is shown in figs. 88, 89, where it is clear that the cells and the fibres they produce are embedded in an amorphous ground substance, which itself is also a product of the cells. The ground substance varies in viscosity from a fluid sol to a viscous gel depending upon the age, activity and physiological condition of the tissue. This ground substance is not the same thing as the tissue fluid although its contents are in equilibrium with the latter. All the functions of the ground substance are not known for certain but it is thought to influence the diffusion of nutrients from the capillaries to the tissues and the growth of many organs, especially bone, cartilage and certain tumours.

Hyaluronic acid, which consists of long chain polysaccharides, is an important constituent of the ground substance.

The fibres in areolar tissue are of three main kinds, elastic fibres, collagen fibres and reticular fibres.

ELASTIC FIBRES are long fine branching fibres which form a loose network throughout the areolar tissue. They are easily stretched but when the tension is removed they readily return to their former length. They can easily be stretched to 150% of their former length by a strain of only 30 kilogram/cm² before they break. When these fibres are abundant and closely packed as they are in ligaments they have a yellow appearance. Chemically they are made of an albuminous protein called elastin

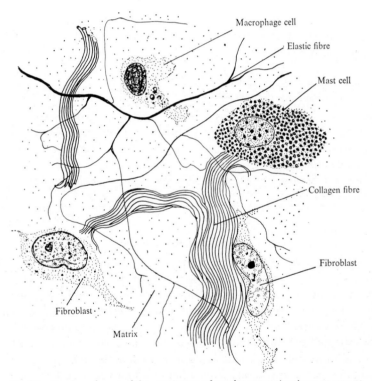

Fig. 88. *Drawing of some of the constituents of areolar connective tissue*, as seen at high magnification. Reticular fibres are not shown.

which is resistant to boiling in acids and alkalies but is digested by protein digesting enzymes like trypsin. The function of yellow fibres is to give elasticity and flexibility but not to confer strength upon tissues.

They are found commonly in ligaments, where they bind bones such as vertebrae together, in the walls of arteries and embedded in cartilage in the external ear. They are sparsely distributed in areolar tissue. The elastic fibres are produced by fibroblasts found scattered in the matrix.

COLLAGEN FIBRES appear as flat and often wavy ribbons of fibres which run through the tissue. The ribbons of fibres may branch but the fibres themselves, unlike yellow fibres do not branch. In unstained tissue the collagen fibres are white and are birefringent—due to the long molecules which run in bundles along their length. The fibroblast cells which make these fibres are closely applied to the fibres. Contrast this with elastin fibres where the fibroblasts are separate from the fibres. Collagen is a protein which is soluble in pepsin and swells when in contact with mineral acid. After boiling in water collagen produces a glue like substance called gelatin. Collagen fibres are flexible, inextensible and strong, capable of withstanding a strain of 120 Kg/cm^2 without breaking. They are therefore useful in strengthening tissues and helping them to keep their form. The direction of the fibres coincides with the lines of stress. Collagen fibres are found in huge quantities in tendons where they serve to attach muscles to bones, e.g. in the Achilles' tendon which transmits the force of the large gastrocnemius muscle to the heel.

RETICULAR FIBRES are extremely fine fibres which are not usually seen in connective tissue unless the tissue is specially stained to show them, using silver salts; hence they are sometimes called argyrophile fibres. Networks of these reticular fibres are found around blood vessels, nerves and muscle cells and in the basement membrane where epithelia joins connective tissue. These fibres are not elastic and are not digested by trypsin; in these two properties they obviously resemble collagen. They may be considered to be immature collagen fibres and are sometimes called precollageneous fibres. They differ from collagen in their form, and in their reaction to stains.

The cells of areolar tissue are of several main kinds, which are however very difficult to distinguish under the microscope unless the tissue is specially stained and the observer has had a great deal of experience. Each cell type has several varieties and its appearance may alter with the different physiological states of the tissue. Any comments made here refer to the normal appearance of the average type of this cell. The student should certainly not expect to see all these cells in the first few slides of areolar tissue examined.

FIBROBLASTS are the most common cells found in areolar tissue, and are flattened or spindle shaped cells often with sharp processes projecting from the cytoplasm. The outline of the cell is often not clear and many cells often seem to fuse together to form a spongelike network throughout the tissue. The nucleus is oval shaped or slightly irregular and contains fine dustlike chromatin granules. Fibroblasts are actively motile cells and are often to be seen dividing but they do not have psuedopodia. They move towards injuries in the skin or tissues or towards sites of parasitic infection and play an important role in making new fibres to seal off the wound from invading bacteria or parasites. The actual process by which collagen fibres are formed by fibroblasts is not yet fully understood but the fibres seem to be a crystallization from the fluids which are secreted between two fibroblasts. The formation of elastin fibres is even more obscure, but they are formed by fibroblasts.

MACROPHAGES are almost as common as fibroblasts and may be more common in areas well supplied by blood. They may be distinguished from fibroblasts by a smaller nucleus with a heavily folded membrane and by coarser granules in the nucleus and inclusions in the cytoplasm. Macrophages are normally non-motile cells and are often called fixed macrophages or histiocytes. Their function is to engulf bacteria in the connective tissue. In inflammation the macrophages become free and actively motile towards the site of infection. Their function is defence by ingesting any bacteria which penetrate the tissues. These cells along with others such as dust cells which ingest particles in the lungs, and the reticular cells of the lymphatic system, make up what is known as the reticulo-endothelial system.

MAST CELLS are found near blood vessels and they produce the ground substance and perhaps a compound called heparin which helps to prevent the clotting of blood in the blood vessels. They stain well with basic dyes.

UNDIFFERENTIATED MESENCHYME cells look like small fibroblasts and are found near blood vessels and where there are reticular fibres, and in parts of areolar tissue which are not highly differentiated. They are a reserve of cells which can develop into several different cell types if the need arises.

PLASMA CELLS are rare but they do occur in blood forming tissue and in lymphatic tissue in large numbers. They are small cells which stain with basic dyes and have a characteristic pale area next to the eccentrically placed oval nucleus. These cells are thought to be the site of production of antibodies. They accumulate in tissues at the site of chronic infection (p. 576).

FAT CELLS are an important constituent of areolar tissue which serve as a food reserve and as a heat insulator. In certain parts of the body, e.g. around the kidneys and below the skin, the fat cells may become very numerous and form what is called adipose tissue. There are always a certain number of fat cells in areolar tissue even in the thinnest of people. The cells can multiply, and can also increase the amount of fat that each cell can hold, if an excess of food is available, producing large shining spherical cells.

PIGMENT CELLS contain the black pigment melanin and are more common in certain dense connective tissues than in areolar tissue. They are very common in the pia mater of the medulla oblongata and in the choroid layer in the eye.

Functions of areolar tissue
1. *Mechanical.* It penetrates the organs and envelops them, giving them form and flexibility. At the same time it allows the movement of the various organs one upon the other. The spaces between organs are packed with this tissue. The mechanical functions are determined by the fibres in the tissue.

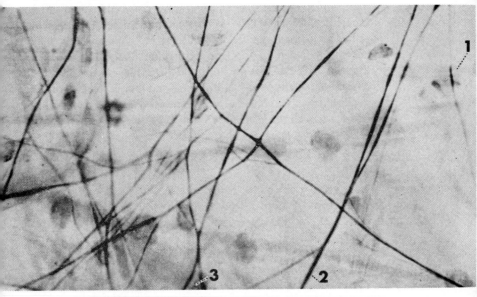

Fig. 89. *H.P. photomicrograph of areolar connective tissue.*
1. faintly staining collagen fibre. 2. elastic fibre. 3. fibroblast nucleus.
(*By courtesy of Dr. S. Bradbury.*)

2. *Nutritional.* The areolar tissue penetrates deep into organs; the cells in muscle for instance are wrapped up in small bundles by a connective tissue envelope called the perimysium whilst each cell is wrapped by the connective tissue endomysium. Thus the ground substance of the areolar tissue is in close contact with the tissue fluids and can affect the nutrition of the cell.

3. *Defence.* We have seen how the fibroblasts can help to isolate disease producing organisms and that the plasma cells produce antibodies. The macrophages are very important in ingesting bacteria. In inflammation many cells of connective tissue become phagocytic and ingest bacteria—these cells are then called polyblasts and they all assume a similar appearance.

Classification of connective tissues

(*a*) CONNECTIVE TISSUE PROPER.

    i. Loose areolar tissue.

    ii. Dense or regular connective tissue. Here the fibrous element is dominant, e.g. collagen fibres in tendons and elastin and collagen in ligaments.

    iii. There are many other kinds of connective tissue depending upon which of the elements of loose connective tissue predominates. Thus fat cells dominate adipose tissue and pigment cells the pigment tissue. Mesodermal cells at surfaces may become flattened to form an epithelium like tissue. Such a tissue is called a mesothelium and lines the cavities like the coelom of the abdomen (peritoneum) and thorax (pleura) or the pericardium. These mesothelia are called serous membranes and contain all the elements of areolar tissue.

(*b*) VASCULAR CONNECTIVE TISSUE. Areolar tissue penetrates into structures like spleen and bone and the cells become modified to produce both the red and white cells of blood. The white cells produced in bone marrow have small granules in the cytoplasm and are called granular leucocytes. The non granular leucocytes are produced by the connective tissue cells of the lymphatic system in the lymph nodes and lymph organs. The lymph, the cells of the lymphatic system and the fluid and cells of blood are all part of the connective tissue, but the vessels which carry these cells and fluids are not tissues but organs since they are composed of so many different tissues. The nature of the cells of the blood and the process of blood formation will be described later (p. 201).

(c) SKELETAL CONNECTIVE TISSUE. (See also Chapter XIII.) We have learned that areolar tissue has a skeletal function by virtue of possessing fibres. In cartilage the amorphous ground substance is of a cheese-like consistency and is called chondrin, and it is well suited to withstand compression forces. If there are no visible fibres* present it is called hyaline cartilage. Yellow fibres run through the chondrin in elastic cartilage, e.g. in the tip of the nose and the pinna. Collagen fibres are present in the non elastic cartilage between the bodies of the vertebrae and in the cartilages in the knee.

Another modified connective tissue which can stand tension as well as compression is bone. Bone often arises in connective tissue to form membrane bone, e.g. some bones of the skull and the clavicle. Mesenchyme cells or fibroblasts in the connective tissues are differentiated to form osteoblasts which produce a fibrous protein similar to collagen and then calcium phosphate and carbonate are laid down, the connective tissues thus becoming ossified.

Many bones like the long bones of the limbs are preformed in cartilage and only later turn to bone. Modified mesodermal cells are responsible for the transformation and for the secretion of the calcareous intercellular substance. The details of bone formation are discussed in Chapter XIII.

Tissues have been defined as collections of cells specialized in a similar fashion to do one main job. This definition is obviously true for epidermal, muscular and nervous tissue, but it needs to be interpreted very freely to embrace all connective tissue. In areolar tissue we have suggested three main functions not one. Connective tissue is a collection of cells of a certain type which are designed to carry out a limited range of functions.

ORGANS AND THE CONCEPT OF THE ORGANISM

Structures like the heart, liver, kidney and stomach are called organs. They are all complicated structures made up of several tissues which are combined in such a way that they can cooperate in the performance of duties which are essential for the life of the whole organism. The heart for instance consists of muscle, connective tissue and epithelial tissue and is supplied by nervous tissue. All four tissues cooperate so that blood may be pumped round the body efficiently for the benefit of the whole organism.

From the earlier part of this chapter we learned how in multicellular organisms the body is split into millions of special cells each specialized

*There are reticular fibres which need special staining processes to demonstrate them.

to carry out one function very well. We learned that loss of some functions has to accompany specialization and therefore there arises the need for integration. We have seen now how the cells are grouped into functional units called tissues and as we read other chapters it will become clear how the tissues cooperate in the work of the organs and how eventually all these highly specialized cells work together in the life of the organism.

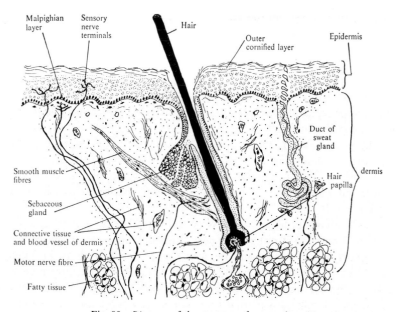

Fig. 90. *Diagram of the structure of mammalian skin.*

Structure of the skin. Skin is a diffuse organ consisting of several different tissues which are integrated to carry out many very important functions on behalf of the whole organism. From fig. 90 it can be seen that skin consists of two main parts, an outer epidermis and an inner dermis. The epidermis is a stratified epithelium which is continuously replacing itself. The lowest layer of the epidermis (the Malpighian layer) is always dividing in a plane parallel to the surface of the skin and thus new layers of cells are always appearing to replace those rubbed off the outside. It is interesting to think that the cells which are in the epidermis of your hands at the moment will not be there in a month's time, and yet the skin will appear to be very much the same. The skin is to be thought of as an extremely active organ rather than an inert layer, it is constantly being rebuilt from new materials into the same pattern as before. This self maintaining, self regulating, dynamic pattern of events is a typical feature of life. As the cells produced by the Malpighian layer get older and nearer to the surface they die and become flattened into squames. These squamous cells are made of a protein called keratin, which is the material from which nails and horns are made. The outer layer of the skin is therefore often called the stratum corneum or the horny layer. Underneath the epidermis is the dermis which is composed of many tissues including nervous tissue, muscle tissue and connective tissues of several types. The dermis contains the sweat and sebaceous glands, the blood vessels, the hair follicles, muscles and the nerve endings.

Function of the skin. As we consider the many functions of the skin we shall see that like all organs the skin is responsible for doing several things on behalf of the whole organism.

1. PROTECTION. The epidermis stretches as a continuous cover over the whole surface of the body and it also continues into the several orifices of the body e.g. the mouth, anus and vagina. If an area of skin is rubbed off and we do not take the precaution of seeing that the wound is kept clean and invading bacteria are killed, then we are often reminded as the wound goes septic that the intact epidermis had been keeping these bacteria out. When the surface of the skin is intact the only places where germs can penetrate are the sweat ducts and the hair follicles (pores). The apertures of the alimentary, respiratory, urinary and genital tracts present the only other portals of entry. If bacteria do somehow get through the epidermis the connective tissue in the dermis is a very good second line of defence. The process of inflammation develops which combats the infection (see p. 562). Pigment cells lie in the dermis just below the Malpighian layer protecting the dermis and all the underlying tissues from the harmful effects of solar radiation. Most people have experienced sunburn early in the summer and many have doubtless prided themselves later on their ability to bask in the sun unharmed once they have acquired their tan. In many mammals the pigment cells are also protective because they help to camouflage the animal either by *cryptic* (e.g. rabbit) or *disruptive* colouration (badger or zebra). Cryptic colouration allows the animal to look like its normal background. Disruptive colouration helps to break up the normal outline of the animal.

2. RECEPTION OF STIMULI. In the dermis are many afferent nerves which convey information about the changes occurring in the outside world to the central nervous system. These nerves may start as finely

branched dendrites or in more highly organized receptors like Meissner's corpuscles, which are responsible for pressure detection. Skin which is hairless like the finger tips of man or the lips of most mammals tend to have more highly organized sense receptors than does hairy skin. It is not possible at present to identify the function of the nerve by the structure of the receptor since nerves that we know to have different functions may start at what appear to be identical receptors. It is thought that a particular neurone can respond to only one kind of stimulus, i.e. there are special pathways for temperature, pressure, touch and pain information. Certain areas of skin are more sensitive to stimuli than other areas. You test the temperature of the bath with your hand and it feels comfortably warm but if you step in and particularly if you promptly sit down you often prove conclusively to yourself that some areas are more sensitive than others. If you hold two dissecting needles in one hand so that the points are about 2 mm. apart and then you gently prick the other hand with these two points simultaneously, you will find that some parts of the hand enable you to distinguish the double prick whereas only a single prick can be felt in other places. Thus we have a picture of the whole surface of the body covered by receptive spots for touch, heat, cold and pain. At any one time thousands of these receptive spots are being stimulated and it is from the general pattern of these stimulation fields that the central nervous system allows the whole organism to be 'aware' of the changing conditions around it. Thus the skin is very important in adapting the animal to its surroundings.

3. TEMPERATURE REGULATION. Protection and reception are characteristic features of the skins of all animals but this third function of temperature regulation is found only in birds and mammals, animals which are consequently referred to as homoiothermic. The maintenance of a constant temperature, often one much higher than the surroundings, is an expensive way of living but it has given homoiothermic animals decided advantages over other animals, which are called poikilotherms. Homoiothermic animals can live in the tropics or the arctic, in the midday sun or the cool of midnight, whilst poikilotherms may be too hot or too cold to move. Some mammals do hibernate when conditions are too cold, e.g. dormouse, or too dark for hunting, e.g. polar bear, but the majority live an active life throughout the year. It is significant that the hibernators are either small or live in very difficult climates. Thus this property of homoiothermy has enabled mammals to extend their range much further than the reptiles could ever have gone and the evolution of this property was necessary before living in air was fully mastered by animals. One of the difficulties of living in air as opposed

to living in water is that the range of temperatures encountered is much greater, and the speed at which the changes occur is much more rapid. An extension of the range for mammals meant that the early mammals could move away from places that were overcrowded by their reptilian competitors or they could feed at times, such as the night when the temperatures were unsuitable for poikilotherms. Thus homoiothermy allowed the mammals to avoid competition and to extend their range to new habitats. A constant internal temperature could also conceivably be a great advantage in a very complex animal, for the chemical reactions can proceed at a more constant speed than is possible in a fluctuating temperature. Where one set of reactions has to be geared into many others this is an obvious advantage.

The skin plays an all important role in temperature regulations as its large surface area is a potential site for heat loss. Therefore the first requirement of the skin of a mammal is that it should be a good heat insulator, and this is usually fulfilled by a layer of fat beneath the skin and by the hairs. The fat layer is built up in autumn in wild mammals of colder climates so that they will have a food store for the winter months and also to act as an heat insulator. In the fur trade trappers speak of 'blue' pelts when they refer to animals that have been killed in the summer. The skin looks bluer than a winter skin because the fat layer has broken down. In the autumn and winter the skin becomes clear, cream coloured and supple, as fat is redeposited. Hair is of importance in temperature regulation because air is trapped under it. Still air is a bad conductor of heat but moving air is a good convector. Heat losses are kept to a minimum by surrounding the body by a layer of still air. In a mammal like a fox there are two main kinds of hair; the guard hairs are long and stiff and serve to make water run off the fur easily, whilst the under fur or 'fur fibre' is shorter and more suited to trapping air. If the under fur gets wet all the air is lost and the hair mats together. So both kinds of hair are useful in preventing heat loss. The power to regulate the amount of heat lost depends on the ability of the mammal to raise and lower the hairs and thus vary the thickness of the layer of air which surrounds the body. This is why a cat looks bigger on a cold day than it does on a warm one. The cooler the day the more the cat contracts the smooth muscle attached to the hairs (see fig. 90). It does not have to do this consciously of course for the smooth muscle is innervated by the autonomic nervous system. The layer of air is increased and the cat loses less heat. The climate affects the nature of the hair, the colder the climate the more fur fibre is made, whilst a damp climate produces long silky guard hairs. The seasons also bring differences in the fur as the hair is shed in the spring. During the shedding time the skin undergoes a cellular change becoming sinewy, tough and reddish—a condition known as springy. It is interesting to note that water animals like beavers have the best pelts in spring when the water is coldest. The colour of the skin may have some influence on body temperature in that it will affect the amount of heat lost by radiation. We would expect a light coloured animal to lose less heat by radiation than a darker animal. It is a fact that many animals in the arctic like the arctic hare do turn white in winter—but this may be an adaptive colour change to camouflage them against the snow.

A third and very important fact governing the amount of heat lost is that the amount of heat supplied to the skin can be varied by altering the diameter of the blood vessels in the skin. The blood carries heat around the body and if the smooth muscles in the walls of the arterioles contract then a reduced quantity of blood and heat is present in the skin. Thus in cold weather the arterioles contract and the skin looks blue or white, and heat is conserved. Conversely when the body is too hot a reflex action mediated by the autonomic system, dilates the arterioles and a greater volume of blood flows through the skin and heat is lost from the body. Therefore when we are too warm we appear flushed. Blood vessels at this time appear to be nearer the surface of the skin, this is because they are dilated. They do not migrate bodily from a lower level to the surface. Thus we see that vasoconstriction and vasodilation control the amount of heat entering the skin whilst the subcutaneous fat and the hair help by insulating the body from the environment.

Sweating is a very rapid method of losing heat from the skin. Sweat is a watery fluid containing small quantities of dissolved substances like sodium chloride and the excretory product urea. When the body is too hot the sweat glands are stimulated to pour out sweat onto the surface of the skin; the water evaporates using body heat in the process. The latent heat of vapourization of water at body temperature is about 580 calories for every gramme of water vapourized. In this way a large amount of heat can be lost very rapidly.

There are other minor ways of regulating body temperature, some animals, e.g. dogs, have no sweat glands and they lose heat by panting thereby allowing water to evaporate from the tongue. All mammals lose some heat in the breath. Behaviour helps to control heat loss. Vigorous movement generates heat within the body and general lassitude tends to keep the body cool; in the morning when we wake up our body temperature is often about two degrees F. lower than it is at midday. The surface area presented to cooling influences is also important since heat is lost from the surface. Therefore parts of the body like ears get cold quickly and on cold days the cat curls up on the rug like a ball, thereby presenting the smallest possible area for cooling. But on hot days it stretches out on its side with limbs extended exposing the maximum amount of surface.

We see then that in this organ, the skin, several tissues cooperate to regulate the temperature. The whole body often lends support to the action of the skin and intelligent behaviour may be employed as when man lights fires and uses ventilation systems. This is a good example of the integration of tissues into organs and of the concerted action of many organs all working harmoniously for the good of the whole organism. How is this integration achieved, how do the sweat glands know when to sweat and the hair muscles when to contract? The somatic nervous system is responsible for the control of breathing movements, posture and shivering, whilst the autonomic (sympathetic) nervous system is controlling the blood vessels, sweat glands and erector muscles of the hairs. The coordination of these two parts of the nervous system is done by a part of the fore-brain called the hypothalamus. This is a region lying just behind the optic chiasma and above the pituitary gland. (See fig. 140.) The hypothalamus functions as a physiological thermostat, one part of it sets up reactions in the body which cause heat to be lost, whilst another part of it is responsible for heat conservation. The hypothalamus is the most vascular part of the brain and contains cells which are sensitive to the temperature of the blood.

Although all mammals are homoiothermic when active, some species are able to suppress their temperature regulating processes and become almost poikilothermic. The metabolic rate falls under these conditions and the animal becomes inert and may appear to be dead. In this condition the animal may be able to withstand much worse environmental conditions than normal, and can tolerate lack of food and low temperatures. Some mammals such as bats undergo these changes for short periods of several hours of the day when they are inactive. Other mammals such as golden hamsters, hedgehogs, dormice and marmots hibernate or aestivate for much longer periods. In the hibernation of these animals the temperature controlling mechanisms are not completely in abeyance and if the environmental temperature falls near to freezing then the animal wakes up and becomes homoiothermic again. Further, the poikilothermic state in some mammals is interrupted at intervals for feeding. Stores of food may be laid up in or near the nest for these periods.

In the hibernating animal the body temperature is low, respiration and the heart rate is slow and the animal slowly loses weight as it draws upon food reserves in the form of glycogen and fat. The onset of hibernation is probably brought about by internal factors which are not at the moment understood. Some animals have been kept under constant conditions of light, temperature, feeding and water supply and yet still hibernate at the time of normal animals subjected to the seasonally changing environmental conditions.

4. WATER CONSERVATION is the fourth major function of the skin. The dead cells of the stratum corneum are heavily keratinized forming a waterproof layer which keeps water in the body. This property of the mammalian skin is of extreme importance as it allows the animal to live in dry places. By contrast the skin of a frog has very little keratin and it loses water rapidly through its skin and is consequently compelled to frequent only damp habitats. The frog can of course breathe through its skin whereas animals with heavily keratinized skins cannot. In mammals the sebaceous glands pour out a lipoid secretion which may help in waterproofing the skin.

5. SUCKLING THE YOUNG. One of the characteristic features of mammals is that they feed their young on milk. The milk is supplied by the mammary glands of the mother. These glands are derived from the skin.

6. SCENT GLANDS are derived from the skin and play an important role in the lives of some mammals. They may be used for defence as in the skunk, or for sexual recognition e.g. dog, or for marking out the territory. The deer has scent glands in the inguinal region near the anus, and antelopes have similar glands near the eyes, whilst other herbivores have scent glands on their feet. These scents allow the mammals to recognize members of their own species and may be an important factor in the development of their social life.

7. NAILS, hooves and claws are all formed from the skin.

Chapter Five

The Circulatory System

Introduction. During the course of evolution animals have not only become more complex but they have become larger. In a very small organism, all of its protoplasm is near the outside world, and the exchanges between the animal and its environment that are necessary for life (e.g. the absorption of oxygen, the getting rid of waste products) can proceed easily.

In small animals the surface area is large compared with the volume and is adequate for all the exchanges which are necessary. When animals become larger, their volume increases at a faster rate than does their surface area (see p. 162) which is no longer adequate for the exchanges between the animal and its environment. In other words, as animals increase in size, parts of the animal body become far removed from the outside world, too far for the process of unaided diffusion to be an adequate link between the internal cells and the outside world.

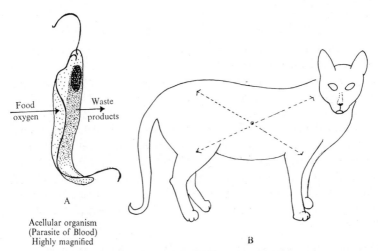

A

Acellular organism
(Parasite of Blood)
Highly magnified

B

A. Shows a drawing of an acellular organism (trypanosome—a parasite of the mammalian blood). Because of the size of the organism direct exchange of substances with the environment from all parts of the organism is possible.

B. A mammal. The dot in the centre of the animal is a representation of a highly magnified cell. This cell is remote from external environment but is connected to all parts of the organism by means of the circulatory system, in particular to those organs which are specialised to relate the animal to the environment e.g. the lungs, kidneys, alimentary tract.

197

These problems were solved, in part, by the development of a transport system, so that substances from the outside world could be transported into the interior of the body and vice versa. This transport system is the circulatory system—a system of hollow tubes containing fluid, circulated by means of a pump, the heart. But the circulatory system not only links each part of the body with the outside world, but it also links the parts of the body with one another. This became very important when special organs were evolved to perform particular functions; the circulatory system plays an important role in uniting the specialist organs into one working unit. Thus, the stomach and intestines, specialized as they are for the digestion and absorption of food-stuffs, supply the raw materials of life to all the other organs of the body by means of the circulatory system. The lungs, specialized for the intake of air and the exchange of gases with the blood, supply the vital element oxygen to all cells of the body by means of the circulatory system. The endocrine glands which produce their chemical messengers, the hormones, are able to influence many other organs in the body, only by means of their connection with the blood stream.

Thus the circulatory system links together the individual organs of the body and is the ultimate connecting link between each cell and the outside world.

CIRCULATION AT THE CELLULAR LEVEL

The circulatory system, as described above, consists of a series of hollow, fluid filled tubes, and a pump, the heart which circulates the fluid. Each cell of the body is not, however, in contact with a blood vessel, and further, the cells are separated from the blood by the wall of the blood vessel itself. The connecting link between each cell of the body and the blood in the vessels is by means of a fluid surrounding all the cells, called tissue fluid. It is through this tissue fluid that the exchanges between the cells and the blood stream occur; substances needed by the cells e.g. oxygen, glucose, diffuse from the blood stream into the tissue fluids and so to the cells, and substances produced by the cells (waste products, hormones etc.) gain access to the blood by the reverse course.

Formation and composition of tissue fluid. Tissue fluid is not a stagnant medium but is being continually replaced. It is formed at the arterial end of the capillaries where the contained blood is at relatively high pressure. The wall of the capillary acts as a semipermeable membrane and through it are filtered out some of the constituents of the blood. Under normal circumstances most of the cellular elements of the blood are retained within the capillaries and only some of the fluid elements pass out into the tissue fluid, particularly those substances of lower molecular weight viz. water, oxygen, salts, glucose, hormones and proteins of lower molecular weight. Only when the walls of the capillaries are damaged, as in inflammation, do the proteins of higher molecular weight and the cellular elements gain access to the tissue fluids.

The tissue fluids thus contain less protein than does blood and so have a lower osmotic pressure. Thus there is a tendency for a diffusion of tissue fluid into the blood stream; this is overcome at the arterial end of the capillaries by the hydrostatic pressure of the blood. Where the hydrostatic pressure is low, that is at the venous end of the capillaries, tissue fluid passes into the blood stream. There is thus a circulation of tissue fluid; it is formed at the arterial end of the capillary and drains back into the capillary at the venous end.

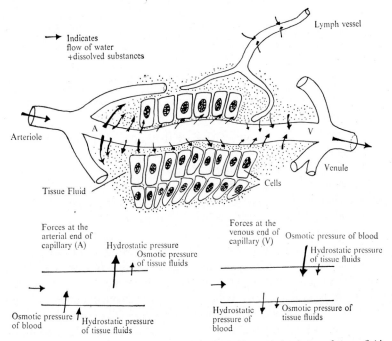

Fig. 91. *Diagrammatic representation of the formation and circulation of tissue fluid.*

The hydrostatic pressure of the tissue fluid and tissues themselves also plays a role in the formation and flow of tissue fluid; at the arterial end of the capillaries the hydrostatic pressure of the tissue fluids is overcome by the higher hydrostatic pressure of the blood. But as the pressure in the capillaries gradually falls along its length, so the tissue fluids drain into the venous end of the capillaries, under the influence of both the hydrostatic pressure of the tissue fluid itself and the higher osmotic pressure of the blood.

The factors concerned in the formation and drainage of the tissue fluid may be summarized thus:

Forces causing fluid to enter the tissues	=	Hydrostatic pressure of blood + Osmotic pressure of tissue fluids.
Forces causing tissue fluid to enter the capillaries	=	Hydrostatic pressure of tissue fluids and osmotic pressure of blood.

It is because of the variation in these forces along the length of the capillary that a circulation of tissue fluid results.

A simpler explanation of the formation and circulation of tissue fluid depends upon the fact that in any particular tissue some capillaries are wide open with a fast flow of blood through them (high pressure capillaries) whereas in other capillaries the flow of blood is reduced by the constriction of the arterioles supplying them (low pressure capillaries). The ratio of high pressure to low pressure capillaries depends upon the activity of the tissue; thus the flow of blood through muscle is much greater during and following muscular activity than at rest, because of the relaxation of the arterioles with the consequent 'opening up' of the capillary bed. Tissue fluid may be produced from the high pressure capillaries and drain back into the low pressure capillaries.

Tissue fluid is also drained by means of the lymphatic vessels which ramify through most of the tissues of the body, excluding the central nervous system.

The importance of the various factors concerned in the formation and drainage of tissue fluid may be well illustrated by referring to certain disease processes. The importance of the osmotic pressure of the blood as a factor in the drainage of tissue fluid is seen in diseases where there is a fall in the amount of protein in the blood, such as in severe malnutrition. Because of the reduced osmotic pressure of the blood there is less tendency for tissue fluid to drain into the capillaries and increasing amounts of fluid collects in the tissue spaces; this excess fluid tends to gravitate downwards to produce swollen ankles and legs, or to collect in various body cavities e.g. pleural and peritoneal cavities.

The role of the lymphatics in the drainage of tissue fluid is seen in a disease such as filariasis (elephantiasis), where grossly swollen legs are the result of blockage of the lymphatic channels by small nematode worms. Any extensive injury to soft tissues damages many lymphatics and so interferes with the drainage of tissue fluid; this is seen particularly in the legs where at the best of times there is some difficulty in returning all the tissue fluids into the circulation against the forces of gravity.

Whenever there is a rise in the hydrostatic pressure of the blood in the veins, this is transmitted back to the venous end of the capillaries and so slows down or stops the drainage of tissue fluid into the capillaries; this is seen classically in what is known as congestive heart failure, where there is a failure of the heart to deal with the blood returning to it via the veins, which leads to the engorgement of the veins with a rise in hydrostatic pressure of the venous blood, and the development of accumulations of excess tissue fluid, which is called oedema.

Recent work using proteins labelled with radioactive iodine (which enables them to be followed within the body) illustrates that the above concept of the formation and circulation of tissue fluid is not entirely correct. The capillary wall is in fact permeable to larger molecules than had previously been supposed and in fact proteins and lipids do leave the blood at the arterial end of the capillary along with water, mineral salts and glucose. Some of the fluid is reabsorbed at the venous end of the capillary but the remainder, together with the larger molecules which cannot re-enter the venous end of the capillary (because of lack of any great driving force) is returned to the blood by way of the lamphatic system. The walls of the lymphatics are more permeable to lyrge molecules than the walls of the capillaries. In the course of a day at least 50% of the total plasma proteins are lost from the capillaries and this is normally returned to the blood in the lymph which drains into the large veins at the root of the neck.

COMPOSITION OF BLOOD

Blood consists of a fluid called plasma in which are suspended the various cellular elements viz. red cells, white cells and platelets. These two constituents of blood, plasma and cells, are easily seen if a tube of blood, which has been prevented from clotting is allowed to stand; the heavier cellular elements gradually settle to the bottom of the tube leaving the buff coloured plasma above. About 45% of blood is made up of cells, the remaining 55% by plasma.

The red blood cells (erythrocytes)

There are about 4,500,000 red cells in each cubic millimetre of blood in the adult human male, slightly less in the female. Variations in the number of cells occurs with age, disease, exercise, exposure to high altitudes etc.

Each erythrocyte has the appearance of a biconcave disc, about $7 \cdot 5 \ \mu$ in diameter. When blood is smeared on a microscope slide some of the red cells stick to one another, appearing like piles of saucers called rouleaux.

If you wish to see these for yourself follow these instructions. The first task is to obtain a sample of your own blood. Roll a handkerchief into a small ball in the palm of your hand but leave about six inches which you can wrap lightly round the base of your thumb. Now swing your arm round and round as if you were a fast bowler at cricket. You will feel the blood being forced into your hand. Now stop and immediately tighten the loop of handkerchief around the base of the thumb. If you now bend the thumb you will see that the flesh at the base of the nail is gorged with blood (you have incidentally demonstrated something about the effects of centrifugal force on the amount of blood in the tissues. Such considerations become very important if you are a pilot in a very fast flying aircraft for you may force blood into or from the capillaries of the brain if you try to turn too quickly or pull out of a dive too fast. You may 'red out' or 'black out' depending on whether you are forcing blood into or out of the brain.) The handkerchief wrapped around the base of the thumb serves to compress the veins which drain blood from the thumb, and by using the handkerchief properly you can keep plenty of blood in the thumb. All you have to do now is to puncture the skin. To ensure that no bacteria are introduced in this operation the skin should be wiped with cotton wool soaked in 75 % alcohol and the stabbing needle should be passed through the flame of a Bunsen burner and allowed to cool. Take the clean needle and jab it into the flesh at the base of the nail which is engorged with blood. This is a relatively painless process because there are few pain sense endings here. Make a firm jab first time and you will get about a $\frac{1}{4}$ ml. of blood. If your thumb was dry the blood will remain as a big drop, but if your thumb is wet the blood will run over the finger and will not be much use. Transfer the drop of blood to a clean microscope slide and by using another slide held at 45° make a very thin smear of blood. When the second slide touches the drop of blood a thin film of blood will run along its base; if the second slide is now pushed firmly along the first slide a very thin film of blood can be obtained. Allow the smear to dry in the air, to allow the cells to stick to the slide. Then add about six drops of Leishman's stain to the smear and leave for one minute. Then add six drops of water and rock the slide so that the stain mixes with the water. Leave for fifteen minutes and then rinse the stain off with water. By this method the red cells are stained a pink colour, and the nuclei of the white blood cells, and the platelets, are stained a blue colour.

Erythrocytes and the carriage of oxygen. The red cells are of importance because they enable the blood to carry more oxygen. They can do this because they contain a special pigment, called haemoglobin, which is red and gives the characteristic colour to blood. Haemoglobin belongs to a group of pigments called respiratory pigments. Respiratory pigments are chemical substances which can form a reversible combination with oxygen.

$$\text{Haemoglobin} + \text{oxygen} \rightleftharpoons \text{oxyhaemoglobin}$$

In the lungs the haemoglobin in the red cells combines with oxygen to form the compound oxyhaemoglobin; as blood takes up oxygen it becomes a brighter red in colour. This oxygenated blood is returned to the heart by way of the pulmonary veins and is then distributed to the various parts of the body by way of the aorta and its branches. In the tissues some of this oxyhaemoglobin breaks down and oxygen is released and leaves the blood for the tissues. When blood gives up its oxygen it becomes a blueish-red in colour.

It is easy to demonstrate these colour changes. Obtain about one ml. of blood from your thumb and dilute it until you have a test tube full. Divide this into three equal volumes in three separate test tubes. Shake one vigorously and the solution will turn a brighter red in colour as oxyhaemoglobin is formed. Add a little sodium hydrosulphite to the second without shaking and this will show you the colour of the deoxygenated blood. Blow a little coal gas into the third and the solution will turn a cherry red colour as the compound carbon monoxy-haemoglobin is formed; carbon monoxide has a higher affinity for haemoglobin than has oxygen and in individuals exposed to atmospheres containing carbon monoxide then carbon monoxy-haemoglobin gradually replaces oxyhaemoglobin in the blood and death occurs due to anoxia i.e. lack of oxygen. Carbon monoxide occurs, for example, in coal gas and in the exhaust fumes of cars. Haemoglobin is a complex molecule consisting of a protein, globin, to which is attached the coloured pigment haem, which is made of a porphyrin compound and iron. Porphyrins are complex organic substances containing carbon, hydrogen and nitrogen. Haemoglobin can be thought of as a big molecule with a little hook at the end. This hook is made of iron and is just big enough to attach to one molecule of oxygen; the hook is shaped so that the molecule of oxygen can easily be unhooked. If carbon monoxide becomes attached to the hook it is not easily removed, so preventing the haemoglobin molecule from carrying further oxygen.

THE OXYGEN DISSOCIATION CURVE OF HAEMOGLOBIN. Large amounts of oxygen can be carried by the blood haemoglobin. 100 mls. of water in contact with alveolar air holds $\frac{1}{3}$ ml. of oxygen, whereas the same quantity of blood holds 20 mls. of oxygen. The average human body contains about six litres of blood and can therefore carry 1,200 mls. of oxygen. The amount of oxygen held by the blood depends upon the partial pressure* of oxygen in the surrounding medium i.e. the alveolar air, the tissues and tissue fluid. This relationship between the amount of oxygen held by the blood and the partial pressure of oxygen in the surrounding medium is an important one. It has been studied by placing known amounts of blood into bottles and then admitting air containing known partial pressures of oxygen. The bottles are suspended in a water bath at constant temperature and rotated so that the blood comes into close contact with the air. After a fixed time samples of blood are taken from the bottles to determine the oxygen content.

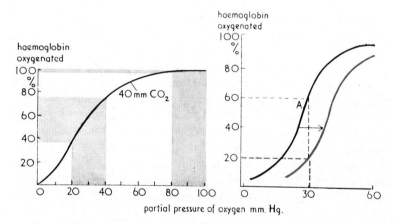

Fig. 92.   *The left hand figure shows the dissociation curve of oxyhaemoglobin.* Compare the effect of a fall of 20 mm. in the partial pressure of oxygen on the % saturation of the blood with oxygen at the two levels marked. *The right hand figure shows the effect of a rise in carbon dioxide concentration or a rise in temperature on the dissociation curve of oxyhaemoglobin.*

When the percentage saturation of the blood with oxygen is plotted against the partial pressure of oxygen in the air, a characteristically S shaped graph is produced (see fig. 92) called the dissociation curve of oxyhaemoglobin. The shape of the dissociation curve is highly significant and illustrates the function of the respiratory pigment haemoglobin.

In the lungs of man the partial pressure of oxygen is 100 mm. Hg and at this level the blood in the lungs becomes almost saturated with

* *Partial pressure.* In a mixture of gases each gas exerts a partial pressure proportional to its percentage in the mixture. Thus at sea level the atmospheric pressure is 760 mm. and contains 20·9% of oxygen. The partial pressure of oxygen is, therefore, 20·9/100 × 760 = 159 mm. Hg.

oxygen (95% saturated, see footnote*). The flat upper part of the curve at oxygen pressures above 80 mm. Hg, means that the arterial blood will remain almost saturated with oxygen in spite of relatively wide variations in the oxygen in the air. If there was a fall of 20 mm. in the partial pressure of oxygen in the air, the blood would still be almost completely saturated with oxygen at these partial pressures; thus although the partial pressure of oxygen falls progressively with altitude oxygen masks are not needed until about 4000m above sea level (see p. 481).

In the tissue fluids the partial pressure of oxygen is much lower than that of arterial blood (p.p. of oxygen of tissue fluids ranges from 5 to 30 mms. of Hg) and here the haemoglobin-oxygen combination breaks down and oxygen is given up to the tissues. At the partial pressures operating in the tissue fluids the blood gives up relatively large amounts of oxygen for relatively small decrements in the amount of oxygen in the tissue fluids—hence the steep part of the dissociation curve. Thus, whereas at 100 mm. of partial pressure of oxygen in the surrounding medium a fall of the partial pressure by 20 mm. of mercury results in the giving up of very little oxygen by the blood, at the partial pressures of oxygen operating in the tissues such a fall of 20 mm. in the partial pressure of oxygen would result in the giving up of large amounts of oxygen by the blood (see fig. 92).

The dissociation curve of haemoglobin is influenced by a variety of factors including carbon dioxide and temperature. When there is a rise of carbon dioxide or of temperature, the haemoglobin will hold less oxygen for a given partial pressure of oxygen; this is indicated on the graph (fig. 92) by a shift of the curve to the right.

In the hypothetical example in the graph on page 204 there is a shift of the dissociation curve from point A to B because of a rise in the carbon dioxide content and temperature of the blood. Thus at 30 mm. partial pressure of oxygen at A the blood will be 60% saturated with oxygen; when there is a shift of the curve to the right the blood will only be 20% saturated—with a loss of oxygen to the tissues.

This effect of a rise in the carbon dioxide content or of temperature is physiologically important; thus in muscular exercise there is a local rise in the temperature and carbon dioxide content of the muscle tissues. This alteration in the local environment facilitates a local increase in the rate of supply of oxygen to the tissues, at a time when this is needed.

* The oxygen tension at which haemoglobin is 95% or more saturated is called the loading tension or tension of saturation. The tension at which the pigment is 50% saturated is called the unloading tension or tension of half saturation.

Respiratory pigments in other animals. All the vertebrate animals have red blood cells containing haemoglobin. Some Annelid worms also use the respiratory pigment haemoglobin although in them the haemoglobin is dissolved in the blood plasma and is not concentrated in special blood cells; much less haemoglobin can be carried in this way and the dissociation curve, for haemoglobin dissolved in the plasma, is much flatter than that of haemoglobin concentrated in the red cells of the vertebrates.

Other respiratory pigments occur in different animals; in Crustacea e.g. crab, crayfish, there is a respiratory pigment called haemocyanin, and in many molluscs e.g. squid, ram's horn snail (Planorbis) there is a respiratory pigment called erythro-cruorin containing the metal copper.

Blood Groups. It is appropriate to consider blood groups whilst we are concerned with red blood cells for it is because of the chemical properties of the surface of the red blood cell that the whole problem of blood groups arises. It is well known that if for any reason a patient has to be given a transfusion of whole blood then it is essential that he be given blood of the right group. There are many blood groups and sub-groups but the best known ones are groups A, B, AB, and O, and the rhesus groups. When blood samples from two persons with the same blood groups are mixed together, no change occurs in the blood, and the two bloods are said to be compatible. However, when two specimens of blood from different groups are mixed together there may be important and potentially dangerous changes in the red cells; the red cells may stick together to form clumps, and the damaged red cells may break down to release haemoglobin into the plasma. If this occurs during a transfusion of blood then the kidneys may be seriously injured by deposits of haemoglobin obstructing the renal tubules. If clumping of the red cells occurs when two specimens of blood are mixed together then the bloods are said to be incompatible. This incompatibility is the expression of an antigen-antibody reaction which is taking place in the blood.

The surface of the red blood cell has on it certain protein substances called antigens. An antigen is a chemical substance which causes the body to produce another chemical called the antibody, which neutralizes or diminishes the effect of the antigen. It is only possible then to define an antigen in terms of the antibody. If a chemical substance causes the body to make antibodies against it then such a chemical irritant is said to be antigenic. It is only possible to give a few examples of the many types of antigens. Many bacterial proteins are antigens and stimulate the body to produce antibodies which are very important in the defences of the body against bacteria (see p. 567). In some persons vegetable proteins such as pollen behave as antigens and cause the body to produce antibodies; these antibodies tend to be concentrated in the lining cells of the respiratory system (viz. nose, bronchi and bronchioles) and when these cells loaded with antibodies meet a dose of antigen, as when air containing pollen is breathed in by a susceptible individual, the antigen-antibody reaction which occurs on the surface of these cells irritates the cells in some way and so gives way to the symptoms of hay fever or of asthma. As described above, a transfusion of incompatible blood may cause the body to produce antibodies which will unite with antigens on the surface of the incompatible red cells, which damages and destroys them. From the above examples it will be seen that the cells on which an antibody-antigen reaction is taking place are damaged or destroyed. A further interesting example of this reaction may be described which is encountered in skin grafting. When a large area of skin has been damaged, for example by a burn, the rate of healing of the skin from the healthy margin is so slow that thin slices of skin from another part of the body are laid by the surgeon in patches on the raw area; and these patches, by means of the growth of new skin cells from their edges, gradually form a new skin cover for the injured area. Now if the injuries are very extensive the patient may not have enough healthy skin from which to take patches to cover the wounds. If the skin from another person is used then the patches will remain healthy on the injured areas for a week or two but then they rapidly die. This is because the patient has made antibodies to these patches of 'foreign' skin, whose proteins have acted as antigens; the antigen-antibody reaction destroys the grafted patches of 'foreign' skin. Grafts of skin and other organs from one person to another usually fail unless the donor and the recipient are identical twins, in which case the proteins of the two individuals, donor and recipient, have identical structure.

Returning now to the red cells, there are two main kinds of antigen on human red cells called A and B. You may have antigen A on your cells and then you are said to have group A blood; if you have antigen B then you have group B blood. If your red cells have both antigens, then you have group AB blood and if you have neither antigen on your red cells then you have group O blood. In the case of blood groups the antibodies to these antigens are naturally occurring and are not the result of introduction into the body of antigens, although abnormal antibodies may develop in this way. These naturally occurring antibodies are dissolved in the plasma. Of course one cannot have antibodies in the plasma which are specific to antigens on the red cells; if one has group A blood then one cannot have antibody a in the plasma or ones own red cells would be destroyed, but one may have other antibodies in the plasma. Thus in group A blood there is antibody b in the plasma and in group B blood there is antibody a in the plasma. In group AB blood there is neither antibody present, and in group O blood both antibodies are present.

Let us now look at some possible transfusions and see which bloods are incompatible. Group A blood can obviously be given to other group A persons since the blood of the latter does not contain antibody a. However group A blood cannot be given to persons of group B or group O because both of the latter contain antibody a in the plasma which would cause clumping of the red cells of the donor. You can work out what happens in the other possible mixings for yourself and check your results against table 2.

If you study this table carefully you may be confused by the fact that you can mix group A with group AB but not AB with group A. It is important to know which is the donor blood and which is the receiving blood. Thus blood of group A can be given to a recipient of group AB in spite of the fact that the donor blood contains antibody b; this is because the plasma containing the antibody b is rapidly diluted when given to the recipient group AB. For this reason persons

Table 2

		ANTIBODIES in RECEIVER'S PLASMA			
		Group A (b)	Group B (a)	Group AB (none)	Group O (a and b)
ANTIGENS	Group A (A)	safe	clot	safe	clot
ON DONOR'S	Group B (B)	clot	safe	safe	clot
RED	Group AB (A and B)	clot	clot	safe	clot
CELLS	Group O (none)	safe	safe	safe	safe

of group AB have been called universal recipients, that is they can receive blood of any of the AB groups without ill effect. In practice however blood is always cross matched in the laboratory with a specimen of the receiver's blood before a transfusion is given; the presence of antibodies and antigens other than the A–B type make this essential. People of blood group O have been called universal donors, that is their blood can be given to persons of any of the A–B groups without ill effects; this is because their red cells have none of the A–B antigens, whilst their plasma containing antibodies a and b is rapidly diluted in the recipients own plasma. In England the most common blood groups are A (40%) and O (45%) whilst group B (10%) and group AB (5%) are less common.

RHESUS GROUPS. There is another antigen present on the red cells of some individuals called the rhesus antigen or rhesus factor. It is

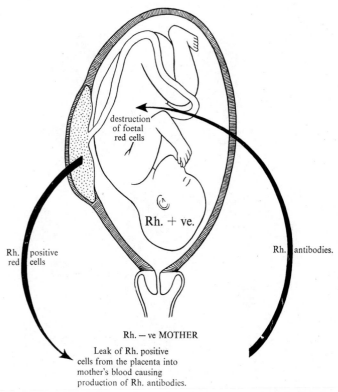

destruction of foetal red cells

Rh. + ve.

Rh. red cells | positive cells

Rh. antibodies.

Rh. – ve MOTHER

Leak of Rh. positive cells from the placenta into mother's blood causing production of Rh. antibodies.

Fig. 93. *Diagram illustrating a possible complication of the Rh. blood blood groups*—Rhesus incompatibility between mother and foetus.

called rhesus factor because it was first discovered in the blood of rhesus monkeys.

Unlike the A–B antigens there is no naturally occurring antibody to the rhesus factor. Those persons containing the rhesus factor on their red cells are called rhesus positive, whilst those without the rhesus factor are called rhesus negative. Antibodies to the rhesus antigen only develop under certain unusual circumstances; firstly they may develop if an Rh-negative person is given a transfusion with Rh positive cells. The Rh positive cells stimulate the production of rhesus antibodies in the receiver's blood. The second way in which rhesus antibodies can develop is in the case of some pregnant women. The rhesus factor is inherited as a dominant gene and thus it may happen that if a rhesus positive man marries a rhesus negative woman, the rhesus negative woman may bear rhesus positive children. Now in the majority of cases of rhesus negative women bearing rhesus positive children there is no effect on the mother's blood. But in some cases it appears that

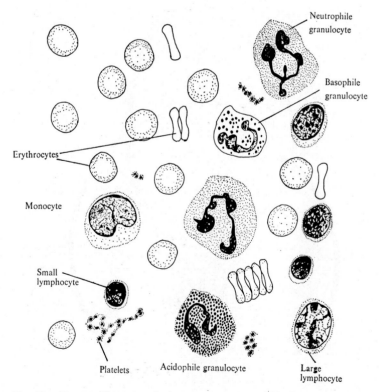

Fig. 94.　*Diagram illustrating the various constituents of blood, as seen at high magnification.*

small leaks of blood from the foetus, through the placenta, into the mother's blood cause the production by the mother of rhesus antibodies. Because of their molecular size these antibodies may gain access to the circulation of the Rh positive foetus through the placenta. When this happens the red cells of the foetus are progressively destroyed because of the antigen-antibody reaction on their surfaces. This effect becomes more important in successive pregnancies of an Rh negative woman bearing Rh positive children and after several Rh positive pregnancies there may be sufficient antibody produced by the mother to produce such a degree of an anaemia in the foetus that the foetus dies in utero. It becomes very important then to know the Rh group of blood before it is transfused to a woman, particularly if she is Rh negative. If by chance Rh positive blood is given to an Rh negative woman it will stimulate the production by her of rhesus antibodies which may damage the infant if the infant so happens to be Rh positive.

The white blood cells

There are about 8,000 to 10,000 white blood cells per cubic mm. of human blood. There are two main kinds;

(i) lymphocytes
(ii) leucocytes.

Lymphocytes. These are produced in the lymphatic tissues (see p. 234) and can be recognized under the microscope (see figs. 94, 95, 96) by the fact that they have a round deeply staining nucleus surrounded by a thin rim of cytoplasm in which no granules can be seen. There are three main kinds of lymphocyte; the most common one is called a small lymphocyte, having a diameter of about 8 μ. 25% of all white blood cells are small lymphocytes and their functions are discussed on page 567. Very similar to the small lymphocytes are the large lymphocytes, with a diameter of about 15 μ; they make up about 3% of all white blood cells. The third type of white blood cell which is lymphoid in origin is the monocyte. Like the small and large lymphocytes the monocytes are actively amoeboid and they are also active phagocytes (i.e. they can ingest bacteria). Like other lymphocytes they have a short life. Strictly it is not a blood cell; it really belongs to a very diffuse network of 'defence' cells called the reticulo-endothelial system which is spread as a network through many tissues including lymphoid tissue, spleen, bone marrow. These monocytes are such active migrants that they wander into the blood and account for 3% of the white blood cells. They are seen particularly in areas of inflammation where they ingest bacteria and cell fragments.

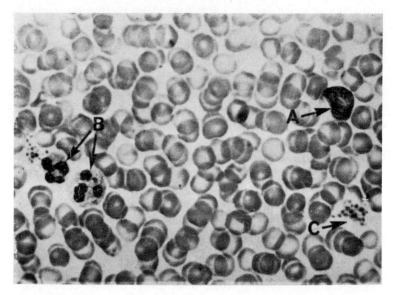

Fig. 95. *Photomicrograph of a film of blood showing erythrocytes, monocyte* (A), *polymorphonuclear leukocytes* (B) *and platelets* (C). (*By courtesy of Dr. S. Bradbury.*)

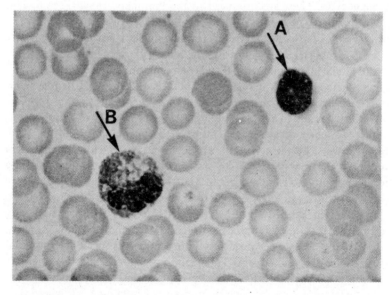

Fig. 96. *Photomicrograph of a film of blood showing erythrocytes, small lymphocyte* (A) *and an eosinophil polymorphonuclear leukocyte* (B). (*By courtesy of Dr. S. Bradbury.*)

Leucocytes. These cells are also called polymorphonuclear granulo-cytes on account of the facts that their cytoplasm contains various granules and they have a distinctive nucleus consisting of several lobes, which increase in number as the cell gets older. These cells are pro-duced in the red bone marrow; in adult man this is situated in the upper end of the femur, in the ribs and bodies of the vertibrae.

The granulocytes are about 10 μ in diameter. The commonest type of granulocyte is the neutrophil granulocyte which accounts for over 70% of the white blood cells; their cytoplasm contains very fine pink staining granules. These neutrophil granulocytes are actively phago-cytic and are seen in large numbers at the site of infection (see p. 564). Pus in fact consists of vast numbers of these neutrophil granulocytes together with the bacteria they have consumed, and cell debris.

There are two other types of granulocytes, found in much smaller numbers in the blood; the acidophil granulocytes with red staining granules in their cytoplasm (these account for 2% of white cells) and the basophil granulocytes containing blue staining granules in their cytoplasm (0·5-1% of white cells). It is necessary to look with great care at stained films of blood in order to see these two types of granulocytes because of the infrequency with which they occur. Neither of the two types are actively phagocytic—a point of contrast with the neutrophil granulocyte. The acidophil granulocyte is found in the blood in larger numbers when there is an infestation with certain parasitic worms, and they may also increase in number in certain cases of asthma; their exact function is not known. The functions of the basophil granulocytes is also uncertain.

In summary one may think of the white cells as the defensive elements of the blood; the most common ones (the neutrophil granulocytes) eating up any invading bacteria, and the lymphocytes engaging in chemical warfare on behalf of the body.

Blood platelets

In addition to red and white blood cells, blood contains some minute colourless corpuscles called blood platelets. Each platelet is a rounded or oval disc about 3 microns in diameter. When seen in stained films of blood the centre of the platelet consists of a group of darkly staining granules, but this is not a cell nucleus in the usual sense of the word as it does not contain chromatin material. The number of platelets in blood is approximately 250,000 per cubic millimetre.

They are produced from large cells in the bone marrow, called mega-karyocytes, by a process of budding off small portions of the cellular protoplasm.

The surface of the platelets is very sticky and when blood is shed the platelets tend to stick together in clusters onto the surface on which the blood is poured; here they rapidly break down and strands of fibrin appear to radiate from these groups of degenerating platelets. It has long been known that the platelets play some part in the process of clotting, but their exact role is not understood. Certainly the platelets are not indispensable to the clotting of blood and plasma which has been freed of platelets clots nevertheless. It is thought that when the platelets disintegrate they release thromboplastic substances which assist in the conversion of prothrombin to thrombin (see p. 215).

THE PLASMA AND THE MECHANISM OF CLOTTING

When blood escapes from the body, as into a wound for example, it quickly changes from the fluid state into a thick jelly like material called a clot. This process of clotting, or coagulation as it is called, is an essential process, in that it prevents the whole of this very valuable fluid from draining from the body from the slightest cut.

We have now to study the way in which fluid blood is so quickly converted into a clot. The process of clotting depends upon the change in state of a protein constituent of plasma called fibrinogen. This protein fibrinogen is normally dissolved in the plasma but on shedding of blood there is a change in the state of the fibrinogen into long fibrous molecules called fibrin, which gradually contract to form a firm clot. As the clot shrinks a clear yellow fluid escapes from it called serum which consists of plasma minus the protein fibrinogen, which has now been converted into fibrin. Within the meshes of the clot are trapped the various cellular elements of the blood. On the surface of the body this clot gradually dries to form a scab which forms a mechanical covering to a wound.

The process of clotting is a highly complex one and is dependent upon the presence of a great variety of factors, but the essential process, as described above, is the conversion of the protein fibrinogen from its corpuscular state into its fibrous state, called fibrin. When blood escapes from a blood vessel into a wound a variety of changes in the blood occur; first the blood is escaping from a series of smooth walled vessels into an area exposed to the air, with roughened, often dirty, surfaces and escaping into the blood are tissue fluids containing substances derived from damaged cells in the area. Some elements of the blood break down upon the surfaces of the wound, particularly the elements called platelets. During this process there is the conversion of an inactive enzyme present in the blood called prothrombin into its active form, thrombin, and it is this thrombin which triggers off the change in the state of fibrinogen. The various substances which are liberated into shed blood from damaged cells and platelets are called thromboplastic factors and it is these factors which produce the change

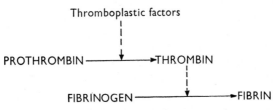

Fig. 97. *Summary of the clotting process.*

of the inactive enzyme prothrombin into thrombin. These changes are summarized in the diagram above (fig. 97). Prothrombin is converted into the active form thrombin under the influence of thromboplastic factors, and the thrombin then initiates the change in state of the protein fibrinogen. The process is not as simple as described above and some of the other factors concerned in the process of clotting will be discussed later.

Prothrombin. We can now investigate in more detail the various factors concerned in the process of clotting. Prothrombin like the other plasma proteins is manufactured in the liver.

VITAMIN K. For the manufacture of prothrombin vitamin K is necessary, although it appears that vitamin K does not form part of the molecule of prothrombin. Anything interfering with the absorption of vitamin K interferes with the synthesis of prothrombin and may lead to some failure in the clotting mechanism. The presence of bile is necessary for the absorption of vitamin K from the bowel because it is fat soluble (see p. 393); thus, when the flow of bile is obstructed, for example by a stone in the bile duct, there is a failure to manufacture adequate amounts of prothrombin. The new born infant may also suffer from a deficiency of vitamin K which may lead to the appearance of spontaneous bleeding from various parts of the body. In part this deficiency of vitamin K in the new born is due to the fact that some of the vitamin K absorbed is derived from bacteria in the bowel and the young mammal does not have a full complement of bacteria in the bowel when it is born; it gradually acquires these. A further example of the importance of these bacteria in the bowel in the manufacture of vitamin K is seen in the treatment of patients with large doses of wide spectrum antibiotics, which tend to eradicate many of the organisms in the bowel, and so may lead to reduced prothrombin levels in the blood.

DICOUMARIN. A substance called dicoumarin, which is present in spoiled clover, interferes with the manufacture of prothrombin in the liver. This then is a dangerous feeding stuff for cattle. Use of di-coumarin-like-substances is made in certain diseases, in which there is an increased tendency to form clots within the blood stream, such as coronary thrombosis, and here, regular doses of these drugs reduces the clotting power of the blood by its effect on prothrombin synthesis, and so may halt the progress of the disease. Prothrombin is a very potent substance and 20 mg. in 100 ml. of blood is more than adequate to clot all the fibrinogen in the blood.

Prothrombin to Thrombin. It is not sufficient merely to have enough prothrombin in the blood but there must also be present a variety of factors which play a role in the conversion of prothrombin into thrombin.

CALCIUM IONS. Firstly there must be an adequate supply of free calcium ions. When blood is taken from donors it is collected into bottles containing a solution of sodium citrate; this precipitates out the calcium ions as insoluble salts and prevents the blood from clotting. If this blood is given in large amounts to a patient who is already bleeding then it is often necessary to give injections of calcium salts to enable the citrated blood to clot. Other salts will also prevent blood from clotting, including oxalates and fluorides, however both are poisonous substances and are only used in laboratory tests.

THROMBOPLASTINS. Another group of substances necessary for the conversion of prothrombin to thrombin includes thromboplastic substances, sometimes called thromboplastins. They have been identified chemically as phospholipids and can be obtained in extracts from a great variety of tissues. In the clotting process, however, they are usually derived from injured cells and probably also from the break-down of blood platelets in the area.

GLOBULINS AND HAEMOPHILIA. A third type of substance necessary is included in the globulin fraction of the plasma proteins, and there are several types of globulins concerned in the conversion of prothrombin into thrombin which are given various names, such as accelerator globulin, anti-haemophilic globulin and Xmas factor. Haemophilia is inherited as a sex linked recessive trait, and the gene responsible is carried on the terminal portion of the X chromosome. The human female has two X chromosomes but the human male has only one X chromosome which pairs with a shorter chromosome called the Y chromosome. In the male the terminal portion of the X chromosome is unpaired since the Y chromosome is shorter than the X. It is in this unpaired region of the male X chromosome that the recessive gene responsible for haemophilia is carried.

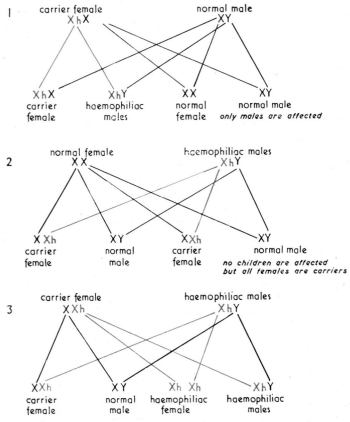

Fig. 98. *Illustrations of the mode of inheritance of haemophilia. Xh represents the X chromosome carrying the recessive factor.*

If a recessive gene is carried on this unpaired portion of the X chromosome in the male, the character will be expressed in the phenotype. In the female however, since the X chromosome is paired along the whole of its length with another X chromosome, a recessive gene will always be paired with another gene; thus the female will only have the disease haemophilia if she has a double dose of the haemophilia gene (i.e. the homozygous recessive state), one recessive gene on each X chromosome. If there is only one recessive gene present then the normal allele, or other gene partner, will be present on the other X chromosome, and this normal gene will govern the production of anti-haemophilic globulin, and the disease will not manifest itself. However, although a female with a single dose of the gene responsible for haemophilia (i.e. heterozygous state) does not suffer from haemophilia she is still able to pass the disease on to male descendants. Females with haemophilia are very rare, because to produce such a female it would mean that a female carrier (or female haemophiliac) would have to marry a male haemophiliac; since the disease is relatively rare such a combination is highly unlikely. But the disease is not transmitted merely by inheritance in this way; there is a fairly high spontaneous mutation rate of the normal gene on the X chromosome, and thus it would be impossible to eradicate the disease by control of marriages. The mode of inheritance of haemophilia is illustrated in fig. 98.

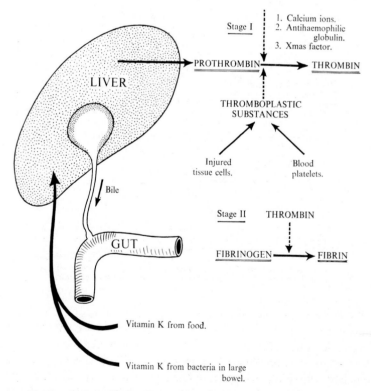

Fig. 99. *Diagram illustrating the factors concerned in the clotting process.*

Another type of haemophilia, sometimes called haemophilia B or Xmas disease is due to the absence of another factor from the plasma, necessary for the conversion of prothrombin into thrombin; this factor is called the Xmas factor. It is inherited in a similar way to classical haemophilia (or haemophilia A) and is due to a sex-linked recessive gene. Another rare type of haemophilia, called haemophilia C, is due to the deficiency of another plasma factor. Unlike haemophilia A and B it is inherited as a Mendelian dominant trait, and therefore affects males and females alike. In haemophilia there is a tendency to spontaneous bleeding or of bleeding on trivial injuries; thus minor injuries to joints or muscles may lead to the appearance of large collections of blood in these situations. Continuous bleeding may even occur after minor operations such as dental extraction. The treatment of these bleeding episodes consists in giving the patient plasma containing antihaemophilic globulin. Because normal freeze-dried plasma or normal stored whole blood loses its anti-haemophilic globulin it is necessary to use fresh-frozen plasma which retains its anti-haemophilic globulin.

HEPARIN. There are several substances present in normal plasma whose actions tend to prevent the development of thrombin within the blood stream and so the appearance of blood clots within the blood vessels. One of these factors is called heparin. Heparin is produced from special mesenchymal cells present in the connective tissues called mast cells, and the liver is particularly rich in such cells. Heparin is a polysaccharide substance and is used in medicine as an anticoagulant and can be injected into the blood stream in cases where clots have appeared within the blood vessels and so helps to prevent extension of the clotting process.

Other factors have been mentioned which tend to prevent the clotting of blood within the vessels; one of these is the continuous smooth surface of the vessel walls to which the blood is continually subjected. When these surfaces become roughened by the degenerative processes of old age then clots may develop along the roughened surfaces; and when this happens in the coronary arteries supplying the heart muscle with blood the result can be disastrous.

Summary. We can now summarize the process of clotting in a little more detail, as shown in fig. 99. The process of clotting is divided into two stages. In the first stage there is the conversion of prothrombin into thrombin, at which stage various factors exert their effects, including antihaemophilic globulin, Xmas factor, accelerator globulin, thromboplastic substances and calcium ions. The second stage is the conversion of fibrinogen into fibrin, under the influence of the thrombin. The presence of heparin within the blood stream helps to prevent the formation of thrombin in the blood; and it is only when the equilibrium of the blood is disturbed, as during injury, when thromboplastic substances are liberated, that thrombin begins to appear locally at the site of injury.

PLASMA AND THE CARRIAGE OF CARBON DIOXIDE

85% of the total amount of carbon dioxide carried by the blood is carried by the plasma as sodium bicarbonate. A summary of the whole process is seen in fig. 100. Carbon dioxide produced in respiration in the cells diffuses out into the tissue fluids and into the blood in the capillaries. Here it dissolves in the blood and combines chemically with water in the red blood cell in the presence of the enzyme carbonic anhydrase. This is a reversible reaction which in the presence of plenty of carbon dioxide and the enzyme, quickly produces carbonic acid. This acid dissociates to produce hydrogen ions which are positively charged (H^+) and bicarbonate ions which are negatively charged, (HCO_3^-). (In passing we might note that this is typical behaviour, for an acid is a substance which gives free hydrogen ions in solution.)

As carbon dioxide is being taken up by the blood from the tissues, oxygen is being given up from the erythrocytes to the tissues. The oxyhaemoglobin of the erythrocyte dissociates into oxygen and reduced haemoglobin because of the lower partial pressure of oxygen in the tissues. The oxygen released from oxyhaemoglobin diffuses through the membrane of the erythrocyte into the plasma, through the capillary wall into the tissue fluid and the cells. Haemoglobin is a remarkable protein not only because of its reversible binding of oxygen but also because the character of the protein changes markedly on oxygenation or deoxygenation. Haemoglobin, like the proteins of plasma, is a basic substance which tends to accept hydrogen ions (p. 103). Reduced haemoglobin is a much stronger base (i.e. it accepts hydrogen ions more avidly) than oxyhaemoglobin. This change in the character of haemoglobin occurs at a time when the concentration of hydrogen ions inside the erythrocyte is increasing from dissociation of carbonic acid. These hydrogen ions are buffered by the stronger base, reduced haemoglobin (see p. 511 for definition of buffer). In fact if hydrogen ions were not supplied to the reduced haemoglobin the blood would actually become more alkaline in the tissues. The character of haemoglobin permits the carriage of large amounts of carbonic acid from the tissues to be excreted by the lungs without any significant change in pH of the blood.

The concentration of bicarbonate ions in the erythrocytes increases as carbon dioxide is taken up from the tissues. The bicarbonate ions now diffuse into the plasma along a concentration gradient. However the erythrocyte is relatively impermeable to cations (positively charged ions such as sodium and potassium) and these cannot accompany the bicarbonate ions out of the erythrocyte to maintain electrochemical neutrality. In order to maintain electrochemical neutrality there is a diffusion of chloride ions from the plasma into the erythrocyte, a phenomenon called the 'chloride shift'.

Thus much of the carbon dioxide taken up from the tissues ultimately appears as bicarbonate ions in the plasma. 10% of the total carbon dioxide is transported by chemical combination with haemoglobin to form carbamino-haemoglobin. A further 5% is carried as dissolved CO_2 in the plasma.

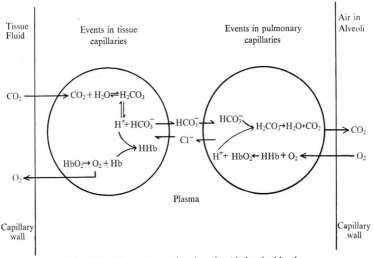

Fig. 100. *The carriage of carbon dioxide by the blood.*

When the blood reaches the lungs the reverse sequence of events occurs. Reduced haemoglobin takes up oxygen from the high partial pressure of oxygen in the pulmonary alveoli. As reduced haemoglobin becomes oxygenated it becomes a much weaker base and it gives up the hydrogen ions which it accepted in the tissues. These hydrogen ions combine with bicarbonate ions in the erythrocyte to form carbonic acid. Because of the lower partial pressure of carbon dioxide in the pulmonary alveoli this carbonic acid dissociates into carbon dioxide and water, and the carbon dioxide diffuses out of the erythrocyte down the concentration gradient into the pulmonary alveoli where it is excreted from the body. The concentration of bicarbonate ions in the erythrocyte declines as carbonic acid is formed, and now bicarbonate ions which left the red cells in the tissues diffuse back into them down a concentration gradient. More carbonic acid is formed and more carbon dioxide passes into the pulmonary alveoli for excretion. The reversible events which occur in the erythrocyte are summarised diagrammatically in fig. 100.

THE HEART AND CIRCULATION THROUGH THE GREAT VESSELS

The capillary bed has been described as an important part of the circulatory system where the blood is in close contact with the tissue fluids; the function of the heart and blood vessels is to provide this capillary bed with an adequate supply of blood. The mammal with its high body temperature, rapid turnover of substances within the body cells and fast rate of movement requires a good supply of blood to the capillary bed. In the fishes the blood leaving the heart has to pass to the capillary bed of the gills before it reaches the rest of the body, where it is at relatively low pressure. By various means the reptiles, birds and mammals have separated off the blood supply to the lungs from the rest of the body circulation so that blood at high pressure can be delivered direct to the capillary bed of the remaining organs.

The capillary network of the body is a vast system and if this system were all open at one time it would take up more blood than there is in the whole circulatory system; the animal would virtually bleed to death into its own capillary system. At any one time only a fraction of the capillary bed is open, the remainder being closed by means of contraction of the smooth muscle in the walls of the arterioles supplying them with blood. By this means blood is diverted from inactive parts of the body into the more active parts. Thus in the marathon runner the capillary bed of the leg muscles may be wide open, whilst that of the gut may be almost closed down. And after a heavy meal the gut will be supplied with large amounts of blood at the expense of other parts of the body, the skin and muscles. The way in which the circulation is adapted to meet the varying demands upon it will be discussed further on page 231.

The structure of blood vessels

1. **Arteries.** The blood is carried from the heart to tissues of the body in vessels called arteries. These gradually decrease in size and branch

as they pass away from the heart. The sum of the diameter of the various branches increases as one passes from the heart, and thus the blood pressure gradually falls towards the tissues. Like all blood vessels the arteries are lined by a smooth flat pavement epithelium (called the endothelium). The wall of the arteries contains several tissues including elastic fibres, collagen fibres, smooth muscle and nerve fibres; the proportion of elastic tissue to muscle gradually changes as one passes from the larger arteries near the heart to the smaller arteries in the periphery. This results in arteries of two main kinds, elastic arteries containing a predominance of elastic tissue in their walls and muscular arteries with a predominance of muscle in their walls. The elastic arteries are those large vessels situated close to the heart and

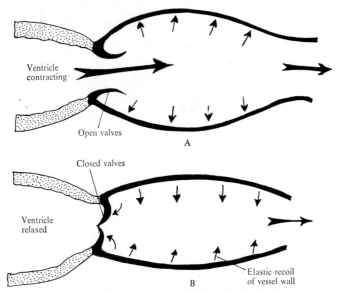

Fig. 101. *Diagram illustrating the role of the large elastic arteries near the heart in maintaining a continuous flow of blood in spite of a discontinuous flow of blood into them from the ventricles of the heart.*

they act as a reservoir of blood; when the ventricles contract and eject their blood into the large elastic arteries, these dilate to take up the increased volume of blood and when the ventricles relax in diastole the elastic arteries decrease in size again and by the rebound of their elastic walls they force the blood along the arteries. (See fig. 101.) Thus even when the ventricles are in the phase of relaxation or diastole there is a continuous flow of blood from the larger to the smaller arteries.

2. Arterioles. The smaller muscular arteries gradually give rise to vessels called arterioles which have smooth muscle in their walls. These vessels, because of the contractile nature of their walls serve to regulate the flow of blood to the various organs. The arterioles supply the capillary system with blood.

3. Capillaries. As described already (page 170) they are thin walled vessels consisting of a single layer of pavement epithelium with a few scattered connective tissue cells along the wall.

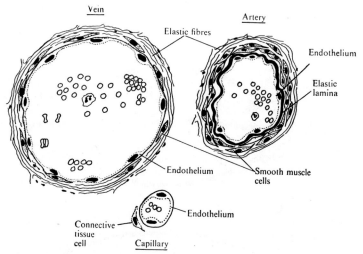

Fig. 102. *Drawing illustrating the structure of a vein, artery and capillary, in transverse section.*

4. Veins. Blood is drained from the capillary bed by a system of veins. These have a structure basically similar to that of arteries with muscle, elastic tissue and collagen fibres in their walls, but the walls are much thinner and the proportion of muscle tissue is much less than in the arteries, and the larger veins near the heart contain very little muscle tissue in their walls. In the larger veins there is a system of valves (see fig. 103) which prevents the backflow of blood. These can be easily demonstrated in the veins of the arm. If the upper arm is constricted to prevent the veins of the arm emptying their blood, these become gorged with blood and the position of the valves can be seen as thickenings along the course of the veins (see fig. 103B). If a finger is stroked along the vein of the arm towards the hand in an attempt to force the blood backwards the positions of the valves are more conspicuous.

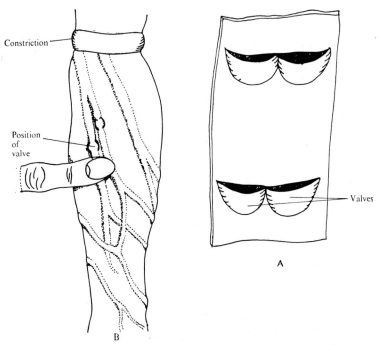

Fig. 103. *A. Diagram of a vein opened to show the valves. B. Drawing illustrating the demonstration of valves in the superficial veins of the fore-arm.*

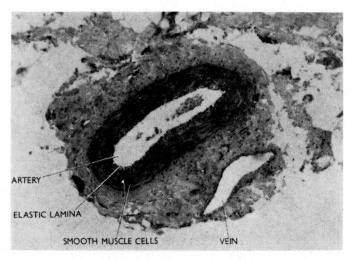

Fig. 104. *Photomicrograph of an artery and vein embedded in connective tissue.*
(By courtesy of Dr. S. Bradbury.)

The structure of the heart and the regulation of the heart beat

Fig. 105 shows a vertical section through a mammalian heart. The two upper thin-walled chambers are the auricles, the left receiving oxygenated blood from the lungs, the right receiving de-oxygenated blood from the rest of the body. Contraction of the heart begins in the auricles and these eject their blood through valved entrances into their respective ventricles. The ventricles are relaxed at the time when the auricles are contracting and so there is little resistance to the flow of blood. There is thus little need for the muscle of the auricle to be powerful. The ventricles then eject their blood into the major arteries, the left ventricle into the aorta and the right ventricle into the pulmonary artery.

The regular beating of the heart with contraction beginning in the auricles and then involving the ventricles is determined by two factors; one, the presence of a region in the auricles which generates the stimulus to contraction and second, specialised tissue which rapidly conducts the stimulus from the auricles to the ventricles. Fig. 106 is a section through the heart of a fish and shows the presence of a chamber, the sinus venosus, which is not present in the mammalian heart. This chamber receives the blood as it returns from the rest of the body. As the sinus venosus becomes stretched by contained blood it contracts and empties itself into the auricle, and the faster it fills the faster it empties itself. The sinus venosus is what we call the pace-maker of the heart and if the sinus contracts rapidly then so does the auricle and ventricle. In the evolution of the mammalian heart the sinus venosus as a distinct chamber has disappeared but it is represented as a vestige in the right auricle. This small patch of tissue is now called the sino-atrial node and it is here that the beat of the heart originates in the mammal. All heart muscle cells show the property of automaticity— the ability to beat rhythmically in the absence of external stimuli. The cells with the most rapid inherent rhythm are the 'pace-maker' cells. These cells lie in the sino-atrial node and show special physiological features. It seems that their membrane is unusually 'leaky' to sodium at rest so that sodium ions diffuse into the cell from the sodium-rich tissue fluid around the cells. This inward diffusion of positively charged ions reduces the electrical potential which exists across the cell membrane to the critical point when rapid changes occur in the permeability of the membrane. The cells rapidly depolarise (see p. 618 for a fuller discussion of these events) and a wave of depolarisation spreads in all directions concentrically from the S.A. node at a rate of about 1 metre/second. This flow of excitation needs no special conducting tissue for the depolarisation spreads from one muscle fibre

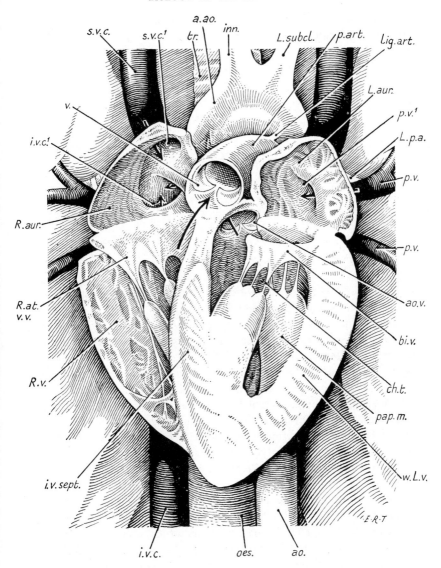

Fig. 105. *The mammalian heart* (rabbit). From J. Z. Young 'The Life of Mammals'. O.U.P.

a.ao. aortic valve. ao. aorta. ao.v. aortic valve. bi.v. bicuspid (mitral) valve. ch.t. chordae tendinae. i.v.c. inferior vena cava. i.v.c.[1] opening of inferior vena cava into right auricle. L. aur. left auricle. L.p.a. left pulmonary artery. Lig.art. Ligamentum arteriosus (remnant of ductus arteriosus). oes. oesophagus. p.v. pulmonary vein. p.v.[1] opening of pulmonary vein into left auricle. pap.m. papillary muscle. p.art. main pulmonary artery. s.v.c. superior vena cava. R.aur. right auricle. R.at.v.v. right auriculo-ventricular (tricuspid) valve. R.v. right ventricle. tr. trachea. v. semilunar valve of pulmonary artery. w.l.v. wall of left ventricle.

to another so that the entirety of the thin auricular muscle is depolarised in about 80 milliseconds. Electrical depolarisation is followed by contraction; spread of excitation is so rapid that all parts of the auricles contract virtually simultaneously.

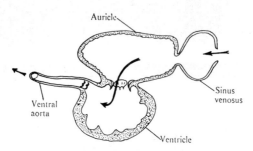

Fig. 106. *Diagram of a longitudinal section through the heart of a dogfish.*

The muscle fibres of the auricles are not continuous with those of the ventricles and the excitation does not spread directly to ventricular muscle. The mass of ventricular muscle is so much larger than that of the auricles that a special conducting system is necessary to rapidly convey the impulse to all parts of the ventricles. This conducting system begins at the base of the inter-auricular septum as a small mass of tissue similar to the S.-A. node. This is the auriculo-ventricular (A.-V.) node. Conduction of the impulse through the A.-V. node is very slow—about 0·1 metres/sec. and this determines that there is a pause between contraction of the auricles and ventricles. From the A.-V. node the impulse is conducted rapidly through special Purkinje tissue (named after the Bohemian physiologist) to all parts of the ventricle—conduction at about 2 metres/sec. From the A.-V. node the Purkinje tissue is organised into a structure called the Bundle of His. This bundle passes a little way down the inter-ventricular septum and then divides into left and right branches which supply the respective ventricles. These branches divide extensively over the inner surface of the ventricular muscle and then penetrate the muscle mass to reach more superficial parts of the ventricles. On its way to the apex of the heart the Bundle of His supplies the papillary muscles situated at the base of the chordae tendinae, inelastic cords which reach from the apex of the ventricle to the margins of the bicuspid and tricuspid valves. The contraction of the papillary muscles a fraction of a second before the main ventricular muscle prevents these valves from being blown inside out. The Purkinje fibres pass round the outer wall of the ventricles after they have supplied the apex of the heart with the consequence that the ventricle beats from the apex towards the auricles. The blood is forced against the mitral and tricuspid valves which are consequently closed and only prevented from being blown inside out by the tension in the chordae tendinae. The semi-lunar valves at the openings of the pulmonary artery and aorta are opened by the fluid pressure. The pressure exerted by the left ventricle is greater than that by the right ventricle since the muscle on the left is about five times as thick as that on the right. This is a good example of the adaptation of structure to function since the left ventricle has to supply blood to the entire body whereas the right ventricle has to supply blood only to the lungs, and even this at low pressure to protect the delicate pulmonary capillaries. The times taken in the various phases of the heart beat in man are shown below.

Phase of the Cardiac Cycle	Time taken in secs.
contraction of the auricles	0·1
contraction of the ventricles	0·3
total systole (Contraction)	0·4
Total auricular and ventricular diastole (relaxation)	0·4
The total time for the whole cycle at 75 beats per minute	0·8

When a muscle is stimulated by putting an electric current through it, the muscle may respond if the current is strong enough. There is a level called the threshold value which is just strong enough to cause a response. At currents below this value nothing happens, but above the threshold the muscle twitches. Once cardiac muscle has been made to twitch further applications of current have no effect no matter how strong they are, until the muscle has contracted and begun to relax. This period of time in which the muscle is insensitive to stimuli is called the refractory period. The characteristic things about the physiology of cardiac muscle is that it can contract strongly and rapidly and does not fatigue but it does have a very long refractory period. Fig. 107 shows the refractory period for cardiac muscle to be as long as the period of contraction. In striated muscle there is a refractory period but it is very short, in fact the striated muscle has recovered from the first stimulus and can be affected by a second one before the muscle has even had time to respond to the first stimulus. Thus in striated muscle the effects of repeated stimulation can build up and the fibre can be maintained in a state of perpetual contraction called the state of tetanus (p. 606). This is not possible in cardiac muscle because of the long refractory period. Thus the muscle of the auricle when stimulated contracts and then follows a long refractory period in which further stimuli are ineffective. It is possible to imagine that herein lies the explanation of the origin of the beating of the cardiac muscle; action followed by enforced rest. A great deal of research has been done on this matter and the origin of the beat is a very complex matter and is not totally explicable in terms of changes in the threshold of stimulation and the period of contraction. It seems certain however that the origin of the beat is myogenic.

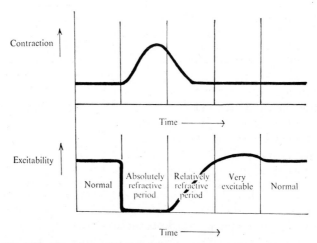

Fig. 107. *The refractory period of cardiac muscle.* The upper figure shows a record of the contraction of the muscle, the lower figure shows the varying excitability of the muscle to stimuli.

We have seen then that the regular beating of the heart is determined by regular electrical changes in cells lying in the S.A. node (i.e. pace-maker activity) and by the properties of cardiac muscle i.e. the absolute refractory period. The cells of the S.A. node spontaneously depolarise and generate a wave of excitation to heart muscle. Following depolarisation the resting electrical potential becomes re-established. Then, because of the 'leaky' cell membrane the cells gradually depolarise again and so the cycle repeats itself. That the regular beating of the heart is determined by the heart muscle itself was realised early in the history of physiology for Harvey showed in 1628 that if strips were cut from a living heart they would continue to beat. Indeed if the heart is rapidly removed from an animal after it has been killed and provided with a supply of oxygen by perfusing the coronary arteries (i.e. those arteries which pass from the base of the aorta to supply the heart muscle itself with blood) with an appropriate oxygenated saline solution then the heart will continue to beat for hours. As we shall see the heart does have a nerve supply but the role of the nervous system is not to *initiate* the heart beat but rather to *modify* the rate and force of contraction according to the needs of the organism.

Adaptation of the circulatory system

At all times the circulation is being adapted to meet the need to supply an adequate quantity of blood at an adequate pressure to the various organs. These needs are for ever changing locally in that the different organs have varying rates of activity and the blood supply must be adapted to meet these changing needs. Adaptation to meet these needs can occur in one or both of two parts of the circulatory system, firstly the heart, secondly the peripheral blood vessels. We will first describe how the heart's action is modified to deliver blood at the correct volume and pressure in spite of changing conditions in the peripheral blood vessels.

The heart. The output of blood by the heart may be modified in two ways; firstly by alterations in the rate of the heart beat and secondly by alteration in the amount of blood put out at each beat.

REGULATION OF HEART RATE. Although the origin of the heart beat is myogenic, modifications in heart rate occur through the influence of the nervous system and endocrine organs. The heart is under the influence of the two divisions of the autonomic nervous system, the parasympathetic and sympathetic nervous systems. The parasympathetic system exerts its effect on the heart by means of the vagus nerve. The sympathetic nervous system can influence the heart in two ways; firstly by means of sympathetic nerves to the heart itself and secondly by means of the hormone adrenaline produced in the adrenal medulla.

In the mammalian heart the vagus nerve spreads extensively in the fibres of the sino-auricular node, in the auricular musculature, auriculo-ventricular node and along the branches of the Bundle of His. The effect of vagal stimulation on the heart is to slow the heart rate and therefore reduce the amount of blood put out by the heart. This depressing action of the vagus nerve is brought into action by means of sensory endings situated in the arch of the aorta and in the carotid sinus. In many mammals these sense endings are in constant activity and by their connections within the brain stimulate the vagus to exert its restraining action on the heart. This restraining action of the vagus varies from one animal species to another and in man it tends to be highest in highly trained athletes who characteristically have a slow pulse rate. The sensory endings in the aorta and carotid sinus are increasingly stimulated by a rise in blood pressure, which produces a reflex slowing of the heart, thus tending to prevent an excessive rise in pressure. This reflex is described in Marey's law of the heart which states that the pulse rate varies inversely with the arterial blood pressure, and is the expression of vagal restraint upon the heart.

The heart is also supplied with branches of the sympathetic nervous system whose effect is augmented by the secretion of adrenaline from the adrenal medulla, which reaches the heart through the coronary arteries. The effect of stimulation by the sympathetic nervous system is to increase the heart rate and the force of each heart beat; this produces an increase in the output of blood and a rise in arterial blood pressure. The whole metabolic activity of the heart is raised, showing itself by an increased rate of utilization of glucose and lactic acid by the cardiac muscle. It will be seen that the effects of the sympathetic and parasympathetic nervous systems are in opposition and by variations in these two controls the heart's action can be adapted to meet a variety of circumstances.

CARDIAC MUSCLE AND CARDIAC OUTPUT. A further way in which the output of blood from the heart can be altered is due to a property of heart muscle itself; when heart muscle is stretched it is capable of working harder. Thus, if there is an increase in the return of venous blood to the heart the output of blood from the heart can be increased because the cardiac muscle is stretched during diastole and therefore works harder. During exercise there is an increased flow of blood returning to the heart due to the effect of contractions of skeletal muscles around the veins, which pushes the blood onward to the heart. In these circumstances there is an increase in the diastolic volume of the heart which stretches the cardiac muscle and so the output of blood increases, thus meeting the increased needs of exercise.

Changes in the Arterioles. The second way in which the circulation can be adapted to meet varying needs is at the level of the arterioles. Like the heart these are under the control of the autonomic nervous system. The effect of stimulation of the sympathetic nervous system is to constrict the arterioles and so reduce the blood supply to the tissues whilst the effect of stimulation of the parasympathetic nervous system is to oppose this effect and to dilate the arterioles. The centre which controls these vasomotor nerves is in the medulla oblongata, with subsidiary centres in the spinal cord. The vasomotor centre itself is sensitive to the carbon dioxide content and pH of the blood passing through it; when there is a rise in carbon dioxide or acidity then the sympathetic division of the vasomotor centre is stimulated, producing a constriction of the arterioles of the skin and gut with a consequent rise in blood pressure. Further the vasomotor centre is connected by nerves to special sense structures situated along the carotid arteries called the carotid bodies; these are also stimulated by a rise in the carbon-dioxide content of the blood, and reflexly stimulate a rise in blood pressure.

During physical activity the muscle tissues produce increased amounts of carbon dioxide. This increased output of carbon-dioxide, by means of the effects on the carotid bodies and vasomotor centre produces a generalized constriction of arterioles. But a local accumulation of carbon-dioxide has a direct effect upon the arterioles causing them to relax. Thus the blood is diverted into the dilated arterioles of the active muscles.

The arterioles are also affected by hormones, particularly by adrenaline. The effect of adrenaline in the circulation is to cause the constriction of the arterioles in the skin and gut, diverting the blood to more important regions.

Adaptation to stress situations. We can now consider some examples of the way in which these mechanisms are brought into action. First we will consider the 'stress' situation which occurs when an animal is confronted by a potential danger. In this situation there is an increased activity of the whole of the sympathetic nervous system and an outpouring of adrenaline from the adrenal medulla. By these means the heart rate is increased together with the cardiac output, and so the circulation is adapted to meet the need for increased activity. The adrenaline also stimulates the contraction of the arterioles of the skin and gut, diverting blood into the more important organs, brain, lungs and muscles.

Adaptation to blood loss. Secondly, we will consider adaptations of the circulation to blood loss incurred for example by an injury. When this occurs there is an immediate fall in blood pressure; this is a dangerous situation in which many vital organs e.g. brain and kidney are being deprived of blood. With the fall of blood pressure the heart is released from vagal restraint (Marey's law) and there is an increase in heart rate which, in itself, tends to promote a rise in blood pressure.

Further there is a reflex contraction of the arterioles in a variety of organs (mediated through the sympathetic nervous system) including the skin and gut, so conserving blood for more vital functions. The effects of these are seen in man in cases of shock due to haemorrhage where the pulse is rapid and the skin pale, cold and clammy due to the constriction of the arterioles to the skin. Formerly, part of the treatment of shock due to haemorrhage consisted of warming the patient; but it will be understood that warming the skin may promote a reflex vasodilation of the arterioles (see p. 466), and so oppose the blood conserving effect of vasoconstriction; this may result in further lowering of the blood pressure.

LYMPH AND THE LYMPHATIC SYSTEM

The lymphatic system begins as very many fine, blind ending hollow tubes, which ramify through most of the tissues of the body; they are about as extensive as the capillary system. These fine tubes drain into larger lymphatic vessels which eventually drain into the great veins in the neck.

The finer lymphatics have a thin wall, consisting of a single layer of pavement epithelial cells. The larger lymphatic vessels have connective tissue in their walls, and have valves which direct the flow of the contained lymph away from the tissues.

The function of the lymphatic vessels is to help to drain away tissue fluids. In particular the proteins and lipids which are filtered out of the circulating blood at the arterial end of the capillaries are returned to the blood stream via the lymph. In one day over 50% of the total plasma proteins may be lost from the blood into the tissue fluids. The capillary wall is much less permeable to large molecules compared with the lymphatic vessels and the filtered protein is removed from the interstitial fluid by the lymphatic vessels. Because of the greater permeability of lymphatics to substances of higher molecular weight much of the hormones secreted by endocrine glands probably reaches the blood stream via the lymph. Thus the lymph draining the thyroid gland is very rich in the large molecules of the thyroid hormone, thyroxine. In the gastrointestinal tract use is made of this permeability in the absorption of fats. Small short chain fatty acid molecules are absorbed by way of the blood capillaries and pass in the hepatic portal vein to the liver. But the long chain fatty acids of higher molecular weight pass into the large lymphatic vessels of the villi of the small intestine, the lacteals, and are transported via the lymphatic system to the root of the neck where they are discharged into the blood stream. The walls of the lymphatics contain no muscle and so are unable to actively propel the lymph; this drains along the lymphatics because of the intermittent pressure of the surrounding muscles, and the flow of lymph is in one direction, away from the tissues, because of the system of valves in the larger lymphatic vessels.

Before the lymph drains into the blood, it passes through a special tissue called lymphatic tissue. This consists of lymphocytes and the cells which produce them, and various connective tissues including macrophages. Lymphatic tissue may be collected in special masses, surrounded by connective tissue, called lymph nodes, or may be scattered diffusely through various organs e.g. in the wall of the gastro-intestinal tract. The lymph nodes are aggregated in special groups

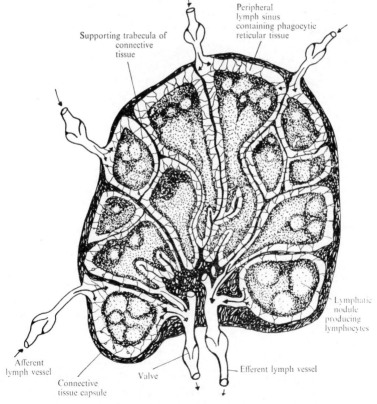

Fig. 108. *Diagram of a lymph node showing several afferent lymph vessels discharging their lymph into the peripheral sinus of the node.* The lymph permeates through the sinuses, which contain phagocytic cells supported by reticular fibres, and is drained away from the node by the efferent lymph vessels.

e.g. in the groins, arm pits, neck, base of the bronchi, and along the larger blood vessels in the abdominal cavity, and lymph drains through at least one set of lymph nodes before it pours into the blood at the junction of the venous and lymphatic systems in the neck. The lymphatics opening into a lymph node pour their lymph into a network of sinuses lined by connective tissue containing macrophages. These engulf dead cells and any organisms which may have been drained away from the tissues, and so help to limit any infection. The course of the lymphatics may often be seen in local infections, when tender fine red

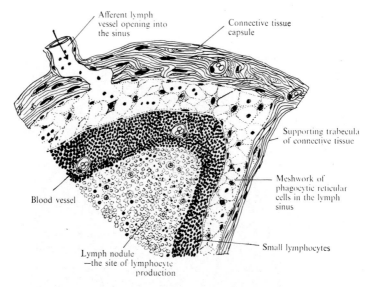

Fig. 109. *Drawing of a portion of a lymph node seen at high magnification* showing the meshwork of phagocytic reticular cells in the sinus.

lines are seen on the skin, radiating away from the infected part; there may be also a swelling of the lymph nodes into which these lymphatics drain, and these may even develop abscesses. Thus lymph nodes in the neck may become enlarged and tender in any infection of the mouth or throat.

In addition to this filtering action of the lymph nodes, they also pour into the lymph large numbers of cells, the lymphocytes, which pass with the lymph into the blood stream, where they have a very short life. Their exact fate and functions are not fully understood, but as discussed in Chapter XIV they have some function in relation to antibodies and immunity. Under conditions of stress, when there is

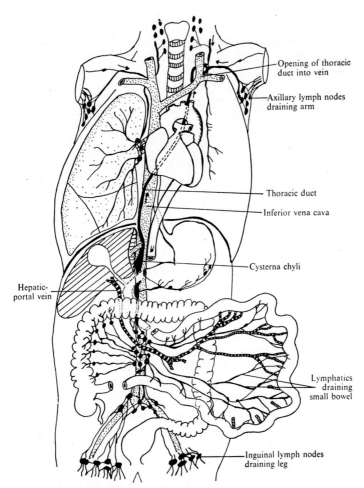

Opening of thoracic
duct into vein

Axillary lymph nodes
draining arm

Thoracic duct

Inferior vena cava

Cysterna chyli

Hepatic-
portal vein

Lymphatics
draining
small bowel

Inguinal lymph nodes
draining leg

Fig. 110. *Drawing illustrating some of the anatomy of the lymphatic system.*
Collections of lymph nodes are shown in the neck, axilla, inguinal region,
around the bronchi, and along the large vessels of the abdominal cavity. Into
these nodes drain lymph from the various tissues. From the nodes the lymph
drains into larger channels. In the abdomen the efferent lymph vessels from the
lymph nodes drains into a sort of reservoir, the cysterna chyli from which a
relatively large duct, the thoracic duct, passes through the thorax to the root
of the neck where, after receiving other tributaries it discharges into the junction
of the jugular and subclavian veins on the left side. On the right side there is no
thoracic duct and the vessels draining the right arm and right side of the head
and neck pass directly into the junction of the right jugular and subclavian veins.

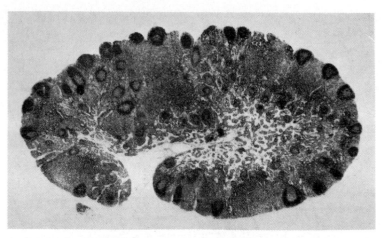

Fig. 111. *L.P. photomicrograph of a section of a lymph node showing the denser cortex, looser medulla, and lymphatic nodules embedded in the cortex.* (*By courtesy of Dr. S. Bradbury.*)

an outpouring of adrenocortical hormones, the lymphocytes are destroyed in large numbers and there is a generalized shrinkage of all the lymphatic tissue of the body; the exact significance of this is not understood.

The composition of lymph varies slightly from one part of the body to another, depending upon the organ it drains; thus the lymph draining the gut tends to be rich in fat globules, particularly after a meal.

EVOLUTION OF MAMMALS AND THE CONSTANCY OF THE INTERNAL ENVIRONMENT

In the story of the evolution of mammals from primitive vertebrates, two of the dominant themes are concerned with the animals' increasing independence of water and increasing power to withstand fluctuations in the temperature of the environment whilst still living an active life. In the change from fishes through amphibian and reptilian grades of organization right up to mammals we see a progressive emancipation from the environment. The fish evolved into an amphibian when the adult stage managed to live in air. The adult fish needs water for breathing and locomotion. Some primitive lung fish evolved so that they were able to breathe atmospheric air by means of an air bladder. Thus arose, 300 million years ago, a group of animals which were air breathing fishes which could squirm around in the muddy banks of the lakes and rivers. This period of Earth's history is known as the Devonian period and it was a time of floods and drought. Many of these air breathing animals would be stranded from time to time and some developed legs from their fins and gave rise to a new stage which is called the amphibian grade of life. The amphibian still had to return to water to lay its eggs and for its larval life, and as an adult it could not stray beyond the wet marshy land for its skin was incapable of conserving the animal's water.

The reptile gradually emerged from amphibian stock as an animal that could breathe air through its lungs, could walk on land far away from water without drying up for it had an impervious scaley skin. Above all it did not have to return to water to lay its eggs. The eggs were provided with a greater amount of yolk than in amphibia, so that the young could spend a longer time inside the egg. There was no metamorphosis and the young emerged from the egg as complex animals—small editions of their parents. This long development in the egg on land was really just as dependent upon water as was the amphibian egg the difference was that in the reptile the water was supplied as part of the organization of the very large and complex 'egg', whereas there was no such internal provision in amphibia and the eggs had therefore to be laid in water. The water supply in the reptilian 'egg' is called the amniotic fluid, and animals which have this type of 'private bathing pool' in which the young begin to develop are called amniotes (reptiles, birds and mammals) whereas vertebrates without the amniotic fluid are called anamniotes (fish and amphibia). In order that the young reptile could breathe during its development inside the egg it was provided with another membrane which fitted close beneath the shell and was richly supplied with blood vessels. This was the allantoic membrane and it also served as a receptacle for the excretory waste of the young reptile. The invention of amnion and allantois were essential before eggs could be laid on land. The independence of water seen in the reptile gave it great advantages over the amphibians and in the Mesozoic period the reptiles increased their range tremendously and were a very successful group.

The reptiles were limited in their range by temperature. As soon as the temperature fell below a certain minimum the chemical reactions in the body slowed down so much that the essential process of respiration could no longer supply energy fast enough for an actively moving life. So the reptiles were confined to warmer habitats as they were poikilothermic—i.e. their body temperature varied with that of the environment.

Mammals evolved the capacity to maintain a constant high temperature and could thereby succeed in living where reptiles failed. They had also other very great advantages including a brain capable of intelligent behaviour and especially the ability of the mother to nurture her young within the uterus. An attachment is made between the allantois and the wall of the mother's uterus and food and oxygen are thus supplied to the developing embryo and waste materials are removed. This very important attachment between foetus (embryo) and mother is called the placenta. Life in the uterus provides food, oxygen, protection, water and a constant temperature. Thus at all stages of the life of a mammal the cells of the body are provided with a constant supply of the things they need. There is a buffering of the fluctuation of the environment, so that the cells in a mammal may live although the conditions outside the body are not good. The famous French physiologist Claude Bernard expressed this idea in his now famous aphorism—'The constancy of the internal environment is the condition necessary for a free life'. The term 'constancy' should be understood as a dynamic equilibrium rather than a static concept. Because the cells inside a mammal have a fluid bathing them whose chemical composition and temperature is very constant these cells are able to function equally well in the tropics or the arctic, in the ocean, in fresh water or in the desert. The ultimate significance of all the varied functions of the circulatory system is that the constancy of the physical and chemical properties of tissue fluids is maintained.

Chapter Six

Integration and Adaptation by the Endocrine System

The endocrine system and the nervous system. The body of a mammal is a very highly differentiated structure and in order that all the specialized organs and tissues can work together harmoniously there must be some control and organization. This control is called integration and ensures that the animal functions as a unit, an organism rather than a collection of separate organs. Integration is achieved by two systems. Firstly by means of electrical impulses passing along specialized conducting elements, the nerves, which pass to and from a central controlling centre, the central nervous system. Secondly by means of chemical substances called hormones which are produced in certain tissues and glands and are distributed throughout the body by means of the circulatory system. The hormones produce their effects by influencing the activities of their target organs. Hormones are the chemical messengers of integration.

These two systems, the nervous system and endocrine system in addition to integrating the various organs of the body also serve to change the activity of the organism in response to changes in the external environment, a function called adaptation.

THE ENDOCRINE SYSTEM

The glands producing hormones are called endocrine glands. Whereas some glands, e.g. the liver and salivary glands discharge their secretions through special ducts, the endocrine glands are ductless and discharge their secretions into the blood stream, which is the agent which distributes the hormones. Thus endocrine glands have a characteristically rich blood supply.

The endocrine system of coordination is a relatively primitive mechanism and endocrine organs have been described in very many animals including molluscs, insects, crustaceans, fish, amphibia, reptiles and mammals. Plants also have chemical substances e.g. auxins, for the purpose of integration. The nervous system has evolved side by side with the endocrine system and the two mechanisms have interconnections. The pituitary gland, which exerts a powerful controlling influence on many other endocrine glands, has intimate connections with the hypothalamus which is the seat of the control of the autonomic nervous system, in the floor of the fore-brain. (See pp. 303, 461.)

An endocrine function has been ascribed to many organs including the pituitary gland, thyroid gland, parathyroid glands, adrenal glands, ovaries, testes, placenta, the islets of Langerhans in the pancreas and to various parts of the digestive tract. Even part of the brain itself, the hypothalamus, produces a hormone, the antidiuretic hormone which controls water reabsorption from the tubules in the kidney. It has been suggested that the pineal body and the thymus are also endocrine glands.

When an endocrine organ is diseased or is removed from the body, a series of symptoms appear which can often be caused to disappear if active extracts of the gland are injected. This is the method of proving that a gland has an endocrine function. There are many disease states in man produced by malfunction of the endocrine organs. Some are described here, not because they are of interest in themselves (except of course to the clinician) but because they give invaluable information as to the normal function of the glands concerned.

The pituitary gland

The pituitary gland is a small round body connected to the floor of the thalamencephalon (fore-brain) by a stalk. Developmentally it has a dual origin, from the floor of the fore-brain and from the roof of the mouth. A projection from the roof of the mouth grows upwards to meet a down-growth from the floor of the brain, the infundibulum. These parts meet and fuse, that from the roof of the mouth (the hypophysis) loses its oral connection and comes to form the anterior glandular part of the pituitary gland. The posterior part retains its connection with the brain and forms the posterior nervous part of the gland. The adult gland is lodged in a depression in the floor of the skull, the sella turcica. It is richly supplied with blood vessels into which the gland discharges its secretions. The posterior lobe provides the connection with the nervous system and nervous influences can effect changes in the secretory activity of the gland.

In the rabbit, nervous stimuli associated with mating cause the pituitary gland to produce a hormone which stimulates the release of eggs from the ovary. If the pituitary stalk is severed, ovulation no longer follows mating. This is a good example of coordination of function carried out by combined nervous and endocrine mechanisms. It might also be noted that the idea of having a mechanism whereby the eggs are released only at mating is a very good one, for it prevents wastage of eggs since it increases the chance that the eggs will be fertilized. In addition to the connection with the nervous system provided by the posterior lobe of the pituitary gland there is a further way in which the nervous system and the pituitary are connected. There is a network of capillary blood vessels in the hypothalamus drained by vessels which lead directly to the anterior lobe of the pituitary where they branch to form a further set of capillaries. These two sets of interconnected capillary beds form the hypothalamico-hypophyseal portal system which provides a means whereby the hypothalamus, by means of substances which pass from the tissue of the hypothalamus into the capillary blood vessels, can influence the activity of the anterior lobe of the pituitary. (See fig. 112.) It is probably the interruption of this portal system that prevents ovulation in the rabbit after section of the pituitary stalk.

The pituitary has been dubbed the master gland or the 'leader in the endocrine orchestra'. These descriptions serve to emphasize the fact that many of the other endocrine organs in the body are under the control of the pituitary. It produces a thyrotrophic hormone which controls the function of the thyroid gland, gonadotrophic hormones controlling the testis and ovary and an adrenocorticotrophic hormone (abbreviated to A.C.T.H.) which controls the adrenal cortex. These trophic hormones will be considered together with the glands they control, i.e. with their target organs. The position of conductor of the endocrine orchestra should be given to the region of the hypothalamus since this is the ultimate controlling centre for many of the activities of the pituitary gland.

THE ANTERIOR LOBE exerts a strong influence on growth, not only of the particular endocrine glands which it controls by trophic hormones, but on the growth of bones and soft tissue generally. If the pituitary is removed from a young animal it fails to grow properly and a dwarf animal results. This is due not only to the reduced activity of the thyroid and adrenal but also to the absence of the pituitary growth or somatotrophic hormone.

The process of the removal of the pituitary gland in an experimental animal is called hypophysectomy, and in an hypophysectomized animal resumption of growth can occur if extracts of the anterior lobe of the pituitary are injected. If very large amounts of the extract are given before the animal reaches maturity, then a giant is produced. In man, disorders of the pituitary in youth, which result in overproduction of growth hormone, produce giants. If excess growth hormone is given after the animal has reached maturity, growth in length of the long

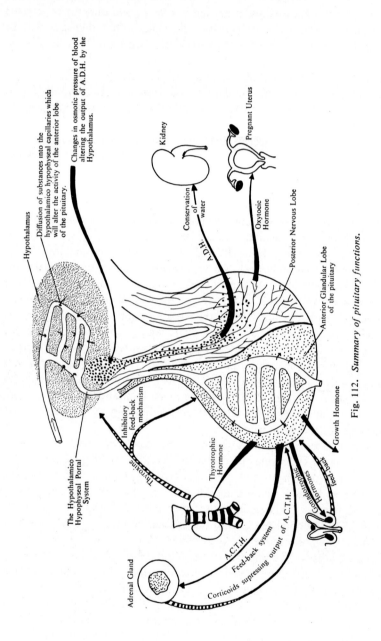

Fig. 112. *Summary of pituitary functions.*

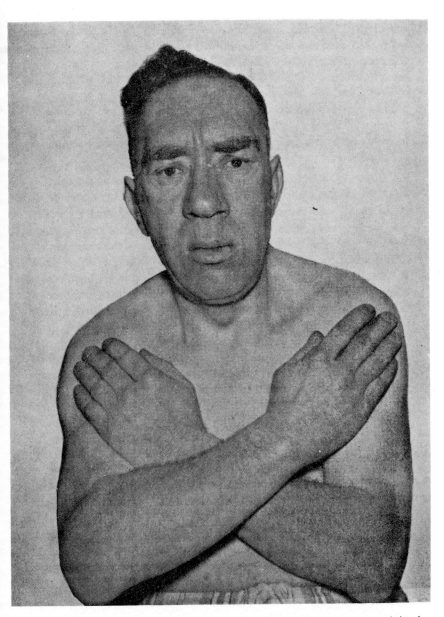

Fig. 113. *Photograph of a patient suffering from acromegaly.* The excess tissue growth involves both bones and soft tissues. This patient shows enlargement of parts of the skull and jaw, thickening of the nose and enlargement of the hands. (*By courtesy of Winifred Hector, 'Textbook of Medicine for Nurses', Heinemann Medical Books Ltd.*)

bones cannot be obtained since this growth occurs at a plate of cartilage near the end of the bone—the epiphysis—and in the mature animal this cartilage has been completely changed into bone. Certain other bones, however can still be stimulated to grow, particularly those at the ends of the body. Therefore overproduction of hormone after puberty causes growth of the jaw, skull, hands and feet, producing a condition known as acromegaly (fig. 113).

The hormones produced by the anterior pituitary gland are complex protein containing substances, some of which have been isolated in a relatively pure state, including gonadotrophic and adrenocorticotrophic hormones.

THE POSTERIOR LOBE of the pituitary gland produces hormones:

i. The oxytocic factor which when injected causes contraction of the uterus. It also causes 'let down' of milk in the mammary gland.

ii. Vasopressor factor which when injected causes a rise in blood pressure by the contraction of the smooth muscles of the blood vessels.

The vasopressor factor is also called the antidiuretic hormone. Removal of the posterior pituitary results in only one marked disturbance—the animal produces a large volume of dilute urine, and because of the continuous water loss the animal has a great thirst. These symptoms are also shown by some human beings suffering from diabetes insipidus, a condition associated with damage or disease of the posterior pituitary or hypothalamus. The anti-diuretic hormone, normally produced by the posterior pituitary and hypothalamus regulates the reabsorption of water from the uriniferous tubule; in its absence there is a reduced reabsorption of water and a large volume of urine is produced. The amount of antidiuretic hormone produced varies with the water intake. If little water is taken, the osmotic pressure of the blood tends to rise and more antidiuretic hormone is produced resulting in an increased reabsorption of water in the tubule, and this vital commodity is conserved. If large amounts of water are taken, less antidiuretic hormone is produced, less water reabsorbed and the excess water is lost in the urine. This mechanism seems to have been evolved by terrestrial vertebrates to enable them to conserve water. The antidiuretic hormone seems to have its effect on the distal convoluted tubules and collecting ducts of the kidney (p. 504).

Thyroid gland

The thyroid gland consists of two main lobes, one on either side of the trachea at the base of the neck. It consists of masses of small follicles, or balls of cells, containing a jelly-like substance called colloid.

The cells of the follicle are flat, cubical or columnar in shape depending on their activity, and they produce the hormone of the thyroid, thyroxine, which is stored in the colloid (fig. 114–116).

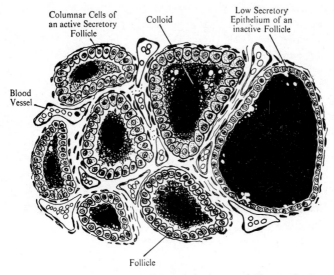

Fig. 114. *Drawing from a section of thyroid gland seen at high magnification.* The drawing shows an inactive follicle with a low epithelium and active secretory follicles with a high columnar epithelium.

The gland is under the control of the anterior lobe of the pituitary which produces a thyroid growth hormone, the thyrotrophic hormone. This stimulates both the growth and secretory activity of the thyroid. Thyrotrophic hormone production by the pituitary is controlled by the amount of thyroxine in the blood; when the blood thyroxine level falls, thyrotrophic hormone production is increased leading to the production of more thyroxine. This increased thyroxine content of the blood however eventually reduces the pituitary gland's output of thyrotrophic hormone. This type of mechanism is termed a 'feed-back' control and is found also in the case of the endocrine activity of the adrenal cortex, ovary and testis.

The thyroid hormone thyroxine has a powerful effect on the metabolic rate of all tissues, on growth, and on amphibian metamorphosis. In mammals removal of the thyroid gland, an operation called thyroidectomy, results in stunted growth, and the failure of sexual maturation when performed on young animals. In amphibia, underactivity of the thyroid prevents the metamorphosis of the tadpole to the adult.

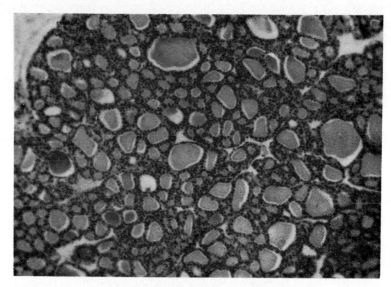

Fig. 151. *L.P. photomicrograph of a section of the thyroid gland of the dog. (By courtesy of Dr. S. Bradbury.)*

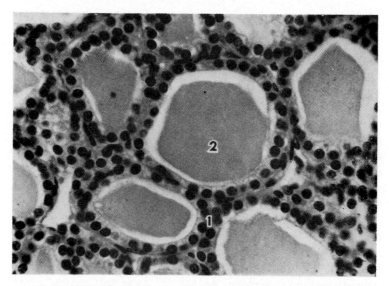

Fig. 116. *H.P. photomicrograph of a section of the thyroid gland of the dog.*
1. secretory epithelium.   2. colloid.   (*By courtesy of Dr. S. Bradbury.*)

These effects are reversible if thyroxine is administered. In man, thyroid underactivity in the infant results in a condition called cretinism; a cretin fails to attain adult stature or sexual maturity and is often grossly obese and is mentally subnormal (fig. 117). Thyroid under-activity in the adult produces a condition called myxoedema, character-ized by a peculiar puffiness of the face and a thickening of the skin. The individual becomes slow witted and loses his normal interest in the environment, and because of the lowered metabolic rate he becomes very sensitive to cold (fig. 118).

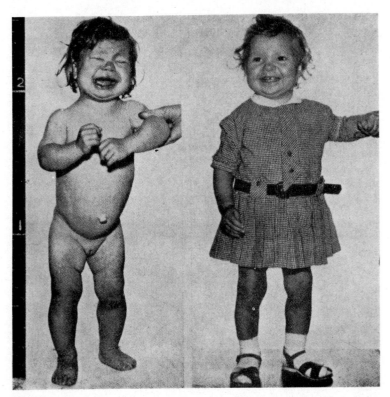

Fig. 117. *This little girl is a cretin.* The effect of six months treatment with thyroxine is shown on the right. (*By courtesy of Winifred Hector.*)

Thyroid overactivity also occurs, often seen in middle aged women, producing the condition of thyrotoxicosis or exopthalmic goitre. Here many of the symptoms are caused by the raised metabolic rate and include loss of weight, nervousness and irritability, quickened heart rate and insensitivity to cold. There is a swollen thyroid in the neck

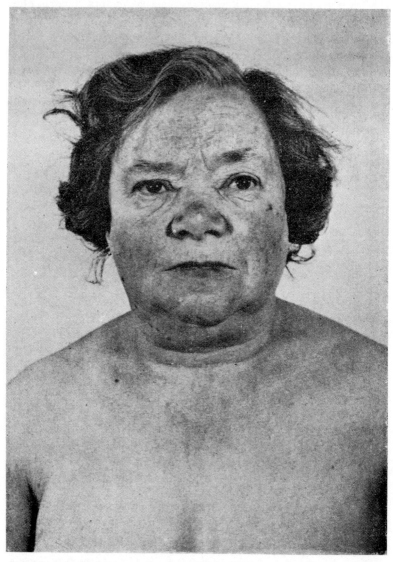

Fig. 118. *Photograph of a patient suffering from myxoedema.* The face shows puffiness and the skin and hair are dry. The rather dull expression is fairly typical of the disease. (*By courtesy of Winifred Hector.*)

and the eyes protrude because of the accumulation of fatty material in the orbit behind the eyeball. The condition is probably due to the failure of the feedback mechanism; the pituitary is no longer so sensitive to the circulating thyroxine and therefore large amounts of thyrotrophic hormone are produced despite a high level of thyroxine in the blood.

In addition to thyroxine there is another active substance produced by the thyroid and stored in the colloid, called tri-iodothyronine. This, like thyroxine, is released from the colloid when necessary and passes to the tissues of the body dissolved in the blood plasma. But tri-iodothyronine is a much more potent substance, weight for weight, than thyroxine, and has a much faster effect on metabolism when it is injected into the body, and it is thought that thyroxine may be converted to tri-iodothyronine in the tissues.

The thyroid hormone thyroxine, contains iodine as an essential part of its molecule. Iodine is thus an essential part of the diet and if the intake is insufficient, thyroxine production is reduced. The pituitary gland becomes released from the inhibitory influence of thyroxine and increasing amounts of thyrotrophic hormone are produced. This causes growth of the thyroid gland which is still unable to produce thyroxine because there is insufficient iodine. This type of enlargement is called 'simple goitre' and was common in mid-continental areas e.g. parts of north America and Switzerland where the soil has a reduced iodine content due to leaching. Simple goitre was also common in Derbyshire, and is sometimes referred to as 'Derbyshire neck'. Table salt is now often iodized to ensure an adequate intake of iodine in the diet.

Because of the marked effect of the thyroid hormones thyroxine and tri-iodothyronine on metabolic rate the thyroid gland forms an important component of temperature regulating devices. Low environmental temperatures stimulate increased secretory activity of the thyroid gland, an effect probably mediated by way of the hypothalamus and anterior pituitary gland. In temperate latitudes the thyroid of many species increases in size and activity in the winter.

The way in which thyroid hormones have their effect on metabolism is not known with certainty. Administration of the hormone to animals increases the rate of oxidation of succinic acid to fumaric acid, perhaps by acting upon the various enzymes concerned e.g. succinic dehydrogenase, cytochrome C, cytochrome oxidase and electron carriers. The hormone might also produce more body heat by uncoupling the energy releasing processes of the cell from the mechanisms for converting the energy into metabolically useful forms such as energy rich phosphate compounds. Thus the energy liberated in the cell from the metabolism of foodstuffs is not as efficiently converted into energy rich phosphate compounds, and some of this energy appears as heat.

The Adrenal gland

The adrenal glands are paired endocrine glands, each situated near the upper pole of the kidneys in man and other mammals. Each gland consists of two parts, an outer cortex and an inner medulla; these two parts are quite distinct in their structure and function and in their origin in the embryo.

In development the cells of the cortex of the adrenal are derived from the same ridge of coelomic epithelium as the gonads, whilst the medulla on the other hand is derived from primitive nerve cells of the spinal cord. These primitive nerve cells (neuroblasts) migrate to the kidney area where they become covered by cells which later develop into the cortex of the adrenal. Some of these neuroblasts never reach the kidney and some migrate even further than the kidney so that there are scattered particles of medullary-like tissue throughout the abdomen particularly on the large arteries. In lower vertebrates, e.g. amphibia and fishes, the two components of the mammalian adrenal gland, the cortex and the medulla, are quite separate.

The activities of the two parts of the gland are under different control. Secretory activity of the adrenal cortex is controlled by a hormone from the anterior pituitary gland called the adrenocorticotrophic hormone. The adrenal medulla however is controlled by part of the nervous system called the sympathetic nervous system.

The Adrenal Medulla. This consists of a grey looking tissue forming the core of the adrenal gland. It consists of groups of modified nerve cells which secrete a hormone called adrenaline (with a smaller amount of a similar hormone called nor-adrenaline). Nerve fibres of the sympathetic nervous system pass into the medulla and end around the secretory cells. The cells of the medulla produce much larger amounts of adrenaline when under conditions of fright and fear. In these circumstances the sympathetic nervous system is very active and the adrenal medulla is stimulated to action by the sympathetic nerves which supply it, resulting in an outpouring of adrenaline into the blood stream. This adrenaline has a variety of actions in the body but the sum total effect is to prepare the body for 'fight or flight'. Adrenaline has a powerful effect on the heart and the blood vessels of the body. The heart muscle is stimulated and so pumps out blood at a faster rate, and the many small blood vessels of the skin and gut are constricted so that blood is diverted into more important channels—to the brain, lungs and muscles. Adrenaline also causes a breakdown of some of the glycogen stores of the liver causing the blood sugar level to rise. The whole nervous system is also stimulated so that the entire organism is prepared for extra effort.

Adrenaline also has a stimulating effect on the anterior pituitary, which begins to produce larger amounts of adrenocorticotrophic hormone which stimulates the adrenal cortex. The significance of this will be appreciated later when the role of the adrenal cortex in stress situations has been discussed.

Adrenal cortex

The adrenal cortex forms a layer of tissue surrounding the medulla. It is yellow in colour because of the lipid material inside the cells. These lipids probably form the raw material for the synthesis of the various hormones. The cells of the cortex are arranged in three layers which are readily recognizable when sections of the gland are examined under the microscope. The outer layer, called the zona glomerulosa, has cells which contain little lipid. This layer produces the hormone aldosterone which is concerned in regulating the body's sodium balance. The inner layers of the cortex are called the zona fasciculata and the zona reticularis. The cells of the fasciculata probably act as a storehouse

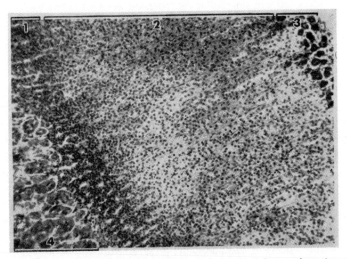

Fig. 119. *Photomicrograph of an L.S. of an adrenal gland showing the various layers of the cortex and the medulla.* 1. zona reticularis. 2. zona fasciculata. 3. zona glomerulosa. 4. medulla.

of raw material for hormone synthesis and contain large amounts of lipid. The inner zona reticularis produces the hormones cortisol and corticosterone, together with small amounts of sex hormones, fig. 119.

Functions of the adrenal cortex

Unlike the adrenal medulla which can be removed without seriously affecting an animal the adrenal cortices are indispensable for life. Without adrenal cortices an animal eventually dies and death is accelerated if the animal is exposed to any form of 'stress' such as heat, cold, anxiety, injury or infection. The hormones of the adrenal cortex influence a large number of body processes, some of which are listed below.

1. Metabolism of protein, carbohydrate and fat.
2. Salt and water balance.
3. Kidney function.
4. Maintenance of blood pressure.
5. Resistance to stress.
6. The processes of inflammation and repair.

We can best look at the functions of the adrenal cortex by considering individual hormones. The various hormones which can be extracted from the cortex have been classified into three main groups according to their major actions, but it should be emphasized that there is considerable overlap of actions between the classes.

1. Glucocorticoid hormones e.g. cortisol, corticosterone.
2. Mineralocorticoid hormones e.g. aldosterone.
3. Sex hormones.

GLUCOCORTICOID HORMONES

Cortisol and corticosterone are classed under this heading because of their marked effect on the metabolism of carbohydrate. But their sphere of influence is much larger than this. One clear effect of injecting glucocorticoid hormones into an animal is to cause an increase in the amount of glucose in blood. This occurs even when the hormones are given to a starving animal showing that the extra glucose in blood must arise within the body. In fact the glucose is 'new' glucose manufactured in the liver from the products of protein metabolism. In fact cortisol promotes the breakdown of tissue protein and the amino acids become available to the liver for the synthesis of glucose. In the liver cortisone has another effect—it stimulates the production of the various enzymes which are needed for the complex process of the conversion of protein to glucose. The rise in the concentration of glucose in blood is not entirely due to an increased rate of generation of glucose in the liver. Cortisone also prevents many cells of the body from utilizing glucose— thus conserving the critical amount of glucose leaving the liver for the needs of the central nervous system.

This action of cortisol is vital for the mammal during periods of fasting when supplies of glucose, which are so essential for the metabolism of the central nervous system, have to be manufactured by the body. Animals from which the adrenal glands have been removed tolerate starvation very badly and often die of convulsions— the product of a brain deprived of adequate amounts of glucose.

Other functions of glucocorticoid hormones are more nebulous conferring as they do resistance to all forms of 'stress'—ranging from severe muscular exercise, injury and infections to exposure to heat, cold or even psychic stress (e.g. in overpopulation, restricted territories etc.). We do not clearly understand how the hormones confer resistance to all these forms of stress. Part of the protection given by the hormones is probably due to their effect on the internal distribution of salt and water between cells and tissue fluid, and also due to an interaction of the hormones with the autonomic nervous system and blood vessels— which facilitates those responses of change in heart action, blood pressure and blood flow which occur during severe exercise, exposure to heat or cold etc. The hormones also have powerful effects on inflammation and repair. They are widely used in medicine for this purpose— in the supression of inflammation of a variety of tissues including the eye, skin, joints etc. But if the hormones are used for this purpose e.g. to ease the pain, swelling and limited movement of an arthritic joint then the benefits have to be weighted against the disadvantages of the effects of the hormones on other metabolic processes.

The effects of the hormones on protein metabolism may be so marked that bones lose much of their protein, become brittle and fracture without any external force.

The output of cortisone from the adrenal cortex is controlled by the adrenocorticotrophic hormone (ACTH) released from the anterior pituitary gland. The anterior pituitary gland is in turn controlled by the hypothalamus, which releases a chemical factor (corticotrophin releasing factor) which reaches the pituitary gland by way of the hypothalamico-hypophyseal portal system of blood vessels. A great variety of stimuli, pain, trauma, infection, emotions, produce their effects on the adrenal cortex by acting through this common pathway- hypothalamus-anterior pituitary-adrenal cortex. There is a feed back mechanism in that the amount of cortisone in the blood controls the activity of the hypothalamus and anterior pituitary gland. Increased amounts of cortisone in the blood depress the output of ACTH from the anterior pituitary.

MINERALOCORTICOID HORMONES

The most potent naturally occurring mineralocorticoid hormone is aldosterone. This hormone is produced by the outer layer of the adrenal cortex, the zona glomerulosa.

The hormone acts at various sites in the body—kidney, sweat glands, salivary glands. At all these sites the hormone influences the transport of sodium and potassium ions across cell membranes. An important site of action is the distal convoluted tubules of the kidney where the hormone stimulates the reabsorption of sodium from the urine into the blood in exchange for potassium which is secreted into the urine. The significance of this action is in the regulation of the body's sodium balance and this is explored in more detail in chapter XII. A deficiency of aldosterone produces a disturbance of these regulating mechanisms causing a critical loss of sodium (and water) in the urine which results in a shrinkage of the volume of body fluid, particularly blood. This is the usual cause of death after removal of the adrenal cortices from an animal. It is only if such an animal is maintained alive by giving it free access to salt and water or an artificial supply of the hormone that disorders appear which are due to a lack of glucocorticoid hormones.

Although the pituitary hormone ACTH influences the secretion of aldosterone by the adrenal cortex (the secretion of aldosterone falls by up to 50% after removal of the anterior pituitary gland) the zona glomerulosa remains structurally and functionally intact after removal of the anterior pituitary gland from an animal whilst the inner layers of the cortex undergo virtually complete atrophy. The factors which regulate the secretion of aldosterone are discussed in detail in chapter XII. Here we can say that the output of aldosterone is regulated, in part, by a hormone called renin which is produced by the kidney—an organ which senses the state of the body's sodium balance.

SEX HORMONES

Both androgens (masculinizing) and oestrogens (feminizing) are produced by the adrenal cortex in both males and females. The effects of these hormones may be disproportionately in evidence in the case of certain tumours of the adrenal cortex when virilizing changes (e.g. growth of beard) may be seen in women, and occasionally feminizing changes in men.

SUMMARY. Although the functions of the different adreno-cortical hormones has been rigidly subdivided into gluco-corticoid, mineralo-corticoid and sex hormone functions there is really no such clear cut division of function, and cortisone, for example, whilst having predominantly glucocorticoid functions also has effects on the renal tubules causing retention of salt and water.

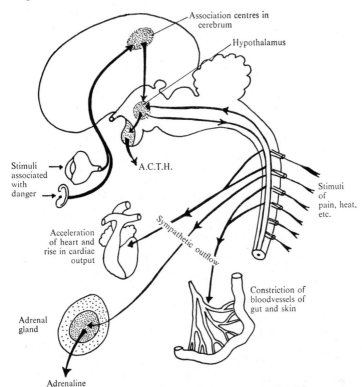

Fig. 120. *Stage 1 of the stress reaction.* The organism is receiving a variety of stimuli, including pain, heat and is becoming aware of a danger situation. These stimuli bring about a reflex activation of the sympathetic nervous system by way of connections with the hypothalamus. There is a generalized sympathetic outflow producing a quickening of heart rate, vasoconstriction of the blood vessels of skin and gut, and a stimulation of the adrenal medulla. The pituitary gland begins to produce increasing amounts of A.C.T.H.

The functions of the cortical hormones may be thought of in one word—*conservation.* By the effects on blood sugar production and utilization, wound healing and the metabolism of cells in general, and by the effects on the conservation of sodium and water, the animal is prepared to withstand the damage and danger to which life subjects it.

An animal deprived of its adrenal cortex can be kept alive provided it is kept warm and supplied with generous amounts of salt, but it is unable to meet any emergencies with certainty, and even a night out in the cold could kill it.

The functions of the adrenal cortex and medulla and the way in which these are integrated to adapt the animal to withstand stress situations is shown diagrammatically in figs. 120–122.

The functions of the adrenal cortex may be illustrated by diseases in which there is underactivity or overactivity of the cortex.

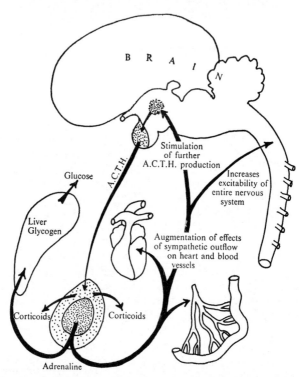

Fig. 121. *Stage 2 of the stress reaction.* The general sympathetic out-flow is now being reinforced by the secretion of increasing amounts of adrenaline by the adrenal medulla. Adrenaline reinforces the action of the sympathetic nervous system on the heart and blood vessels, but it also brings about a breakdown of glycogen in the liver, the release of larger amounts of A.C.T.H. by the anterior pituitary gland and an increased excitability of the entire nervous system. The animal is now fully prepared for fight or flight.

Addison's disease is a condition of underactivity of the adrenal cortex, sometimes due to a destruction of the glands by tuberculosis. There is a tendency to pigmentation, particularly of the face, mucous membranes and hands. The pigmentation of Addison's disease has now been proved to depend on the fact that melanocyte stimulating activity is built into the adrenocorticotrophic hormone molecule. There is general muscular weakness, wasting and a low blood pressure. The life of patients with Addison's disease is threatened by crises: these can be brought about by infections, injury and fatigue. Because of the lack of the conserving functions of the cortical hormones these people are unable to withstand these stresses and develop crises in which there is vomiting, dehydration and collapse. There is a fall of blood pressure and blood sugar level. The cause of many of these changes can be discovered by referring back to the functions of the individual adrenocortical hormones.

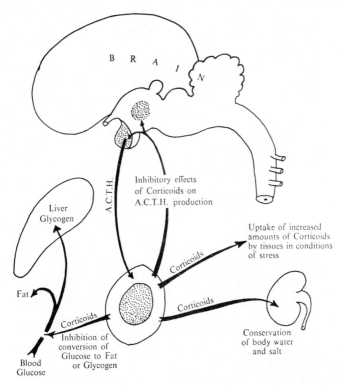

Fig. 122. *Stage 3 of the stress reaction.* Adreno-cortical hormones are being produced in larger quantities, and these, by their various actions on the organism, are producing their effect of CONSERVATION. The extent to which this reaction is brought into play is somewhat determined by the extent of damage to the organism incurred in the stress reaction.

The basis of treatment of Addison's disease consists in the replacement of the hormone deficiencies by the administration of cortisone, and desoxycorticosterone if necessary. In infections and injuries the doses of these hormones is increased because at these times of stress the hormones are utilized by the body at a rapid rate.

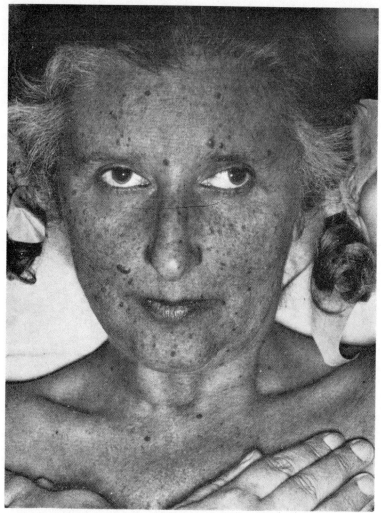

Fig. 123. *Photograph of a patient suffering from Addison's disease.* The main point to notice is the increased pigmentation of the skin. This is due to an increased production of the hormone A.C.T.H. from the anterior pituitary gland in an 'attempt' to stimulate the secretory activity of the diseased adrenal glands. (*By courtesy of Winifred Hector.*)

In *Cushing's disease* there is an outpouring of large amounts of adreno-cortical hormones, because of a tumour of the adrenal or the pituitary gland. There may be an increased production of all three types of hormones. Increased output of gluco-corticoid hormones may produce a high blood sugar (due to increased protein breakdown and the synthesis of sugar from the products of protein breakdown). And because

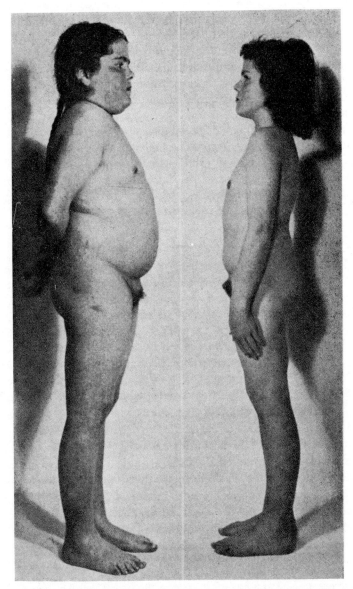

Fig. 124. *Photographs of a girl with Cushing's syndrome aged* 13 *years, before* (L) *and after* (R) *treatment.* The main features to be noticed here are the obesity—due to excess fat storage and water retention—and the 'moon face' which is typical of the disease. This patient does not show abnormal hair growth or acne which are seen in many women with the disease. In this patient the disease was due to the excess production of A.C.T.H. by the anterior pituitary gland. Treatment consisted in irradiation of the pituitary glands with X-rays. (*From 'Textbook of Paediatriacs', by Collis, Heinemann Medical Books Ltd.*)

of this protein breakdown there may be muscular weakness and a weakening of the bones due to the dissolving of their protein matrix (this may be so extreme that spontaneous fractures occur). Increased output of mineralo-corticoid hormones results in the retention of abnormal amounts of salt and water in the body which results in a rise in blood pressure. Because of increased sex hormone production from the adrenal cortex there may be signs of masculization in females e.g. growth of hair on the face.

Treatment of Cushings disease is usually surgical and consists in removal of the tumour. Symptoms similar to those seen in Cushings disease may be produced by overdosage of patients with cortical hormones. These are being used with increasing frequency today in the treatment of a variety of diseases, particularly because of the powerful 'anti-inflammatory' effect of cortisone.

The control of blood sugar level, an example of integration

The blood contains an ever ready source of energy for the tissues of the body in the form of glucose. The amount of this glucose is constantly fluctuating, depending on how fast the tissues are abstracting it from the blood, and on whether or not glucose is being absorbed from the intestine. Although the level of glucose fluctuates it must not fall below a certain level; the brain does not store glucose (and only $0 \cdot 1 \%$ glycogen) and is therefore almost completely dependent on blood glucose for its supply of energy. Thus if the blood glucose falls below a critical level, nervous function is impaired and the whole activity of the organism is disturbed. This is a further example of the importance of the relative constancy of the internal environment. It is also a good example of the sort of thing which is constantly occurring in a complex organism such as a mammal. Cells are highly differentiated and a sequel to this is that they lose some of the functions that a less specialized cell would retain. The highly differentiated cell performs its special function admirably but it is very dependent upon the supporting functions of other cells. Thus in this case the highly specialized cell of the brain is dependent for its working upon all the mechanisms in the body which are involved in the regulation of the blood sugar level.

Sugar gains. The amount of glucose in the blood may be increased by three means:

   i. From the absorption of glucose by the alimentary tract
   ii. By synthesis of glucose from the products of protein breakdown
   iii. By the conversion of glycogen into glucose.

Sugar losses. Sugar can be lost from the blood in three ways:

i. By the utilization of glucose in tissue respiration
ii. By the conversion of glucose into fat
iii. Temporarily lost in the form of glycogen stores in the liver and muscles.

These processes of loss and gain can be summarized conveniently as seen in fig. 125.

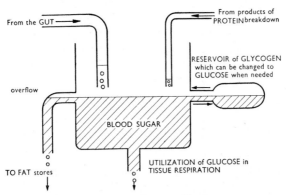

Fig. 125. *Losses and gains in blood sugar.*

Hormones influencing sugar level. In fig. 125 the blood sugar is represented as a tank of fluid. Glucose is supplied to the tank by a large pipe bringing glucose from the gut and a smaller pipe from the products of protein breakdown. Out of the tank there is a constant outflow into the tissues where glucose is constantly used as a source of energy. The reservoir tank contains glycogen which can be reconverted into glucose when the level of glucose begins to fall. An overflow is also provided whereby excess glucose (from excess eating) can be piped off and converted into fat. It will be obvious that a relatively constant level of sugar in the tank can be obtained by varying the rate of inlet and/or outlet. What is required is a set of taps to regulate the flow in the pipes. In the body this is brought about by means of the action of several hormones. There are many hormones which are thought to play a role in the metabolism of glucose but the best understood ones are insulin, adrenaline and the glucocorticoid hormones. The role of these may now be summarized.

1. ADRENALINE. It is produced in the adrenal medulla and stimulates the breakdown of glycogen into glucose. We have seen how this helps in the 'fight or flight' reaction by providing the body with a readily

available source of energy. The role of adrenaline under normal conditions is not as important as the role of insulin or glucocorticoid hormones.

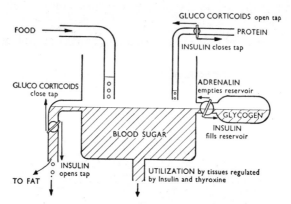

Fig. 126. *The influence of hormones on the losses and gains in blood sugar.*

2. GLUCOCORTICOIDS. These hormones cause an increased rate of breakdown of proteins and from these breakdown products glucose is synthesized. In addition these hormones also slow the rate at which glucose leaves the blood e.g. in the form of fat. We have defined the function of the corticoid hormones as 'conservation' and in the control of the blood sugar it is obvious that by increasing the rate of glucose production and by reducing the rate at which it leaves the circulation, the body's supply of glucose is conserved for important functions.

3. INSULIN. Insulin is a hormone produced by groups of cells, called the islets of Langerhans, which are scattered throughout the substance of the pancreas. It is discharged into the blood stream. The hormone is a polypeptide containing 51 amino acids. The cells of the islets respond directly to the changing concentration of glucose in the blood. When the concentration of glucose in the blood rises insulin is released. When the concentration of blood glucose falls below a certain critical level insulin secretion stops.

Insulin has widespread effects on metabolic processes including,
(1) a lowering of the concentration of glucose in the blood,
(2) an increase in the rate of synthesis of fat, glycogen and protein by tissues.

The action of insulin on uptake of glucose by cells. In a resting animal when no food is being absorbed from the gut, the glucose in the blood is derived from the liver, which produces glucose from glycogen break-down and from some of the products of protein breakdown. The output of glucose by the liver in a fasting animal is strictly limited, and most of this glucose is used in the metabolism of the brain and heart.

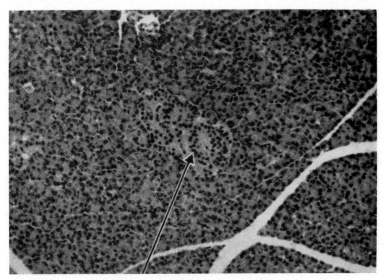

Fig. 127. *Photomicrograph of a section of pancreas showing an islet of Langerhans embedded in exocrine secreting tissue.*

During fasting most of the tissues of the body, especially muscle and fat, seem to be prevented from taking up glucose from the blood. A barrier exists at the cell wall preventing the entry of glucose. This barrier can be reduced in two ways, by giving glucose to the animal, and by muscular work. Giving glucose to an animal stimulates the release of insulin from the islets of Langerhans. Insulin reduces the barrier at the cell membranes of fat and muscle cells allowing them to take up glucose. Increased muscular effort has a similar effect, allowing muscle cells to take up glucose from the blood. This effect of insulin has been shown directly. In the fasting state when little insulin is being released, muscle cells contain little glucose. But on giving insulin (or glucose, to stimulate the release of insulin) glucose begins to appear in the cells. Earlier theories about the action of insulin, in particular those concerned with the effect of insulin upon the enzyme hexokinase, have been superseded by the present concept which regards insulin as a factor allowing cells to take up glucose. Not all the tissues of the body have a membrane barrier to the penetration of glucose, for example the cells of the liver and central nervous system. The central nervous system is almost entirely dependent on carbohydrate as a source of energy and in nerve cells the uptake of glucose is independent of the action of insulin. Muscle cells rely more heavily upon fatty acids as a source of energy and perhaps carbohydrate metabolism may be more of an emergency source of energy.

In the absence of insulin the uptake of glucose by fat and muscle cells is greatly reduced. Meanwhile glucose continues to be absorbed from the intestine, and the liver continues to produce glucose from the products of protein metabolism, so that glucose begins to accumulate in the blood and tissue fluids. Increasing amounts of glucose are filtered off in the glomeruli until the capacity of the renal tubule cells to reabsorb glucose is exceeded. Glucose then begins to appear in the urine and this is one of the first signs of the disease called diabetes mellitus which is due to a deficiency of insulin production.

The action of insulin on fat and protein metabolism. We can now consider the effect of a lack of insulin on fat and protein metabolism. Insulin is necessary for the synthesis of both fat and protein. When insulin is lacking there is an increase in the rate of breakdown of fat and protein. Amino acids from protein breakdown are no longer resynthesized into body protein but instead they undergo transamination in the liver (p. 441) to form pyruvic acid from which glucose is formed, thus helping to increase the amount of glucose in the blood of the diabetic. In fat cells there is a constant turnover of fat which is being broken down to fatty acids which are normally re-esterified to neutral fat by glycerophosphate. Glycerophosphate is formed from carbohydrate metabolism, which is markedly depressed in diabetic fat tissues. Thus the breakdown of fat continues, resynthesis is deficient, and an increased amount of fatty acids is freed from the fat depots into the blood. Some of these are used by the muscle in its metabolism. Much of the remainder is oxidized in the liver to aceto-acetyl Co A from which ketone bodies, aceto acetate and acetone, are formed. These can to some extent be metabolized by muscle but in severe diabetes they accumulate in the blood and appear in the urine. The accumulation of these acids leads to a depletion of body bases (p. 510) and leads to a condition of acidosis. The excretion of increasing amounts of glucose in the urine causes loss of water in the urine, since glucose cannot be excreted without water. The volume of blood and tissue fluid falls and severe thirst and circulatory failure develops. If untreated, the diabetic progresses into coma with severe dehydration and acidosis. The disease is treated by regular injections of suitable amounts of the pure hormone which is extracted from the pancreas of animals, e.g. cows and pigs. The hormone cannot be given by mouth because it is a protein and would be digested in the alimentary tract with loss of its biological activity.

There is an interesting condition which is just the reverse of that found in diabetes mellitus. There are certain rare tumours of the islets of Langerhans which produce abnormally large amounts of insulin. Here the blood sugar level, particularly in between meals, falls to dangerously low levels (because most of the blood sugar is being turned into glycogen or fat) and coma develops because the brain is being deprived of a sufficient source of energy in the form of glucose. A similar condition may develop in the diabetic who inadvertently takes too much insulin, or who does not eat sufficient food.

Summary. We can now envisage how by means of a varying production of hormones, the blood sugar can be maintained at a fairly constant level at all times, even during periods of relative starvation. Other hormones affect the utilization of glucose by the tissues e.g. thyroxine, but their role in the regulation of blood sugar level is negligible.

The mode of action of hormones

We have seen that hormones produce diverse effects in the animal affecting as they do the metabolism of carbohydrate, fat and protein, the transport of ions and glucose across cell membranes, the secretion of other hormones, contraction of smooth muscle etc. For many years biochemists and physiologists have, rather unsuccessfully, sought answers to the question of how hormones produce their effects on the target cells. The impact of modern studies in molecular biology on these searches has been profound and at last we are able to provide at least partial explanations for the mode of action of various hormones.

There seem to be *at least* three distinct sites in the cell at which hormones act:

1. they influence the functions of the cell membrane,
2. they influence the activity of enzymes,
3. they influence the genetic material of the cell nucleus causing changes in the character and/or amounts of messenger RNA produced on the nuclear DNA. We will briefly consider examples of these three sites of action.

Insulin and the cell membrane

We have seen that the hormone insulin has an important effect in determining the availability of glucose to many body cells (mainly muscle and fat cells). When an animal is deprived of external supplies of glucose e.g. during fasting, the glucose requirements of the central nervous system are provided by the liver which generates glucose from other substrates e.g. aminoacids and glycerol. In order that this limited supply of glucose is available for the C.N.S. many other body cells are prevented from utilizing glucose. These cells seem to have a membrane barrier to the free movement of glucose from blood and tissue fluid into the interior of the cells. It has been suggested that the transfer of glucose into these cells depends upon a special 'carrier mechanism' located in the cell membrane. The carrier molecules are thought to enter into a reversible combination with glucose molecules. The carrier is visualized as oscillating across the cell membrane and association with the carrier at the outer surface of the cell membrane and dissociation from it at the internal surface of the cell membrane seems to be the only way in which glucose can cross the cell membrane (fig. 128). The effect of insulin is to alter the properties of the carrier mechanism so as to increase the efficiency of glucose transport across the cell membrane. When supplies of glucose are scarce little insulin is secreted by the islets of Langerhans and the transport of glucose into

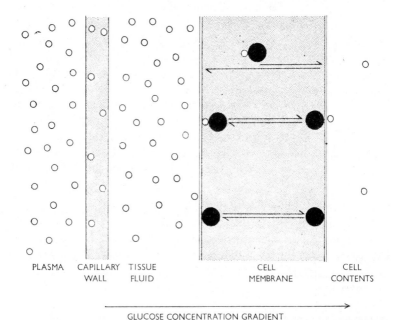

PLASMA   CAPILLARY   TISSUE   CELL   CELL
          WALL   FLUID   MEMBRANE   CONTENTS

GLUCOSE CONCENTRATION GRADIENT

Fig. 128. *Diagram showing the membranes separating glucose in plasma from the cell contents. (From Clegg and Clegg, 'Hormones, Cells and Organisms', Heinemann Educational Books Ltd.)*

muscle and fat cells proceeds only at a slow rate. This conserves the glucose for the C.N.S. If now a source of glucose is eaten then the concentration of glucose in blood gradually rises. This directly stimulates the islets of Langerhans to secrete more insulin. This insulin now increases the efficiency of glucose transport into muscle and fat cells and the extra glucose is diverted into fat in fat cells or into glycogen in muscle.

These conclusions of the mode of action of insulin on fat and muscle cells have been built up from many studies of the distribution of glucose and other sugars in tissues in the presence and absence of insulin. These studies have shown that in the absence of insulin there is very little free glucose in the interior of fat and muscle cells. The addition of insulin increases the amount of glucose that one can detect in these cells. The evidence for the existence of a carrier mechanism with the properties we have described is thus indirect and as yet no such carrier mechanism has actually been isolated from the cell membrane.

The fact that a hormone produces effects on the cell membrane manifested by changes in the distribution of ions, sugar etc., across the cell membrane is no evidence that the hormone is acting directly upon the cell membrane. Although the hormone aldosterone produces changes in the properties of cell membranes, studies with radioactive aldosterone show that the hormone becomes concentrated in the nuclei of cells and not in the cell membrane. The changes in the properties of the cell membrane produced by aldosterone would then appear to be secondary to effects on the nucleus of cells.

We do have a variety of techniques which permit a localization of the site of action of a hormone. An action on the DNA of the chromatin of the nucleus can be blocked by an antibiotic called Actinomycin D. This antibiotic binds itself to the base guanine in the helix of DNA; it sticks out of the molecule and prevents the enzyme called polymerase from progressing along a stretch of DNA in the synthesis of RNA. This action effectively blocks the synthesis of RNA in the nucleus; if a hormone produces its characteristic effects on a cell in the presence of Actinomycin D this is good evidence that the action of the hormone is not dependent upon changes in the synthesis of nuclear RNA and hence on the synthesis of new proteins.

Similarly a site of action in the cytoplasm of the cell can be excluded if the hormone produces its effects on the cell in the presence of various metabolic inhibitors or if it still operates at low temperatures which slow down metabolism and enzyme action. Another antibiotic called puromycin inhibits the growth of the polypeptide chain on the ribosome.

If a hormone acts in the presence of puromycin then its action is regarded as being independent of the synthesis of new protein in the cytoplasm.

Insulin does alter the properties of cell membranes (it increases the rate of transport of glucose into cells) and this effect is regarded as direct action of the hormone on the membrane because it can be obtained in the presence of the various metabolic inhibitors we have discussed and the hormone operates at low temperatures. Other hormones such as oxytocin, anti-diuretic hormone and noradrenaline seem to have direct effects upon the membrane of their target cells.

It must be stressed that insulin has diverse effects in the mammal e.g. it increases the rate of glucose transport into many cells, it increases the rate of incorporation of amino acids into protein and it depresses the rate at which glucose is generated in the liver (gluconeogenesis from protein). We have analysed here only one of these effects and it seems clear that these other effects of insulin are produced in quite different ways.

Adrenaline and the enzyme phosphorylase

Some hormones appear to exert their effects on enzymes located either in the cell membrane or in the interior of the cell. One clear example of such an action is the effect of the hormone adrenaline on the enzyme phosphorylase of liver cells.

The enzyme phosphorylase is responsible for the hydrolysis of glycogen stores in the liver which ultimately results in the liberation of free glucose into the circulating blood. This action is a rapidly acting emergency measure whereby the body is provided with a supply of glucose if the amount of glucose in blood falls below a critical level. This is only a temporary measure because supplies of glycogen are strictly limited. Phosphorylase activates the reaction in which glycogen reacts with inorganic phosphate, breaking the α-1,4, links between component glycosyl units in the glycogen molecule (page 92). This reaction produces glucose-1-phosphate which under the influence of another enzyme, phosphoglucomutase, is converted to glucose-6-phosphate. Glucose-6-phosphate is then dephosphorylated in the presence of an enzyme phosphatase which results in the formation of free glucose which can enter the blood stream.

The amount of phosphorylase in liver cells is in a constant state of flux depending upon the rates at which the enzyme is inactivated and reactivated. Adrenaline acts upon this system by increasing the amounts of the active form of the enzyme. It does this by activating yet another liver enzyme. But we will leave the details of the action here but stress that under the influence of adrenaline there is an increase in the amount of active phosphorylase in liver cells which can be detected by classical biochemical techniques.

Cortisone and DNA

We have already seen that an important effect of the hormone cortisone is to increase the rate of production of glucose in the liver. This action is of special importance during fasting. There are two components of this effect.

1. Cortisone promotes the breakdown of protein in the tissues, supplying amino acids which form the raw material for the synthesis of glucose by the liver.

2. Cortisone also increases the amounts of various enzymes in the liver which are involved in the complex conversion of protein to glucose.

We do not yet know how cortisone produces the first effect but there is evidence that the effect of cortisone on liver enzymes is due in part to an action of the hormone on the DNA of liver cells.

We have seen (page 136) that much of the DNA in the nucleus of a given cell is not available to act as a template for the synthesis of messenger RNA i.e. the genes are repressed. One way in which a hormone might affect the cell is to cause an uncovering of some of these 'hidden' stretches of DNA i.e. to derepress certain genes and so permit expression of the genes in terms of the synthesis of new cytoplasmic proteins. In certain giant chromosomes of some insects (Diptera) this effect has actually been visualized. Shortly after administering a hormone (ecdysone) to these insects certain section of the giant chromosomes previously inactive in the synthesis of RNA become greatly swollen and active in the synthesis of RNA. Some of this RNA has the characters of messenger RNA which leaves the nucleus to instruct the ribosomes in the synthesis of new kinds of protein (fig. 129).

In mammals we have to use more indirect techniques to determine that a hormone is influencing genes. Indirect kinds of evidence include the following.

1. An early increase in the amount of RNA in the nucleus of cells after treatment with the hormone.

2. The appearance of increased amounts of cytoplasmic RNA and new kinds of protein after treatment with the hormone.

3. The effects of the hormone can be prevented by inhibitors which
   (a) prevent DNA from acting as a template for the synthesis of RNA e.g. Actynomycin D,

(b) prevent the synthesis of new protein on the ribosomes e.g. puromycin.

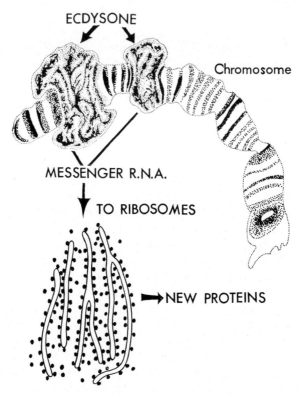

Fig. 129. *The mode of action of ecdysone.* (*From Clegg and Clegg, 'Hormones, Cells and Organisms', Heinemann Educational Books Ltd.*)

The action of cortisone on liver cells fulfils these criteria. Shortly after administering the hormone to animals one can detect increases in the amount of nuclear and cytoplasmic RNA and the appearance of new proteins (enzymes). The proteins which appear are literally 'new' and this can be shown by the simultaneous administration of cortisone and radioactive-labelled amino acids to an animal. The enzymes which appear in the liver cells are also radioactively labelled showing that they have been synthesized under the influence of cortisone. The effect of cortisone on liver cells can be readily blocked by either Actinomycin D or puromycin.

Many other hormones also seem to act in this way by acting as derepressors of genes e.g. oestrogens, testosterone, growth hormone. In one species the administration of the sex hormone oestrogen causes the derepression of genes and the production of proteins which normally are never produced throughout the life span of the individual. In the hen oestrogen stimulates the liver cells to produce proteins which will ultimately appear in the yolk of the egg. The liver of the cock bird normally does not produce these egg yolk proteins but even he will produce them after treatment with oestrogen.

We have discussed three ways in which hormones may influence cells. Other modes of action will no doubt be discovered; growth hormone, although it appears to act as an agent of gene derepression also seems to influence the process of protein synthesis at the level of ribosomal translation of the message of RNA.

Chapter Seven

Integration and Adaptation by the Nervous System

Introduction. In Chapters V and VI we have shown that the specialized cells of the mammalian body are united to form one functional whole organism and we have called this unification, integration. The nervous system plays a very important part in integration and also in adapting the organism to its environment. We are crossing a field and see an enraged bull bearing down upon us whereupon we flee into safety. In so doing we have used thousands of specialized cells in coordinated action and we can only do this by virtue of the activity of our nervous system. What is more this action has saved us from harm and perhaps death, the action was an adaptive one. The function of our nervous system is both integrative and adaptive.

From the example just quoted of the bull in the field, it is obvious that the nervous system is conducting messages around the body at times of great action. It should be understood clearly however that the nervous system is carrying messages all the time. Even when we are relaxing as completely as we know how to, there are still many messages being carried in the system, in sleep even this is still so, for our heart, under nervous control continues to beat and breathing movements continue, tension remains in many of our muscles, all of which are controlled by our nervous system. There is a ceaseless pattern of activity then in the nervous system, and as the world around us changes the pattern of activity in the nervous system also changes. The nervous system has been likened to a telephone system since the nerves seen in dissection are like wires, and this is a useful analogy, but we must remember that the wires are living wires and the activity incessant.

A plan of the system. The basic cell type of the nervous system is the neurone which has been described on page 180. These neurones are built up into nerves which may be either medullated or not (see page 181). The whole system consists of three main parts

(1) The sensory system
(2) The central system
(3) The motor system.

The sensory nervous system supplies information about the outside world (from eyes, ears and the sense organs of smell, taste, touch, temperature etc.) and from the inside world (from cells sensitive to pressure, tension, chemicals etc.).

The central nervous system consists of the brain and spinal cord and the messages are conducted towards it by the sensory system and away from it by the motor system. The central nervous system can be regarded as the place where the sensory imput is sorted out and appropriate action initiated by way of the motor system.

The motor system carries the messages from the central system to the appropriate glands or muscles where action is taken. It is sometimes referred to as the efferent system whereas the sensory system is called the afferent system. The motor system has two components the cerebrospinal nerves and the autonomic nerves. The former is sometimes called the somatic motor nervous system for it takes messages to the somatic (striated) muscle and it controls, by means of nerves of the head (cranial) and spine, activity which *may* come under the will e.g. walking. The autonomic system or visceral efferent system controls activity which is purely visceral and not under the control of the will e.g. the contraction of blood vessels, the secretory activities of glands, peristalsis.

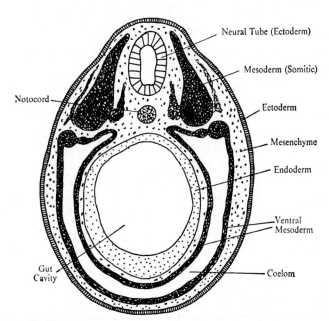

Fig. 130. *Diagram of a transverse section of an embryo showing the three basic germ layers, ectoderm, endoderm and mesoderm.*

The word visceral means associated with the internal organs of the body e.g. heart, lungs, gut, etc. These organs are largely derived from the visceral mesoderm (see fig. 130).

THE REFLEX ARC

The mammalian body like that of all the animals with backbones (and some others too like earthworms and insects) is what is called technically 'metamerically segmented'. This means that the body is built up of a series of units called segments, which are all constructed on a similar basic plan. Fig. 130 shows this plan in cross section. In an animal like an earthworm the pattern of segments is obvious, as it is in the tail and body of a fish. You can eat fried fish segment by segment very easily. Since all the segments are similar in basic plan and all of more or less the same age they are called metameric segments. There are some animals like tapeworms which have segmented bodies but the segments are very different from each other; the tapeworm is not metamerically segmented but the earthworm and the vertebrate animals are. In a mammal this basic repeating pattern of similar units is difficult to see since the animals have a very complex pattern of surface muscles which obscures the segmentation. Evidence that the mammalian body is metamerically segmented comes from the way the embryo develops. In the embryo the segmentation is obvious. (See fig. 131.) In the adult there are clear signs of the segmental pattern in the segmented nature of the backbone and in the segmental nature of the nerves which leave the spine. Many of the activities of the body are carried out within the one segment of the body. The other segments are informed what is going on but effective action is taken within the one segment. The posture of the body is mainly carried out in this way by what are called segmental reflexes. When we go to the doctor for a medical examination he wants to check that our nervous system is in good order and he does so by checking whether a selection of our reflexes are working. The best known of these tests is the knee-jerk reflex. If the legs are crossed so that one foot is off the floor and then a sharp tap is given to the upper of the two legs at a point just below the knee cap, the lower part of this limb swings upward. Let us look at this reflex action and see what has happened. The sharp tap applies tension to the tendon of the quadriceps muscle and stimulates stretch receptors within the muscle (see p. 610). The stretching of the muscle sense organs constitutes the 'stimulus'. A stimulus is any change in the environment which affects a sensitive cell. Thus sound waves in the air are stimuli to us, unless we are deaf when they cease to be stimuli.

We have no receptor cells in our bodies which are sensitive to radio waves so we remain oblivious of the fact that radio waves are present in our environment, unless we switch on a radio set which transforms them into sound waves which we can hear. The sensitive cells which are affected by environmental changes are called receptors. These cells change the energy of the stimulus into a kind of electrical effect which

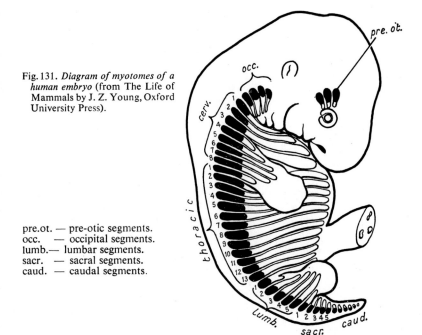

Fig. 131. *Diagram of myotomes of a human embryo* (from The Life of Mammals by J. Z. Young, Oxford University Press).

pre.ot. — pre-otic segments.
occ.   — occipital segments.
lumb.— lumbar segments.
sacr.   — sacral segments.
caud.   — caudal segments.

passes along the sensory nerve to the central nervous system. This wave of electricity in a nerve is called an 'impulse'. The nerves from the various receptors all carry a similar impulse; that is, it is not possible to distinguish an impulse caused by a smell from one caused by a sound or the tap on the knee. Now we know what an impulse is we can follow its course. Fig. 132 shows the route taken by the impulse. The sensory neurone carrying the impulse has its cell body in the dorsal root ganglion. A ganglion is simply the technical term used to describe a swelling on a nerve. The swelling on the sensory nerve is caused by the fact that all the cell bodies are at this site. The axon of the sensory neurone takes the impulse on into the more central parts of the spinal cord. There comes a point called the synapse where the sensory nerve

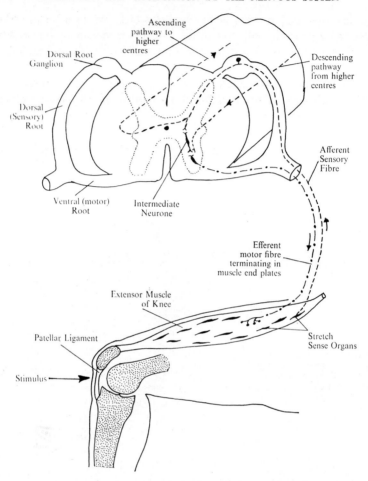

Fig. 132. *The knee jerk reflex. To be accurate this particular reflex is monosynapic i.e. there is no intermediate neurone between afferent sensory and afferent motor neurones. However, since most reflex arcs do have one or more intermediate neurones an intermediate neurone is included in this figure to show the position they occupy in reflex arcs.*

ends and the impulse is handed on to another neurone called the intermediate or connecting neurone. The dendrites of the connecting neurone are not in actual physical contact with the end of the axon of the sensory neurone but are separated by a small space called the synapse. The connecting neurone conveys the impulse to the motor or efferent neurone from which it is separated by a synapse. Note that the cell body of the motor neurone is reached by very short dendrites and then the long axon leaves the cord via the ventral root. The ventral root leads back into the spinal nerve up which the impulse passed on its way to the cord. Note however that although this is a nerve which carries both motor and sensory impulses these impulses are carried in very distinct neurones. The neurones, and there may be many, which are carrying the motor impulses from the cord are going to the skeletal (striated) muscle in the thigh and when they reach this muscle they end in what are called motor end plates. The structure of a typical end plate is shown in fig. 80. The muscle is stimulated to contract by the impulse which arrives in the axon of the motor neurone and the lower leg moves. The strength with which the leg moves depends on the number of muscle cells which are caused to contract, and this in turn depends on the number of motor neurones which are carrying the impulse. It should be mentioned at this point that the impulses are all of equal 'strength' and they cause the muscle cell where they end to contract with a certain strength. The cell either contracts or it does not contract when it gets the impulse. It is rather like pressing down the switch of the electric light in your room, the light goes on full or not at all. It is possible to get graded responses from muscle of course, we all know that from our own experience, and this is done by varying the number of muscle cells used. In a similar way if you want more light in your room you switch on more bulbs.

Even when the eyes are closed one knows that the knee has been tapped, and that the leg moves in response to this tap. This must mean that the brain is informed when impulses pass up the sensory neurones from the knee to the spine. It will be seen, if fig. 132 is carefully inspected, that a branch of the sensory nerve goes to the brain. At the synapse between the segmental connecting neurone and the motor neurone there is a return neurone from the brain, which can influence the impulses which pass to the muscles of the leg. Thus although the segmental reflexes can respond to stimuli all within the segment many other segments of the body, including the brain, are informed. Here again we see a good example of the nervous system integrating many special cells and also informing the whole organism. Reflex actions are well developed in vertebrate animals but in many of the very lowly groups of animals, e.g. sea anemones, they are almost absent. Reflexes are very useful and extremely economical for they ensure that effective action is taken without the whole organism having to go into action. Thus if you touch a hot object you reflexly pull away your hand quickly, you do not need to step back to remove your hand from the hot object. Similarly if an object passes your eye quickly you reflexly blink your eyelids, so protecting the delicate cornea. Thus reflexes are spontaneous automatic actions which are made in response to definite stimuli, the same stimulus constantly eliciting the same response, usually from only a part of the body. The action taken by this part of the body may serve to protect the body from danger i.e. it is a biologically advantageous reaction, which is inborn and does not have to be learned.

Conditioned reflexes. The experiments which the Russian scientist Ivan Pavlov (1849–1936) performed at the beginning of the twentieth century on the factors affecting the secretion of digestive juices in dogs are now classical experiments in psychology. They hold this honoured position on account of the fact that these experiments were investigating the mind of the dogs used. Pavlov and his workers noticed that the flow of saliva in a dog's mouth was a reflex activity and when food was placed in the mouth there was a flow of saliva which helped to lubricate the throat and assist swallowing. Saliva was made when it was required. The taste of meat would cause saliva to flow but the smell of the feeder's hand also caused the flow of saliva even before the meat was put into the dog's mouth. Pavlov could tell when the saliva flowed because he had put a small brass tube called a canula into the duct of the salivary gland so that the saliva dripped into a bottle strapped onto the dog's chest instead of into its mouth. It was thus an easy matter to see when saliva was caused to flow. Pavlov saw that stimuli (e.g. smell of meat), which were associated with the main stimulus (taste) of the normal saliva producing reflex, could also produce the flow of saliva. The stimulus of smell of meat had in some way become connected with the motor nerves which went to the salivary glands. This sort of thing could only happen in a system that had synapses where new connections of neurones, or at least where new routes for impulses could be made. The two stimuli, of taste and smell are said to be associated in this example. Pavlov went on to show that if the dog was allowed to smell the meat but not allowed to eat it, the response of secreting saliva to the stimulus of the smell alone became weaker and weaker until the dog did not salivate when he smelled meat. If the smell was always followed by the taste of meat the response to smell became stronger. Thus it was possible to 'teach' the dog when to respond to an associated stimulus and when not to respond. When the dog responded to the smell alone, after a period of presenting both the stimuli of smell and taste together, the saliva producing reflex was said to be 'conditioned'. In this case the salivary reflex had been conditioned to respond to the smell of food, but it could equally be conditioned to respond to other stimuli, the ringing of a bell for example. This mechanism of conditioned reflexes is useful to the animal in nature for it helps him to forecast what is going to happen and to take the appropriate action in readiness. The dog could start to salivate when he saw the keeper approaching from a distance at mealtime. This kind of conditioning gives the appearance of intelligent activity.

Using his idea of the conditioned reaction Pavlov could explain how the behaviour occurred without bringing in 'psychic' qualities. He had reduced psychology to physiology. This started a controversy which still exists; is the mind the product of various connections within our nervous system or is it something different, has it got some special psychic force and merely expresses itself through the nervous system? The interesting point for us to note here about conditioned reflexes is that they demonstrate how the simple segmental reflex can be connected to sources of information in other segments and coming into the body by different routes and perceived by different receptors. This connecting of the reflex arc with other circuits is done via connecting neurones which pass up and down the spinal cord in the white matter. Fig. 132 shows where these connections are plugged into the reflex arc. We see then in reflex action economical biologically useful activity of specialized parts of the body operating automatically and without the need for thought, but at the same time the isolated reflex is connected with the whole organism by neurones which run up and down the spinal cord, to and from the brain. Once again the nervous system is seen as an integrating and adapting mechanism.

Threshold stimulus and the receptor. When a receptor cell is stimulated by a change in the environment we say that it has responded to a stimulus, and the result of this stimulation is that an electric current is sent along the sensory nerve to the central nervous system. The change in the environment must be sufficiently marked for the receptor cell to be stimulated or else nothing happens in the nervous system; there are many sounds in the world which we do not hear for they are not loud enough to stimulate the ear. We get a big surprise when a sensitive microphone and amplifier is used to let us hear what a fly sounds like when it is about to take off, or what a caterpillar sounds like when it is chewing. The point at which the sound is just loud enough to be audible is called the threshold level of stimulation. There is a threshold level for all the receptor cells below which they do not respond and above which they do respond. So we can say that provided the stimulus is above the threshold of the receptor concerned that receptor will send an electric current to the spinal cord or the brain.

The receptor cells of our bodies are very specialized cells and are designed to receive one kind of stimulus only. Our bodies are not capable of receiving and interpreting radio waves even though the air is full of these waves, we have to have a radio set or a T.V. set to interpret these changes in the environment. There are many changes in the environment of which we are not aware either because we have no suitable receptor or because the stimulation is not above the threshold level of our receptors. Some moths can smell the female of their species from a distance of many miles whilst we cannot detect the smell of the moth at all. Dogs can hear 'silent whistles' but we cannot because the pitch of the sound they emit is beyond the range of our sensitivity.

A very interesting situation arises when the intensity of the stimulation is just about on the level of the threshold of our senses. Some people claim that bats do not make a noise whilst others can hear the high pitched squeak that they make when they fly. This squeak is used as an echo sounding device by the bat and it enables it to navigate through small openings at high speed even in intense darkness. The important point for us to note at present is that the pitch of this note is just within the hearing range of some folk and just outside the hearing range of others.

The impulse. We have seen that stimuli of many different kinds can cause various sense organs that are sensitive to that kind of stimulus e.g. eyes to light and ears to sounds, to send an electrical charge down a sensory neurone to the brain. This electrical charge is called a nerve impulse. The very surprising fact is that a nerve impulse passing from the ear as a result of stimulation by a sound, is identical to one in the optic nerve caused by a visual stimulus. All impulses are similar and it is impossible to distinguish a pain impulse from a sight impulse or a taste impulse. In fact there are no such terms known to science and I have just invented them. Remember that all impulses are identical no matter what kind of stimulus causes them to come into being. It is said that if it were possible to take a neurone from the eye and one from the ear and cut them both and then join up the wrong ends it would be possible to see sounds and hear sights—this serves to illustrate the universal nature of the impulse. The sorting out of the nature of the changes occurring in the environment is done by having a large range of sense organs each sensitive to a particular part of the environment. It is rather like having a whole series of door bell pushes on your door, one for the postman, one for the milkman, one for your friends and one for the vicar. These bell pushes are wired to lights in your room so that you can see at a glance who is at your door. The labelled bell pushes have sorted out the environment for you—similarly your sense organs sort out the environment for you.

We have said that if a touch receptor is touched very gently so that the pressure is not big enough to reach the threshold level no impulse will be sent up the sensory neurone, but when the touch is made heavier then the threshold is reached and an impulse flows along the sensory nerve. When the touch is made even heavier the impulse that flows is just the same. How are we able then to distinguish light touch from heavy touch and all the grades in between? There are two methods employed by the body; first there are special receptors for light touch and heavy touch and these receptors have different thresholds, secondly there can be an alteration in the frequency with which the impulses flow along the neurone. Let us imagine that we can stick a pin into the flesh with nine different pressures which we will denote by the numbers one to nine.

No. 1	below threshold	no response
2	below threshold	no response
3	on threshold of the light pressure receptor	impulses pass along the neurone from light pressure receptor to the brain at 2 per sec.
4	above threshold of the light pressure receptor	impulses pass along the neurone from the light pressure receptor at 10 per sec.
5	above threshold of light pressure receptor	impulses pass along the neurone from L.P.R. to brain at 20 per sec.
6	on threshold of heavy pressure receptor	impulses pass along the neurone from the heavy pressure receptor at 2 per sec.
7	above threshold of H.P.R.	impulses pass at 20 per sec.
8	pain threshold is reached	impulses pass along the neurone from the pain receptor at 2 per sec.
9	above the pain threshold	impulses pass along the neurone coming from the pain receptor at 20 per sec.

The brain can interpret the environment by having impulses from known sources and by interpreting the frequency with which these impulses arrive. In nerve impulse conduction there is a law called the all or nothing law this means that below the threshold level there are no impulses, above it there are impulses. We can measure the intensity of the stimulus not by the size of the impulse, for they are all alike, but by the frequency of the arrival of the impulses.

Charge on the membrane and the conduction of the impulse. The chemical composition of the interior of cells differs markedly from that of the interstitial fluid bathing the cell. The intracellular fluid is characterized by a much higher concentration of potassium ions and a lower concentration of sodium ions than the interstitial fluid. The anions which accompany the sodium ions of the interstitial fluid are mainly chloride ions. Inside the cell the anions probably consist mainly of proteins in anionic form. Separating these two fluids, intracellular and interstitial fluid, is the cell membrane, a double layer of lipid molecules sandwiched between two films of protein molecules. The diffusion of substances across this cell membrane is much slower than that of diffusion in water. However there is a net tendency for potassium ions to diffuse out of the cell and for sodium ions to diffuse into the cell, the ions moving along their concentration gradients. The membrane is much more permeable (x100) to potassium than sodium ions, perhaps because of the smaller size of the hydrated potassium ion. Positively charged potassium ions thus diffuse out of the cell, through the cell membrane. They are not accompanied by negatively charged anions from inside the cell because these are too large to pass through the cell membrane. As potassium ions diffuse out of the cell the interior of the cell becomes increasingly electrochemically negative to the outside, because of the loss of the positively charged particles. This electrochemical gradient created by the diffusion of potassium ions out of the cell actually tends to attract potassium ions back into the cell. Thus the potassium ions which diffuse out of the cell along their concentration gradient are held close to the cell membrane by electrochemical attraction. An equilibrium is set up between the tendency for potassium ions to leave the cell along their concentration gradient and the tendency for them to return to the cell along the electrochemical gradient. At this equilibrium there is a potential difference across the cell membrane which amounts, in muscle cells, to 90 millivolts, the interior of the cell being electrochemically negative.

Although the cell membrane is much less permeable to sodium than to potassium ions there is a net tendency for the intracellular sodium to come into equilibrium with the extracellular sodium. Sodium ions enter the cell along both a concentration gradient and along an electrochemical gradient, generated by the outward diffusion of potassium ions. The intracellular sodium concentration is, however, kept at low levels because of some active mechanism, requiring energy, called the sodium pump, which extrudes sodium ions from the cell into the interstitial fluid. The sodium pump is also linked to potassium, and as sodium is extruded from the interior of the cell, potassium ions are taken into the cell from the interstitial fluid, thus maintaining the high intracellular concentration of potassium ions.

The above features are characteristic of all cells. However, there are certain excitable cells, nerve cells, muscle and gland cells in which there are additional features. In these cells some change in the external environment of the cell of a chemical or electrical nature, can bring about an alteration in the permeability of the membrane to ions and this leads to a change in the membrane potential.

In muscles and nerves this change in the properties of the membrane is rapidly propagated to adjacent regions of the cells—the phenomenon of conductivity. This propagated alteration of the features of the membrane is called the impulse and is manifested electrically as the action potential. In the resting state nerve and muscle cells maintain a fairly steady electrical potential across the cell membrane. The absolute value of the resting potential varies from tissue to tissue. When the excitable cell is stimulated, either by chemicals or by passing an electrical current outward through the membrane there is a marked change in the membrane potential. The membrane potential rapidly falls to zero and for a short time the potential is reversed so that the inside becomes positive to the outside. Following this the membrane potential slowly returns to the resting level. If the muscle or nerve fibre is thus stimulated at one point, there is a fall in the transmembrane potential at this point and this change is propagated along the fibre as the impulse. Behind the front of the impulse the normal transmembrane potential is restored.

The fall in the membrane potential to zero value in the generation of the action potential is due to a sudden increase in the permeability of the membrane to sodium ions (500 fold increase). Sodium ions move across the cell membrane from the interstitial fluid along their diffusion gradient. They are also attracted inwards electrochemically because of the negative charge of the interior of the cell. The movement of positively charged sodium ions into the cell abolishes the electrochemical negativity of the interior of the cell and for a brief period the interior of the cell becomes electrochemically positive because of a net excess of positively charged cations. The increase in the permeability of the membrane to sodium ions is a very brief event, lasting only a millisecond or so, and the recovery is associated with a change from a sodium-permeable to a potassium-permeable membrane, the normal state of the resting cell. In the recovery process excess sodium ions are extruded from the cell by the sodium pump mechanism.

The energy for the action potential is already present in the cells before stimulation. The basis of this energy is the concentration

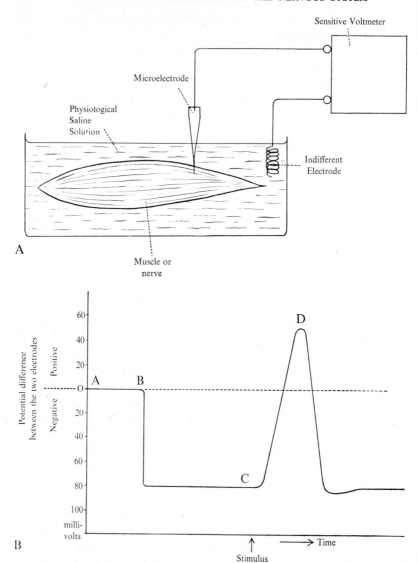

Fig 133.   A. Scheme of apparatus for recording the transmembrane potential and action
potentials of excitable tissues (muscle, nerve). The stimulating electrode is not shown.
B.   Record of the transmembrane potential and action potential.   At point A the microelectrode has not penetrated the tissue and there is no potential difference between the
two electrodes, which are bathed in the same solution. At point B the microelectrode
penetrates a cell and the microelectrode becomes 80 m. volts negative to the indifferent
electrode. At point C the tissue is stimulated by passing an electric current from the inside
to the outside of the cell. After a latent period there is a rapid fall of the potential difference
between the two electrodes to zero as sodium ions enter the muscle cell. There is also an
overshoot of zero potential to D, the inside of the cell and thus the microelectrode
become positively charged compared with the indifferent electrode. The potential difference then is restored to the original testing level.

gradient of sodium ions established by the sodium pump, and the electrochemical gradient developed by the diffusion of potassium ions from the interior of the cell. As soon as the membrane becomes permeable to sodium ions then these move along the already existing electrochemical and concentration gradients. It is the recovery process which is energy consuming, requiring the re-establishment to the electrochemical and concentration gradients.

One way of investigating the transmembrane potential and action potential of excitable cells is by the use of microelectrodes. These are made by pulling out a piece of glass tubing to a very fine point, less than one micron in diameter. These are filled with an electrolyte solution of potassium chloride. The microelectrode and another indifferent electrode are connected to a sensitive voltmeter. A piece of tissue to be studied is suspended in a bath containing a solution of mineral salts at physiological concentrations. The indifferent electrode and microelectrode are then placed into the solution and at this point there is no potential difference recorded between them. The microelectrode is then advanced towards the piece of tissue, the progress being regulated and observed by means of a micro-manipulator. The microelectrode is pushed into the tissue to penetrate a cell. As soon as the microelectrode tip penetrates the interior of the cell a potential difference between the two electrodes is recorded by the voltmeter, because the interior of the cell is electrochemically negative to the external solution (fig. 133B).

If the excitable cell is then stimulated, for example by means of an electric current, the recorded potential difference between the two electrodes falls to zero as positively charged sodium ions enter the cell. There is an overshoot, the inside of the cell becoming positively charged compared with the external solution. The recovery process begins after a millisecond or so and the resting potential difference is re-established. The fall of the transmembrane potential to zero and its recovery is recorded as the action potential (fig. 133B).

Propagation of the action potentials along a nerve or muscle fibre. When a point on a nerve or muscle fibre is stimulated to cause an alteration in the permeability of the cell membrane to sodium ions, the transmembrane potential at this point falls as sodium ions move into the intracellular fluid and the inside of the cell becomes transiently positive to the outside of the cell (A, fig. 134). A flow of current occurs from adjacent parts of the cell membrane to the active region, the current flowing through the interstitial fluid to the active region, through the cell membrane into the intracellular fluid and then out through the cell

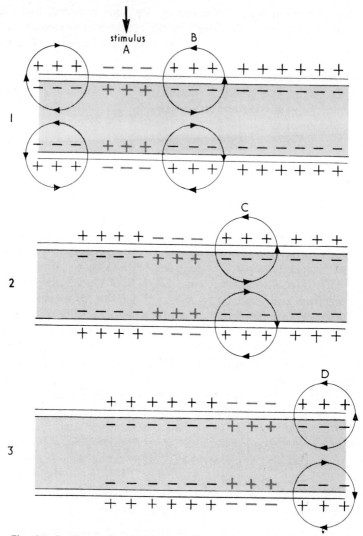

Fig. 134. *Conduction of impulse along a new fibre.*

1. The nerve is stimulated at point A with reversal of transmembrane potential, the interior becoming positive to exterior. Local current now flows (from + to −) in from adjacent regions, the current flowing into the fibre in the active zone and out of the fibre in adjacent inactive regions (B).

2. The current flowing out of B has now increased the sodium permeability with reversal of the membrane potential. Thus the active zone has moved up the fibre. The active zone of (1) has now restored the resting transmembrane potential. Current is now flowing out of the fibre in the adjacent inactive region C. (For simplicity the impulse is shown travelling in one direction only).

3. Region C has now become active and current is flowing out of the fibre in region D which will be the next to become active.

membrane (B, fig. 134). This outflowing current through the inactive region reduces the membrane charge in this region and when this falls to a critical value, known as threshold, the permeability of the membrane to sodium in this region rapidly increases, with the associated disappearance of the transmembrane potential (C, fig. 134). Thus the action potential induced at a particular point on the cell membrane induces activity in adjacent areas of the cell membrane because of local flow of current, and the action potential is propagated along the length of the elongated nerve or muscle cell. Following in the wake of the action potential the membrane permeability to sodium ions rapidly falls and the resting transmembrane potential is re-established. Until the transmembrane potential has been restored to the resting level the nerve fibre in this region will not respond to a stimulus, the so called refractory period. During the development of the action potential at a point on the fibre this region is completely refractory to a second stimulus. When the fibre is repolarising, only stimuli of greater strength than the original stimulus will induce a second action potential, and only when the transmembrane potential is fully restored does the sensitivity of the fibre to stimulation return to normal. If the neurone is medullated then the impulse travels more quickly because the fibre is insulated, except at the nodes of Ranvier. The depolarization spreads very rapidly from one node to another. Thus in man the medullated nerves of large diameter may conduct at more than a hundred metres per second whereas in the finer non-medullated fibres conduction may be as slow as one metre per second.

In the evolution of the nervous system one of the significant advances has been the increase in the length of the neurones and the consequently fewer number of connections needed. The point at which two neurones connect with one another is called a synapse. In the lower groups of animals, e.g. the sea anemones and jellyfish, the nerve cells are very short and so many synapses are needed that the nervous system is spoken of as a nerve net. Conduction is very slow ($0 \cdot 1$ metres per sec.).

The synapse. In mammals the relations of neurones with one another may be very complex. Each neurone in the central nervous system may be related to many other neurones by way of synaptic connections. Fig. 135 shows a neurone with some synaptic connections with other neurones. The terminals of the branching dendrites of the neurones making their synaptic connection with the neurone A, end in small swollen processes which abut directly onto the surface of neurone A. There is no direct continuity of the protoplasm of the synaptic endings with the protoplasm of A, and they are separated by the membranes of the two neurones. This gap in continuity acts as a block to the nervous impulse which is passing along a dendrite to neurone A. The nervous impulse causes a local electrotonic depolarization of the neurone and if this is of sufficient intensity then it may result in the propagation of a wave of depolarization down the neurone. Because each neurone is connected to several other neurones by way of synaptic connections it is easy to envisage how the individual effects of the various dendrites impinging upon a neurone can have additive effect and so overcome the threshold of the neurone when a single impulse would have no effect. This effect is called summation and is an important property of the activity of the central nervous system. The neurone, because of its synaptic connections, can thus be linked by means of nervous impulses, with activities in many segments of the body, thus enabling information from one part of the body to influence the activities of another part of the body. And it is because of widespread synaptic connections that conditioning of reflexes is possible.

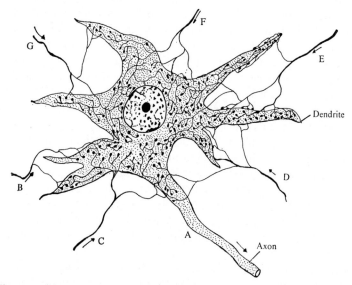

Fig. 135. *Diagram of a neurone* (A) *showing synaptic connections with other neurones* (B-G), indicating the complexity of the relationships between the neurones.

The physical break in continuity between neurones which occurs at the synapse is believed to be bridged by means of a chemical substance, liberated at the terminals of the dendrites, called acetyl choline (see also p. 608). This substance, liberated by the synaptic endings, is thought to cause local changes in the permeability of the membrane of the cell body and so initiate those electrical changes responsible for the initiation of the impulse. The acetyl choline is prevented from having long lasting effects by the presence of an enzyme cholinesterase which rapidly destroys the acetyl choline. Not all impulses which impinge at the synaptic connections of a neurone are stimulatory, some are inhibitory. The way in which some impulses can cause an inhibition is not understood.

We can think of the synapses as a complex and numerically vast series of points where information in the form of nerve impulses, some stimulatory, some inhibitory, is sorted out, the net effect depending upon the number, frequency and nature of the impulses impinging upon the synapse.

THE CENTRAL NERVOUS SYSTEM

The Spinal Cord

In the vertebrate animals the spinal cord would be better called the spinal tube, or the neural tube as it is a hollow structure with a central canal filled with fluid. By contrast of the invertebrate animals e.g. insects and worms, have a solid nerve cord which is ventral in position whereas the neural tube in vertebrates is dorsal and protected by the backbone. The difference between the structure of the nerve cord in animals is due to their different mode of origin during the development of the embryo. In invertebrate animals certain cells are set aside for the construction of the nerve cord which come to lie inside the animal along the ventral midline and give rise to a solid structure. In vertebrate animals however the story is very different for the nerve tube is derived from the ectoderm of the embryo. Two ridges arise in the mid-dorsal line of the embryo and the area between them is depressed. As the area between the ridges sinks inwards the two ridges rise as folds and meet and fuse in the mid-dorsal line, enclosing a hollow tube of ectoderm from which the brain and spinal cord develop.

When the spinal cord is exposed in dissection by removal of the neural arches of the vertebrae, the cord is seen as a long white structure tapering in the lumbar region where it terminates. It is surrounded by three layers of connective tissue which are penetrated by the nerve roots. The outer membrane is a capacious sheath of tough connective tissue, the dura mater. As the spinal nerve roots pierce the dura mater they receive a sheath of membrane which surrounds them as they pass out into the intervertebral space. The dura does not lie completely free in the vertebral canal and is loosely attached to the inner surface of the vertebrae, but in the skull the dura mater is closely attached to the inner surface of the skull. On the inner side of the dura mater there is a thin vascular membrane, the arachnoid mater which is separated by a space, the sub-arachnoid space, from the third sheath of membrane, the pia mater, which forms a fine vascular membrane closely adherent to the surface of the brain and spinal cord. The sub-arachnoid space contains the cerebro-spinal fluid which circulates around the surface of the brain and cord. This fluid provides a shock-absorber and a supporting medium for the delicate tissues of the brain and cord. It is formed within the cavities of the brain from special collections of vascular tissue, the choroid plexuses, which project inwards into the ventricles from the surface of the brain. The fluid formed inside the ventricles passes to the surface of the brain through an aqueduct in the roof of the fourth ventricle. The cerebro-spinal fluid is being continually drained from the sub-arachnoid space as it is absorbed through special thickenings of the arachnoid mater on the surface of the brain.

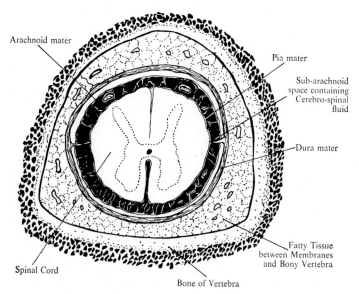

Fig. 136. *Diagram of a transverse section through the spinal cord.* The sub-arachnoid space containing cerebro-spinal fluid is indicated in black.

In transverse section (see fig. 136) the spinal cord is seen to consist of two different kinds of material, a thick outer coat of white matter and a central core of grey matter surrounding the fine central canal. The white matter consists of the medullated axons of neurones which pass up and down the cord, to and from the brain. Some of these are sensory fibres carrying impulses responsible for the sensations of touch, temperature, pressure, pain, tension etc. from the various segments of the body to the brain. Other fibres are motor fibres from the higher centres in the brain, carrying impulses which will stimulate neurones lying in the grey matter of the spinal cord, from which impulses will pass to the various motor organs of the body. The grey matter is H-shaped and consists of a dorsal and ventral horn on each side. If these horns are traced outwards they are seen to be connected to the dorsal and ventral roots of the spinal nerves. The grey matter divides the white matter up into three columns on each side, the dorsal, lateral and ventral columns. Within the grey matter there is functional separation of neurones, the anterior (ventral) horn contains the neurones concerned with the somatic motor system and the visceral motor system whilst the posterior (dorsal) horn contains the cells of the visceral and somatic sensory nerves. This arrangement is what one would expect considering the routes taken by the neurones and the position of the synapses in the cord.

The brain

Early in development the swollen end of the neural tube shows a division into three parts, a fore-brain, mid-brain and hind-brain (see fig. 137). In the more primitive vertebrates, e.g. fishes, these three divisions can clearly be seen in the adult brain, where the three divisions

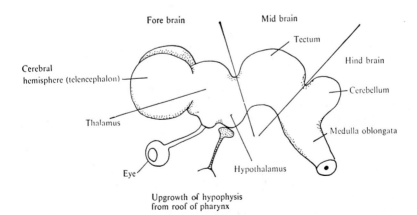

Fig. 137. *Diagram of the appearance of the brain early in development* showing the fore-, mid-, and hind-brain. The thalamus and mid-brain are largely obscured later by the tremendous overgrowth of the cerebral hemispheres.

of the brain correspond to the major groups of sense organs. The olfactory nerves pass to the fore-brain and in fishes this part of the brain is concerned mainly with the sense of smell. And in those fishes in which the sense of smell is very important in the detection of food, e.g. the dogfish, the fore-brain is found to be expanded. The optic nerves have central connections with the mid-brain and fishes like the trout which hunt by sight are found to have swollen optic lobes of the mid-brain. In fishes the roof of the mid-brain, called the tectum, is also concerned with the coordination of a variety of impulses from many parts of the body, and is for example concerned with balance. The hind-brain receives impulses from the sense organs of balance, taste and hearing (from the lateral line system of sense organs).

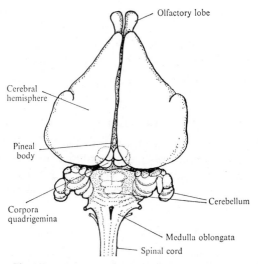

Fig. 138. *Diagram of dorsal view of brain of rabbit.*

In higher vertebrates, e.g. birds and mammals, this fundamental tripartite plan tends to be somewhat obscured by an extensive overgrowth of certain parts of the brain; the roof of the fore-brain is, for example, enormously expanded in mammals and in man it covers almost the entire brain. The roof of the hind-brain is also expanded in mammals and birds to form the cerebellum. The cerebrum and cerebellum are association areas where sensory input from a variety of sense organs can be correlated and appropriate action initiated by way of the motor system. The tectum of fishes is an association area but it is small and its functions relatively limited when compared with the large association areas of the higher vertebrates.

The fore-brain (figs. 137–141). The fore-brain is developed from the most anterior of the swellings of the hollow neural tube. There are two hollow lateral pouches of this swelling which develop into the cerebral hemispheres, the telencephalon. The cerebral hemispheres communicate with the unpaired portion of the fore-brain, the diencephalon or thalamencephalon, by way of their hollow cavities (ventricles) which drain into the venticle of the diencephalon.

The two cerebral hemispheres are connected with one another by means of transversely running commissures of which the corpus callosum is the largest, a thick sheet of fibres running between the hemispheres.

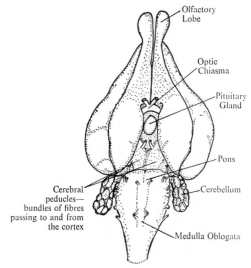

Fig. 139. *Diagram of ventral view of brain of rabbit.*

THE TELENCEPHALON

The telencephalon consists of the cerebral hemispheres which are incompletely divided by a longitudinal fissure in the floor of which runs the corpus callosum, the transversely running fibres of which connect one hemisphere with the other. The high development of the cerebral cortex is characteristic of mammals and especially of the order of mammals called primates, to which the monkeys, apes and man belong.

The cerebral hemispheres are covered by a layer of grey matter enclosing the central white matter. The cortex shows a complex pattern of folds which greatly increase the surface area of the cortex. These cortical foldings are called convolutions. The grey matter consists of a vast number of nerve cells arranged in several layers.

As described above the fore-brain in lower vertebrates is mainly concerned with the sense of smell but as one ascends the evolutionary scale there is an increasing number of other sense organs which send their messages to the fore-brain. In mammals the original 'smell brain' is obscured by the large number of connections of the cerebral cortex with all the sense organs of the body. Passing through the dorsal wall of the diencephalon (the thalamus) to the cerebral hemispheres are fibres carrying information from the lower centres of the brain and from all the sense organs of the body.

This information, in the form of nerve impulses, passes to various parts of the cortex where it is synthesized with other sources of information, including that which arises from the memory of previous experience. Appropriate action may be initiated by way of fibres which pass from the cortex to the lower centres of the brain and to the spinal cord.

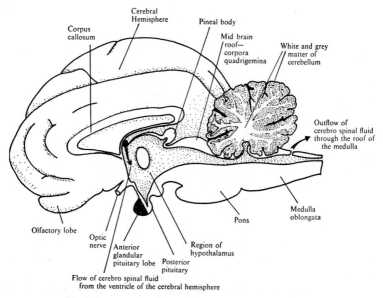

Fig. 140. *Diagram of longitudinal section of brain of sheep.*

LOCALIZATION OF FUNCTION IN THE CEREBRAL CORTEX. In man, and other animals, many of the functions of the cortex have been localized to certain areas, shown in fig. 141. These have been discovered in various ways, e.g. by noting the behaviour of animals and men with disease of certain areas of the cortex or in which certain areas of the cortex have been removed by operation. The brain itself is insensitive to pain and some brain operations may be performed on a conscious patient. In these operations electrical stimulation of the brain may also give rise to useful information. By stimulating small parts of the cortex concerned with movement, the movement of individual parts of the body may be noted. Stimulation of other areas may be followed by seeing

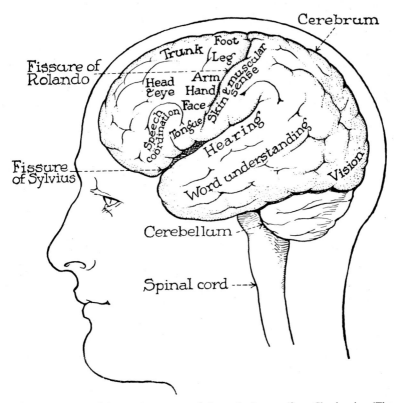

Fig. 141. *Diagram of the association areas of the cerebral cortex* (from Clendenning, 'The human body', Heinemann).

remembered sights, or hearing noises or music. Although the functions of the cortex appear to be fairly localized each part of the cortex is connected to other parts by means of association fibres. It is interesting to note that the amount of area of cortex devoted to each function gives an indication of the amount of information passing to and from that area. If we were to draw a picture of a human being as reflected in the sensory functions of the cortex we obtain a picture shown in fig. 142 in which the sensory areas of the fingers and lips appear of great importance.

THE MOTOR PATHWAYS FROM THE CEREBRAL CORTEX. From the motor areas of the cortex two sets of motor fibres pass inwards, the fibres of the pyramidal system and those of the extra-pyramidal system. The pyramidal fibres form a pair of thick bundles of fibres called the cerebral peduncles which pass in the floor of the mid-brain to the spinal cord where they terminate to synapse with the cells of the somatic motor fibres in the anterior (ventral) horn of the spinal cord.

Fig. 142.   *Drawing of a man, the size of parts indicating the importance of the part as reflected in its sensory representation on the cerebral cortex.*

The extrapyramidal fibres pass inwards from the cortex to collections of grey matter deeply placed within the hemisphere called the basal ganglia. These basal ganglia constitute the corpus striatum, which will be further discussed below.

These two motor systems, the pyramidal and extrapyramidal systems tend to have opposing actions, and the interplay between the two permits normal coordinated movements and normal muscle tone. If one of this pair of opposing systems is damaged the function of the other will be seen exaggerated. Thus after a haemorrhage in the basal ganglia, a form of 'stroke', the relatively unopposed action of the pyramidal system is seen; the limbs stiffen as the tone of the muscles increases and the reflexes become exaggerated, because the normal antagonistic and inhibitory effects of the extrapyramidal system is not functioning properly. In cases of damage to the pyramidal system the muscles become flaccid as their tone is lost, due to the unopposed inhibitory effect of the extrapyramidal system. In the anterior horn of the spinal cord a great variety of influences act upon the somatic motor cells whose fibres innervate the voluntary muscles of the body, and the net effect of these influences depends upon the strength of the various factors which include impulses carried by pyramidal fibres, extrapyramidal fibres and the fibres relaying local segmental reflexes.

INTELLIGENCE. Intelligent behaviour is a striking feature of mammalian behaviour, particularly of the primates. In intelligent behaviour there is no fixed sequence of events but actions are performed appropriate to the task in hand, and if a usual pattern of events in the environment changes then the intelligent animal should be able to adjust its behaviour accordingly.

There is no fixed area of the cortex in which intelligence is localized but intelligence depends upon the associative activity of the entire cortex, the richer the association paths the greater the intelligence. The anterior poles of the cerebral hemispheres, the frontal lobes, have been called the 'silent areas'; electrical stimulation of these areas produces little or no change in the animal and it is considered that these are areas of rich association, contributing their part to intelligent behaviour.

ELECTRICAL CHANGES IN THE CEREBRAL CORTEX. We do not yet know how the cerebral cortex allows us to think, to see the relationship between objects, or to store up information. In the investigation of these problems much work has been done on the measurement of electrical changes in the cortex in a variety of situations. Measurements of the electrical discharges from the brain in man show that there is an incessant rhythmic activity. These rhythms are electrical brain waves and are of several kinds. When we see a uniform visual field our brain waves are of a variety called the alpha type. When the visual field changes to a patterned type there is a change in the electrical activity to a beta type. When we sleep a different set of brain waves is recorded. These currents in the brain are apparently not affected by cutting the tracts to and from the cerebrum and they are thought to be a product of the activity of the cortex itself. Interesting models have been made using valves and electronic equipment which have been taught to remember certain things e.g. model toys which can learn to avoid obstacles. We are only at the very beginning of our search for useful scientific knowledge about the functioning of the brain.

CORPUS STRIATUM AND INSTINCT. The base of the telencephalon is also thickened in the mammal to form two very important swellings, one on each side, called the corpora striata. The corpus striatum is so called because it appears to be striped grey and white when it is sectioned. These striated bodies are important as they are the seat of the instinctive

actions of the animal. Birds are par excellence the animals which behave instinctively and they are also the animals with the largest corpora
striata. At first when the brain of a bird and a mammal are seen side
by side they present a superficially similar appearance for both of
them have the large swellings in the fore-brain area. If the roof of the
fore-brain is removed however in the case of the bird it is seen to be
very thin whilst in the mammal it is thick. Under the thin roof of the
cerebrum in a bird is the very big thickening of the floor—the corpus
striatum, whilst in the mammal although the floor is thickened the
striatum is not as big as it is in the bird. Thus in both these very highly
advanced vertebrates there is a great development of the association
centres in the telencephalon, in the bird the emphasis is on the corpus
striatum and instinctive behaviour whilst in mammals the stress is on
the cerebral cortex and intelligent behaviour. It is necessary to be quite
clear at this stage what we mean by the terms instinct and intelligence.
Instinctive behaviour can be very complex behaviour as for instance
nest building and territory defence in birds. Birds seldom get to fighting
each other, for when a cock bird enters the territory of another cock
the latter makes a display of aggressiveness and the former eventually
flees. This is a very sensible thing to do for it prevents birds of the
same species fighting each other and also more or less guarantees that
each bird will have a big enough territory and therefore enough food
to feed the young when they are hatched. Now all this complex behaviour is not learned but is as it were built into the bird. He does not
need to have seen any other bird or ever taken up a defensive posture
and yet he can take up the exactly appropriate posture when required
to do so. When he is threatened and he is in someone's territory he
knows it is his turn to run away. The instinctive action is something
that is of great biological importance to the bird, in this case it is concerned with the rearing of the young. The instinctive act is triggered
off by what is called a sign stimulus, that is by some significant pattern
in the environment. For instance a robin in the breeding season will
fiercely attack an egg-shaped piece of wood on which is painted a red
area if it is placed in his territory but he would take little notice of a
stuffed robin provided that the red breast had been painted brown.
Obviously the thing that causes a fighting reaction from a robin is a
red area which does not run away when it is displayed at. The sign
stimulus is red-circle-on-brown in this case. Thus we may think of an
instinctive action as a complex action which is inborn in the animal
and triggered off by a sign stimulus. It usually fulfils some important
biological purpose for the animal and the action is usually accompanied
by a fairly large amount of energy or 'drive'. The action cannot be

modified to any great extent, for instance in the case of this robin it is obviously stupid to go on displaying at what to an intelligent animal is a piece of wood with a red circle on it, but the robin will go on and on and on for the bird cannot easily modify its instinctive behaviour. In the natural world it would be a very sensible action on the bird's part, it is only because some human has come along and artificially confused the issue by placing artificial sign stimuli about that the action seems stupid. Man has not many very obvious instinctive re- actions, or if he has it is difficult to pick them out because he is always likely to apply his intelligence and inventiveness to any situation and modify the instinctive action. The rounded chubby features of a human baby's face or any other object which has a similar outline, especially if it has a large head and a relatively small body, will act as a sign stimulus to call out the pattern of events that we call maternal behaviour in a woman. This probably accounts for the popularity of the teddy bear and explains why human dolls are not shaped like snakes. It is obviously biologically useful if the form of the human infant elicits motherly love from the mother for then she will hold the baby to her breasts and allow it to suckle, an action which incidentally it can do without having to be taught, for it is instinctive. Small baby kittens push their noses into the mother's fur before their eyes open and when they feel the nipple on their lips they suck. They can do this a few hours after being born, what is more they push the mammary gland around the nipple with their paws and this helps to release the milk from the nipple. These activities associated with feeding of the new born are far too important to have to be learned for the baby needs feeding at once and time for learning can not be afforded, so these feeding reactions are instinctive.

THE CORPUS STRIATUM IN MAN. Little is known for certain about the function of the corpus striatum in man, but we do know that it is im- portant in the control of muscle tone and in steadying muscular move- ment. Patients who suffer damage to the corpus striatum sometimes have paralysis agitans or chorea.

THE THALAMENCEPHALON OR DIENCEPHALON

The thalamencephalon is the more posterior part of the fore-brain and connects the cerebral hemispheres with the mid-brain. The roof of the thalamencephalon consists of a thin plate of non-nervous vascular tissue from which projects the pineal body or epiphysis. In lower ver- tebrates e.g. lamprey and some reptiles the pineal body has the structure of a third eye but in mammals its function is unknown although it has been suggested that it has an endocrine function (see page 216).

THE THALAMUS. In mammals there are large collections of grey matter in the lateral walls of the thalamencephalon constituting the thalamus. Through the thalamus large amounts of information pass from the lower centres of the brain to the cerebral cortex. But the thalamus is not merely a relaying station for the cortex and fibres pass from the cortex to the thalamus where some form of primitive integration is carried out. This part of the brain is often referred to as the

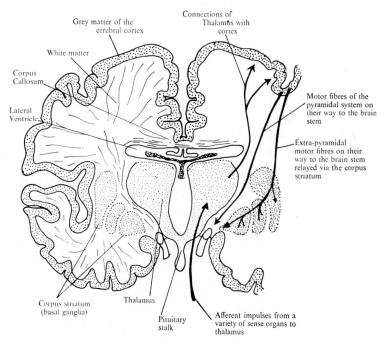

Fig. 143. *Transverse section through human fore-brain showing cerebral hemispheres, corpu striatum and thalamus.* The afferent (sensory) pathways to the cortex via the thalamus, and the two motor pathways (pyramidal and extra-pyramidal) from the cortex are indicated.

pain-pleasure-brain, since pain, extremes of temperature and rough contacts with the environment are appreciated in the thalamus. The thalamus gives us a crude sort of consciousness whereas the cerebral cortex gives a more sensitive impression. The cerebral cortex can inhibit the activity of the thalamus, and if this inhibitory influence is removed the animal then reacts in a manner suggesting displeasure, even to stimuli which are only mildly irritating to an animal with an intact cortex. A cat with no cortex is often found snarling with claws extended.

THE HYPOTHALAMUS. The thickened floor of the thalamencephalon forms the hypothalamus. A projection from the floor of the hypothalamus, the infundibulum, is continuous with the posterior nervous lobe of the pituitary gland and is an important channel for the co-ordination of nervous and endocrine activity.

The hypothalamus is the head of the autonomic nervous system; visceral afferent fibres reach the hypothalamus and after coordination with each other, efferent fibres carry impulses down the brain stem to lower subsidiary centres and eventually down the spinal cord and out along the spinal nerves. There seem to be two anatomically discrete centres in the hypothalamus one controlling the activity of the sympathetic nervous system, and the other controlling the parasympathetic nervous system.

In addition there are regions of the hypothalamus which initiate and control quite complex patterns of behaviour, e.g. sleep, feeding, aggression. The hypothalamus also contains certain sense organs, called osmoreceptors, which are sensitive to the sodium ion concentration of the blood. The axons of these cells pass down the infundibular stalk to the posterior pituitary gland, and granules of neurosecretory material migrate down the axons to the posterior pituitary where they are liberated (see p. 246). In addition to this effect of a rise in the salt concentration of the blood the hypothalamus initiates somatic motor activity in the form of searching for water. Correlated with this sensitivity of certain cells of the hypothalamus to changes in the blood chemistry, the hypothalamus has a rich blood supply, probably the richest of the entire nervous system.

The hypothalamus also exerts a powerful controlling influence on the activities of the anterior pituitary gland.

The mid-brain or mesencephalon. In the lower grades of vertebrate organization the roof of the mid-brain is thickened to form the tectum, of which there are two enlargements forming the optic lobes. The tectum is the 'heart' of the fish brain and to it are relayed impulses from all the receptor organs of the body; the information is integrated in the tectum and motor activity initiated. In the mammal there are other association centres which over-ride the tectum. The tectum in the mammal is divided into four lobes, the corpora quadrigemina. The anterior pair of lobes correspond with the optic lobes of the fish brain and some visual reflexes are located in them. The posterior pair of lobes has developed in association with the cochlea and some auditory reflexes are located here.

The floor of the mid-brain houses the pyramidal fibres from the motor area of the cerebral cortex which are passing down towards the spinal cord. Two thick bundles of fibres run on either side of the floor of the brain stem. These thick bundles are called the cerebral peduncles (or crura cerebri) and they come to occupy the whole of the floor of the mid-brain.

In summary we can think of the mid-brain as being concerned with some visual and auditory reflexes and being an important link between the cortex and the rest of the brain.

The hind-brain or metencephalon. The hind-brain consists of two important parts, the cerebellum and the medulla oblongata.

CEREBELLUM. The cerebellum is a large and complex association centre which dominates the appearance of the hind-brain in mammals. It receives large numbers of afferent fibres from the spinal cord and other parts of the nervous system, including the cerebral cortex, and from it efferent fibres pass to different parts of the brain stem. Many of the afferent fibres reaching the cerebellum are sensory, from a variety of sense organs, and a large number of these are from proprioreceptor organs. In spite of a large sensory input to the cerebellum none of these sensations reach consciousness and the activity of the cerebellum is unconscious. The sensory data supplied to the cerebellum is integrated there and used in the control of voluntary movement initiated by the cerebral cortex. The cerebellum does not initiate movement, rather it coordinates movement, ensuring that each muscle engaged in a particular movement contracts at the right time and to the right degree. Some parts of the cerebellum are specially concerned in the maintenance of equilibrium. Damage to the cerebellum in man may produce an unsteady gait, and jerky uncoordinated movements, and there may be also changes in muscle tone.

THE MEDULLA OBLONGATA. The medulla oblongata is formed from the posterior end of the hind-brain swelling in the embryo. It is not sharply demarcated from the spinal cord into which it merges. The roof of the medulla is non-nervous and consists of a vascular choroid plexus under which lies the ventricle of the medulla, the fourth ventricle of the brain. In the roof of the fourth ventricle there are three pores, a central one and two lateral ones through which cerebro-spinal fluid passes from the ventricles of the brain into the subarachnoid space.

The medulla contains a number of nuclei (collections of nerve cells) of the more posterior cranial nerves, which supply the head and body (e.g. the vagus nerve). In addition there are large numbers of fibres passing through the medulla from the spinal cord on their way to higher centres of the brain, and from higher centres (e.g. pyramidal fibres from the cerebral cortex) on their way to the spinal cord.

The medulla is the centre of control for certain vital functions including the control of respiratory movements (for the activities of the medullary respiratory centre see p. 477) and the control of the tone of blood vessels. The vagus nerve, the nucleus of which is located in the medulla, is concerned with reflexes of the heart and gastro-intestinal tract. An animal can live even when the brain is severed from its connections with the spinal cord at the level of the mid-brain. But if the medulla is damaged, even with the rest of the brain intact, the animal dies, either because of a cessation of breathing movements or a loss of tone in the blood vessels.

THE AUTONOMIC NERVOUS SYSTEM

The general functions of the autonomic nervous system. The autonomic nervous system is a part of the nervous system which is concerned with the regulation of visceral functions, glands and smooth muscles, contrasting with the somatic nervous system concerned as it is with the functions of voluntary muscle, and with receiving information from the external environment. The somatic nervous system integrates and adapts the animal to the external environment whilst the autonomic system is concerned with the maintenance and modifications of the internal environment. This division of function is not as clear cut as it might seem for both systems may be integrated and act at the same time. Thus whilst the somatic nervous system may be receiving information from the external environment, by way of the sense organs, to which the animal responds by rapid movement (e.g. hunting prey) initiated through the somatic motor system, the autonomic nervous system, by means of modifications of breathing, heart beat, the tone of smooth muscle in walls of arterioles (see also p. 231), is initiating those changes of the internal environment which make the response of the somatic motor system possible. Because these functions are involuntary and not normally under conscious control the autonomic nervous system is sometimes called the involuntary nervous system. Because of the connections of the autonomic with the somatic motor system at many points this is not always strictly true. Thus in micturition there is an element of conscious control, although the bladder muscle and internal bladder sphincter are under the control of the autonomic system.

The anatomy of the autonomic nervous system. The centres which integrate and control the activities of the autonomic nervous system lie within the central nervous system where there are many connections between the autonomic (visceral) and somatic functions. Centres of integration of autonomic activity occur at various levels in the central nervous system and include the cerebral cortex (although the autonomic functions of this area are not well understood), the hypothalamus, medulla oblongata and spinal cord. The higher centres of integration are connected to the cell bodies of neurones lying within the brain and the lateral grey column of the spinal cord. Axons from these neurones pass out of the central nervous system to terminate in ganglia (containing collections of nerve cells) which lie at varying distances from the central nervous system. Some of these ganglia lie close to the spinal

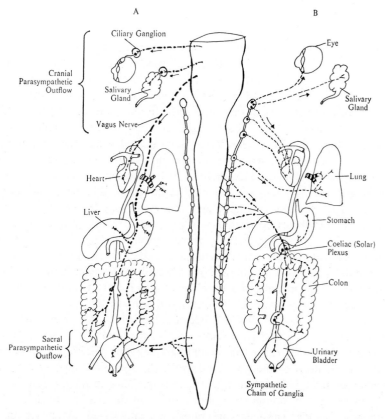

Fig. 144. *Diagram illustrating the general arrangement of the autonomic nervous system.*
A. *The Parasympathetic division* composed of fibres leaving the C.N.S. in the region of the brain stem and sacral region of the spinal cord.

B. *The Sympathetic nervous system* composed of fibres leaving the cord in the thoraco-lumbar region and relayed by way of the sympathetic chain of ganglia directly to the organs innervated or to more peripherally situated ganglia.

cord, others are actually situated in the organs innervated. The pre-ganglionic fibres arising within the central nervous system synapse with post-ganglionic neurones, the cell bodies of which lie in the ganglia. The post-ganglionic fibres then leave the ganglia and pass to the organs innervated. The length of the post-ganglionic fibre depends on the site of the ganglion, it is long if the ganglion is situated close to the central nervous system, short if it is situated in the wall of the organ innervated. Thus whilst there is only one neurone between the central nervous system and the effector organ in the somatic motor system, in the autonomic system there are two neurones which synapse with one another in a special structure called a ganglion. The pre-ganglionic fibre is a medullated fibre, the post-ganglionic fibre non-medullated.

So far as we have described the autonomic nervous system it consists of higher controlling centres, with subsidiary centres connected by motor pre-ganglionic fibres to a series of autonomic ganglia from which post-ganglionic fibres pass to the various visceral structures. This is the classical description of the autonomic system and it will be seen that it is entirely motor in function, i.e. a visceral efferent system. The classical description does not include a visceral afferent or sensory system. This afferent system must obviously exist to supply the higher autonomic centres with information on the basis of which modifications in the pattern of visceral activity can occur. The rectum and bladder contract only when full; this would be impossible without a system of visceral afferents from these organs. These visceral afferents are not anatomically separated from the somatic afferents and pass into the spinal cord in the dorsal roots of the spinal nerves, from which they are projected upwards in the cord to the higher autonomic centres.

The sympathetic and the parasympathetic divisions of the autonomic nervous system

The autonomic nervous system is divided into two major divisions, the sympathetic and the parasympathetic systems. These are anatomically distinct and each system opposes the other in its functions.

The sympathetic nervous system. The sympathetic nerves consist of those autonomic nerve fibres which leave the spinal cord in the thoracic and lumbar regions. These groups of medullated pre-ganglionic fibres leave the thoraco-lumbar region of the cord by way of the ventral root of the spinal nerve (fig. 145), and leave the ventral root in a bundle of fibres called the white ramus communicans. The white ramus leads to a ganglion where the pre-ganglionic fibre synapses with the post-ganglionic fibre. In the thoracic and lumbar regions of the cord these ganglia form a chain on either side of the anterior (ventral) surface of the vertebrae, each ganglion being connected to the one in front and behind it. The non-medullated post-ganglionic fibres leave the ganglion and rejoin the ventral root of the spinal cord by way of the grey ramus communicans. The post-ganglionic fibres are then distributed to the effector organs by way of the spinal nerves.

However, not all the pre-ganglionic fibres have their synapse in the para-vertebral sympathetic ganglia, and some pass through the ganglia on their way to more peripherally situated ganglia.

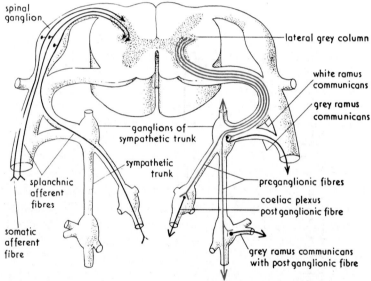

Fig. 145. *Diagram showing the course of sympathetic afferent fibres* (black) *and sympathetic efferent fibres* (red). The cell bodies of the efferent fibres are situated in the central grey matter of the spinal cord where they are activated by means of impulses passing down the cord from higher centres. Four fibres are shown leaving the cord. They pass out in the ventral root of the spinal nerve and pass to the sympathetic ganglia by way of the white ramus communicans. Some fibres synapse here and a post-ganglionic fibre then re-enters the ventral root of the spinal cord by way of the grey ramus. Other fibres pass through the ganglion without synapse and pass to higher or lower ganglia in the para-vertebral chain before they synapse or leave the spinal cord to more peripherally situated ganglia, e.g. the coeliac plexus of ganglia.

From the upper end of the thoracic sympathetic chain pass pre-ganglionic fibres which terminate in ganglia in the neck, the superior, middle and inferior cervical ganglia, which supply post-ganglionic fibres to the structures of the head and the heart. The cervical ganglia have no direct connection with the spinal cord.

From the lower parts of thoracic chain of paravertebral ganglia pass pre-ganglionic fibres which go to the abdomen where a complex network of ganglia is situated around the major branches of the aorta in the abdomen, forming the coeliac plexus, situated around the coeliac artery, with subsidiary ganglia around the other arteries of the abdomen. In these ganglia the pre-ganglionic fibres terminate and synapse with the post-ganglionic fibres which leave to innervate the gut and other organs.

The parasympathetic nervous system. This division of the autonomic nervous system constitutes those pre-ganglionic fibres which leave the central nervous system via the cranial nerves and by way of the sacral nerves. Whereas the pre-ganglionic fibres of the sympathetic nervous system terminate in ganglia along the vertebral column or in ganglia situated in the neck or abdomen, the pre-ganglionic fibres of the parasympathetic nervous system pass directly to the organs they innervate, on or in the walls of which they make their synapse with the post-ganglionic fibres, which are characteristically very short.

In the head the oculomotor nerve carries pre-ganglionic fibres which terminate in the ciliary ganglion, situated in the orbit, from which post-ganglionic fibres run to the eye to serve the ciliary and pupillary muscles. Through the ciliary ganglion pass post-ganglionic fibres from the superior cervical sympathetic ganglion, which supplies sympathetic fibres to the eye. There are other ganglia in the head, the spheno-palatine ganglion supplying secretory fibres to the lachrymal gland, the otic ganglion supplying fibres to the parotid salivary gland, and the submandibular ganglion supplying fibres to the submandibular and sub-lingual salivary glands. All these ganglia receive pre-ganglionic para-sympathetic neurones by way of the cranial nerves. In addition the vagus nerve arises in the head which carries pre-ganglionic parasym-pathetic fibres supplying the lungs, heart and gastro-intestinal tract. The pre-ganglionic fibres of the vagus have their synapses with the post-ganglionic fibres on or near the walls of the structure they inner-vate. The post-ganglionic neurones of the gut actually lie in the wall of the gut.

The remaining outflow of pre-ganglionic parasympathetic fibres comes from the sacral region of the spine. These fibres terminate in a series of ganglia around the organs of the pelvis, rectum, bladder and internal genital organs, and some fibres pass up into the abdomen to supply the lower reaches of the gastro-intestinal tract with a para-sympathetic nerve supply.

The functions of the autonomic nervous system

The two divisions of the autonomic nervous system, the sympathetic and the parasympathetic, have been described as having opposing actions on the structures they innervate. Thus stimulation of the sympathetic nerves supplying the muscle of the iris causes the pupil to dilate, whereas stimulation of the parasympathetic nerves causes the pupil to constrict. The heart rate is quickened in response to sympathetic stimulation, slowed by parasympathetic (vagal) stimulation. In the bladder the muscles of the bladder wall (the detrusor muscle) relax and the sphincter of the bladder contracts in response to sympathetic stimulation whereas the detrusor contracts and the sphincter relaxes, thus achieving micturition, when the parasympathetic is stimulated. The effects of sympathetic and parasympathetic stimulation on a variety of structures is shown in table III. In some organs, e.g. salivary glands, the two systems reinforce each other. Some structures e.g. sweat glands are innervated by only one system.

Table III. *The effect of sympathetic and parasympathetic stimulation on various structures.*

Structure	Sympathetic Stimulation	Parasympathetic stimulation
Heart	Quickening of heart rate	Slowing of heart rate
Blood vessels of skin and gut	Constriction	Dilatation
Coronary arteries	Dilatation	Constriction
Iris of eye	Dilatation of pupil	Constriction of pupil
Bladder	Relaxation of detrusor muscle and contraction of bladder sphincter	Contraction of detrusor muscle and relaxation of sphincter
Bronchial muscle	Relaxation	Constriction

But the functional differences between the two divisions of the autonomic system are wider than this dual control of various visceral structures. The sympathetic nervous system, under certain circumstances, tends to be activated as a whole, producing widespread effects upon the organism. This is in part due to the fact that the sympathetic fibres ramify to a far greater extent than those of the parasympathetic system. And in the sympathetic ganglia a single pre-ganglionic fibre may be linked to as many as twenty or thirty post-ganglionic fibres; one pre-ganglionic fibre when stimulated can give rise to widespread effects. But the spread of sympathetic stimulation is also due to the fact that the hormones adrenaline and nor-adrenaline produced by the medulla of the adrenal gland produce effects on visceral structures which mimic those of sympathetic stimulation. The adrenal medulla is innervated by pre-ganglionic fibres of the sympathetic system from the coeliac plexus, and these fibres terminate around the secretory cells of the medulla. These medullary cells are considered to be modified post-ganglionic neurones which arise in development from the tissues of the spinal cord, from where they migrate to the abdomen where many of them become surrounded by adrenocortical tissue to form the adrenal medulla. The terminals of the post-ganglionic fibres of the sympathetic nervous system are thought to produce their effects on visceral structures by the release of the hormones nor-adrenaline and adrenaline. The post-ganglionic fibres which come to form the secretory cells of the adrenal medulla are specialized in this respect and pour out large amounts of these hormones into the circulation when the sympathetic nervous system is activated.

The function of widespread sympathetic activity is to prepare the animal for 'fight or flight' and this function is discussed in Chapter VI. We can summarize here the effects of a generalized sympathetic discharge:

1. There is a constriction of the arterioles of the skin and gut so diverting blood into those regions which may need a greatly increased blood supply during the effort of fight or flight—the voluntary muscles and the lungs. This widespread vasoconstriction leads to a rise in blood pressure which is available to force more blood through the dilated arterioles of the active muscles.

2. The pyloric sphincter contracts so that food does not pass into the lower digestive tract which has had its blood supply reduced.

3. The coronary arteries, supplying the heart muscle with blood dilate. This permits a greater cardiac output of blood by supplying the heart muscle with increased amounts of oxygen and glucose. Adrenaline increases the rate of conduction in the heart and increases the ability of the heart muscle to utilize glucose as a source of energy.

4. The glycogen stores in the liver, under the influence of adrenaline, are converted into glucose, providing the muscles and brain with a readily available source of energy.

5. In some species of mammal the spleen contracts, pouring out reserves of red blood cells into the circulation.

These responses of the organism to sympathetic stimulation prepare the organism for a sudden burst of activity.

The parasympathetic nervous system, with its post-ganglionic fibres located on or in the walls of the structures which it innervates permits a much more precise response to stimulation. Its activities, concerned as they are with such things as movements of the bowel, evacuation of the rectum and bladder, the secretory activity of the digestive glands, and its restraining influence on the heart, are more concerned with maintaining the status quo and conserving the organism. The terminals of the post-ganglionic fibres of the parasympathetic liberate the chemical substance acetyl-choline at their endings and this substance acts upon the effector organ (see also p. 608). Acetyl-choline is rapidly destroyed in the body by enzymes called cholinesterases, and even if acetyl-choline is injected intravenously there are minimal effects because it is so rapidly destroyed. Thus the parasympathetic activity is rendered precise (anatomically limited) not only by the organization of its fibres but also in its chemical mediator acetyl-choline. Adrenaline enjoys a far longer life when released into the general circulation from the adrenal gland, and it is a dangerous substance to inject intravenously because of its potent effects on the activity of the heart.

THE EYE

The eye is one of the major special sense organs in most mammals, especially in man. It is light sensitive and capable of supplying information to the brain about the size, shape, and in many cases the colour of objects in the environment, and also about the direction and intensity of light.

Structure of the eye. It is a complex organ whose structure is shown in fig. 146. The eye is a spherical hollow organ in which is suspended a lens system which focuses light onto the back of the eye. The lens system separates the eye into anterior and posterior chambers. (In medical texts the small space between the lens and the iris is often called the posterior chamber, but here the term is used meaning the whole of the chamber posterior to the lens.) The anterior chamber is full of a watery fluid called the aqueous humour while the posterior chamber contains a more viscous vitreous humour. The wall of the anterior chamber is transparent and is called the cornea, which is covered by a layer of squamous epithelium on its outer surface. At the margin of the cornea this layer of epithelium is continuous with the conjunctiva, a loose layer of epithelium covering the visible portion of the sclera and reflected forwards to form the inner layer of the eyelids. The cornea contains a great number of sense endings, so that one becomes painfully aware of the smallest particle resting on the surface of the cornea. The posterior chamber has a three-layered wall, the outer layer of which is a tough membrane called the sclera; it contains many white fibres and the external eye muscles which are responsible for moving the eye within the bony orbit of the skull are attached to it.

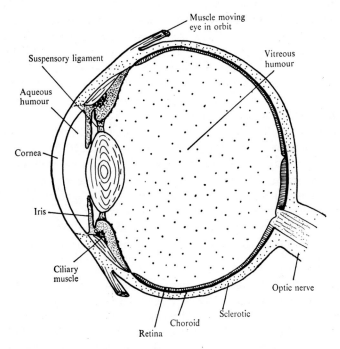

Fig. 146. *Diagram of horizontal section of human eye.*

The visible portion of the sclera forms the white of the eye. The inner most layer of the wall of the posterior chamber is the retina, which is the light sensitive layer of the eye and corresponds to the film in a camera. Between the retina and the sclera is a soft brown vascular tissue called the choroid, responsible for the nutrition of the retina. The brown colour is due to pigment cells called melanophores which absorb the light after it has passed through the retina thus preventing internal reflection within the eye. At the junction of the sclera and cornea is a ring of tissue called the ciliary body which forms the anterior edge of the brown choroid. Within the ciliary body are the ciliary muscles which are attached to the sclera near the corneo-sclerotic junction. The fibres of the muscle run in an anterior-posterior direction. The posterior part of the ciliary body forms the ciliary processes from which the suspensory ligaments arise. These ligaments form a circular disc of tissue in the centre of which the crystalline lens is held. The suspensory fibres in this circular ligament run radially and in the centre are continuous with the outer capsule of the lens. The lens is a cellular structure and the columnar cells at the centre of the lens are called lens fibres. Their protoplasm is fluid so that the lens behaves like a water filled balloon and when the suspensory ligaments are not putting the lens under tension it assumes a more or less spherical shape. The anterior margin of the ciliary body continues in front of the lens as the iris, which is a pigmented disc with a central aperture, the pupil. The iris has muscles which allows the pupil to vary in size and shape. The colour is due to pigment cells in the iris, and it is this structure which makes the human eye brown or blue.

The eye is protected by the eyelids. In man there are only two of these but in some mammals e.g. cat, there is a third eyelid which moves from the nasal to the temporal side of the eye. This is called the nictitating membrane and is very thin and transparent and can best be seen when a cat is just waking from a sleep. The eyes are kept free from dust by the lubricating action of the tears which are formed by the lachrymal gland, the duct of which opens under the upper eyelid. The well-known blink reflex also serves to protect the eye by causing the eye lids to close.

Passage of light through the eye. The cornea is curved and has a higher refractive index than the surrounding air (air R.I. = 1·0, cornea R.I. = 1·38) so that parallel light entering the eye converges onto the retina. The effect of the high refractive index of the aqueous humour lens and vitreous humour is to cause the light to continue to converge towards the retina, where the light stimulus is converted into electrical impulses which pass along the optic tract to the brain. The nerve fibres run along the inner surface of the retina, adjacent to the vitreous humour, and converge to a point where they penetrate the capsule of the posterior chamber to reach the brain; at this point there are no light sensitive cells and therefore it is called the blind spot. In some animals e.g. the squirrel the blind area forms a narrow band in the retina. This is an advantage to the squirrel because if the image of a branch to which it is jumping, falls on the blind area, it can, by a very small movement of the eye in a vertical plane, move the image onto a light sensitive area. If the blind spot were circular, as in man, and the image fell on the centre of the circle the eye would have to move through a greater distance in order to reach a light sensitive area.

Structure of the retina. The retina is the inner light sensitive layer of the posterior chamber of the eye, and under the high power of an ordinary microscope it is seen to consist of several layers (see fig. 147). This layered appearance is produced because there are three main kinds of cells in the retina; the visual elements, called rods and cones, the bipolar cells and the ganglion cells. The nuclei of these cells and the connections of one type of cell with another cause the layered appearance.

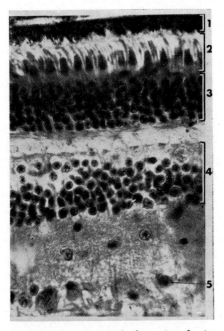

Fig. 147. *Photomicrograph of a section of retina showing the various layers.* 1. pigment layer of retina. 2. rods and cones. 3. nuclei of rods and cones. 4. nuclei of bipolar cells. 5. ganglion cell.

It will be seen from fig. 147 that light has to pass through the nerve fibres, ganglion cells and bipolar cells before it reaches the rods and cones. This is very surprising when it is realized that the rods and cones are the light sensitive cells and the bipolar and ganglion cells are merely conducting elements. The vertebrate retina is said to be inverted because it seems to be back to front. The reason for this inversion is explained later (p. 321).

THE ROD is the most common light receptor in the human retina and it has been estimated that there are 130,000,000 of them in a pair of human eyes. There are some mammals e.g. the opossum and some bats which have pure rod retinas. In nocturnal mammals the rod is the most important light receptor for it is extremely sensitive and sends out nerve impulses to the bipolar cells even when light of very low intensity falls upon it. Since there are only about 450,000 nerve fibres in each optic tract this means that each nerve fibre must supply about 150 rods. It is seen that several rods are connected to each bipolar cell and several bipolar cells to each ganglion cell, whose axon is the nerve fibre of the optic tract. Experiments have shown that when a lot of rods are stimulated it is possible to cause the ganglion cell which unites these rods to discharge an impulse to the brain, whereas if only a few rods are stimulated the ganglion cell may not discharge. In technical language we say that there is summative interaction between the rods stimulated. The rods are interconnected to a greater degree around the periphery of the cup of the retina in the human eye than at the base of the cup near the back of the eye and this is perhaps one of the reasons why the periphery of the cup of the retina is very sensitive to dim light. It is well known that you can see better in dim light if you do not look directly at an object but look a little to one side so that light falls on the rod rich area of the retina. We have seen that rods discharge impulses when light of low intensity falls upon them and that these impulses are summated i.e. added together, so that they can overcome the threshold of a ganglion cell, ensuring that the latter discharges an impulse to the brain. Thus the brain is informed of an object even when it is lit by very dim light. In short the eye is sensitive. However, because one ganglion cell is connected to about 150 rods, 150 points of light from the object will only be represented by one impulse to the brain and the brain can only form a vague picture of the object. Thus if an animal had only rods in its retina it would be able to see in dim light but it could never see an object in any great detail, even in the brightest light.

THE CONE. The second type of receptor in the retina is called the cone and it is the cone which affords us great visual acuity. Acuity means accuracy and if we have acute eyes we should be able to inspect objects and see them in very great detail. It is a great advantage for man to be able to see details for he is a tool maker and in order to make and use tools and to manipulate the environment it is necessary to be able to see in detail. Cones have a high threshold of stimulation and can only function well in conditions of high light intensity. Many cones have their own bipolar cells and do not share them with other cones and each of these cones is connected to its own ganglion cell. Thus if light is of sufficiently high intensity to stimulate the cones to discharge an impulse the brain will receive one impulse for each cone stimulated. Each point source of light which affects the cone will therefore be reflected as a point of stimulation on the visual area of the cerebral cortex. Therefore using cones it is possible to get a very accurate image of the object in the brain, i.e. the cones provide us with acute vision. The disadvantage is that high light intensity is needed for this acute vision. You should be reading this book in good illumination otherwise the cones cannot operate and you will be attempting to read using the more sensitive rods which cannot provide you with a sharp image. Since you get a blurred image you try to use your eye muscles to focus the eyes and still you cannot see clearly. You are obviously not using these very delicate instruments under the proper conditions

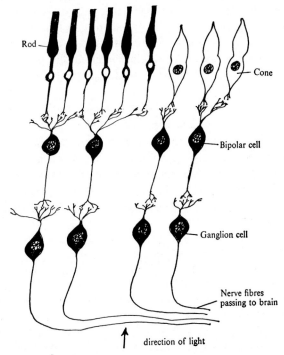

Fig. 148. *Diagram illustrating the connection of rods and cones with bipolar and ganglion cells.* Note that several rods are connected to one bipolar cell.

and perpetual misuse of your eyes like this will probably give you a headache.

In order to explain the principles of acute vision we have described how each cone has its own ganglion cell but this is only so at certain special points in the retina e.g. at the yellow spot (fovea). There are about 8 cones in the retina for every neurone in the optic tract since each eye has about 3,500,000 cones and only 450,000 neurones in each optic tract. It is obvious therefore that each cone cannot have its own fibre in the optic tract and there must be some connections between the cones as their dendrites join the bipolar cells. It is known that at some places in the retina many cones do share one bipolar cell, and there are in fact some cells called horizontal cells which link rods and cones together. The situation is obviously more complicated than the simple case which has previously been put forward. It is clear that the old idea that each cone had an individual nerve supply is inaccurate but since there are many fewer cones than rods the general principle already stated holds true that cones are capable of allowing much more acute vision than are rods. The fovea is obviously an area in which very precise images can be formed, and passed on to the brain. The position of the fovea is so arranged in man that when an object is held in the hand at about one foot from the nose the image falls on the fovea. We automatically hold things in about this position for close examination for we have learned by trial and error that this is how we are able to see them best.

We have seen that the human eye has two kinds of receptor, the rod suitable for night vision, and the cone suited for day vision; the rod gives us a sensitive eye and the cone an acute eye. The rod and cone are so called because it is sometimes possible to distinguish them by their appearance under the microscope (see fig. 148). We must not imagine that all retinas have the same distribution of rods and cones as we have, the retinas of different species of animals are as different as are their bodies.

THE BIOCHEMICAL BASIS OF COLOUR VISION

The presence of light sensitive pigments within the cells of the retina allows the retina to respond to light energy and ultimately to send nerve impulses to the brain. Electron micrographs of the retina show that the ends of both rods and cones are filled with thin membraneous sacs which contain the light-absorbing pigments. These light sensitive pigments are called retinal or chromophores. These chromophores are combined with fatty protein molecules called opsins. In the rods the chromophore-opsin complex is called rhodopsin and this is not associated with colour vision in man and is sensitive to white light. The cones are responsible, in man, for colour vision and they contain chromophore-opsin complexes called iodopsins. There are 3 different proteins associated with the chromophore in human cones and hence there are 3 different iodopsins, but each cone only contains one kind of iodopsin. These 3 kinds of iodopsin are sensitive to blue, green and yellow light respectively. Colour vision in man is thus said to be trichromatic as it depends upon 3 populations of cones each containing a pigment sensitive to either blue, green or yellow light.

Study of purified chromophore-opsin combinations has shown that each kind of combination has a different affinity for light of a particular wavelength. The 3 opsins in human cones show peak absorption of light at 450, 525 and 550 nanometers—which are the blue, green and yellow regions of the visible spectrum respectively (see fig. 149). When a pattern of coloured light meets the retina, this pattern is reflected in

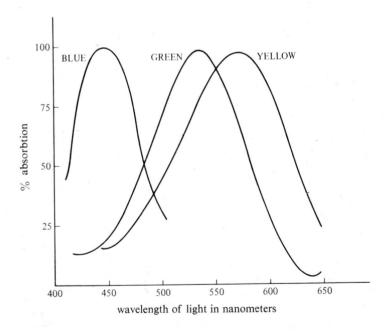

Fig. 149. *Colour vision in man depends upon the presence of three different opsins combined with the chromophore retinal in the cones. There are thus three different iodopsins (opsim + chromophore) and they absorb visible light at three different wavelengths. Each cone contains only one kind of iodopsin so that there are three populations of cones, one sensitive to blue light, one to green light and one to yellow light. The above graph shows the amount of visible light absorbed by the three iodophopsins at different wavelengths. Colour vision is thus said to be trichromatic.*

the nerve impulses which leave the retina because of the presence in the retina of these 3 different types of cone cells, each sensitive to light of a particular wavelength.

The molecular changes induced by light in the chromophore

On excitation by light all the 4 human opsins change in the same way. They only differ in the kind of light to which they respond.

The molecule of the chromophore (retinal) is characterized by the presence of alternating single and double bonds, in chemical terms a conjugated system. Such molecules are structurally fairly stable because the groups of atoms which are attached by double bonds cannot rotate around these bonds. However, the molecules, if adequately extended, have a low energy state and can be excited by light. The effect of excitation by light is to 'relax' the double bond character so that the molecule can change from a cis configuration to a trans structure. (Fig. 150.)

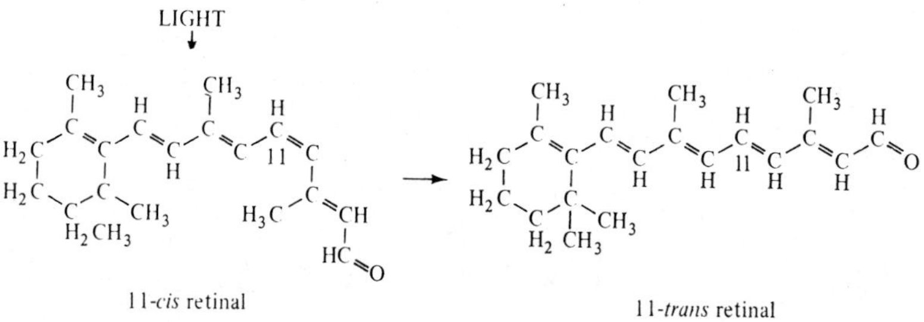

Fig. 150. *Vision depends upon the presence in the retinal rods and cones of a light sensitive pigment, a 'chromophore' called 11-cis retinal. The only stage in the biochemistry of vision which is dependent upon light is the transformation of 11-cis retinal into 11-trans retinal; this change triggers off a complex series of events in the protein molucules (opsins) with which the chromophore is associated. This change, from cis- to trans-configuration occurs in the first thousandth of a second after excitation by light.*

This change in the molecular configuration of the chromophore (retinal) is the only one which light is involved, and it occurs in an unbelievably small fraction of a second. Following this cis-trans transformation in the pigment are many complex chemical events in the associated protein opsin but these are not light dependent and can occur, as it were, in the dark. The change of the chromophore from cis to trans configuration sets in train this complex series of changes in the structure of the associated protein opsin and these changes finally lead to the release of the chromophore from the opsin.

The released all-trans chromophore is later reduced to the alcohol form and then oxidized back to the aldehyde form to regenerate all-cis chromophore which is ready to respond again.

The study of the rapid events that take place when light impinges upon chromophores has been revolutionized by low temperature chemistry. Solutions of chromophores can be held at the temperature of liquid nitrogen ($-140°C$) and their responses to irradiation by light of different wavelengths and intensities can be studied at rates which are far slower than they actually occur at body temperature.

Colour blindness

Normal human vision utilises 4 different opsins, 1 in the rods to make rhodopsin and 3 in the cones for the colour vision pigments. Each of these must be specified by a different gene. Normal colour vision is trichromatic, depending as it does on the presence of 3 kinds of opsins in the cones. Most colour blindness is dichromatic, resulting from an absence or deficiency of one kind of opsin in the cones, producing blue-, green-, or red-blindness. Red- and green-blindness is caused by a sex-linked recessive mutation. About 1% of men are red colour blind, about 2% green colour blind, whereas both conditions are usually very rare in women. Blue-blindness is very rare, not sex-linked and affecting only 1 in 20,000 persons.

INVERTED RETINA. One of the very surprising things about the retina is that the rods and cones seem to be pointing the wrong way, as it were, for the light has to pass through the nerve layer then the bipolar cell layer and only finally does it reach the receptor cell. Such a retina is said to be inverted (see fig. 147). The reason for this peculiar design is to be found in a study of the way in which the eye develops. One of the important ways in which vertebrate animals differ from those without backbones is that the vertebrate animals have a central nervous system which is hollow and dorsal whereas the invertebrate has a ventral solid nerve cord. The vertebrate nerve cord is a tube which is formed by a rolling up of the skin (see fig. 143). Note how in the diagram the outer layer of the skin becomes the inner lining layer of the nerve tube. Fig. 152 shows how the fore-brain bulges out to form buds which develop into the optic cup. If the layers are carefully followed on the diagram you will see how cells which were once on the outer skin of the animal come to lie in the position of the rods and cones in an inverted retina. The outer wall of the optic cup becomes pigmented and forms the pigment layer which backs the retina. This black layer in man serves to absorb the light after it has passed through the receptor layer. If the layer behind the receptors was shiny then light would be reflected back through the receptor cell layer and the cones might be restimulated and confusion could arise because of this internal dazzle effect. Some nocturnal animals e.g. cats do have a shiny backing to the retina called the tapetum. The shiny tapetum does

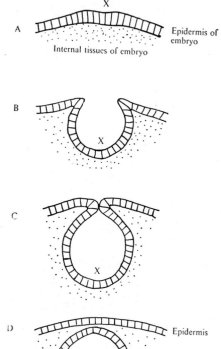

Fig. 151.   *Development of the nerve tube.* A. Shows the epidermis of the embryo. B. and C. show successive stages in the rolling up of the epidermis to form the nerve tube. D. shows the nerve tube complete. Note how the outer surface of the embryo at X comes to lie in the inside of the nerve tube.

reflect the light back into the receptors and the rods are restimulated, there is no dazzle of course for the cat is a nocturnal animal and is only using this tapetum in dim light on a rod rich retina. If such a dark adapted animal is caught in the beam of a car headlight the reflected light can easily be seen. The unfortunate animal will probably be dazzled and unable to see anything for the internal reflection inside the eye.

Before leaving the developmental story of the eye it must be mentioned that the lens is formed in a very interesting way. The epidermis overlying the developing optic cup is said to be organized to form the lens (see fig. 152). Chemical substances are made by the developing optic cup and as these diffuse outwards they effect a change in the surrounding tissues, the epidermis folds in to form a lens and the new

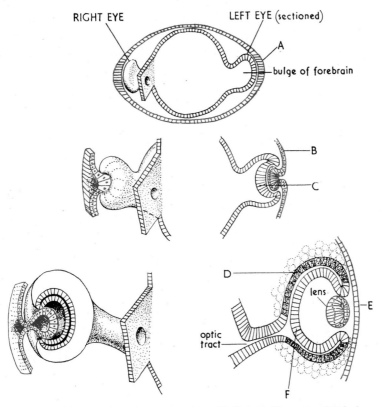

Fig. 152. *Diagrams to show how the optic cup is formed.* The upper sketch shows a section across the embryo in the region of the eyes. A is the epidermis which is influenced by the bulging nerve tube to form a lens placode (C in middle sketch). B is the epidermis which will form the conjunctiva (E in lower sketch). In the lower sketch D represents cells which will form the pigment layer of the retina. This is surrounded by mesodermal cells which will form the choroid and sclera. F represents cells which will form the retina (rods, cones etc.).

outer layers become the transparent cornea. Mesodermal cells congregate around the cup bringing blood to the developing retina. Incidentally here is a physical reason why the inverted retina may have been so successful, for by inverting the retina the receptors are brought very near to the blood supply. Other connective tissue cells form the fibrous protecting sclerotic around the outside of the eye. This story of eye development was one of the early examples of organization that was discovered. We now know that the direction which the cells of our body take as they differentiate is governed by the position that these cells have on a particular diffusion gradient. A differentiating organism is a very complex nexus of diffusion gradients.

Accommodation. Ciliary muscles have their origin at the junction of the cornea and sclerotic and are inserted into the ciliary body, which is attached to the choroid layer. The suspensory ligaments are stretched

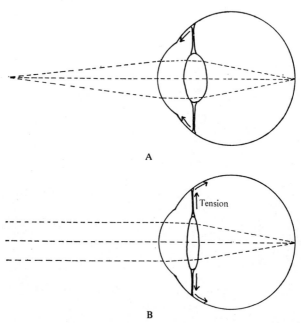

A

B

Fig. 153. A. *Eye with lens adjusted for focusing the image of near objects.* The ciliary muscle is active, and by drawing forwards the front of the choroid the tension on the suspensory ligament is reduced and the elastic lens assumes a more spherical shape. B. *Eye adjusted for focusing the image of distant objects.* The ciliary muscle is inactive, and the intra-ocular tension increases the tension on the suspensory ligament, so flattening the lens.

from the processes of the ciliary body to the capsule of the lens, with which they are continuous. When the ciliary muscle contracts the ciliary body and the choroid are pulled forward to a point at which the diameter of the eye is smaller. Because the suspensory ligaments are now stretched over a smaller distance there is less tension in them and therefore there is less tension on the lens, which because of its elasticity, becomes more spherical. As it assumes a spherical shape its focal distance shortens and it becomes capable of turning light through a greater angle; it is adjusted for seeing near objects, from which the light rays are divergent (see fig. 153A). When the ciliary muscles relax, the tension in the stretched choroid is transmitted to the suspensory ligament and the lens assumes a less spherical shape, its normal resting position. It can now only bend light through a small angle and is suited for dealing with light from far distant objects, since light from these objects is parallel rather than divergent (see fig. 153B). It is a surprising fact, but a true one, that when the ciliary muscle contracts, tension falls in the suspensory ligament and vice versa. This is the method of accommodation in the mammal, but it must not be thought that all animals accommodate in this way. Many fishes accommodate by moving the lens backward and forward in the eye by means of a special muscle called the retractor lentis. Birds and some reptiles accommodate by the ciliary muscle actually squeezing the lens and making it more convex.

ACCOMMODATION, VISUAL ACUITY AND SIZE OF EYE. Accommodation is the method used to focus the eye accurately so that objects can be seen with greater precision, but precise focusing is worthless unless the retina is capable of recording a detailed image. An image can only be seen with great precision in a cone containing retina for reasons already explained on p. 317. The image must also be a large one so that it falls on many cones. Photographers know that if they want to make negatives which have to be enlarged many times they have to have a fine grain emulsion on their film, since the negative must be very precise if it is to be enlarged. It is the same with the retina; the image if it is to be precisely seen must fall on a fine grained retina, that is one containing many cones. For the light to fall on many cones the retina must be large. Therefore acuity of vision demands large cone rich retinas, and a lens system which is capable of accurate focusing and able to throw a large image on the retina. If the eye is small as in a mouse or any small rodent the image formed will also be small and the number of cones stimulated will be small. Therefore in small rodents acuity of vision is not possible and it is interesting to note that these small animals cannot accommodate. They are nocturnal creatures and their eyes are used only to indicate the direction and intensity of light, not to see in accurate detail. To these small nocturnal rodents the sense of smell is more important than sight. In the primates, a group which includes monkeys, apes and man, the eyes are relatively large and capable of throwing a large image onto the retina. Accommodation is now worth while and we have seen something of the method in man.

In mammals the ciliary muscles are poorly developed in rodents and show an increasing development in herbivores, carnivores, until they reach a maximum development in the primates. In general the extent of accommodation is low in mammals except in primates.

The Herbivore eye. The eye is important in these animals for safety. In the typical case of the horse the eye is useful for detection of the

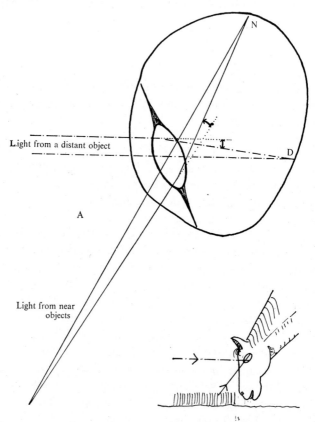

Fig. 154. *The ramped retina of the horse.*

movement of possible predators as the animals graze in the open plains. The horse is active during the day and has a cone rich retina and a large eye so that a large detailed image can be formed, and the movement of a stalking predator can be easily seen. The horse has a special device to ensure that almost the whole field of vision including near and far distant objects can be in focus without accommodation. The posterior surface of the eyeball is not equidistant from the lens at all points but is much nearer to the lens at the point where images of far distant objects fall. The shape of the eye with its sloping posterior wall is shown in fig. 154A.

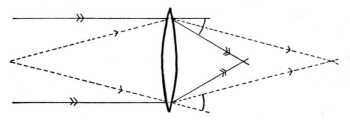

Fig. 155. *Diagram illustrating the relationship of the position of the object and the position of the image with a convex lens of fixed focal length.*

The horse has very little ability to alter the shape of the lens so that it cannot make the lens more convex for the focusing of images of near objects. Therefore the lens is not capable of having a variable focal length and the only way in which near and far objects can be focused on the retina is by having different distances from the lens to the retina where the image falls from different distances. Fig. 155 shows that objects at different distances from the lens are focused at different distances behind the lens if the focal length of the lens remains constant. There is a formula in optics which expresses this:

$$\frac{1}{v} - \frac{1}{u} = \frac{1}{f}$$

u = distance of object from lens
v = distance of image from lens
f = focal length of lens

When the horse's head is bent down as it eats, the image of the grass focuses in the upper part of the retina at point N (fig. 154B), whilst distant objects focus at point D in the lower part of the retina. The light from near and distant objects is bent through the same angle but both objects are in focus simultaneously because of the sloping retina. Such a retina is called a ramped retina. The ramped retina replaces the ciliary muscles functionally.

The eyes of the horse are large and set high in the front of the head, with the result that the horse can see what is happening all round it except for a small area the width of its head behind its body. The visual fields of each eye overlap in front of the head (see fig. 156). The horse has a broad, horizontal oblong pupil, so that it can take advantage of the possibility of all round vision, even when the pupil is small in very bright light.

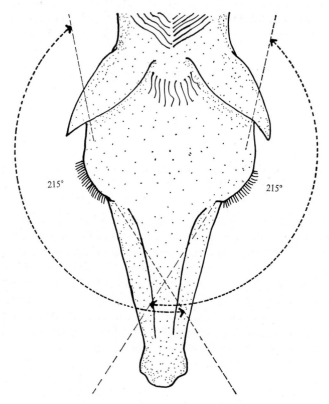

Fig. 156. *The visual field of the horse.*

The horse's eye has a very small anterior chamber and a large posterior one, thus giving a large image. Movement of the object is detected by movement of the image on the retina. The image will move further in a large eye than in a small one. Thus in the large eye of ungulates movement is easily detected, so that the animal easily sees the predator coming and can flee.

It is necessary for distant vision to have sensitive cells in the retina, as the intensity of light from a distant object may be small. The ungulate eye is large enough to pack in rods, which are sensitive, between the cones.

In ungulates the cone rich area of the retina (area centralis) is surrounded by an area where the blood vessels of the choroid run perpendicular to the surface of the retina. The areolar connective tissue of the choroid between the blood vessels is fibrous giving parallel rows of fibres around the vessels. These fibres reflect light back through the retina, making a tapetum fibrosum. The dim images from a distance are thereby brightened. Therefore animals like the horse have well adapted eyes for life in open grassland. They are best adapted to bright daylight but can see better than man at dawn and dusk.

Nocturnal eyes. The development of very sensitive eyes capable of seeing in dim light is considered to be a secondary specialization. Many carnivorous animals, e.g. cats, have adopted nocturnal habits for they stand a better chance of catching their prey in dim light, if their prey is adapted to the high light intensities of daylight. Many carnivores hunt mainly by scent and hearing, and sight is only used to estimate the accuracy of the final pounce.

Nocturnal eyes differ from diurnal eyes (i.e. daylight adapted eyes) in the following ways:

(a) large spherical corneas which take up a large percentage area of the perimeter of the eye. This enables the eye to have a large surface through which light can enter; big windows let in a lot of light.

(b) the lens is spherical and the posterior chamber is small relative to the anterior chamber. The nocturnal eye must not be merely large, it must be disproportionately so. It needs to be large to let in a lot of light but it must not have large distance from lens to retina, for the intensity of light diminishes as the square of the distance; that is if you double the length of the eye you double the distance light has to travel from lens to retina and the intensity is diminished by four times. If the eye becomes three times longer, the illumination intensity is nine times smaller. Now the eye capable of seeing in dim light must make the image as bright as possible so it must have a short posterior chamber. If the posterior chamber is small then the lens must be powerful enough to bend the light through a large angle in order to focus light on the retina. The spherical lens of the nocturnal mammal is capable of this.

(c) The retina of a nocturnal animal must be sensitive and contain many rods whose effects are summated. Bats which fly at night have pure rod retinas.

(*d*) Tapetum. In many nocturnal animals there are layers of cells either in the retina, or more usually in the choroid which reflect light back through the light sensitive retina. Such layers are called tapetal layers. Cats eyes shine green at night, as the light is reflected back through the eye by a tapetal layer in the choroid. As the white light is reflected through the choroid and retina, red light is absorbed by the red pigment in blood and so the light appears green, since green is the complementary colour to red. In the cat the epithelium proliferates around the blood vessels in the choroid to form thin tiers of cells. The joints between these coincide and blood vessels run straight between the piles of cells. The piles of cells reflect the light and form what is known as a tapetum cellulosum. The cells in these piles contain crystalline threads in the protoplasm. These threads are very fine and form bundles running in varying directions, and it is these refractile threads which reflect the light.

Twenty-four hour eye. We have seen that some eyes are adapted for use in strong light or in dim light. Since animals live half their lives in light and half in dim light or darkness it is not surprising to find that most animals can see under a great variety of light intensities although they often have a very good performance under only a small range of light conditions. The opossum is incapable of seeing if he is disturbed in bright light—he can only see in the dark, but animals like the opossum are exceptions. Man can see well in bright light and fairly well in the dark and he manages this by having a special area of the retina called the fovea (yellow spot) which is very rich in cones for daylight use and other areas of the retina which are rich in rods for nighttime use. Many mammals are like man in that they possess dual purpose retinas.

The iris diaphragm is a most useful organ for regulating the amount of light which enters the eye. It is a circular sheet of muscle with a central aperture, the pupil, size and shape of which can be altered to control the passage of light into the eye. The eye of a cat at night has a large circular pupil as the iris is allowing all the available light to enter. In bright light the cat's pupil becomes a very narrow vertical slit so that only a little light can enter, and the sensitive rods of the retina are protected. The iris thus allows the cat to live successfully during all light conditions.

Eye defects.

1. FAR SIGHTEDNESS. As we get older the lens becomes progressively more inelastic which means that as the ciliary muscles contract during

accommodation for near vision the lens assumes a less spherical shape, and so is less able to converge the diverging rays of light from near objects. The closer the object to the eye the more divergent are the rays of light from it and the more converging the lens has to perform. There is a maximal amount of convergence that can be brought about by the lens system and there comes a point when an approaching object can no longer be focused properly on the retina. The point of maximum capacity of convergence is called the near point, and with increasing age the near point recedes progressively. A youth should be able to focus adequately up to about ten inches from his nose end but in older persons the near point is often three feet away. This defect may be easily corrected by fortifying the eye's powers of convergence of light by supplying spectacles containing biconvex (converging) lenses. People suffering from this defect can see distant objects well with the unaided eye and the defect is thus called far sightedness.

2. LONG SIGHTEDNESS. Persons with long sightedness can only see distant objects clearly. This is not due to a loss of elasticity in the lens, as in the above case, but is due to a defect in the shape of the eye. The eye of a long sighted person is too short, the retina being too near the lens, and the maximal powers of accommodation are unable to converge the light from near objects to focus them on the retina. The rays of light are in focus behind the retina and thus the image is blurred. The defect may be corrected by spectacles containing biconvex (converging) lenses (see fig. 157).

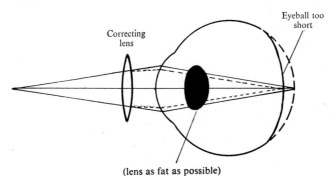

Eyeball too short

Correcting lens

(lens as fat as possible)

Fig. 157. *The correction of long sightedness.*

3. SHORT SIGHTEDNESS. In this condition the eyeball is too long so that the retina is further away from the lens than in a normal eye. This places far less strain on the powers of accommodation and the light from near objects can be focused onto the retina with less activity of the ciliary muscles. A short sighted person can see objects clearly

when they are only a few inches from the eye. However the images of distant objects are focused in front of the retina even when the accommodation processes are completely at rest, because of the increased length of the eyeball. The person with short sightedness can be made to see distant objects clearly by the use of spectacles with divergent lenses (biconcave) (see fig. 158).

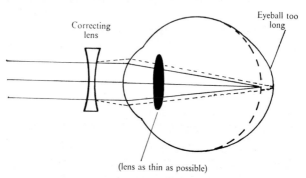

Fig. 158. *The correction of short sightedness.*

4. ASTIGMATISM (fig. 159). This is a defect of the eye in which there is an irregularity of the corneal surface so that a point source of light may be focused on the retina if the rays pass through one part of the cornea, yet be out of focus if the rays pass through another part of the cornea. Irregularity of the corneal surface, which is normally spherical, can also lead to a distortion of the image on the retina. Provided that the irregularity is a regular one e.g. a flattening of the surface in one

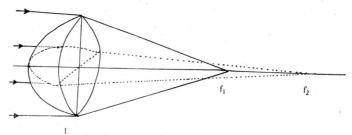

Fig. 159. *To illustrate the principle of astigmatism* by showing light passing through a glass lens in which the curvature of the vertical meridian is greater than that of the horizontal meridian. Light passing in the vertical meridian is thus refracted greater than that passing in the horizontal meridian and is brought to focus at a point nearer to the lens (f_1) than light passing in the horizontal meridian (f_2). Thus it is impossible to get a point focus of light. This defect can be corrected by the use of an additional appropriately shaped lens which will equate the refractive powers of the horizontal and vertical meridians. In the eye regular astigmatism which can be corrected by means of an optical lens is due to a cornea with different curvatures in the horizontal and vertical meridians.

meridian the defect can be corrected by using a cylindrical lens so that the lens and cornea have the same effect as a normally shaped cornea would have.

THE EAR

Principles of the mechanism of hearing. We can perhaps best understand the complex process of hearing by comparing the sensory apparatus of the ear to a piano sounding board. Whereas the strings of the piano are made to vibrate when they are struck by hammers the strings in the ear apparatus are made to vibrate indirectly by means of sound vibrations in the air.

The production of vibrations in a string by sound waves travelling through the air is called sympathetic resonance and can easily be demonstrated. Two strings, which if plucked would produce sounds of the same pitch, are attached side by side to a board. If one string is now made to vibrate, the other will soon be vibrating in exactly the same manner. It is as if sound waves from the first string were actually plucking the second string.

This sympathetic resonance will only occur when the two strings are capable of producing notes of the same pitch. We have now to enquire what it is in the strings which determines the pitch of the note produced. There is a formula which connects the relevant facts together:

$$n = \frac{1}{2l}\sqrt{\left(\frac{t}{m}\right)}$$

where
l = length of string
t = tension on string
m = mass of unit length of string
n = number of vibrations per second.

The pitch of the note is determined by the number of vibrations per second (n); the greater the number of vibrations per second the higher the pitch and vice versa.

Now the number of vibrations per second in the string is determined by various qualities of the string, its length, tension and mass. From the formula which connects these factors it is evident that to produce a note of high pitch the string should be relatively short, taut and light. To produce a note of lower pitch the string should be longer, slacker or heavier.

Thus in the piano the strings on the right hand side, producing notes of high pitch, are relatively short, taut and fine. The strings on the left hand side are longer, less taut and are bound round with copper wire to make them heavier.

The structure in the ear which we have compared to the piano sounding board is called the cochlea. Its essential structure is that of a fine membrane, surrounded by fluid and embedded deep in bone. This membrane may be likened to a lot of strings closely applied side by side. As in the piano sounding board the qualities of the strings vary from one end of the membrane to the other. At one end the fibres are shorter, more taut and are probably less heavily loaded with fluid (i.e. less mass) than at the other end.

We can now obtain a picture of how this membrane functions. Sound waves passing in the air reach the ear apparatus and set up vibrations in the fluid surrounding the membrane of the cochlea. Parts of the membrane now begin to vibrate, but only those parts that vibrate with the same frequency as the sound waves reaching the ear. Nerve fibres attached to the part of the membrane which is actually vibrating are now stimulated and electrical impulses pass to the auditory areas of the brain.

Structure of the hearing apparatus. We can now discuss how the ear apparatus is designed to carry out these activities. For descriptive purposes the ear apparatus may be divided into three parts, the outer, middle and inner ears (fig. 160).

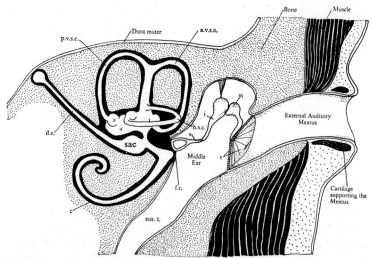

Fig. 160. *Diagram of the structure of the hearing apparatus.* The coils of the cochlea have been simplified. The black surrounding the membranous labyrinth of the inner ear represents the fluid perilymph.

a—ampulla.
a.v.s.c.—anterior vertical semicircular canal.
c.—cochlea.
d.e.—ductus endolymphaticus
eus.t.—eustachian tube.
h.s.c.—horizontal semicircular canal.
f.r.—fenestra rotunda.

i.—incus.
m.—malleus.
p.v.s.c.—posterior vertical semicircular canal.
s.—stapes.
t.—tympanic membrane.
sac—sacculus.
u—utriculus.

The outer ear consists of the pinna, the 'ear' of everyday speech and a tube, the external auditory meatus which leads into the skull. The pinna deflects waves of sound into the external auditory meatus. In some animals e.g. dog and rabbit, the pinna is movable and can be directed independently from the head towards the source of particular sound waves. In man the pinna faces more or less forwards and is immovable and thus he is compelled to look and hear, as it were, in the same direction. Animals without a pinna e.g. amphibia are at a distinct disadvantage.

The external auditory meatus conducts sound vibrations inwards to the middle ear. It is separated from the middle ear by a taut membrane, the ear drum or tympanum, which is set in vibration by the sound waves.

The middle ear is an air filled cavity in bone, closed from the outside by the tympanum and leading into the inner ear through an oval aperture in the bony wall, the fenestra ovalis. Three bones, the ear ossicles, are suspended across the air filled cavity of the middle ear; the malleus (hammer), incus (anvil) and stapes (stirrup). These three bones are articulated to one another, with synovial joints between them. The arm of the malleus is attached to the tympanum and the stapes abuts onto the fenestra ovalis.

The bones serve to transmit the vibrations from the air in the external auditory meatus to the fluid which bathes the delicate apparatus of the inner ear.

The ear ossicles are suspended from the wall of the middle ear cavity by ligaments and the tension across the bones is regulated by means of two small muscles.

The area of the tympanum is large compared with that of the fenestra ovalis so that the pressure (i.e. force per unit area) acting upon the fenestra ovalis is greater than that on the tympanum. This increase in force is available for transmitting vibrations from the air of the middle ear to the more dense medium (i.e. fluid) of the inner ear.

The air in the middle ear is kept at atmospheric pressure by means of a connection to the back of the throat, the eustachian tube. Thus the pressure on the two sides of the tympanum is kept equal, at atmospheric pressure. The aperture of the eustachian tube in the throat is opened during swallowing. If the tube is blocked, as it often is by mucus plugs when one has a cold or catarrh, oxygen is absorbed from the air of the middle ear by the blood; the tympanum now bulges inwards because of the lower pressure in the middle ear, producing the familiar ringing and buzzing sounds.

THE INNER EAR consists of a delicate hollow membranous structure, the membranous labyrinth filled with fluid called endolymph, bathed externally by the fluid perilymph and embedded deep in the temporal bone. The membranous labyrinth performs two functions, that of balance and orientation in space, and that of hearing. The part of the labyrinth called the cochlea is responsible for the sense of hearing.

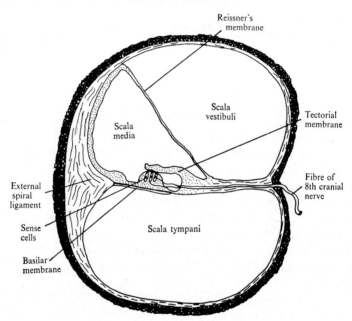

Fig. 161. *Diagram of a transverse section through the cochlea.*

The cochlea is a membranous structure wound spirally like a snail shell inside a bony canal. Running along the whole length of the bony canal and stretched from one side wall of the canal to the other are two membranes. The part of the canal enclosed by the two membranes, the scala media, contains the fluid endolymph which is continuous with the endolymph of the rest of the membranous labyrinth. It is virtually a closed cavity. Of the outer two cavities the upper one is called the scala vestibuli because it communicates with the fluid filled cavity, the vestibule, into which the stapes abuts. The lower cavity is called the scala tympani; it connects with the scala vestibuli at the extreme tip of the cochlea (the helicotrema) and it connects to the middle ear cavity by another closed window, the fenestra rotunda (see figs. 160–162).

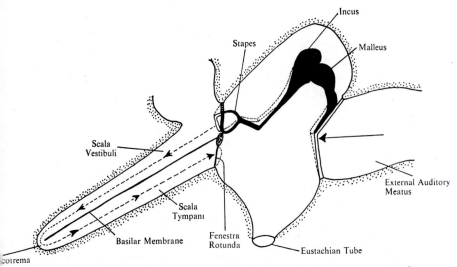

Fig. 162. *Diagram to show the course of vibrations in the ear.* The cochlea is represented as a straight tube. The movement of the ear ossicles is represented by dotted lines. The stapes rocks in the fenestra ovalis and the vibrations pass in the fluid perilymph of the scala vestibuli and into the scala tympani by way of the helicotrema. A rise in pressure of the perilymph of the scala tympani is compensated by a bulging of the fenestra rotundae.

When vibrations are set up in the fluid perilymph of the vestibule by the action of the stapes rocking in and out of the fenestra ovalis, they pass down the scala vestibuli via the helicotrema, into the scala tympani and end at the fenestra rotunda. The course of such vibrations is indicated by the dotted lines in fig. 162. Fluid cannot be compressed so when the stapes rocks into the fenestra ovalis the pressure wave set up in the perilymph causes the fenestra rotunda to bulge into the middle ear. When the stapes moves out of the fenestra ovalis the fenestra rotunda moves inwards into the perilymph. The fenestra rotunda is a compensating mechanism which protects the delicate cochlea from the effects of pressure.

The upper membrane of the cochlea called Reissner's membrane is very delicate and can be disregarded in considering the sensory function of the ear. It is the lower membrane, called the basilar membrane which is the important one, and it is this membrane which we have compared to the piano sounding board. The basilar membrane is firmly attached at each side of the bony canal by fibrous tissue, that on the outer side being called the external spiral ligament (see fig. 160). The membrane is widest at its apex; here the external spiral ligament is a very delicate structure compared with that at the base i.e. at the vestibular end of the basilar membrane. From these facts it is evident that the constituent 'strings' of the basilar membrane are longer and less taut at the apex than at the base. From our study of the relationship between the pitch of a note and the length and tension of the string that produced it, we must conclude that notes of high pitch set the basal part of the basilar membrane in motion whereas notes of low pitch set the apical part of the membrane vibrating. This has been confirmed experimentally.

The basilar membrane possesses rows of sense cells which because of the fine hairlike processes coming from their surfaces are called hair cells. These sensory hair cells are part of the organ of Corti. These processes are embedded in a ribbon of jelly called the tectorial membrane. The hair cells receive a very rich nerve supply from the auditory branch of the eighth cranial nerve. The sensory hair cells of a particular sector of the membrane are stimulated when that sector vibrates. Thus since particular parts of the membrane vibrate only to notes of a particular pitch, particular nerve cells are stimulated only when the ear receives notes of that pitch. This is the basis of the 'place theory' of hearing.

Balance and orientation in space

We have seen that the cochlea is the part of the membranous labyrinth concerned with hearing, now we must look at the remaining part of the labyrinth, consisting of the semicircular canals and the otolith organs (saccule and utricle).

Semicircular canals. The membranous semicircular canals are embedded in the bony semicircular canals, the space between them being filled with the fluid perilymph. Inside the membranous canals is the fluid endolymph which is continuous with the fluid in the utricle and saccule, and thence with the endolymph in the cochlea.

The canals lie in three planes which are at right angles to each other; there is a horizontal or lateral canal, a posterior vertical canal and an anterior vertical canal (see fig. 160). At one end of each canal, where it passes into the utricle there is a swelling called the ampulla, containing a sense organ, the crista acoustica. The crista consists of a mound of cells containing sensory hair cells, the fine processes of which project from the mound to become embedded in a cone of jelly (the cupola). The sense cells are joined to nerve fibres of the eighth nerve (see fig. 163).

FUNCTION OF THE SEMICIRCULAR CANALS. When the head moves from a resting position the endolymph in the semicircular canals is set in motion. According to the plane of movement of the head so the endolymph movement is restricted to a particular canal in that plane. Thus during an old-time waltz it is the horizontal semicircular canal in which the endolymph is mostly moving. There will be slight movements in the other two of course.

When the endolymph of a canal moves, the cupola in an ampulla is displaced and this stimulates the hair cells of the crista acoustica. From the pattern of stimuli coming from the three ampullae of each labyrinth the mammal is aware of the movement and the kind of movement of the head.

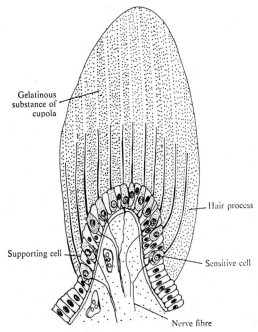

Gelatinous substance of cupola

Hair process

Supporting cell

Sensitive cell

Nerve fibre

Fig. 163. *The ampullary sense organ, the crista acoustica.*

Otolith organs (fig. 164). When the head is still the fluid in the semi-circular canals is stationary and the sense organs in the ampullae are unstimulated. These provide no information about the position of the head at rest. This information comes from the sense organs in the utricle and perhaps the saccule. The sense organs of the utricle and saccule are called the maculae. The macula of the utricle consists of a thickening of the wall containing sensory hair cells surrounded by supporting cells. A collection of small crystals of calcium carbonate, called otoliths, adheres to the surface of the macula. The hair cells are supplied with nerve fibres and under the influence of gravity the hairs are distorted and the nerves stimulated due to the weight of the crystals. If the animal is upside down the crystals fall away from the hair processes and stimulation is reduced. From this sort of information the animal is aware of the position of its head in space at rest. Movement of the head also stimulates the maculae of course, even movement in a straight line will do so. (It would not stimulate the cristae of course.)

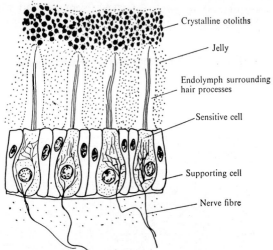

Fig. 164. *Part of the macula sense organ of the utricle.*

Summary of the function of the labyrinth.

   i. Cochlea—hearing.

   ii. Crista acoustica of the ampullae—sensitive to angular acceleration—kinaesthetic sense.

   iii. Macula of the utricle—sensitive to gravity and linear acceleration—static sense.

Other mechanisms for orientation. We have seen that the membranous labyrinth is a source of information about the position of the body in space but there are at least three others. The way that a mammal orientates itself in space can perhaps be best understood by giving a simplified account of some experimental work.

THE ROLE OF THE EYES IN BODY ORIENTATION. If a normal cat is held upside down by its legs and then dropped, it always lands on all four feet. However, if a blindfolded labyrinthectomized cat is dropped in a similar manner it falls on its back. When the blindfold is removed the labyrinthectomized cat lands on all four feet like a normal cat. It is obvious that the eyes have supplied information about the position in space.

THE ROLE OF BODY CONTACT IN BODY ORIENTATION. A blindfolded labyrinthectomized cat lying on its side on the ground has the ability to turn its head and body to a more vertical position. Remember that this was not possible when the cat was allowed to fall through the air. Some information must have been gained from a contact of the animal's body with the ground. If a weighted board is placed upon the cat's body whilst it is lying on its side it cannot now turn its head, although the head be free to move. Why is this? The explanation must be that the turning of the head depends upon an unequal stimulation of the flanks of the body.

Let us now turn our attention to a normal cat and describe the series of events that occur when it stands up from lying on its side. Lying on its side the cat is aware of the position of the body in space because of information received from the eyes, utricle of the labyrinth and from body contacts. Using this information appropriate action can be taken. First the head is slightly raised and twisted into a near vertical position. During this action some of the muscles of the neck connecting the head and body are twisted and stretched. Stretching of the muscles stimulates proprioceptors within them, providing information about the position of the head in relation to the body. Appropriate action can now bring the head and body into alignment.

As soon as the head has started to move from the lying down position the sensory apparatus of the semicircular canals is brought into action providing the animal with information about the angular movement and acceleration of the head.

Summary of orientation mechanisms. We now know that the mammal gets information about the position and movement of the body in space from the following sources.

1. Membranous labyrinth; semicircular canals and utricle.

2. Eyes.

3. Body surfaces.

4. Proprioreceptors.

PART III

ORGAN SYSTEMS

Chapter Eight

Nutrition

Food chains. In order to grow and live as a healthy organism the animal has to select certain substances from the environment and incorporate them into its own body, using some for the provision of energy for the working of its body. This whole process is called nutrition and it involves, if necessary, hunting the food, eating it, digesting it into particles small enough to be absorbed into the blood stream, then using these substances in cellular metabolism either in anabolism (body building) or in breaking down the substances further to release their energy (catabolism). The point to be grasped here is that the term nutrition is a very wide term embracing a number of biological processes.

The body contains organic chemicals of great complexity, and these compounds contain a lot of carbon, and the procuring of carbon is a very important task for the mammal. In any natural community the interrelationships between the animals and plants are very complex. Some animals are predators, like owls or stoats, whereas others are herbivores feeding on the green plants. If you make a close study of any natural community and work out what the various animals feed on you will find that in the end their source of food is in a green plant. A list of animals arranged in order, so that you can see at a glance what any of the animals is feeding on, is called a food chain. For instance you can think what you had for breakfast and perhaps if you are an Englishman you will have had bacon, eggs, toast and marmalade and tea (with sugar and milk). Now think where these things came from and you will be able to build up a picture something like this:

Bacon......pig......potatoes and cereals (green plants)
Eggs.......hens.....corn (green plants)
Toast.......wheat (green plants)
Marmalade......oranges and sugar (green plants)
Tea......(green plant)
Milk......cow......grass (green plant)

From this simple investigation you will see that man is an omnivore and he is at the top of a chain of animals and plants. The base of this food chain, and all other food chains too, is a plant which can synthesize its own food from inorganic sources, in particular it can make organic substances from carbon dioxide. Work using radio active isotopes of carbon in carbon dioxide has shown that the gas is quickly turned into phosphoglyceric acid in photosynthesis. We shall see later that this substance is a very important one in the metabolism of cells. There has been research in recent years which seems to show that animals too can convert some carbon dioxide into organic substances, but they can only do this to a very limited extent, so that our generalization that animals all depend upon green plants for their food is not invalidated. The old adage, somewhat modified, has it that all flesh is grass and all fish diatom.

If you study a community of animals and plants you will find that several animals have the same food source and the food chains cross at many points to produce what is called a food web. This is a very important fact for professional biologists who are concerned with pest control for very often they find that the best way to control the pest is to break the food chain, for example by removing the food plant. For example the potato root eelworm is a nematode parasite of the potato which is very hard to kill in the potato field, for the female worm produces very resistant eggs which are enclosed in resistant cysts and almost the only way to attack this pest is to stop growing potatoes in that field for about the next twenty years, so starving out the pest. There are many now famous cases of pest control by the introduction of animal predators. Perhaps one of the most famous is the introduction into Australia of the moth Cactoblastis cactorum in order that its caterpillar might eat up the prickly pear cactus (Opuntia) which was spreading rapidly, forming a dense scrub which could not be grazed by the sheep. The moth caterpillar ate away the prickly pear more quickly than the latter could grow. The moth reproduced and its population size increased and before two years were up the cactus was eaten away and grass pastureland could be re-established. Fortunately before the introduction of the moth, exhaustive tests had been done to find out which plants were at the base of its particular food chain. It had been found that even after weeks of starvation these moth larvae would rather die than eat anything other than prickly pear cactus. So when the cacti were eaten the size of the moth population diminished. Now there are a few prickly pears being eaten by a few Cactoblastis caterpillars and there is a balanced population.

In an ecological survey of plant and animal feeding relationships, if you count the numbers of plants and animals involved you will see that they form what is called a pyramid of numbers (see fig. 165). From this brief excursion into ecological theory you should have gathered that man is dependent upon green plants eventually for all his food. The balance of the various numbers of organisms involved in the food web is very important and if the population density of any particular animal gets too great then the whole pattern of the food web has often to be readjusted and often many animals must die. Such problems are very important to man and a good example may be seen in the problems of the herring industry in the 1920's. After the invention of the power

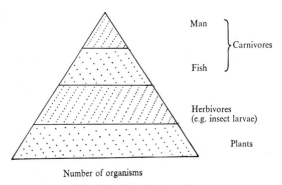

Fig. 165. *Pyramid of numbers.*

driven fishing boat the fisherman could obviously travel faster and catch more fish. He could now fish in weather which previously had been unsuitable. This increase in the fisherman's efficiency brought problems, for it became more and more difficult to catch fish as the North Sea became overfished. Official government action was taken to limit the size of the fish which were to be landed. So the smaller fish were spared in order to provide a good catch at a later date when they had grown bigger. The balance of nature had been disturbed but the balance was readjusted.

The balance of the ratios of the numbers of animals and plants in the world is most important for the future of man. There is a growing fear in the minds of many scientists that as the human population increases, partly accounted for by the increase of medical skill which prevents folk dying, there will be too many mouths to be fed. If the number of food plants in the world cannot be increased then the human population of this planet must be reduced by world starvation. There are two solutions to the problem: one is to find ever new sources of food, perhaps by tapping the resources of the sea, or by growing food more efficiently to feed man rather than his food pests. The other solution is to limit the size of the total human population by sensible family planning on a world scale. There are religious objections to the latter course, and this is only one of the difficulties that would face the people who attempted to control world population. There are obvious political problems and also problems of human nature and sexual customs in different societies to deal with. However, the problem is very serious and if our race is not capable of dealing with the problem then its fate is sealed.

In thinking about the nutrition of man we should try to have a dynamic picture in mind rather than a static one. Organic materials are constantly being used up by man, as he replaces and repairs worn out tissues and uses carbohydrates as fuel which he breaks down to release energy. The animal body may be pictured as the banks of a stream through which flows a river of organic material. The individual animal can be recognized by the shape and nature of the banks and by the speed of the river, but if the water is not constantly supplied from the source then the life of that river is at an end and the banks are useless —in terms of the analogy the life flow stops and the animal is dead. The river of nutrients comes eventually from green plants across a network of rivulets called a food web and after it has flowed through man it goes on to be returned to the source by green plants and bacteria in a very complex series of reactions called the carbon and nitrogen cycles. When you look at a field or the seashore with the eye of an informed biologist you should see this organic stream of nutrients and feel it in yourself. Nothing in the natural world is standing still, all is in dynamic flux, including you.

DIET

The materials of the diet and their functions

The diet should contain materials from which the body can obtain energy, body building materials and all the necessary raw materials for a healthy life (with the exception of oxygen which is absorbed through the lungs). The food contains carbohydrates, fats, proteins, water, mineral salts, vitamins and roughage, and all these substances have a definite part to play in the life of the organism.

Calorific value of the diet. It has long been known that the diet must contain enough energy for the body's needs. This energy is measured as heat energy and is expressed as a number of kilocalories (or Calories with a capital C). The number of Calories required depends upon the age, activity, size and sex of the body. Thus a growing human male of 16–20 years will need about 3,800 Calories compared to a man of similar physique and occupation of 25 years who will use about 3,200 Cals/day. Small mammals, such as the shrew, have a very large surface area in relation to their volume and lose body heat rapidly; such an animal will need a larger number of Calories per unit weight than an animal such as a cow. Indeed such small mammals have to spend much of the twenty four hours of the day eating to provide an adequate number of Calories. (1 Cal $=4\cdot18$ kilojoules $=0\cdot00418$ MJ.)

The first requirement of the diet then is that it must contain an adequate number of Calories for the individual.

A balanced diet. In a balanced diet the Calories must be properly distributed between carbohydrates, proteins and fats. A balanced diet would not be one in which the entire calorific needs are supplied by bread alone. And in addition to the carbohydrates, fats and proteins, the diet must also supply the other food substances, mineral salts, water and vitamins.

An adequate diet. A diet may be balanced in that all the necessary factors are present in the correct proportions but it is also essential that an adequate amount of each substance be present. In the human adult this means about 3,000 Calories provided by about 450 gms. of carbohydrate, 80 gms. of fat and 70 gms. of protein. In addition there should be sufficient vegetables and fruit to supply vitamins, mineral salts and roughage. The diet must also contain sufficient water, either as liquid or in the food, to supply the needs of the body.

Protein

Proteins in foodstuffs supply the raw materials for the growth and repair of body tissues and for the manufacture of various enzymes and hormones. Some of the amino acids of which proteins are constructed can be synthesized by the body and these are not a vital component of the diet—the so called non-essential amino acids. Other amino acids cannot be synthesized by the body and must be supplied in the diet—these form the essential amino acids. The essential and non-essential amino acids are listed below.

1. Table of essential and non-essential amino acids

Non Essential	Essential
Alanine	Histidine (for many mammals e.g. rat but not essential for man).
Aspartic acid	Isoleucine
Arginine	Leucine
Hydroxyproline	Lysine
Proline	Methionine
Glycine	Threonine

1. Table of essential and non-essential amino acids (contd.)

Non Essential	Essential
Serine	Tryptophan
Glutamic acid	Phenylalanine
	Valine
	Tyrosine (if enough phenylalanine is not present. Not essential if the former is present in large enough quantity).
	Cysteine ⎱ These can be made from methio- Cystine ⎰ nine if there is enough present. They are not therefore essential in the presence of methionine.

Note that arginine may be essential for the optimal growth of some young animals, e.g. rat but it is not essential for man. For many animals it is essential e.g. chicken.

Protein can also be used as a source of energy—and indeed is the main energy source of carnivorous animals. It the diet does not supply sufficient energy for the body's caloric needs then proteins of body tissues are used as a fuel source. The liver is the chief organ which converts protein into a fuel which can be used by all body cells (i.e. in the forms of glucose). This process of conversion of protein to glucose is called gluconeogenesis.

The body of an average adult human (70 kg) contains about 10 kg of protein and the bulk of this is in muscle tissue. Studies using amino acids which have been labelled with a radioactive tag of heavy nitrogen (N^{15}) have shown that the proteins of the body are constantly being broken down into amino acids and resynthesized again. During the resynthesis of body protein some of the amino acids come from those released by breakdown of protein elsewhere in the body i.e. they are 'secondhand' amino acids. But the body's amino acid economy is not complete and when proteins are resynthesized about 20% of the amino have to be provided by the diet. If there is no protein in the diet the body's amino acid economy becomes more efficient but even so there is always some loss of amino acids which become converted into urea which is excreted in the urine.

The body also loses protein in the form of cells shed from the lining of the bowel wall, from skin and in nails and hair. These losses are in the order of 28 G of protein each day. An adult has to take in at least this amount of protein each day to prevent him 'wasting away'

But even if the diet contains enough Calories to prevent protein being used as a source of energy the protein in an average British diet is only about 70% utilized. This means that to replace the inevitable losses of nitrogen as described above an adult needs to eat $28 \times \dfrac{100}{70} = 40$ G of dietary protein/day. Table 2 shows the protein content of some common foodstuffs. Reference to this table shows that this amount could be supplied by a day's diet which included ¼ lb steak, 1 egg and 7 oz of bread. However the daily requirement could be provided by bread alone, provided enough was eaten to satisfy total caloric needs.

Table 2

100 G of edible portion	G Protein	G Fat	G Carbohydrate	Calories
Apples (sweet)	0·3	0·4	15·0	58
Bananas (1 banana 6in. = 100 G)	1·3	0·4	24·0	94
Grapes	0·8	0·4	16·7	74
Cabbage	1·6	0·1	5·7	25
Carrots	1·1	0·2	9·1	46
Lettuce	1·3	0·2	2·8	15
Peas (fresh)	6·7	0·4	17·0	80
Potatoes (fresh)	2·0	0·1	19·1	85
Almonds	18·6	54·1	19·6	640
Walnuts	15·0	64·4	15·6	702
Bread white	8·5	2·0	52	266
Cornflakes	7·9	0·7	80·3	359
Cane sugar (refined)	0	0	96	384
Butter	0·6	81	0·4	716
Eggs (1 medium size)	6·9	6·2	0·4	85
Fresh cows' milk	3·3	3·7	4·9	65
Beef grilled steak	16·9	25·0	0	293
liver	19·8	4·2	3·6	131
corned beef	24·4	15·0	0	232
tripe	19·1	2·0	0	94
Chicken (grilled)	20·2	12·6	0	194
Cod (fresh)	16·5	0·4	0	70
Herrings (fresh)	19·0	6·7	0	136
Beer	0·6	—	4·0	50

(100 G = 3·5 oz)
(1 Cal = 4·18kJ)

A man needing a supply of 2800 Cals/day would need to eat about 2½ lb of bread to provide these Calories—and this would provide adequate protein with enough essential amino acids.

In this country it is rare for protein deficiency to occur because of a low intake of protein in the diet. This is so even for infants, pregnant and lactating women who need more protein because of growth, supplies for the foetus and milk secretion respectively. Shown below are the protein and calorie requirements of various normal humans. The Calories specified refer to a fairly sedentary occupation.

Age	Calories/day	Dietary protein G/day
Birth	408	10·2 G
	(120/kg body weight)	(3 G/kg body weight)
6 months	770	15·4
	(100/kg body weight)	(2 G/kg body weight)
8 years	2000	30
Adult female	2500	35
Pregnant female	2500	40
Lactating female	2700	50

Caloric and protein requirements at various ages. The figures for the new born and 6 month infant are based on assumed weight of 3·4 kilos and 7·7 kilos respectively.

The quality of dietary protein

Animal proteins usually contain a high proportion of essential amino acids in a balanced mixture (see table 3). For this reason they have been classified as 'first class' protein. Not all proteins of animal origin have such a first class composition. Gelatin for example, a component of jellies, sweets etc. is completely deficient in tryptophan and is low in methionine. Vegetable foods are not only relatively low in protein content but the amino acids present may not be in the proportion which is characteristic of animal tissues (see table 3). This means that if

Table 3

Essential amino acid content of animal and vegetable proteins and absolute minimum daily requirements of essential amino acids for adults (when the diet contains enough nitrogen for synthesis of non-essential amino acids). Essential amino acids as given a % of total weight of protein.

	Iso-leucine	Leucine	Lysine	Methio-nine	Phenyl-alanine	Threo-nine	Trypto-phan	Valine
Minimum requirements	G 0·7	G 1·1	G 1·1	G 1·1	G 1·1	G 0·5	G 0·25	G 0·8
Content of various foods	%	%	%	%	%	%	%	%
Cows' milk	5·8	9·2	8·5	2·6	4·7	4·3	1·3	7·1
Beef	5·5	7·6	8·6	2·5	3·9	4·8	1·1	5·4
Egg (whole)	6·7	9·2	7·3	3·1	5·4	5·3	1·6	7·1
Peas	4·6	6·6	5·7	0·7	4·7	4·0	0·8	5·0
Whole wheat	3·7	6·2	2·4	1·1	4·8	2·5	1·17	4·4
Whole maize	4·1	9·6	2·3	1·4	4·3	3·5	0·5	5·3
Carrots	4·4	5·6	3·1	0·9	4·1	4·0	0·6	5·7
Brussel sprouts	4·4	7·2	5·6	2·6	3·7	5·0	1·6	6·4

one vegetable food forms a major part of the diet then deficiencies of essential amino acid may develop, and this may arise however much total protein is contained in the diet. In some parts of the world maize forms the bulk of the caloric and protein supply of the diet. This seed is notoriously deficient in tryptophan and an individual living entirely on maize may develop a negative nitrogen balance (i.e. will lose more body protein than is taken in the diet) even when enough maize is eaten to satisfy caloric needs. However, if vegetable foods are combined so that the overall amino acid composition is well balanced then they can be an adequate source of protein as eggs or meat—provided they are eaten in adequate amounts.

Protein and special diets

There is a great deal of prestige value in eating a high protein diet. First class protein is expensive and much in demand—particularly by athletes, shot-putters and boxers who imagine that it will increase the size and power of their muscles. There is however no evidence that extra protein can be incorporated into muscle tissue merely by eating a high protein diet. Certainly muscles increase in size when they are subjected to increasing exercise—e.g. in the training of athletes and weight lifters. During such training muscles may increase in size by 50%. But the increase in the amount of protein in the muscles is only in the order of 10%, which could be readily provided by an average British diet.

Another misconception about protein is its significance for slimming diets. Diets aimed at producing slimming contain less Calories than the individual requires. The deficit of Calories is made up by the body using body tissues as a fuel. Some slimmers imagine that when body tissues are used as a fuel to make up the caloric needs it is only fat stores which are broken down. This is not true. When a diet fails to provide sufficient Calories then both body fat and protein are used as a fuel. A negative nitrogen balance develops i.e. the body loses more protein than that taken in the diet. This negative nitrogen balance develops even if the diet consists of pure protein. The caloric requirements of the body take first precedence and it will use all the diet of pure protein to supply energy needs, and if these supplies are not adequate then the body tissues (protein and fat) are used to make up the deficit. There is thus no advantage in eating expensive proteins during slimming. Table 4 shows the estimated protein losses of an adult who changes to a caloric deficient diet at various levels of protein intake. On a 900 caloric diet the same amount of body protein is lost

Table 4

The loss of body protein ($-$G/day) when a normal adult changes his diet to one deficient in Calories and/or protein.

Protein intake of diet (G/day)	Calorie intake (Cal/day) 900 Cal	(Cal/day) 2800 Cal
0	$-$ 45 G	$-$ 40 G
20	$-$ 30 G	$-$ 20 G
40	$-$ 30 G	0
60	$-$ 30 G	0

(30 G/day) whether he eats 300 G of roast lean beef (providing 60 G protein) or only 100 G roast lean beef (providing 20 G protein)—or 100 G of cod!

Fat

Fat is a palatable and energy rich part of the diet. Without fat—and thus without fried food, cream, cheese, eggs etc., a diet would seem very uninteresting to Western palates. Weight for weight fat provides more than twice the number of Calories than does protein or carbohydrate. 1 G of fat releases 9 Calories when oxidized; thus when large numbers of Calories have to be provided e.g. for people with strenuous occupations, fat provides a convenient source of Calories in that it reduces the bulk of food needed.

Because fat is slowly digested and absorbed (it reduces the rate of stomach emptying) it makes a meal satisfying, and hunger pains do not return as quickly as with a meal composed predominantly of carbohydrate or protein. But because of this delay in emptying of the stomach fat is not a suitable food for supplying a rapidly available source of energy.

In addition to acting as a rich source of Calories, fats (lipids) are incorporated as constituents of protoplasm and cell membranes. In some foods the fat contains the fat soluble vitamins e.g. A, D, K—see page 404.

The body can synthesize fat from excess carbohydrate or protein in the diet. However some animals—particularly the young—need a supply of polyunsaturated fatty acids (linoleic or arachidonic) in the diet for physical well being. The commonest source of polyunsaturated fatty diet is vegetable oil e.g. corn or olive oil. If these fatty acids are absent from the diets of rats then disturbances of growth and of the skin appear. They are probably also needed by human infants.

In addition to these functions, fat has a role to play in regulation of body temperature. Concentrations of fatty tissue under the skin of mammals act as heat insulators. This layer of fat is particularly prominent in aquatic mammals such as the whale where it forms the blubber.

Carbohydrates

Carbohydrates are taken in many forms in the diet including monosaccharides (glucose, galactose and fructose), disaccharides (maltose, lactose, sucrose) and polysaccharides (mainly starch but also glycogen, insulin etc.).

During the process of digestion (see later) the carbohydrates appear mainly as a mixture of disaccharides—maltose, isomaltose, lactose and sucrose in the lumen of the small intestine. These disaccharides then enter the cells of the intestinal mucosa where there are special enzymes—disaccharidases—which convert these disaccharides into monosaccharides—glucose, galactose and fructose. Thus the final stage of digestion is intracellular. These are the forms in which the carbohydrate of the diet reaches the circulating blood. However in the body these monosaccharides are readily interconvertible so that it does not matter which carbohydrate is present in the diet.

In the typical British diet about 50% of the total Calorie requirement is provided by carbohydrate. Only about 5% is derived from protein, the remainder being supplied by fat. About 40% of dietary carbohydrate comes from cereals, 40% from sugar, 10% from potato and about 5% from milk.

When carbohydrate is removed from a diet and Calories replaced by protein and fat, individuals suffer from fatigue, loss of weight and ketosis (an acidosis due to accumulation of products of incomplete fat metabolism). It seems that the rate of formation of glucose within the body from protein is not normally fast enough to meet the body's requirements for glucose. Some carbohydrate in the diet would seem to be important—although some races e.g. eskimos seem to have been able to adapt themselves to a diet consisting predominantly of fat and protein.

Water

The importance of an adequate supply of water is obvious. Water is the most abundant component of the animal body—in man it forms 45-75% of body weight. This wide variation in water content is due to difference in fat content of individuals. Fatty tissue contains much less water than lean tissue; thus fat individuals contain less water (as expressed as % of total body weight) than lean individuals. But in contrast to this wide variation in water content between people, individuals have a remarkably constant water content.

This constancy of water content is achieved by mechanisms which match water intake with water losses. We can obtain an estimate of the daily water needs by looking at the ways in which water is lost by the body.

1. *Loss by evaporation of water in the respiratory passages and lungs*

In the respiratory passages and lungs atmospheric air (which is relatively dry) is exposed to the warm moist respiratory membranes. These respiratory membranes lose, on average, about 400 ml of water to the atmospheric air. This amount can be considerably larger if the air is very dry or if the rate of ventilation is increased.

2. *Loss via the skin*

Although the skin of mammals is relatively waterproof, enabling them to live on dry land, some water is lost from the surface of the skin. This is a generalized diffusion of water vapour through the layers of heratin from the underlying tissues. These losses are invisible—the so called insensible loss of water. The amount of water lost in this way varies greatly, depending upon such factors as air temperature and humidity, air currents, and body temperature.

Water is lost from the skin in a second more tangible way—by the secretion of fluid onto the surface of the skin by sweat glands. Sweat is not pure water but contains minerals such as sodium. The evaporation of this water takes up heat from the skin—as the latent heat of evaporation. The body makes use of this by increasing the secretion of sweat when body temperature rises. The activity of the sweat glands are regulated by nerves. The controlling centre in the nervous system which determines the degree of activity of sweat glands lies in the hypothalamus (see p. 467).

People living in temperate climates lose up to 1000 ml of water from the skin in the form of insensible losses and sweat. If body temperature rises e.g. in strenuous exercise then these losses may be much greater than this. The maximum losses from both lungs and skin may be greater than 5% of body weight/hr—i.e. 3·5 L/hr in a 70 kg man!

In normal conditions in temperate climates the combined losses from lungs and skin in man amount to 1–1·5 L/day. Note that this loss is virtually pure water and it is an inevitable loss in that it does not come under the influence of the body's water regulating mechanisms.

3. *Faeces*

Small amounts of water and electrolytes are lost in the faeces. This loss is normally only 60–150 ml/day. However there is a vast turnover of fluid and salts in the gastro-intestinal tract; in one day a man secretes large quantities of the various digestive juices (saliva, gastric juice, bile, pancreatic juice etc.) containing water and mineral salts. A total of over 8 litres of fluid is secreted and reabsorbed in one day (note the volume of blood plasma is only 3·5 L). It can be readily appreciated that if these secretions are not fully reabsorbed—e.g. vomiting and diarrhoea—then this will lead to severe disturbance of water and salt balance. In infants losses of fluid and salts by vomiting and diarrhoea in the various forms of infectious gastro-enteritis can be rapidly fatal.

4. *Urine*

The daily losses we have described are inevitable. Losses of water from the kidneys in the form of urine are not inevitable and it is by varying the volume of urine produced that the mammal maintains water balance (see also chapter XII). After the supply of water to the body has satisfied the needs of the inevitable losses then any excess is available to the kidneys for excretion as urine water. The kidney has the task of excreting nitrogenous waste (mainly urea) and in 24 hrs the kidney has to excrete at least 35 G of waste products and any excess minerals. Of course these wastes are excreted in solution and about 500 ml of water are needed for this purpose. However if the function of the kidneys is at all impaired by age or disease then much larger amounts may be needed to excrete the wastes. A more generous allowance is thus normally made for the needs of urine water— amounting to 1·5 L/day.

Balance sheet for man

We can now summarize water losses from the body:

1. Skin and pulmonary losses   1500 ml
2. Urine water   1500 ml

This results in an optional intake of about 3000 ml/24 hrs.

We have ignored the small faecal losses because some water is actually released in the body during the oxidation of foodstuffs. This water of oxidation is more than enough to cancel out losses of water in the faeces.

Regulation of water balance

We have created a balance sheet for man of normal losses of water for the body, which shows that about 3 litres of water is an optimal daily intake. That amount may have to be considerably modified if there are abnormal losses from the body—vomiting, diarrhoea, heavy sweating, rise in body temperature, increased ventilation rate. But except for individuals who are ill or unconscious we do not have to make these measurements of losses and calculation of water requirements. The body performs this task automatically.

The body controls water balance in two ways.

1. By regulating the osmotic pressure of body fluids. This is achieved by excreting or retaining either water or salts. Here we are concerned with the regulation of water. Changes in the osmotic pressure of blood are appreciated by special receptors lying in that vital part of the forebrain called the hypothalamus. A rise in the osmotic pressure of blood produce for example too excessive losses of water from the skin and lungs or due excess salt intake, activates these receptors. These in turn set in motion the complex mechanisms of thirst and the searching for water. Stimulation of these hypothalamic osmoreceptors by a rise in the osmotic pressure of blood also causes the release of anti-diuretic hormone from the posterior lobe of the pituitary gland. This hormone causes the kidney to retain more water, thus the urine produced is more concentrated (see p. 502).

2. Regulation of the volume of fluids. If the above mechanism were the only one operating it would be possible to produce a progressive expansion of the body fluids—and the body!—by drinking a solution of salts in water (e.g. 0·9% NaCl) isosmotic with body fluids. But the body also possesses pressure receptors which constantly monitor the volume of fluid (particularly the volume of blood). Changes in volume of fluids results in reflex changes in kidney function which oppose the volume change—by either excreting more or less urine (see p. 503).

Specialized mechanisms in water balance

As far as water balance is concerned man is a rather unspecialized creature. We will look briefly at two animals which show special physiological features which allow them to go without a supply of water for long periods.

Camels live in deserts and steppes where they are exposed to the hot sun and dry air. These animals show an unusual tolerance to these hot drying conditions. A man fed on dry food at an air temperature of 38°C would survive for about two days. Under the same conditions a camel could live for at least seventeen days. What special features give the camel this advantage over man? Camels survive well in the desert because they are very tolerant of changes in body temperature and body water content. Although the camel has sweat glands they do not secrete sweat if the body is deficient in water and this conserves a good deal of water. But because of this adaptation of the sweating mechanism the animal has to tolerate rises in body temperature—and the camel will tolerate a rise of body temperature amounting to 6°C whereas a man would suffer 'heat stroke' if body temperature rose 4°C above normal levels. During the night the body of a camel loses the heat stored during the day and indeed, because of dilatation of blood vessels in the skin, it loses more heat than was stored the previous day so that it starts a new day with an abnormally low body temperature.

In addition to tolerance of extreme changes in body temperature the camel can also tolerate extreme losses of body water. It will tolerate a loss of water amounting to 30% of body weight.

The kangaroo rat is even more specialized for life in the desert. This animal needs *no* dietary water. It is able to reduce its water losses to such an extent that they are less than the amount of water which is generated within the body during the oxidation of foodstuffs. The special features of the kangaroo rat include the following:

1. During the hot part of the day the animal lives in a burrow which is relatively cool (< 30°C) and which has a high relative humidity. This cool humid air reduces the amount of water which is lost from the respiratory tract and skin.

2. The animal has no sweat glands.

3. Produces very dry faeces.

4. Produces highly concentrated urine (page 498).

5. Tolerates a high body temperature—up to 41°C which is 6°C above normal body temperature.

Mineral salts

The basic minerals which are needed in the diet include sodium, potassium, calcium, magnesium, phosphorus, chlorine. The many different functions of these elements are considered in various sections of the book and reference to the index should be made. Here we are concerned only with generalities of function, distribution, and food sources of these major elements. If we examine a particular body tissue or fluid for its mineral content we find that its composition may bear little relationship with that of other tissues. Bones and teeth for example contain enormous quantities of calcium and phosphorus compared with other tissues. And if we compare the composition of the fluid inside body cells with that of the fluid which bathes them (tissue flood and blood plasma) we find that inside cells the predominant

360 NUTRITION

cation is potassium whereas in the fluid around the cells sodium is predominant cations (see fig. 166). These differences in the distribution of elements is closely related to their different functions.

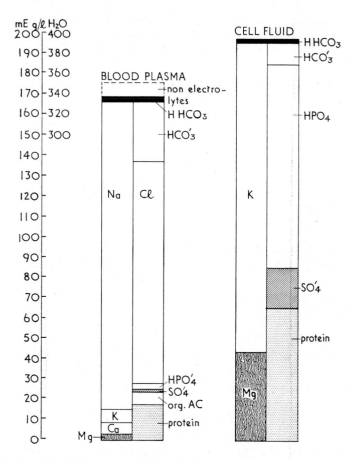

Fig. 166. *Stick graphs showing the composition of blood plasma and cell fluid.* The section labelled non-electrolytes consists of nitrogenous waste products, glucose, etc.

Before we go on to look at the minerals of the body we must touch the subject of units of measurement. It has been the custom in recent years to express the amount and concentrations of electrolytes in body tissues and fluids in terms of absolute weight of substance i.e. G% or mg%. This, however, disregards the fact that chemical substances react together not in terms of absolute weight but in terms of equivalent weight. If we express the composition of a body fluid in terms of equivalent weight of components the information becomes more significant. The unit of measurement for this purpose is the milli-equivalent (m. equiv.) and m. equiv/L. One m. equiv. is 1/100th of the equivalent weight expressed in grams.

The two units are readily interconvertible:

$$\text{milli-equivalents} = \frac{\text{mgm. percent} \times \text{valency}}{\text{atomic weight}}$$

SODIUM

Sodium is mainly found outside body cells and it forms the most abundant cation of blood and interstitial fluid (the tissue fluid which bathes all cells). It is the most important osmotically active ion in body fluids. Because of this a gain in the amount of body sodium results in a gain of water and a loss in sodium leads to a loss of water—both of these changes arising from the vital need to maintain the osmotic pressure of body fluids within normal limits. We can best illustrate the role of sodium by looking at a fairly common disorder due to critical losses of sodium and water during excessive sweating as occurs in exertion in hot conditions e.g. in mines. Sweat contains less sodium than blood so that after copious sweating the body loses proportionately more water than sodium chloride. Loss of this hypotonic fluid produces changes which are seen first of all in blood. The volume of blood falls and its sodium content increases. The cells of the body are now surrounded by a fluid which is hypertonic to them and there is an osmotic movement of water from the cells into the blood so that blood volume gradually rises. The rise in the sodium concentration in blood also stimulates thirst and an individual usually drinks an amount of water similar to that lost in sweating.

The drinking of pure water in the above state leads to further disorders of fluid distribution in the body. Much of this water enters the cells of the body which have previously lost water to blood. Proportionately less of the water remains in the blood stream. We now have a state in which there is still a deficit of blood volume and more water in the cells.

The above condition due to excessive sweating and the drinking of water is seen in 'miner's cramps'. The disturbance in distribution of water and salts leads to spasms of muscle, particularly the fingers. In addition there may be a shrunken appearance of skin and eyes, loss of appetite and sickness, fatigue and dulling of mental abilities. The condition is readily treatable by giving salt solutions instead of pure water to replace the losses due to sweating. Of course now that the condition is a well recognized danger of working in hot conditions, prevention by giving salt tablets and water is the ideal solution to the problem.

Sodium then sharply influences not only total body water but also the internal distribution of water. A 70 kg man contains about 4,200 m. equiv. of sodium, half of which is present in blood and interstitial fluids. Only 10% of total body sodium is in the interior of the cells, the remainder being in bones.

The normal daily requirement of sodium (as sodium chloride) is 7–15 grams. When the body is losing abnormal amounts of sodium then this amount will have to be increased. When sweating is profuse 1 G of sodium should be taken for every litre of water drunk over 4 L.

Normal sweat has about 2–3 G sodium chloride/litre. But on continued exposure to hot conditions the body adapts by reducing the sodium content of sweat to about 0·5 G/litre. After adaptation much less additional salt will be needed.

Sodium is very widely distributed among foodstuffs and some e.g. seafoods are particularly rich in the element. The addition of salt to food may be more in the interests of palatability rather than sodium requirements and the body readily excretes excess sodium. Some foods however may have such a high sodium content that they produce unpleasant although temporary disturbances. A 'chop-suey' syndrome has been described in which some individuals experience numbness, weakness and palpitations after eating out in Chinese restaurants. This temporary disorder, it has been suggested, is due to the high sodium content of Chinese food (soya sauce, sodium glutamate etc.).

POTASSIUM

The body contains a total of about 3000 m. equiv. of potassium (in a 70 kg man). Nearly all of this is inside the body cells—only 2% is present in blood and interstitial fluids. No other ion can replace the functions of potassium inside the cell and an adequate supply of potassium is necessary for normal cell function. In particular this distribution of potassium is related to the electrical potential across cell membranes (see p. 284). Disturbances of potassium balance are thus seen primarily in excitable cells where normal cell function depends upon normal electrical features of the cell membrane. Disorders of skeletal, smooth and cardiac muscle are prominent in disorders of potassium metabolism. The excitable properties of heart muscle are so dependent upon normal distribution of potassium that accidental or thoughtless injection of even small amounts of a potassium salt directly into the blood can lead to fatal changes in heart action.

Like sodium, potassium is widely distributed in food stuffs; all cells, both plant and animal, are rich sources of the mineral. If extra supplies are needed because of abnormal losses in disease then fruit juices (i.e. cell contents) provide a rich supply.

CALCIUM AND MAGNESIUM

Both of these elements are essential for life. The physiological functions of calcium are dealt with elsewhere (page 640). Magnesium, like calcium, is necessary for the normal function of the neuromuscular system. The element is also vital for a variety of enzyme systems e.g. those involved in the synthesis and destruction of acetylcholine. If animals are deliberately deprived of magnesium then death eventually occurs preceded by disturbances of heart rhythm, spasms of muscles and convulsions—disorders very similar to those caused by a deficiency of calcium.

Magnesium is widely distributed in foodstuffs and disorders due to magnesium deficiency are rarely if ever seen in the absence of disease. In some parts of the world magnesium deficiency in the soil can lead to disorders in grazing animals.

TRACE ELEMENTS

Just as some organic foodstuffs are needed in such small amounts to warrant their classification into a special group (of vitamins) so some minerals which are needed in minute daily amounts are grouped together as trace elements of the diet.

Some mineral elements are present in such small amounts in the body that it is only recently that sophisticated techniques such as atomic absorption spectrophotometry and neutron activation have permitted accurate measurements of the amounts present. For this reason the biological significance of some of these minerals has only recently been appreciated.

The definition of a trace element is obviously arbitrary. The substances must be present in small quantities. But how small? Some authors define an element as a trace element if it makes up less than 0·01% of body content. There are at least 25 elements which fulfil this definition, although for many of them we have virtually no information about their biological significance.

Table 5 shows the amounts of some trace elements which are present in an average 70 kg human being. For purposes of comparison some of the commoner elements of the body are also included.

Table 5

Some *Trace Elements*	Amount in 70 *kg body* (% composition)
Iron	5,000 mg (0·007)
Zinc	2,300 mg
Copper	150 mg
Iodine	30 mg
Manganese	20 mg
Cobalt	5 μg/100 G of liver (wet weight)

Some *Common Elements*	Amount in 70 *kg body* (% composition)
Oxygen	54,500 G (65%)
Carbon	12,600 G (18%)
Hydrogen	7,000 G (10%)
Nitrogen	2,100 G (3%)
Calcium	1,300 G (1·9%)
Sulphur	175 G (0·25%)
Potassium	105 G (0·2%)
Sodium	80 G (0·11%)

IRON

Of all the trace elements iron is by far the most abundant in the body, an average 70 kg body containing about 5000 mg of the element. Iron forms a vital component of haemoglobin, muscle myoglobin, cytochrome oxidase and other enzyme systems.

The daily requirements of iron is of the order of 12 mg/day for men and 14 mg/day for women. The differences between the sexes is due to the loss of blood and hence iron (of haemoglobin) during menstruation. Even larger amounts of iron are needed during pregnancy for the synthesis of haemoglobin and enzymes in the foetus. Not all foods provide an adequate supply of iron. Liver is the richest source of iron but meat, egg yolk and cereals are good sources of the element. Milk is particularly low in iron and for this reason supplements of iron are often added to dried milk for babies to provide additional supplies of iron until the infant graduates to mixed feeding. Pregnant women are also routinely given extra iron (as tablets of ferrous sulphate or some other salt). Note that excess iron can be highly toxic, particularly to children, and many children have died of iron intoxication by mistaking for sweets the iron tablets of their pregnant mothers.

The body exercises a strict iron economy. When worn out red blood cells are broken down their iron content is stored in reticulo-endothelial cells particularly in the spleen and liver. The iron is stored in combination with a protein (apoferritin) to produce a substance called ferritin. When iron is needed the element leaves this complex and is available for the synthesis of new haemoglobin or enzymes. The mucosal cells of the intestine also contain apoferritin. If body stores of iron are full then this apoferritin is in the form of ferritin (iron and protein). The presence of ferritin in these cells 'blocks' the absorption of iron from the intestine. Thus iron is not absorbed until the body requires it.

Because of the importance of iron for haemoglobin synthesis, the most marked effect of iron deficiency is anaemia—a deficiency of haemoglobin in red blood cells. This state reduces the oxygen carrying power of blood and in order to maintain normal supplies of oxygen to the tissues the circulation rate has to increase and tissues have to abstract an abnormally large percentage of the oxygen reaching them in the circulating blood.

Iron was used empirically for the treatment of anaemia by the Greek physicians and the ancient Hindus. In 1681 Lydenham, a great English physician, recommended 'iron or steel filings steeped in cold Rheinish wine' for patients with pallor and fatigue, a condition then called chlorosis. This was before it was recognized that iron was a component of blood.

ZINC

Quite large amounts of zinc are present in the human body (2,300 mg/70 kg adult). The element is present in various enzymes e.g. some dehydrogenases and carbonic anhydrase. The highest concentrations of zinc present in the animal body occur in the pancreas where the element is associated with the hormone insulin stored in the islets of Langerhans. Zinc is essential for the growth of plants and animals and since most foods contain the element no disease due to zinc deficiency has been described in man. Animals fed artificial diets deficient in zinc show marked disorders with loss of weight, failure of growth and abnormalities in the nervous system. There is a fatal disease of pigs—porcine parakeratosis—due to zinc deficiency. The animals develop dermatitis, diarrhoea, weight loss and eventual death. The condition is readily treated by giving zinc supplements.

An excess intake of zinc, and many other trace elements, can lead to serious disorders. An excess intake of the element is usually associated with exposure in the metal industry—e.g. inhalation of zinc fumes in 'brass founders ague'. Excess amounts of zinc may also be taken in with acidic food prepared in galvanized iron pans.

COPPER

Copper is an essential element in the diet since it is needed for the synthesis of various enzymes and for the formation of haemoglobin. The element is present in water, soil and all animals and plants. It is required by micro organisms, plants and animals. For these reasons it is difficult to provide a diet which is too poor in copper. A natural deficiency of the element is thus rarely seen. Milk is one food which contains little copper and it has been used as a diet in experimental animals to produced copper deficiency. If animals such as sheep, pigs and rats are maintained on a milk diet for some time a severe copper deficiency develops which is associated with a severe anaemia (small red blood cells with a deficient haemoglobin content). With a more prolonged copper deficient diet other disorders appear—disturbance of bone growth, hair and skin.

Only small daily amounts are required by normal man (1–3 mg/day). This amount is readily provided by a normal diet.

COBALT

Human beings and many other animals take in the essential trace element cobalt after it has already been incorporated into vitamin B_{12}, which contains about 4·5% cobalt. The daily human requirement of vitamin B_{12} is 1–2 micrograms (0·001 mg) which contains about 0·005 micrograms of cobalt.

Ruminants such as the sheep acquire their supplies of B_{12} from microorganisms in the upper alimentary tract. A supply of elemental cobalt is needed by these animals to provide enough for B_{12} synthesis by the microorganisms. About 0·1 mg cobalt/day is needed by sheep.

The functions of vitamin B_{12} and the effects of deficiency are described on page 412 and will not be pursued here. A deficiency of elemental cobalt (i.e. not incorporated into B_{12}) is felt mainly by ruminant animals —for reasons we have described above. In some parts of the world the soil is deficient in cobalt and a deficiency of the element may develop in ruminants which feed on plants in this region. In Western Australia there is a fatal wasting disease of sheep and cattle due to deficiency of elemental cobalt in the soil and plants. The animals develop a severe anaemia, loss of appetite, diarrhoea, a rough coat and scaly skin. Because of the failure of normal red blood cell development and haemo-globin formation the iron supplies of the body are not fully used and heavy deposits of iron appear in the liver and spleen. This disease is readily treatable with cobalt. In Southern Australia another deficiency disease occurs in grazing sheep, the so called 'coast disease'. This is due to a deficiency of both copper and cobalt and is treated by giving both of these elements.

IODINE

Only small amounts of iodine are present in body fluids. Some of this is present as inorganic iodide but much (80 %) is organically bound iodide. In the latter, iodide is present in the molecules of thyroid hormones—thyroxine and to a lesser extent triiodothyronine. This is the main function of dietary iodine—as a component of thyroid hormones. The thyroid tissue has a great affinity for iodine and can trap the element from the circulating blood and concentrate it into thyroid tissue where its concentration may be several hundred times that in the circulation. In the thyroid gland iodide is oxidized to iodine which is then incorporated into the amino acids which constitute the thyroid hormones.

Only small amounts of dietary iodine are needed to maintain iodine balance and normal function of the thyroid gland. A few micrograms of iodine per kg/body weight is all that is needed by normal individuals. Certain regions of the world, particularly in the centre of large land masses, have a deficient amount of iodine in water, soil and plants. This leads to disturbance of thyroid function which may be seen especially at period of growth and activity—in infancy, childhood, puberty and pregnancy. Because the thyroid gland cannot manufacture sufficient amounts of the vital hormones the pituitary gland increases its stimulus to the thyroid (thyrotrophic hormone) and the thyroid gland enlarges in 'an attempt' to produce more hormone (see also p. 251). However the 'attempt' is doomed to failure because iodine is an indispensable part of the molecule of thyroid hormones. These enlarged glands are called goitres. Goitres used to be a frequent disorder in Derbyshire (Derbyshire neck), Switzerland and the central States of North America. In infants and children severe disorders may arise from a deficiency of iodine. These include 'endemic cretinism'—a condition in which the importance of the thyroid hormones for growth and development is all too clearly demonstrated by the sexually immature imbecilic dwarf, known as a cretin.

Giving extra iodide can prevent or treat all these disorders. In fact iodine has been used empirically for the treatment of goitres for more than a hundred years. One of the first rational uses of iodide was that of two workers called Marine and Kimball (1917) who used iodine to prevent the development of goitre in a group of school children in Ohio (U.S.A.) a region where iodine deficiency and endemic goitre is common. The success of this work led to the widespread use of prophylactic therapy with iodine in most regions of the world where goitre is endemic. Small amounts of iodide is added to table salt; about 1 part of potassium iodide to 100,000 parts of sodium chloride gives an adequate dose.

TEETH AND FOOD

The structure of a typical tooth.

ENAMEL. The structure of a typical mammalian tooth is shown in fig. 167. That part of the tooth which sticks out above the gum is called the crown. It is capped with a substance called enamel, which is the hardest material of the body. It consists of cylinders of very hard

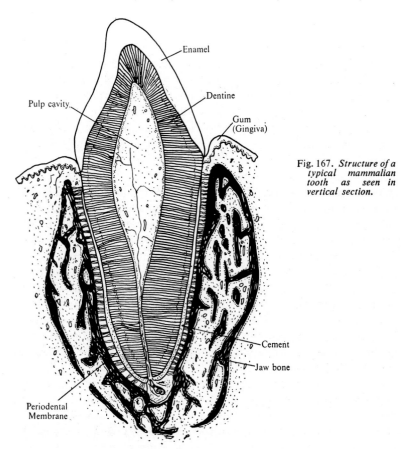

Fig. 167. *Structure of a typical mammalian tooth as seen in vertical section.*

inorganic material which are aligned perpendicularly to the surface. The enamel consists of 96 % mineral matter and is the only part of the tooth formed from the epidermis, all the rest of the tooth being mesodermal in origin. The enamel is completely formed before the tooth erupts from the gum.

DENTINE. This material forms the bulk of the tooth and consists of an organic fibrillar network in which mineral salts, mainly of calcium, are deposited. About 70% of dentine is mineral matter. It is a hard substance, harder than bone and is commonly called ivory. The dentine is perforated by canals, which run at right angles to the surface, in which lie the processes of the odontoblasts, the mesodermal cells which produce the dentine. The odontoblasts line the inner edge of the dentine layer and continue to lay down dentine as long as they are adequately nourished. They are supplied by the blood vessels of the pulp cavity in which the cells lie. In most teeth the base of the pulp cavity, at the root of the tooth, becomes constricted soon after the teeth have reached their full size, so reducing the blood supply to the odonto-blasts. Dentine production now ceases and the tooth stops growing. Some teeth such as the incisors of the rabbit or the molars of a sheep continue to grow throughout life as they are worn away by the friction with the food material and with the opposing teeth. They can do this because the base of the pulp cavity never closes and thus the odonto-blasts receive a rich blood supply. Teeth which have a non-occluded pulp cavity are said to have open roots.

CEMENT. This is a form of bone which surrounds the root and the neck of the tooth i.e. all the parts of the tooth embedded in the gum.

PERIODONTAL MEMBRANE. The tooth develops in the gum at the same time as the bone is growing and the jaw bone grows around the tooth. The cup like depression in the jaw bone into which the tooth fits is called the alveolus and the tooth is held firmly in the alveolus by a strong fibrous connective tissue called the periodontal membrane which connects the cement covered root to the bone of the alveolus.

Development of teeth. Teeth are formed from what are called tooth buds which arise in the developing jaw. The tooth bud consists of a cup like enamel organ which fits over a group of mesodermal cells called the dental papilla. The cells of the papilla produce the dentine forming the bulk of the tooth. The enamel is produced by the cells of the enamel organ.

The first sign of the formation of teeth is the development of a ridge of thickened epithelium along the line of the jaw, called the dental lamina. At regular intervals along the dental lamina groups of epithelial cells proliferate into the mesodermal tissues of the jaw and give rise to the enamel organs. The enamel organs consists of a two layered

1)

(1) The epithelium of the jaw sinks in to form the dental lamina (d.l.).

(2)

(2) The lamina forms cuplike enamel organs (e.o.) around groups of mesoderm cells (m).

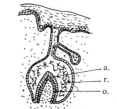

(3)

(3) The inner layer of the cup forms enamel forming cells (ameloblasts a.) but first causes some mesoderm cells to form odontoblasts (o). Inside the enamel organ characteristic reticulate cells (r) are seen.

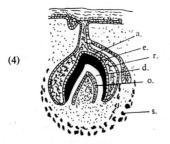

(4)

(4) Ameloblasts form enamel (e). Odontoblats form dentine (d). The bone of the jaw forms a tooth socket (s).

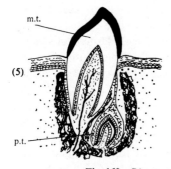

(5)

(5) The root of the milk tooth (m.t.) grows as more dentine is added. The permanent tooth (p.t.) germ is developing.

Fig. 168. *Diagrams of the stages of tooth development.*

cup and within the cavity of the cup are the mesodermal cells of the dental papilla. When there are two sets of teeth, the milk teeth and the permanent teeth, the dental lamina has two rows of tooth buds, the uppermost being the milk tooth buds, the lower row the permanent teeth, which may develop several years later pushing the milk teeth out of the jaw as they grow upwards.

The cells of the inner layer of the enamel organ produce the enamel and induce the cells of the outer edge of the dental papilla to form the dentine. The remainder of the papilla is eventually invaded by nerves and blood vessels to form the tissues of the pulp cavity. The stages in the development of the tooth are shown in fig. 168.

The jaw bone eventually grows round the developing tooth to form the socket or alveolus, to which the tooth eventually becomes firmly attached by way of the periodontal membrane.

Variations in dentition

Food is complex organic material which we have seen has been synthesized in the first place by green plants. This food may be transformed into the protoplasm of animals when the plants are eaten. The protoplasm of plants is surrounded by a cellulose cell wall which during the period known as secondary thickening in the life of the plant, may become chemically altered or impregnated with various substances like lignin. Walls of plant cells are often woody or in the case of grasses they have a high silica content which makes them very hard. Plant material is much harder to chew than the soft juicy flesh of animals.

During the evolution of mammals from reptiles one of the significant trends has been the formation of the false palate in the roof of the mouth, separating the breathing passage from the mouth. The maxillae and the palatine bones have grown to form this shelf. The significance of the palate is that its possessors have the ability to breathe whilst the mouth is full of food. This development may well be associated with the evolution of warm bloodedness and the need for constant breathing to supply energy in respiration. As soon as food can be retained in the mouth without preventing breathing there is virtue in having a battery of teeth which can deal with this food. In all classes of vertebrates below the mammals the teeth are only of use to stop the prey escaping from the mouth and all the teeth are of similar shape—a simple pointed cusp. With the development of the palate in the mammal is associated the development of teeth of different kinds suited to the special kind of food which the animal eats. Special kinds of teeth are absolutely essential to deal with food such as grass, so much so, that one of the reasons given for the sudden eclipse of the giant dinosaurs at the end of the mesozoic era is that the grasses were becoming the dominant vegetation in the cretaceous period (100 million years ago) and the dinosaurs' dentition was not capable of dealing with such hard siliceous material. In order to understand the adaptive radiation which has occurred in mammals let us look more closely at a few different types of mammalian dentition. It is interesting to comment here that in evolution once a great step forward has been taken e.g. when the amphibians came onto land, or when the mammals learned how to maintain a constant body temperature there frequently follows a period of great evolutionary experimentation. Different combinations of genes produce different characters and those which are successful are retained as their possessors live long enough to reproduce and pass on these gene combinations to their offspring. When the palate was formed there seems to have been an opportunity to try several methods of dealing with food. The attempts to exploit this new opportunity for efficiency constitutes a good example of what we call adaptive radiation.

Dentition of the sheep. The teeth of mammals are characterized by being heterodont and di-phyodont. These two technical terms mean that there are teeth of different kinds and that there are two sets of teeth, a set of 'milk' teeth followed by the permanent set. There are four kinds of teeth in the typical mammal and they are called incisors, canines, premolars and molars. The numbers of these teeth present can be expressed in the dental formula e.g. $I.\frac{3}{3}$, $C.\frac{1}{1}$, $P.M.\frac{4}{4}$, $M.\frac{3}{3}$ or more simply $\frac{3.1.4.3.}{3.1.4.3.}$ This formula is derived in the following way. The letters represent the different types of teeth, and the top number after each letter indicates the number of teeth of that kind found in half the top jaw. The denominator of the fraction indicates the number of teeth of this kind found in half of the lower jaw. Thus the typical mammal with the above quoted formula has twelve incisors four canines, sixteen premolars and twelve molars, that is forty four teeth in its permanent dentition.

The formula for the sheep is $\frac{0.0.3.3.}{3.1.3.3.}$, but this does not tell us enough about the dentition so we must draw and describe it. Fig. 169 shows that the incisors of the lower jaw bite against a horny pad on the top jaw. In the upper jaw the canines are absent and a space called the diastema separates the horny pad from the premolars. This space is characteristic of herbivores. The premolars look similar to the molars and together they constitute an effective grinding battery. The teeth in this grinding battery have open roots i.e. they continue to grow throughout life. As they wear away they do so very unevenly because they are made of three different substances. The dentine, enamel, and cement, wear at different rates and leave a crescentic pattern on the

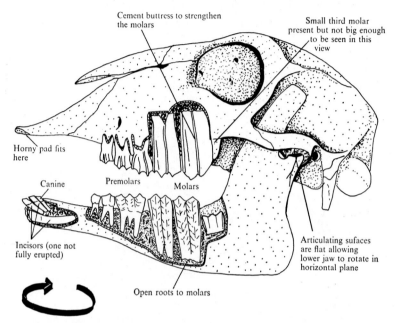

Cement buttress to strengthen the molars

Small third molar present but not big enough to be seen in this view

Horny pad fits here

Canine

Premolars

Molars

Incisors (one not fully erupted)

Articulating sufaces are flat allowing lower jaw to rotate in horizontal plane

Open roots to molars

Fig. 169. *Drawing of the skull of a young sheep, with some bone cut away to show the roots of the teeth. In an older sheep the last molars are as big as the second molars and the lower first incisor erupts. The arrow indicates the plane of movement of the lower jaw.*

surface of the tooth. There are sharp ridges of the hard enamel and slightly softer dentine passing from front to back of the tooth surface and also from side to side of the tooth surface, so that whichever way the bottom jaw moves, grinding is sure to occur. If we look at the jaw joint we see that it is very flat allowing the lower jaw to move in a circular path. If you watch a sheep chewing you will see that this is in fact how the lower jaw works thus exploiting the grinding ridges on the tooth surface. The sheep's jaw is not a strong one and examination of a sheep's skull will show how very easy it would be to dislocate the jaw, but this does not matter to the sheep for the grass does not struggle violently when it is bitten. If you look carefully at the molars and premolars of the sheep you will see that the sides of these teeth are strengthened, especially at the corners, by buttresses of cement. These pillars of cement serve to prevent the edges of the tooth being chipped off by the siliceous food.

The dentition of the dog. The appearance of the dog's dentition indicates at once that it is a flesh eater although many animals which are very closely related e.g. the fox are known to have a very mixed diet often including insects. The formula is $\frac{3.1.4.2.}{3.1.4.3.}$ The dog belongs to the genus Canis and it is not for nothing that the canine is so called. These canine teeth are well developed in the dog and are used as a weapon of defence and attack. They are used to spear the prey. If you watch a dog chewing meat or better still, a bone, you will see how he turns his head on one side and gets the food to the angle of

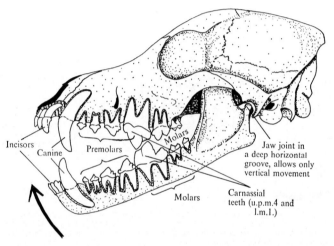

Fig. 170. *Drawing of the skull of a dog with some bone cut away to show the roots of the teeth. The arrow indicates that the lower jaw moves in the vertical plane.*

his jaw. This is where the carnassial teeth are. They are specially designed teeth for cutting flesh and are developed from the last upper premolar and the first lower molar, although in other carnivores different teeth in the molar battery may be involved. Fig. 170 shows the carnassial teeth of the dog; they have vertical surfaces which act like a pair of scissor blades because the jaw joint in the dog does not allow side to side movement but only movement in the vertical plane. The lower jaw is inserted into the skull by the long transversely running condyle of the lower jaw, which is housed in a deep transversely disposed groove in the skull. Fig. 170 will make the structure clear and indicates how movement is restricted to the vertical plane only. The deep jaw joint is essential in order to prevent dislocation of the joint when the prey struggles. The emphasis in the carnivore is on the canine and carnassial teeth, on attack and on chopping up the meat into chunks, which are then quickly swallowed.

Dentition of the rabbit. If the sheep can be called a grinder and the dog a chopper-up of its food, then the word for the rabbit is a nibbler. The formula is $\frac{2.0.3.3.}{1.0.2.3.}$ and the emphasis is undoubtedly on the incisors, which have open roots and grow continually throughout life so much so that if you keep any rodents as pets you should be sure to provide them with something hard to chew like a piece of wood so that they can keep their incisors sharp and prevent them from becoming

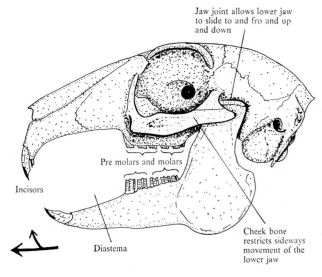

Fig. 171. *Drawing of the skull of a rabbit. The arrow shows that the lower jaw moves to and fro and in the vertical plane.*

overgrown. The front surfaces of rodents' incisors have a very thick layer of enamel and as the curved tooth wears it leaves this sharp edge of enamel exposed, because the enamel is more resistant to wear than the rest of the tooth. (See fig. 171.) The rabbit has the typical large diastema of the herbivore. There is no superficial difference between the molars and premolars, and together they form a very efficient grinding battery. The ridges on the surfaces of the molars are transverse and well designed, when it is realized that the jaw moves to and fro in the vertical plane only. The movement of the jaw is restricted by the method of insertion of the dentary (lower jaw bone). The articulating surface is the top edge of the end of the dentary and it fits into a longitudinal groove in the squamosal bone of the skull, whilst any side to side movement is obviously restricted by the quadratojugal arch of the cheek bones (see fig. 171).

We have seen in the sheep, dog and rabbit three very different sets of teeth, each designed with a jaw joint adapted to the special type of dentition. Each of these three animals is a very specialized feeder and by comparison man has a very unspecialized set of teeth. This may be related to the fact that there is little selective value for man in having specialized teeth, since he has the ability to use tools to help him in his feeding.

DIGESTION AND THE ALIMENTARY TRACT

The principle of digestion. The food ingested by the mammal contains a great variety of compounds which have been incorporated into the tissues of other animals and/or plants. Many of these compounds are highly complex, with high molecular weights and are often insoluble in water. The function of digestion is to alter the ingested food to make it available for absorption into the body, and in order to do this many of the compounds have to be broken down into simpler substances of lower molecular weight, soluble in water and capable of being absorbed through the mucous membrane of the intestine. Thus protein molecules must be broken down to their constituent amino acids before they are absorbed. Each animal species, indeed each individual, builds up its protein from constituent amino acids in a special pattern and if food proteins were absorbed directly into the blood stream they would act as antigens (see p. 567) and call forth the production of antibodies; once antibodies have been formed then continued absorption of the protein into the blood stream could lead to a fatal shock reaction. Thus apart from the physical problems of absorbing large complex protein molecules into the blood, digestion of proteins into constituent amino acids is virtually a biological necessity. Carbohydrates are stored by plants and animals in complex forms viz. starch, cellulose, glycogen, and these must be broken down into simpler substances, e.g. glucose, before absorption takes place. Even the simpler carbohydrates in the form of di-saccharides are not absorbed as such but are converted to monosaccharides. Fats are different in that some fat may be absorbed as such, in the form of very fine particles, and various digestive processes are engaged in producing such fine particles.

The complex food is broken down into simpler substances by the process of hydrolysis, in which a great variety of enzymes are engaged. Although we may describe the hydrolysis of foods as occurring in varying stages, each stage catalyzed by a different enzyme, perhaps in different parts of the alimentary tract, it is necessary to try to view the process as a whole in which many enzymes are acting in integration to achieve the hydrolysis of the food-stuffs.

The digestive abilities of animals vary from one species to another. Herbivores for example employ special mechanisms (p. 398) for the digestion of the carbohydrate cellulose, which is such a predominant part of the structure of plants. Probably no mammal possesses an enzyme capable of hydrolyzing cellulose and they have to rely upon bacteria to do this for them. Carnivores are unable to digest cellulose, and indeed this carbohydrate does not form a part of the diet of such animals.

The alimentary tract

The alimentary tract is a long hollow muscular tube starting at the mouth and ending at the anus. Along its length are well defined regions, buccal cavity, pharynx, oesophagus, stomach, duodenum, ileum, caecum, appendix, colon and rectum. Opening into the alimentary tract are the ducts of several glands which produce secretions concerned with the digestion of food; these glands include the salivary glands, pancreas and liver. The gut is suspended from the dorsal wall of the coelomic cavity by a double layer of the peritoneum which lines the coelom, and between the two layers of peritoneum pass the blood vessels, nerves and lymphatics which supply the gut.

The basic structure of the gut is shown in transverse section in fig. 172. The innermost layer consists of the mucosa. In the buccal cavity and anal canal this consists of a layer of squamous stratified epithelium since these regions were formed in the embryo by intuckings of ectoderm. The remaining portions of the gut (except the oesophagus) are lined by a simple columnar epithelium. In some regions there are intuckings of this epithelium into the deeper layers of the wall of the gut to form glandular structures producing digestive secretions. In the small intestine the mucosal layer is highly folded to form the intestinal villi, which greatly increases the surface area available for the absorption of digested food-stuffs. Lying beneath the mucosa is a loose connective tissue, containing blood vessels and lymphatics, called the submucosa. In most regions of the gut this submucosa contains a thin layer of muscle, the muscularis mucosa.

Outside the mucosa and submucosa are the muscle layers of the gut, an outer longitudinal and an inner circular layer. These layers of muscle consist of smooth muscle fibres except in the upper part of the oesophagus where striated muscle is found. The muscle coats are responsible for propelling the food along the gut in a movement called peristalsis. When the gut tube is stretched by a bolus of food the circular muscles on the oral side (mouth) contract thus pushing the bolus of food onward into a region where the muscles are relaxed.

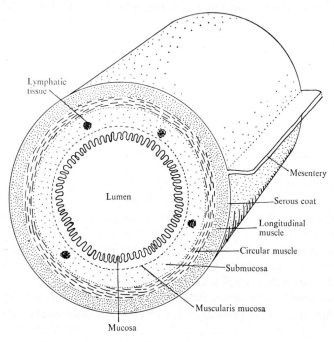

Fig. 172. *Diagram to show the basic structure of the gut as seen in transverse section.*

The outermost layer of the gut is formed by the serous coat, a thin, smooth surfaced layer of peritoneum. When the abdominal cavity is opened this layer of the gut is found to be moist and shining. There is a small amount of free serous fluid within the peritoneal cavity and this serves to lubricate the covering of the intestines so that they can move one on the other freely, without friction.

The movement and secretory activity of the gut is under the influence of the autonomic division of the nervous system. But even if all the nerves supplying the gut are severed it is still capable of peristaltic movement and secretory activity. This is in part due to the fact that the wall of the gut contains its own intrinsic network of nerves. There are two systems within the gut wall, the myenteric plexus of Auerbach lying between the muscle coats and the submucous plexus of Meissner in the submucous layer of the gut.

The mouth and oesophagus. In the mouth the ingested food undergoes changes in its physical and chemical structure. These changes are brought about by the effects of the teeth, tongue and the saliva which is produced by the salivary glands and poured into the oral cavity along the salivary ducts. In addition the mouth is the seat of the sensation of taste which provides the sensory basis for the reflex production not only of salivary juices but of secretions of the gastric mucosa.

The oral cavity is lined by a stratified epithelium which has its origin from an intucking of ectoderm, the stomodaeum, in the embryo. The lining of the mouth is subjected to much friction as the food is masticated and the outer layers of the epithelium are being continually shed and replaced by the deeper layers. If a drop of saliva is examined under the microscope it will invariably be found to contain numbers of flattened squames which represent the outer dead layers of the epithelium of the mouth (buccal cavity).

THE SALIVARY GLANDS AND SALIVA. In man there are three pairs of salivary glands, the parotid, submandibular and sublingual glands. These discharge their secretions by ducts which open into the oral cavity. The openings of the submandibular glands can be seen as two papillae under the anterior end of the tongue where the fold of mucous membrane called the fraenum is attached to the floor of the mouth. If the taste buds have been suitably stimulated e.g. by some lemon juice placed on the tongue, then saliva will be seen to be pouring out of these papillae. The opening of the parotid ducts can be seen on the inner side of the cheek opposite the upper second molar tooth.

There are two ways in which the salivary glands may be stimulated. First, by means of a reflex from the taste buds on the tongue. There are receptors on the tongue which enable us to distinguish bitterness, sweetness, sourness and saltness and the position of these receptors has been mapped out by placing different substances at different sites on the tongue which has been prepared for the experiment by removal of excessive saliva. In this way maps have been made showing the sites of the different sense organs (fig. 173). Saliva is also secreted in response to such stimuli as the sight or smell, sometimes even the thought, of food.

Saliva contains water, mineral salts, mucin and an enzyme called ptyalin. The exact composition varies according to the type of stimulus, thus the saliva contains more mucin and ptyalin following stimulation by bread and meat than by lemon juice. About 1·0–1·5 litres of saliva are produced each day in man. The water and mucin moisten and lubricate the food so facilitating swallowing. The enzyme ptyalin is an amylase and it begins the conversion of starch (amylose and amylopectin) and glycogen into maltose, with the production of a small amount of glucose. Ptyalin works best in the slightly alkaline conditions of the mouth, and the action of the enzyme is helped by the

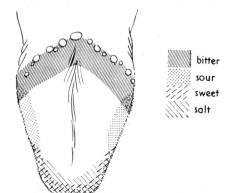

bitter
sour
sweet
salt

Fig. 173. *Diagram showing the distribution of the sense endings of the four primary taste qualities (sour, salt, bitter and sweet) in man. The tip of the tongue is sensitive to all four qualities but particularly to sweet and salt. The sides of the tongue are sensitive to sour stimuli but may also respond to salt. The posterior basal part of the tongue is sensitive to bitter stimuli. Note that most of the upper surface, particularly that near the midline is relatively insensitive.*

presence of chloride ions. Because food does not stay an appreciable length of time in the mouth the salivary amylase does not play a significant part in the digestion of carbohydrates. However, the action of ptyalin continues for a time in the centre of the bolus of food even when it has reached the stomach, until the hydrochloric acid of the gastric juice reaches the centre of the bolus and inactivates the ptyalin, by virtue of the low pH.

MASTICATION AND SWALLOWING. The degree to which food is masticated by the teeth and tongue varies from species to species. Carnivores are notorious bolters of the food. In man the food is masticated to a variable degree to produce a bolus of food, by the action of the tongue against the hard palate. The bolus is then pushed to the back of the mouth by raising the front of the tongue. The bolus is then rapidly and reflexly ejected into the pharynx and the upper end of the oesophagus. During this time the mouth cavity is closed from the pharynx by the tongue pressing against the palate, and the naso-pharynx is closed off by elevation of the soft palate. Breathing movements cease and the larynx is lifted up under the base of the tongue so that the laryngeal opening is protected against the inhalation of food.

THE OESOPHAGUS. The oesophagus is a thick walled muscular tube through which the food passes from the pharynx to the stomach. Most of the oesophagus has no outer peritoneal coat and its outer layer consists of a loose connective tissue, the adventitia. The muscular coats are very prominent particularly the outer longitudinal layer. At the upper end of the tube the muscle fibres are striated, giving way lower down to unstriped muscle fibres. There is a thick submucous layer in which many secretory glands occur, the oesophageal glands. These pour their secretions into the lumen of the oesophagus and so lubricate the epithelial surface to reduce the friction of the passage of food. The epithelial lining of the oesophagus consists of a stratified epithelium. In the resting organ the surface of the epithelium is folded and this permits the accommodation of food.

Liquids and soft food pass down the oesophagus at a fast rate, by the force of the initial act of swallowing and may reach the lower end of the oesophagus in 0·1 sec. Food of this consistency tends to pass directly through the cardiac sphincter into the stomach. More solid food passes down the oesophagus by peristaltic action of the oesophageal muscles. A peristaltic wave takes about five seconds to reach the cardiac sphincter which relaxes to allow the passage of the food into the stomach. There has been much argument as to whether there is a true sphincter at the junction of the oesophagus and the stomach, the cardiac sphincter. Although there may be no anatomical sphincter there seems to be a functional one.

The stomach. The stomach is a muscular bag whose main function is to store the food and convert it into a semi-liquid material called chyme before it passes into the duodenum. Part of the stomach, indeed sometimes almost the entire organ, may be removed surgically, without interfering generally with life. But the patient has to eat small regular meals, emphasizing the normal function of the stomach— which is storage. The human stomach is a J-shaped organ as shown in fig. 174. Its size varies greatly, according to the food contents, but the average capacity is about two pints. Four main regions of the stomach are described:

1. The cardiac region which surrounds the region of the cardiac sphincter.

2. The fundus, an air-filled portion, lying adjacent to the cardiac region.

3. The body of the stomach, the main portion where food is stored and where the food becomes more fluid as the process of digestion begins.

4. The pyloric region, the distal part of the stomach which opens into the duodenum by way of the pyloric sphincter.

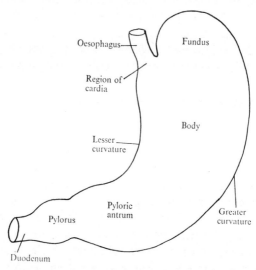

Fig. 174. *Diagram showing the main regions of the human stomach.*

HISTOLOGICAL STRUCTURE OF THE STOMACH. The mucous membrane of the stomach is thrown into large folds which are taken up as food is accommodated; this is made possible by the loose submucous tissue. In addition to these large folds there is a system of very much smaller grooves covering the entire surface of the mucosa. The surface of the mucosa is also pitted by small apertures which open into the lumen of the gastric glands which occupy the thickness of the mucosa. The glands vary in structure from one region of the stomach to another. The surface of the stomach is covered by a layer of columnar cells which secrete an alkaline mucus, and this layer of cells merges into the stratified epithelium of the oesophagus at the cardia and with the intestinal epithelium at the pylorus.

There are four types of cell in the gastric glands figs. 175, 6, 7, the chief or zymogen cells, the parietal (or oxyntic) cells, the mucous neck cells and the argentaffine cells. The zymogen cells produce the pepsinogen which is converted into pepsin when secreted into the stomach. These cells are situated in the lower half of the tubule of the gastric gland. In the resting state they contain granules. The mucous neck cells are situated in the neck of the gastric gland. The parietal cells are situated between the zymogen cells. Their cytoplasm stains red with aniline dyes. They

contain an intracellular canaliculus, where it is thought that the hydrochloric acid is produced, away from the cytoplasm of the cell. The
argentaffine cells are thought to be concerned with the production of

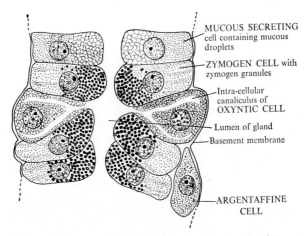

MUCOUS SECRETING
cell containing mucous
droplets

ZYMOGEN CELL with
zymogen granules

Intra-cellular
canaliculus of
OXYNTIC CELL

Lumen of gland

Basement membrane

ARGENTAFFINE
CELL

Fig. 175. *Diagram of part of a gastric gland showing four different
types of cell.*

the intrinsic gastric factor (of Castle) (see p. 412). These cells are shown
in fig. 175 which is a section of the fundus of the human stomach. The
glands of the pyloric region are rather different from those in the fundus

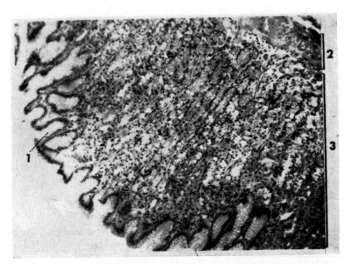

Fig. 176. *L.P. photomicrograph of a section of fundic stomach.* 1. gastric
pits. 2. muscularis mucosa. 3. mucous membrane with gastric glands.

and body of the stomach in that the gastric pit reaches much deeper in the mucous membrane (fig. 178). In addition the glands are more branched and coiled. The secretion of the pyloric glands contains mucus but no enzymes.

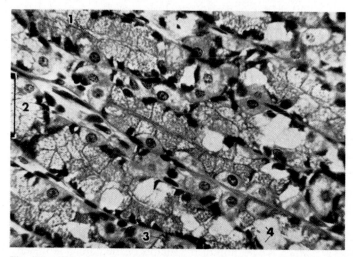

Fig. 177. *H.P. photomicrograph of a section of fundic stomach.* 1. zymogen cells. 2. gastric gland. 3. oxyntic cells. 4. mucous secreting cells.

As in other regions of the gut the stomach wall contains a submucous layer and muscularis mucosa. From the muscularis mucosa strands of fibres extend towards the surface of the mucous membrane between the glands, and the contraction of these fibres probably assists in the emptying of the glands. The external muscle coat of the stomach consists of an outer longitudinal, a middle circular and an inner oblique layer. In the region of the pylorus the middle circular layer forms a thick sphincter, the pyloric sphincter. Externally the stomach is limited by a serous coat.

THE SECRETIONS OF THE STOMACH. The secretion of the stomach, sometimes called gastric juice, contains free hydrochloric acid, mucus and the enzymes pepsin and renin. In addition there may be substances which have been regurgitated back through the pylorus from the small intestine, including bile and some of the intestinal enzymes. Gastric juice also contains the intrinsic factor of Castle which is necessary for the absorption of vitamin B_{12} (p. 412). Large volumes of gastric juice are produced each day and in man this amounts to 2–3 litres.

Hydrochloric acid. Hydrochloric acid has its origin in the parietal cells of the gastric glands. The cytoplasm of these cells has been shown to be neutral and the hydrochloric acid is presumably formed within the intracellular canaliculus (fig. 175). These cells contain the enzyme

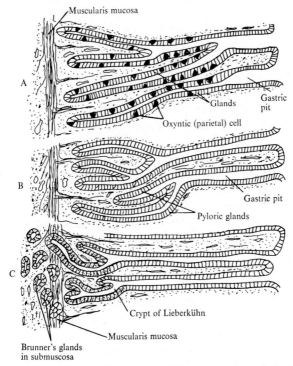

Fig. 178. *A diagrammatic comparison of the glands in* A. *the fundus,* B. *the pyloric region and* C. *the duodenum.*

carbonic anhydrase which catalyses the formation of carbonic acid from carbon dioxide and water. The hydrogen ions of hydrochloric acid are derived from dissociated carbonic acid, the chloride ions from sodium chloride.

The hydrochloric acid released from the parietal cells performs two main functions. Firstly it forms the acid medium in which the enzyme pepsin works under optimum conditions. The pepsin is secreted from the chief or zymogen cells in an inactive form pepsinogen, which is converted into the active form pepsin under the influence of the hydrochloric acid. In addition the acid medium of the stomach prevents the growth and multiplication of bacteria and it kills most of the organisms ingested with the food.

The control of gastric secretion. Three phases of gastric secretion are described, the nervous, gastric and intestinal phases. The first stimulus to the production of gastric juice is a nervous one. When we see, smell or taste food, impulses pass along the vagus nerve to the stomach stimulating the production of gastric juice, and this prepares the stomach to receive the food. This was first demonstrated by Pavlov in dogs. He operated on dogs in order to bring the oesophagus out to open in the neck so that food could not pass directly into the stomach. In the same dogs he prepared pouches of stomach which were brought to the surface of the body so that specimens of gastric juice could be easily obtained. When these animals ate, the food passed out from the opening in the neck but gastric juice was still produced from the stomach. When the vagus nerves to the stomach were cut the production of gastric juice ceased.

The next phase of gastric secretion is called the gastric phase and depends upon the fact that the contact of food with the gastric mucosa stimulates the production of gastric juice. This is not a direct effect but the food causes the mucosa of the pyloric region to produce a hormone called gastrin which circulates in the blood back to the rest of the stomach which responds by the production of gastric juice. The gastric juice produced in response to gastrin appears to be predominantly acid and is not rich in pepsin.

When semi-digested food reaches the small intestine there is a further reflex production of gastric juice, perhaps based upon the production of another hormone from the mucosa of the duodenum.

DIGESTION IN THE STOMACH. There are only two known enzymes produced in the stomach, pepsin and rennin. Pepsin is secreted in the inactive form pepsinogen and activated by hydrochloric acid. Once pepsin is formed it is capable of activating more pepsinogen. Pepsin belongs to a group of protein digesting enzymes called endopeptidases (which include the enzymes trypsin and chymotrypsin produced by the pancreas). These enzymes by acting upon the peptide bonds (p. 104) within the molecules of protein break up the large protein molecules into smaller fragments. These smaller fragments are then further broken down by a group of enzymes, produced by the small intestine, called exopeptidases. These exopeptidases act by breaking down terminal peptide bonds to liberate free amino acids. The differences between pepsin, trypsin and chymotrypsin lie in the fact that each enzyme is capable of breaking down only certain peptide bonds, depending upon the chemistry of the protein molecule adjacent to the peptide bond.

In the stomach then, protein digestion is begun by the action of pepsin, which, by acting as an endo-peptidase, breaks down the molecules of protein into smaller fragments. These smaller fragments, sometimes called polypeptides, are broken down further when they pass into the duodenum and small intestine and meet other enzymes.

There is no fat digesting enzyme produced in the stomach but the condition of the fat is altered by the warmth and churning action of the stomach. Further, globules of fat are liberated from animal tissues when these become softened and partially digested by the enzyme pepsin.

There is no enzyme produced which is capable of digesting carbohydrates, but the action of the enzyme ptyalin, which is mixed with the food, continues until the pH of the gastric juice falls sufficiently to inactivate this enzyme. Some disaccharides e.g. sucrose, are hydrolysed by the presence of dilute hydrochloric acid into monosaccharides. The special features of the stomach of some herbivores for the digestion of carbohydrates is described on p. 397.

Rennin is another proteolytic enzyme and is characteristically found in the gastric juice of young mammals. It is secreted in an inactive form, pro-rennin, which is activated by the hydrochloric acid of the gastric juice. Rennin catalyses the conversion of the protein of milk, caseinogen, into paracasein which is precipitated in the stomach as a calcium salt. The precipitated paracasein forms a firm curd in the stomach. This process ensures that milk stays for some time in the stomach so that it becomes exposed to the action of the proteolytic enzymes.

STOMACH MOVEMENTS AND EMPTYING. The contents of the stomach are mixed by waves of peristaltic contraction which pass along the stomach from cardia to pylorus. The pyloric sphincter is open for most of the time but contracts as a wave of peristalsis reaches it. Food is free to leave the stomach for the duodenum as soon as the consistency is sufficiently fluid and as soon as the peristaltic waves are strong enough to force the chyme out through the bottle neck which the pylorus represents. The rate at which the stomach empties depends upon the type of food contained. Naturally fluids tend to pass out quicker than solid foods. Fats have a characteristic effect in slowing the rate of stomach emptying and this explains why hunger does not return so quickly after a meal containing a good proportion of fat as after a light carbohydrate meal. There is also a possibility that a hormone, called enterogastrone, produced by the duodenum exerts a controlling influence on gastric motility and emptying.

The small intestine. The chyme produced by the stomach is poured into the first part of the small intestine, the duodenum. Here the digestive juices produced from the mucosa of the duodenum itself mingle with the external secretion of the pancreas and the bile produced in the liver, and the process of digestion already started in the mouth and stomach is carried further.

THE STRUCTURE OF THE SMALL INTESTINE. The small intestine is, for the purpose of description, divided into three parts, the duodenum, jejunum and ileum, and while there are differences in their structure, they are basically similar.

In order to increase the area of contact between the food and the wall of the intestine, the mucosa is thrown up into circular folds. Further, the entire mucosa is covered by fine outgrowths, the villi, like a carpet with a fairly close pile. Opening between the bases of the villi are the orifices of glands, the crypts of Lieberkuhn. The surface mucosa consists of a layer of mucous secreting columnar epithelium. This is subjected to considerable wear and tear and the cells are replaced by mitotic division of cells in the crypts of Lieberkuhn. Opening into the crypts are the glands which in the duodenum take the form of highly branched and coiled structures which penetrate the muscularis mucosa (contrasting with the appearance of the stomach where the glands do not penetrate the muscularis) (see fig. 178).

The submucous layer of the intestine contains quantities of lymphoid tissue, varying in amount from region to region. The muscular layer of the wall consists of an outer and inner layer. Although they are usually described as an outer longitudinal and an inner circular layer, both layers are circular, the outer one forming a wide spiral, the inner forming a close spiral.

PANCREATIC SECRETION. Pancreatic tissue consists of a mixture of two glandular elements. The main bulk of the pancreas consists of the glandular tissue which discharges its digestive secretions by way of the pancreatic duct into the duodenum (fig. 179). The other component, the islets of Langerhans, produces the hormone insulin which passes into the blood stream. (See fig. 127.) We are concerned here with the exocrine, digestive secretions.

The secretory activity of the pancreas is controlled in two ways. Firstly the pancreas can be stimulated to produce secretions by nervous influences. A more important factor is the effect on the pancreas of hormones produced by the duodenal mucosa when this is stimulated by the contact of food from the stomach. There seem to be two components of the endocrine secretion of the duodenal mucosa; one hormone, called secretin, is responsible for eliciting a pancreatic secretion rich in sodium bicarbonate, whilst the other hormone, called pancreozymin calls forth a secretion rich in digestive enzymes. Pancreatic juice contains two classes of substance, salts, including sodium bicarbonate, and digestive enzymes. The sodium bicarbonate neutralizes the acid content of the duodenum and produces a mildly alkaline medium in which the digestive enzymes have their optimum activity.

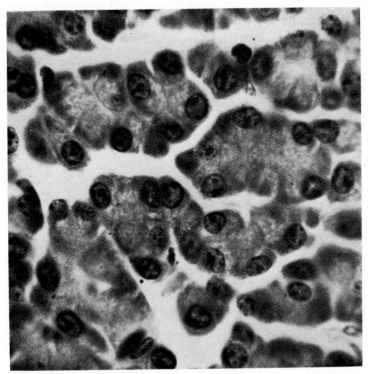

Figure 179. *H.P. photomicrograph of exocrine secreting cells of the pancreas.*

PANCREATIC ENZYMES.

1. *Protein digesting enzymes.* Pancreatic juice contains several protein digesting enzymes. There are two proteolytic enzymes belonging to the class of endopeptidases (p. 386) called trypsin and chymotrypsin. These are secreted in the inactive forms of trypsinogen and chymotrypsinogen and are not activated until they reach the duodenum where they meet a substance produced in the duodenal mucosa called enterokinase. Enterokinase converts the trypsinogen into trypsin which is then able to activate chymotrypsinogen in addition to being able to convert more trypsinogen. Trypsin and chymotrypsin, like pepsin, are endopeptidases and are able to break down the large protein molecules

into smaller fragments. They do this by hydrolyzing peptide bonds within the molecule but each of the above enzymes breaks down the molecule at different places, depending upon the chemistry of the molecule adjacent to the bond. (See fig. 180.)

When the endopeptidases have acted upon the protein molecules the exopeptidases are then able to operate and these attack terminal peptide bonds, liberating free amino acids one by one until only dipeptides

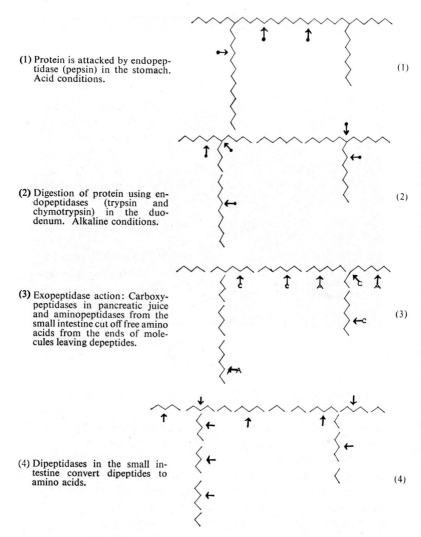

(1) Protein is attacked by endopep-
tidase (pepsin) in the stomach.
Acid conditions.

(1)

(2) Digestion of protein using endopeptidases (trypsin and
chymotrypsin) in the duo-
denum. Alkaline conditions.

(2)

(3) Exopeptidase action: Carboxypeptidases in pancreatic juice
and aminopeptidases from the
small intestine cut off free amino
acids from the ends of molecules leaving depeptides.

(3)

(4) Dipeptidases in the small intestine convert dipeptides to
amino acids.

(4)

Fig. 180. *A schematic representation of protein digestion.*

remain (consisting of two amino acids joined together). These dipeptides are then broken down into amino acids by the action of enzymes called dipeptidases present in the juices of the small intestine. There are two classes of exopeptidases, carboxypeptidases which remove the terminal amino acid when there is a terminal free carboxyl radical (fig. 180) and amino-peptidases which remove the terminal amino acid when there is a free terminal amino group. Pancreatic juice contains, in addition to the endopeptidases trypsin and chymotrypsin, an exopeptidase of the carboxypeptidase class.

2. *Carbohydrate digesting enzyme—pancreatic amylase.* Pancreatic juice contains an amylase which continues the digestion of starch, already started by the salivary amylase, ptyalin, to produce maltose with some glucose.

3. *Fat digesting enzyme—pancreatic lipase.* The digestion of fats does not begin until the chyme produced in the stomach reaches the small intestine where it meets the lipases produced by the pancreas and the intestine. The bile salts, by reducing the surface tension of the fat globules break down the fats into a fine emulsion. This greatly increases the surface area available for the action of the lipases. The lipases break down the fat into glycerol and fatty acid; the fatty acids are removed one by one from the typical triglyceride producing a mixture of glycerol, mono- and di-glyceride.

$$
\begin{array}{l}
CH_2.COOR_1 \\
| \\
CH.COOR_2 \\
| \\
CH_2COOR_3 \\
Fat
\end{array}
\;\xrightarrow{H_2O}\;
\begin{array}{l}
CH_2.OH \\
| \\
CH.COOR_2 \\
| \\
CH_2.COOR_3 \\
diglyceride
\end{array}
\; + R_1.COOH \;\xrightarrow{H_2O}\;
\begin{array}{l}
CH_2.OH \\
| \\
CH.OH \\
| \\
CH_2.COOR_3 \\
monoglyceride
\end{array}
\; + R_2COOH
$$

Fatty acid

$$
R_3COOH \; + \;
\begin{array}{l}
CH_2.OH \\
| \\
CH.OH \\
| \\
CH_2OH \\
glycerol
\end{array}
\; + H_2O
$$

Fatty acid

THE DIGESTIVE SECRETIONS OF THE SMALL INTESTINE. The glands of the small intestine, including Brunner's glands of the duodenum produce a secretion containing enzymes which complete the process of digestion.

There is an exo-peptidase of the amino-peptidase type, also dipeptidases which complete the digestion of proteins by breaking down the dipeptides into their constituent amino acids. These proteolytic enzymes in the juices of the small intestine were formerly known under the one name erepsin but it is now known that this consists of several enzymes.

The starch of the diet has been broken down to maltose by the action of the salivary and pancreatic amylases. In the small intestine an α glucosidase called maltase completes this process by hydrolyzing the maltose into glucose. The other disaccharides are also hydrolyzed, sucrose into fructose and glucose by a glucosaccharase called sucrase, and lactose by a galactosidase called lactase.

BILE. Bile is produced within the cells of the liver and collects in a structure, the gall bladder, on the under surface of the liver, where it is stored and concentrated, and from where it passes intermittently into the duodenum by way of the bile duct. It is a yellow-green, alkaline, mucous fluid containing bile salts, bile pigments, cholesterol and salts.

In the digestive process the function of bile is the emulsification of fats prior to their digestion and absorption. In this the bile salts, sodium taurocholate and glycocholate, are important. After these salts have passed, in the bile, into the intestine and carried out their role in fat digestion, they are reabsorbed and pass in the hepatic portal vein to the liver where they are re-secreted into the bile; there is thus a circulation of bile salts to and from the liver.

Bile also contains excretory products, the bile pigments which are responsible for the colour of the bile. These bile pigments, biliverdin and bilirubin are breakdown products of haemoglobin, after removal of the protein moiety and the iron. Effete red cells are continually being removed from the circulation and their haemoglobin broken down. This process occurs in the reticulo-endothelial system. The iron is stored within the reticulo endothelial cells and the remainder of the haem portion of the molecule is excreted in the form of bile pigment. These pigments undergo further changes in the intestine and give the characteristic colour to the faeces.

In the fasting animal bile is stored in the gall bladder. When chyme reaches the duodenum the gall bladder responds by contracting, and bile pours into the duodenum. There seems to be a hormone produced by the duodenal mucosa which is released when chyme reaches the duodenum. This hormone, called cholecystokinin, passes into the blood stream and stimulates the gall bladder to contract.

THE ABSORPTION OF FOOD SUBSTANCES FROM THE ALIMENTARY CANAL. Most of the food substances are absorbed over the large area of the small intestine, the surface of which is further enlarged by the presence of the villi. Absorption through the stomach is very limited although alcohol is absorbed readily. In the large intestine absorption is mainly limited to water although some vitamins, produced by the bacteria, are absorbed, and in some herbivores e.g. the horse, the caecum absorbs the products of the breakdown of cellulose (see p. 397).

The absorption of fat (fig. 181). Views on the mechanism of fat absorption have tended to vaccilate during the last century. Early in physiological investigation it was found that after giving an animal a meal rich in fat, dilated white vessels (lymphatics) were seen in the mesentery of the small intestine. These were found to contain minute droplets of fat and led to the idea that much, if not all, of the fat in a meal could be absorbed directly without previous hydrolysis into glycerol and fatty acids. Bile salts were regarded as having an important role in emulsifying the fat into particles small enough to allow direct absorption. This early view was variously modified and it became accepted that although some of the dietary fat was absorbed directly in the form of minute globules, some fat was hydrolyzed into glycerol and fatty acids. The fat globules were regarded as being absorbed into the lacteals and lymphatic system, the glycerol and fatty acids being absorbed and transported to the liver by way of the hepatic portal vein. Bile salts in addition to their role in emulsifying fat also assisted the absorption of fatty acids by combining with them to render them water soluble so that they could cross the cell membranes of the intestinal cells. This view is represented in fig. 181. Fat globules are broken down into small particles, called chylomicrons, by the emulsifying action of the bile salts (the fat-bile salt combination being represented by $\rightleftharpoons^{\mathrm{o}}$). After traversing the mucosa of the intestine the bile salts (represented as **...**) are freed from the fat and returned to the liver by way of the hepatic portal vein while the neutral fat ($\underline{\qquad}^{\mathrm{o}}$) is transported away from the intestine in the lymph. Some of the fat is hydrolyzed by the action of lipase into fatty acids and glycerol. The fatty acids, combined with bile salts to render them water soluble ($\rightleftharpoons$), and the glycerol (o) are absorbed, freed from the bile salts and converted into phospholipids ($\underline{\qquad}^{\mathrm{o}}$ –P) in the wall of the intestine and transported to the liver in the blood by way of the hepatic portal vein.

Recent work using fat specially prepared with glycerol 'labelled' with a radioactive isotope, and the fatty acids of a non-metabolizable nature (so that the fate of the fat can be readily followed) has cast some doubt on the above proposed mechanism. It seems now that all of the dietary fat is hydrolyzed to some degree, to glycerol, fatty acids and monoglycerides. The fat globules which appear in the lacteals seems not to be fat absorbed as such from the intestinal contents but new fat synthesized in the gut wall from the products of fat hydrolysis.

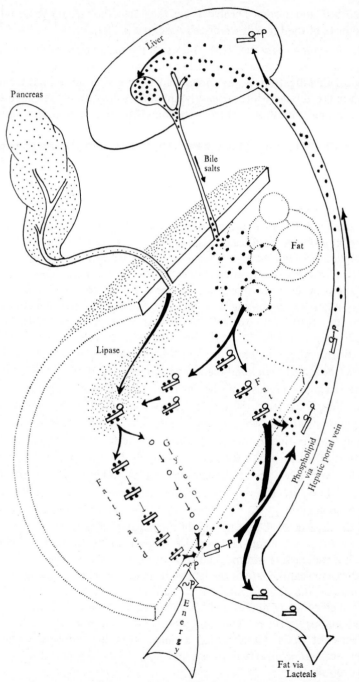

Fig. 181. *Schematic representation of the digestion and absorption of fat.*

The absorption of glucose. During the absorption of glucose into the mucosa the glucose is probably phosphorylated. The absorption proceeds at a quicker rate than can be accounted for on the grounds of simple diffusion and substances which poison the energy-producing systems within the cells prevent the selective absorption of glucose. Glucose can be absorbed from weak solutions in the gut, another point of evidence that the absorption is not a simple diffusion along a diffusion gradient. The glucose passes from the mucosal cells into the capillaries draining into the hepatic portal vein which passes to the liver.

The absorption of proteins. As already described on p. 376 proteins are not absorbed as such but in the form of their constituent amino acids. The absorption of amino acids, like that of glucose, involves active transport mechanisms. There are probably distinct active transport mechanisms for individual amino acids or for small groups of amino acids. These pass into the tributaries of the hepatic portal vein and thence to the liver.

MOVEMENTS OF THE SMALL INTESTINE. It is essential that the chyme entering the duodenum from the stomach be thoroughly mixed with the digestive juices of the pancreas, duodenum and small intestine and with the bile. This mixing is achieved by the movements of the small intestine. Peristaltic movements push the contents of the small bowel onwards to the large bowel. In addition there are segmentation movements of the small bowel brought about by the contraction of the circular muscles at intervals along the length of the bowel. These movements serve to mix the contents rather than propel it onwards. The contraction of the circular muscles at intervals give the appearance of strings of sausages to the segmenting bowel. After a time the muscles at the sites of contraction relax and the bowel then begins to segment at other sites along its length. The pattern of segmentation is altered in the active bowel about twenty times in each minute. In addition to these movements the villi of the mucosa are capable of an independent movement by virtue of slips of muscle which pass up into them from the muscularis mucosa. The movements of the villi help to ensure that the layer of fluid in contact with the intestinal epithelium is not a static one, and also helps to empty the lacteals of the villi into the lymphatic channels of the gut wall.

The large intestine. The importance of the large bowel varies from species to species of mammal. Except in those herbivorous mammals in which the digestion of cellulose occurs here, the large bowel has no digestive function. It is, however, important in the conservation of water. In man about 400 ml. of fluid passes from the ileum into the

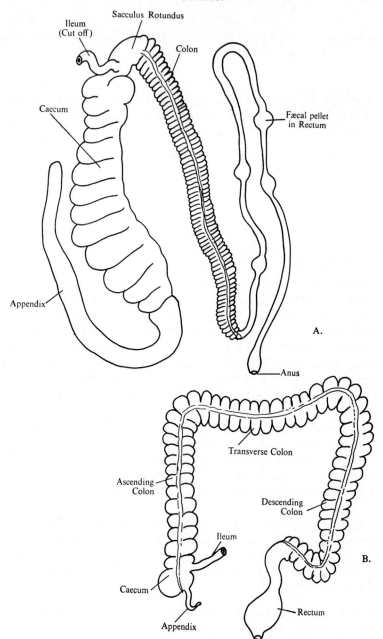

Fig. 182. *Diagram of the structure of the large bowel of* **A.** *rabbit and* **B.** *man. Note the extensive development of the caecum and appendix in the rabbit (the figures are not drawn to scale).*

large intestine each day. Much of this water is reabsorbed by the mucosa of the large intestine.

The regions of the large bowel include the caecum, the colon, rectum and anal canal. In man the caecum is a small cul-de-sac terminating in the vermiform appendix and is of no great importance. In herbivores such as the rabbit it is a much larger structure (fig. 182). The colon in man consists of ascending, transverse and descending portions. The mucosa is a simple columnar type and an alkaline mucus is secreted which permits the drying contents to move along. From time to time strong peristaltic movements occur in the colon pushing the contents onwards into the rectum, a short but expansile structure which can accommodate the faeces before they are evacuated. When the faeces enter the rectum and stretch its walls there is a reflex action in which the muscles of both colon and rectum contract in order to evacuate the faeces. In man a measure of voluntary control is exercised in evacuation by the action of a circular hoop of voluntary muscle, the external anal sphincter.

Modifications of the alimentary tract of herbivores

The natural diet of herbivorous animals consists of grasses and other vegetation containing large amounts of the carbohydrate cellulose. No mammal is known to secrete a cellulase in its digestive tract and the animal has to rely upon bacteria in the bowel to perform this task of the breakdown of cellulose. Under the moist, warm conditions of the bowel with a plentiful supply of fermentable material the bacteria grow and multiply rapidly. The animal obtains its energy from the products of bacterial fermentation and by the digestion of bacterial protoplasm. The site of these symbiotic bacteria in the bowel varies. In the rabbit, for example, these bacteria are located within the caecum, whereas in ruminants such as the cow they are situated within the specialized stomach.

In non ruminant animals such as the rabbit and the horse the bacteria are housed in the large caecum and here they ferment the cellulose. Much of the products of this fermentation are presumably absorbed from the caecum. In the rabbit, however, there is a special habit called refection which permits the material in the caecum to be exposed to the action of the small bowel so that the bacterial products are fully utilized. During the night the rabbit produces faeces which are large and creamy white. The animal sleeps with its mouth close to the anus so that these night faeces are eaten. The faecal pellets produced during the day are dark brown in colour and of a firmer consistency; the material in these faeces has been twice through the alimentary tract.

In ruminants, such as the cow, the bacteria are housed in the specialized stomach (fig. 183). The stomach in these animals consists of four chambers, the rumen, reticulum, omasum (psalterium) and abomasum. After the initial chewing of the food it is passed into the first part of the stomach, the rumen, which is a large bag where juices are added and where bacterial fermentation commences. From the rumen the food passes into a much smaller pouch called the reticulum whose mucosa is folded like a honeycomb, where bacterial digestion proceeds further. The food is then regurgitated back into the mouth for 'chewing of the cud'. When this is reswallowed it is prevented from re-entering the rumen by the closure of a groove on the side of the reticulum which leads the liquid cud into the omasum. This chamber, like the rumen and reticulum has a mucosa of oesophageal type which is folded into vertical ridges. These resemble the leaves of a prayer book hence the alternative name of psalterium. In the omasum the food is churned about and then enters the 'true' stomach or abomasum which produces the typical acid secretion of the stomach.

The products of bacterial fermentation include some sugar but also large amounts of short chain fatty acids, particularly acetic and

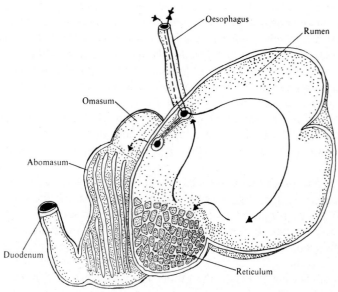

Fig. 183. *Diagram illustrating the structure of the stomach of the sheep. The arrowed lines indicate the course of the movement of food; the solid line indicates that food passes first into the rumen and reticulum from which it is regurgitated into the mouth, the dotted line indicating the passage of food after rumination from the mouth and oesophagus into the omasum by way of a groove in the wall of the rumen.*

proprionic acids, which form the main source of energy for the animal (see also p. 443). In addition any bacteria which pass into the true stomach are killed by the acid secretions and their protoplasm is digested. The bacteria can build up protein from non-protein nitrogen such as urea and amides, and by the digestion of the bacteria which pass into the abomasum these become available to the host animal.

The liver

In the primitive chordates the liver, an outgrowth of the alimentary tract, is a digestive organ but during the evolution of the vertebrates it has acquired a variety of functions and has become a very important and indispensable organ.

The liver is the largest organ in the abdomen. Its convex upper surface fits into the dome of the diaphragm. It is supplied with blood from two sources. The major source of blood supply is by way of the hepatic portal vein which drains the stomach and small bowel; this blood contains the products of digestion, amino acids, glucose, phospholipids, vitamins etc. As the hepatic portal vein enters the liver it breaks up into a series of branches which supply a capillary bed in the liver tissue, from which the blood drains into the hepatic veins which discharge into the inferior vena cava. The blood passing into the hepatic veins has thus traversed two capillary beds, one in the gut wall and another in the liver tissues. This special arrangement of the blood supply reflects the function of the liver as a regulator of metabolites, the blood from the intestine loaded with food materials, and in addition some toxic materials, being subjected to the regulating function of the liver before it is passed into the general circulation of the body. A smaller part of the blood supply of the liver is by way of the hepatic artery, which delivers oxygenated blood from the dorsal aorta. On the under surface of the liver is the gall bladder which stores and concentrates the bile secreted by the liver; the gall bladder contracts at intervals and discharges the bile, via the bile duct, into the duodenum.

HISTOLOGICAL STRUCTURE OF THE LIVER. The liver tissue is made up of innumerable small units, called liver lobules, each having the shape of a polygonal prism (fig. 184). In many mammals, e.g. the pig, each lobule is surrounded by a layer of connective tissue, Glisson's capsule. Running through the centre of each lobule in a longitudinal direction is a branch of the hepatic vein, and blood drains across the tissue of the liver lobule from branches of the hepatic portal vein situated around the periphery of the lobule, into the central vein. As the blood drains across the lobule it comes into intimate contact with the liver cells.

At the periphery of the lobule, in addition to branches of the hepatic portal vein, there are branches of the hepatic artery, and tributaries of the bile duct (fig. 184). The tributaries of the bile duct drain away the bile produced by the liver cells.

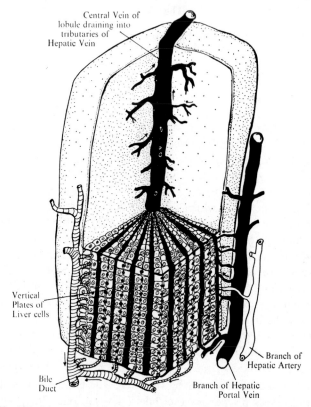

Central Vein of lobule draining into tributaries of Hepatic Vein

Vertical Plates of Liver cells

Bile Duct

Branch of Hepatic Artery

Branch of Hepatic Portal Vein

Fig. 184. *Diagram of the structure of a liver lobule. The wall of the lobule has been shown cut away in various planes to show the vertical plates of liver cells radiating out from the central vein. The vertical tangential sections through the vertical plates shows the bilary canaliculi arising between the liver cells.*

The tissue of the lobule consists of vertical plates of liver cells separated by blood spaces, the sinusoids, through which blood flows from the periphery of the lobule inwards towards the central vein. The sinusoids are lined by mesenchymal cells of the reticulo-endothelial system which are capable of phagocytosis of particles within the blood stream. One special type of phagocytic cell is called the Kupffer cell and is often seen to contain fragments of effete red blood cells and granules of pigment derived from the breakdown products of haemoglobin. The vertical plates of liver cells forming the liver lobules are two cells in thickness and between these cells run fine bile canaliculi which drain the bile produced by the liver cells. The bile canaliculi drain outwards to the periphery of the lobule where they join larger branches of the bile duct system (fig. 184).

THE FUNCTIONS OF THE LIVER

The liver performs a variety of functions but basically there are three main types of function: firstly the regulation of metabolites, including fats, proteins and carbohydrates, secondly the production of bile, Thirdly the detoxification of toxic materials absorbed into the body of the animal. There are in addition other functions which will be discussed below.

(a) THE LIVER AND THE METABOLISM OF PROTEINS, FATS, CARBO-
HYDRATES.

1. *Protein.* When the amino acids produced from the digestion of proteins are absorbed into the body they join the quantity of freely circulating amino acids in the blood, known as the amino acid pool From this pool amino acids are taken up by the various tissues of the body in the process of growth and repair. Now the quantity of amino acid absorbed fluctuates as the protein intake in the diet changes. In well nourished animals the protein intake may greatly exceed the requirements of the body for growth and repair. And even on a fairly low protein intake much of the amino acid produced may not be required by the body for growth because these particular amino acids can be synthesized by the body, that is they are non-essential amino acids. The liver is the organ in which excess and non-essential amino acids can be broken down; the amino group of the amino acid is taken up in the onithine-arginine cycle (p. 516) to produce urea, which is excreted by the kidneys, and the organic acid remaining can be utilized as a source of energy by way of Krebs' cycle, or used in various synthetic processes (p. 517). The liver is thus highly important in carnivorous animals by making the protein of the diet available as a source of energy.

In addition the liver is the site of synthesis of proteins which are dissolved in the blood plasma including prothrombin, fibrinogen and the plasma albumens and globulins.

2. *Carbohydrates.* Excess glucose absorbed from the intestine following a meal can be diverted into two ways of storage, firstly as glycogen and secondly as fat. The main store of glycogen is in the cells of the liver, although muscles also store glycogen. The glycogen of the liver cells forms an easily mobilized store and is converted into glucose when the blood sugar level falls. The deposition and mobilization of glycogen is under the control of the hormones insulin and adrenaline (see p. 262).

3. *Fats.* The liver is intimately concerned in the metabolism of lipids. At times when the animal is metabolizing fat as a main source of energy, e.g. during starvation, large quantities of fat appear in the liver, and its accumulation here seems to be the first stage in its metabolism.

(*b*) THE LIVER AND THE FORMATION OF BILE. The role of the liver in the formation of bile has already been discussed. To summarize, bile is a secretion which serves two functions: first the bile salts are concerned in the emulsification and digestion of fats, and secondly the bile pigments are excretory products derived from the breakdown of haemoglobin. Bile contains no digestive enzymes.

(*c*) THE LIVER AND DETOXIFICATION. Detoxification of a variety of poisonous substances which may be absorbed into the body occurs in the liver, e.g. benzoic acid, picric acid, chloroform.

Detoxification also includes the inactivation of substances produced in the body, particularly of the hormones. Thus the chemistry of the sex hormones is altered in the liver, rendering them biologically relatively inactive, and they may be excreted by the kidney in this form. This function of the liver may be illustrated by referring to a disease of the liver, cirrhosis, in which much of the liver tissue is destroyed and replaced by fibrous connective tissue. The small quantity of surviving liver tissue may not be able to deal with even the relatively small amounts of oestrogen produced by the human male and oestrogens may accumulate in the blood sufficiently to produce signs of feminization, including loss of the beard and development of breast tissue.

(*d*) THE LIVER AND THE METABOLISM OF VITAMINS. The fat soluble vtamin A is manufactured in the liver from carotene precursors (p. 405), and this explains the rich source of the vitamin in fish liver oils. Other fat soluble vitamins are also stored in the liver.

Vitamin B_{12} is stored in the liver, explaining the use of raw liver in the treatment of pernicious anaemia (p. 412).

(*e*) THE RETICULO-ENDOTHELIAL SYSTEM IN THE LIVER. The reticulo-endothelial tissue lining the liver sinsusoids has already been described.

The phagocytic cells of this tissue engulf particles from the blood stream including dead bacteria and blood parasites (e.g. malarial parasite), effete red cells etc.

In the embryo red and white blood cells are also produced within the sinusoids of the liver from the mesenchymal cells of the reticulo-endothelial system. As the embryo grows and develops the bone marrow becomes more important in this respect.

THE VITAMINS

Classification:

1. *Fat Soluble vitamins*

   Vitamin A
   Vitamin D
   Vitamin K
   Vitamin E

2. *Water Soluble vitamins*

   Vitamin C or ascorbic acid.
   Vitamin B Complex: Vitamin B_1, Aneurine hydrochloride or Thiamine hydrochloride.
   Vitamin B_2 Riboflavin.
   Vitamin B_6 Pyridoxine.
   Nicotinic acid.
   Pantothenic acid.
   Vitamin B_{12}, Cyanocobalmin.
   Folic acid.

Introduction. The word 'vitamine' was first used by Funk in 1912 to describe the accessory food factors which were essential in the diet in addition to carbohydrates, fats, proteins, mineral salts and water. The vitamines were essential for a healthy life, indeed for life itself. Funk isolated a crystalline product from rice polishings which contained basic nitrogen and was thought to be an amine and an accessory food factor—hence its name vit-amine, an amine essential for life. We now know that not all accessory food substances are amines and so we drop the terminal e and call them vitamins.

Diseases caused by lack of vitamins are called deficiency diseases and some of these like beri-beri, pellagra, scurvy and rickets, have been recognized for a long time. In the early part of the twentieth century vitamins A, B, C, D, E, were named and associated with certain deficiency diseases. In many cases today the vitamins are called by their chemical names rather than by a letter and a number.

We may define vitamins as organic substances needed for normal animal life; although they do not provide energy or body building material they are essential for energy transformations and regulation of metabolism. They are parts of enzyme systems and are effective in small amounts, operating within narrow limits of temperature and pH. They are usually synthesized by plants or micro-organisms and the mammal is only capable of their partial synthesis e.g. mammals can make vitamin A from β-carotene, and vitamin D by the action of sunlight on substances already present in the skin. It is for this reason that an adequate intake of vitamins is necessary for mammals. Since vitamins are part of enzymes involved in metabolism it is not surprising that the amounts of vitamins needed depend on the relative amount of carbohydrate in the diet, the age, activity and size of the animal and upon the physiological state at the time.

In recent years the distinction between vitamins and other components of the diet has become less clear. Now that more is known about the way in which vitamins play a role in metabolism it is clear that many of them are incorporated into the cellular chemistry as co-enzymes and are therefore body building materials like the essential amino-acids, which cannot be synthesized. The difference lies in the fact that the vitamins, being incorporated into the enzyme systems are needed in much smaller amounts than the essential amino acids; thus the amount of vitamins required is in the order of 10^{-8} to 10^{-11} of body weight per day, whilst requirements of the essential amino acids lie in the order of 10^{-5} of body weight per day.

We have already seen that bacteria and higher plants can synthesize the chemicals which are essential components of the mammalian diet; it is thought that during the course of evolution of the vertebrates they have given up the synthesis of these essential substances, perhaps because they were so readily available in the diet. It is perhaps surprising that the vertebrates have not given up the synthesis of many more compounds, but the reason for this may lie in the fact that they were not provided in an adequate amount or in a sufficiently constant supply in the diet.

Vitamin A

HISTORY. Experiments which led to the discovery of vitamin A were carried out in 1913 when it was found that animals fed on a diet containing lard as the sole source of fat developed a disorder of the skin and cornea which could be relieved by the addition of butter, milk or egg yolk to the diet.

CHEMISTRY. Vitamin A is fat soluble, a primary alcohol with the following structure, which exists in several isomeric forms.

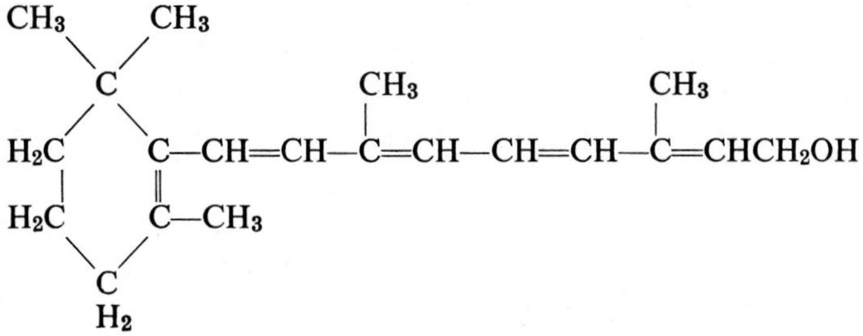

It can be synthesized in the body by splitting the large symmetrical molecule of β-carotene into two parts.

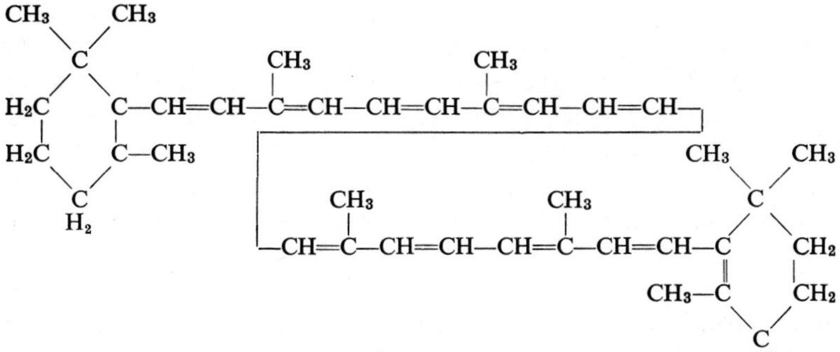

β-carotene

SOURCES. Vitamin A can be obtained from two sources; firstly as the vitamin itself from animal sources e.g. milk, butter, egg-yolk, fish liver oil, and secondly as β-carotene from plants. Carotene is a common plant pigment and is found in the green parts of plants where it is masked by the green pigment chlorophyll and it is also present in other parts e.g. in the red fruit of the tomato and in many yellow flowers.

PHYSIOLOGICAL FUNCTIONS. Vitamin A has two important functions in the body; firstly it is important for vision, particularly night vision, and secondly it is essential for the health of epithelial cells.

Vitamin A and vision. It has long been known that a deficiency of vitamin A produced a condition called night blindness. Adaptation of vision to darkness depends upon the formation of a pigment in the rods of the retina called rhodopsin or visual purple, which breaks down on exposure to light of low intensity and initiates nerve impulses; after breakdown the pigment is resynthesized. In the retina vitamin A is converted into a substance called retinaldehyde by means of enzymes; the retinaldehyde combines with a protein called opsin to produce the visual pigment rhodopsin.

Vitamin A and epithelia. The structure and function of all the epithelial cells of the body are dependent upon an adequate supply of vitamin A. The way in which vitamin A plays its function in the health of epithelial cells is not understood.

DEFICIENCY DISEASE. Many types of epithelia are damaged when there is a severe deficiency of vitamin A. The epithelial cells of the cornea become cornified and ultimately ulcerate, and blindness follows; this condition is called xerophthalmia. Cornification of the epithelial cells of the upper and lower respiratory tract predisposes the animal to respiratory infections; thus vitamin A has been called an anti-infective vitamin but the way in which it fulfils this function is by preserving the integrity of the epithelia, so making them less susceptible to invasion by bacteria and viruses, rather than by any direct effect upon the organisms themselves. The skin as a whole becomes dry and heavily cornified (phrynoderma or toad-skin). There are changes also in the epithelium of the urinary tract which may lead to the formation of stones in the kidneys. Night blindness is a constant finding.

Infants may receive additional supplies of the vitamin in the form of cod-liver oil and in the addition of the vitamin to dried milk powders. Giving too much of the vitamin may be as serious as deficiency and in recent years many infants have been given too much vitamin A causing painful swellings in the skin, excess growth of bone, loss of appetite, dry scaling lips and bleeding cracks at the corners of the mouth. The skin is dry and scaly and the liver is enlarged.

The Vitamin B complex

The vitamin B complex consists of a group of at least twelve vitamins which have different chemical structures and physiological functions. They are grouped together because they are all water soluble and can be obtained from similar sources; they are present in all living cells. The first vitamin to be isolated was called vitamin B and it was only later that this vitamin was found to consist of several substances.

Vitamin B₁. Aneurine hydrochloride or Thiamine hydrochloride.

HISTORY. Takaki (1884) was a naval surgeon and director of the Tokyo naval hospital. He had the idea that a disease called beri-beri which was common among his sailors was due to a deficiency in the diet. European sailors were not subject to the disease and the only difference he could see in their way of life and hygiene and that of

Japanese sailors was in the diet they ate. In a Japanese ship which returned from a voyage round the world, out of a crew of 376, 169 had beri-beri and 45 of these died. Takaki tried an experiment with another ship, sending it round the world on the same course as the first but with a different food supply. He gave the sailors food with a high nitrogen content; there was not a single case of beri-beri, and since then beri-beri has ceased to exist in the Japanese navy. Takaki arrived at the right answer for the wrong reason; in increasing the protein he was increasing the supplies of vitamin B, the deficiency of which caused beri-beri. In 1905 Eijkman published a paper on Vitamin B; this paper was written in Dutch and for this reason did not come to the attention of the scientific world, but his researches are very interesting. He was sent out in 1886 by the Dutch government to investigate beri-beri in the native population of the East Indies. Despite Takaki's effective work on beri-beri in the Japanese navy the mission decided that beri-beri was an infective illness rather than one due to defective nutrition, but Eijkman was not satisfied with the report and he stayed to do further work. He noticed fowls in the hospital yard with limpness of the neck, drooping wings and an unsteady gait, and he recognized that here in the hens was a condition similar to the human affliction beri-beri. He discovered that the fowls had been fed on scraps of food left by patients with beri-beri and thought that the food carried the germs of beri-beri. He decided to carry out bacteriological experiments on the fowls; these experiments were interrupted by officials who were shocked to find that the fowls were being fed on milled rice and the officials ordered that the fowls should be fed on unmilled rice, presumably to save expenses incurred in milling the rice. So the fowls were fed on the unmilled rice, regarded as unfit for human consumption, and Eijkman thought that his experiments were ruined until he noticed that the fowls were getting better and presently became normal. Eijkman studied records of prisoners fed on unmilled rice and from 37 prisons only in one case was beri-beri reported. In 61 prisons polished rice was used and cases occurred in 36 of these prisons. However, such was the influence of Pasteur and Koch and their researches on bacterial disease that Eijkman still believed that beri-beri was due to a germ and that the substance in the bran from milled rice was the antidote to beri-beri. Four years later in 1906 his colleague Grijns showed that beri-beri was due to a deficiency in the diet causing a disorder of the nervous system, and Eijkman came to agree with him and said 'there is a substance present in rice polishings, a substance of a different nature from proteins, fats or salts, which is indispensable for health, the lack of which causes a nutritional polyneuritis'.

Funk in 1913 was able to extract from rice polishings and from yeast a crystalline substance which could cure beri-beri. This compound was found to contain basic nitrogen, and was thought to be an amine, hence the name 'vitamine'.

CHEMISTRY. Thiamine hydrochloride is a complex organic substance containing a pyrimidine and a thiazole nucleus.

$$\left. \begin{array}{c} \underset{\substack{\text{CH}_3.\text{C} \quad \text{C}-\text{CH}_2-\text{N}}}{\overset{\text{N}=\text{C}.\text{NH}_2}{}} \quad \overset{\text{CH}_3}{\underset{\text{CH}-\text{S}}{\text{C}==\text{C}.\text{CH}_2.\text{CH}_2\text{OH}}} \\ \text{N}-\text{CH} \qquad \text{Cl} \\ \text{Pyrimidine portion} \qquad \text{Thiazole portion} \end{array} \right\} \text{HCl}$$

SOURCES. The pericarp and embryos of cereals, yeast, egg yolk, nuts, liver and leguminous vegetables e.g., peas, beans.

PHYSIOLOGICAL FUNCTION. Thiamine has a vital function in carbohydrate metabolism; it is essential in the oxidative mechanisms which involve decarboxylation. In the body, thiamine is active as the pyrophosphate which is a component of the co-enzyme cocarboxylase. This enzyme promotes the decarboxylation of pyruvic acid before it is incorporated in the citric acid (or Krebs') cycle (see p. 434). Thus in deficiency states of thiamine there is a failure of complete oxidation of pyruvic acid, which accumulates in the blood. Therefore the amount of thiamine needed by the body depends upon the percentage of carbohydrate in the diet; those with predominantly carbohydrate intake need a proportionately larger supply of thiamine.

DEFICIENCY DISEASE. Because of the vital role played by thiamine in carbohydrate metabolism it is hardly surprising that deficiencies of this vitamin can profoundly disturb the body. Severe thiamine deficiency leads to the condition known as beri-beri, in which many tissues may be affected.

There are three types of beri-beri—neuritic or dry, cardiac or wet, and cerebral beri-beri with severe mental changes. Many of the nervous symptoms are due to inflammation of the peripheral nerves—peripheral neuritis; thus there may be weakness of the limbs, abnormal sensations or patches of anaesthesia. Ultimately there may be paralysis of the limbs. The central nervous system is also affected producing character changes, tiredness and irritability. In the cardiac form of beri-beri, in which the cardio-vascular system is severely affected, there may be an enlarged heart with breathlessness on effort. There may be accumulations of fluid (oedema) in the legs; this is partly due to a low level of plasma proteins. When a person is consuming a diet deficient in thiamine, invariably there are deficiencies not only in the protein intake but also in the other members of the vitamin B complex and in the treatment of a condition such as beri-beri it is not usually sufficient to replace the thiamine alone; a general improvement in the standard of the diet is necessary.

Vitamin B_2, Riboflavin

HISTORY. The water-soluble substances which had been extracted from yeast and called vitamin B were soon found to contain two factors; one factor which could cure beri-beri was destroyed by heat leaving another heat stable factor with growth promoting properties. This second factor, or vitamin B_2 as it was called had an intense yellow colour, and was found to be identical with yellow pigments which had been previously extracted from a variety of animal tissues.

CHEMISTRY. Riboflavin possesses the following structural formula:

SOURCES. Natural sources of riboflavin include yeast, milk, liver, egg and leafy vegetables.

PHYSIOLOGICAL FUNCTION. Like thiamine, riboflavin must be phosphorylated before it functions in the cells. In the cells it appears as a nucleotide, primarily as flavine adenine dinucleotide. Flavine adenine dinucleotide is a co-enzyme which acts as an acceptor of hydrogen in the citric acid cycle.

DEFICIENCY DISEASE. In man, deficiency of riboflavin in the diet produces a variety of lesions of skin and mucous membrane. The cornea of the eye loses its transparency because it becomes invaded by capillaries. There is cracking of the skin at the corners of the mouth called cheilosis. The lips are dry and may ulcerate. The tongue is purplish red or magenta coloured and there may also be a dermatitis.

Nicotinic acid. Nicotinamide. An American called Goldberger working in the United States Public Health Service established that a disease called pellagra was due to a deficiency in the diet, at that time thought to be vitamin B_2. Later it was discovered that liver extracts were effective in the treatment of human pellagra and the active principle was found to be nicotinic acid amide.

CHEMISTRY. Nicotinic acid is pyridine-3-carboxylic acid; in the body it functions as nicotinamide. The structure of nicotinic acid and nicotinamide are as follows:

$$
\begin{array}{cc}
\text{CH} & \text{CH} \\
\diagup \diagdown & \diagup \diagdown \\
\text{HC} \quad \text{C.COOH} & \text{HC} \quad \text{C.CONH}_2 \\
\| \quad | & \| \quad | \\
\text{HC} \quad \text{CH} & \text{HC} \quad \text{CH} \\
\diagdown \diagup & \diagdown \diagup \\
\text{N} & \text{N} \\
\text{Nicotinic acid} & \text{Nicotinamide}
\end{array}
$$

SOURCES. Nicotinic acid is widely distributed in food substances, including liver, yeast, cheese, milk, eggs and cereals.

PHYSIOLOGICAL FUNCTION. Nicotinamide is an essential part of co-enzyme I (diphosphopyridine nucleotide) and co-enzyme II (triphosphophyridine nucleotide). The coenzymes act as hydrogen acceptors and donors in some of the oxidation-reduction reactions of the citric acid or Krebs' cycle.

DEFICIENCY DISEASE. A deficiency of nicotinic acid in the diet is one of the most important factors in producing the human disease called pellagra and the vitamin has been called the pellagra preventing factor. The symptomatology of pellagra has been expressed as three D's, which are diarrhoea, dementia and dermatitis indicating that the symptoms and signs of the disease refer mainly to three systems of the body, the gastro-intestinal tract, the central nervous system and the skin and mucous membranes. In the skin, those areas which are exposed to light—forehead, neck, hands and feet, become darkened and scale. In the gastro-intestinal tract there is a 'beefy-red' appearance of the tongue, loss of appetite and diarrhoea. In the central nervous system, changes lead to headache, depression, impaired memory, and in severe cases confusion and hysteria. The disease was common in the maize eating regions of the United States; pellagra will develop in these regions even when the dietary intake seems to be adequate. This is probably due to the fact that the body can synthesize a variable amount of nicotinic acid from the amino acid tryptophane and maize protein has a very low tryptophane content.

Vitamin B$_6$, Pyridoxine

CHEMICAL STRUCTURE:

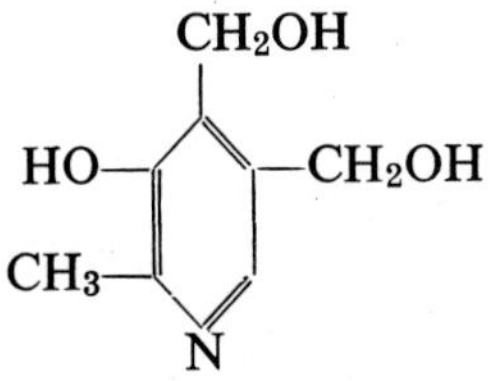

$$CH_2OH$$

HO— —CH$_2$OH

CH$_3$—

N

SOURCES. Egg yolk, yeast, peas, soya bean, meat, liver. (Most green and root vegetables contain little of the vitamin, and milk is a poor source.)

PHYSIOLOGICAL FUNCTION. In the body pyridoxine acts as pyridoxal phosphate which acts as a coenzyme in many aspects of protein metabolism including transamination and decarboxylation. Thus the dietary requirement of pyridoxine depends on the amount of protein in the diet.

In cases of pyridoxine deficiency there may be an excessive production of urea because many of the amino acids which would undergo transamination (see p. 441) are deaminated instead and the free amino groups which are liberated are excreted as urea. Pyridoxine is also essential for the synthesis of certain unsaturated fatty acids and pyridoxine deficient rats develop a condition known as acrodynia, with scaling of the ears, paws and snout, which can also be produced by feeding diets lacking in certain unsaturated fatty acids.

Pyridoxine is necessary for the conversion of the amino acid tryptophane into the vitamin nicotinic acid.

DEFICIENCY DISEASE. It seems that man is able to obtain almost all the pyridoxine needed from the bacteria living in the bowel and deficiency disease can only be produced in man by giving drugs which have an antagonistic action to pyridoxine.

As described above deficiency of pyridoxine in rodents produces the condition known as acrodynia, which is a manifestation of the effect of pyridoxine on the synthesis of certain unsaturated fatty acids.

Pantothenic acid.

HISTORY. Pantothenic acid was first identified in 1933 when it was found to be a substance essential for the growth of yeast. The name of the vitamin indicates its widespread occurrence in nature.

CHEMISTRY. It is a complex organic acid with the following structure:

$$HO-\underset{\underset{H}{|}}{\overset{\overset{H}{|}}{C}}-\underset{\underset{CH_3}{|}}{\overset{\overset{CH_3}{|}}{C}}-\underset{\underset{OH}{|}}{\overset{\overset{H}{|}}{C}}-\overset{\overset{O}{||}}{C}-N-\underset{\underset{H}{|}}{\overset{\overset{H}{|}}{C}}-\underset{\underset{H}{|}}{\overset{\overset{H}{|}}{C}}-\overset{\overset{OH}{|}}{C}=O$$

PHYSIOLOGICAL FUNCTION. Pantothenic acid is part of coenzyme A; this important coenzyme activates both the carboxyl and methyl carbon atoms of acetate (p. 435). It is active in the synthesis of acetyl choline and in catalyzing the first step in the citric acid cycle and derivatives of coenzyme A are concerned in the synthesis of amino acids and in the metabolism of fatty acids.

DEFICIENCY DISEASE. In spite of the vital role of coenzyme A in metabolic processes pantothenic acid deficiency in man has not been described ; this is probably because pantothenic acid is so widespread in nature that even the poorest of diets contain sufficient amounts of this vitamin, and further the bacteria in the bowel may provide a source of the vitamin.

Vitamin B_{12} or Cyanocobalmin

HISTORY. Before 1926 pernicious anaemia was invariably a fatal disease. In this year Minot and Murphy discovered that if sufferers from this disease ate a large amount of raw liver every day their condition improved and they could be kept alive. Later highly active extracts of liver were produced so that injections of small amounts of this extract could replace the almost intolerable daily diet of raw liver which these patients had to suffer. Castle (1936) postulated that in pernicious anaemia there is lacking in the stomach a factor (called the intrinsic gastric factor) which is necessary for the absorption of some factor in the diet (the extrinsic factor); he postulated that there was a reaction between the extrinsic and intrinsic factors with the production of an anti-anaemia principle. It was found that by giving patients powdered hog stomach (containing the intrinsic factor) the anaemia improved.

In 1948 a red crystalline substance was isolated from liver extracts which was highly active in the treatment of pernicious anaemia and this substance was given the name of vitamin B_{12}. This was later shown to be the factor in the diet (Castle's extrinsic factor) which is not adequately absorbed in patients with pernicious anaemia because of the absence of the intrinsic factor of Castle in the stomach.

CHEMISTRY. Vitamin B_{12} is of complex composition and the complete structural formula is not yet known. It is a red crystalline substance containing the element cobalt.

SOURCES. Vitamin B_{12} is almost entirely absent in plant products and herbivores obtain it from the bacteria in their gut. Even in most animal products the concentration of the vitamin is low; in fresh liver, one of the 'richer' natural sources there is only about 500 mg. in a ton of liver. However, on a weight basis the vitamin is the most potent of known vitamins.

PHYSIOLOGICAL FUNCTION. The vitamin has a large number of important and apparently unrelated functions; at the moment its exact metabolic function is not understood and it may be that one fundamental biochemical function will be discovered which will explain its diverse effects. A number of observations indicate that it is important in protein metabolism, particularly in nucleo-protein synthesis. It is also involved in transmethylation reactions and in the metabolism of labile methyl groups.

DEFICIENCY DISEASE. In pernicious anaemia there is a degeneration of the mucosa of the stomach, which in addition to failing to produce hydrochloric acid and pepsin fails to produce the intrinsic factor of Castle which is essential for the normal absorption of vitamin B_{12} from the food. In this disease there are changes in the gastric mucosa, tongue, bone marrow and central nervous system. In the bone marrow there is a failure of the normal maturation of red blood cells; the red blood cells become larger, contain less haemoglobin and are more fragile. There are degenerative changes in the nervous system particularly in the spinal cord and peripheral nerves, which may produce abnormal sensations and weakness of the muscles. Without treatment the disease is invariably fatal; treatment consists of regular injections of vitamin B_{12} which is now obtained from bacterial fermentation.

In addition to this classical pernicious anaemia which is caused by an abnormality of the gastric mucosa a similar condition can occur in many abnormalities of the bowel which disturb the absorption of vitamin B_{12}; two examples from the many causes of this condition include the removal of part of the stomach (because of an ulcer or cancer) and the infestation of the gut with a worm called Diphyllobothrium (this occurs in Scandinavian countries where infection occurs by eating certain fish).

Folic acid or Pteroylglutamic acid

HISTORY. Folic acid was first isolated as a growth factor for certain bacteria which could be isolated from liver or yeast. It was given the name of folic acid by certain workers who isolated it from leafy vegetables. It has now been synthesized.

CHEMISTRY.

$$H_2N \underset{HO}{\underset{N}{\overset{N \quad N}{\diagdown}}} CH_2 \quad NH\diagdown\diagup CO. \quad NH.CH.CH_2.CH_2.COOH$$

| Pteridine residue | PABA residue | Glutamic acid residue |

with COOH below under the Glutamic acid residue.

Pteroyl group

NATURAL SOURCES. Folic acid is found in all green leaves and in many animal tissues such as liver and kidney. It is also present in yeast and milk.

PHYSIOLOGICAL FUNCTION. In the body folic acid is converted into its active form folinic acid, and is dependent on the presence of vitamin C for this conversion. It is an important factor in the synthesis of some purines, pyrimidines and amino acids. Since purines and pyrimidines are important in the manufacture of nucleoproteins, folic acid is an important growth factor. It is therefore the actively growing tissues which are disturbed by a deficiency in folic acid, particularly the bone marrow, which in the normal adult human is producing some 200,000,000 red blood cells every minute.

DEFICIENCY DISEASE. In most animals a deficiency in folic acid intake is difficult to produce because of the production of the vitamin by bacteria in the bowel. In certain diseases of the bowel there is a disorder of digestion and absorption, and bulky fatty stools are produced containing many unabsorbed and undigested food substances; in these diseases folic acid deficiency may result in a severe type of anaemia which can be relieved by giving increased amounts of the vitamin.

Summary of the Vitamin B complex

Name	Sources	Physiological Functions	Deficiency disease
B_1, Thiamine or Aneurine	Yeast, liver, egg yolk, milk, pericarp of cereals peas and beans.	A component of the co-enzyme co-carboxylase.	Beri-beri.
B_2, Riboflavin	Yeast, liver, egg yolk, milk, leafy vegetables.	Acts as flavine adenine dinucleotide—a hydrogen acceptor.	Produces vascularization of cornea, changes in skin and tongue.

Summary of the Vitamin B complex—*continued*

Name	Sources	Physiological Functions	Deficiency disease
B₆, Pyridoxine	Yeast, liver, egg yolk, peas, soya bean. (Milk and green vegs. are poor sources.)	A co-enzyme in protein metabolism (transamination and decarboxylation). Also it is important for the synthesis of some unsaturated fatty acids and for the conversion of tryptophane to nicotinic acid.	Not seen in man because of its production from bowel organisms. Produces Acrodynia in rats because of its effect on fatty acid synthesis.
Nicotinic acid	Is widely distributed e.g., yeast, liver, eggs, milk, cereals.	Is an essential part of Co-enzymes I and II	Pellagra
Pantothenic acid	Very widespread	Part of co-enzyme A	Not seen in man because of the widespread distribution of the vitamin.
B₁₂, Cyanocobalmin	Present in animal products, e.g., liver. Almost absent in plant products.	Important in protein metabolism and transmethylation reactions.	Pernicious anaemia.
Folic acid	Present in all green leaves and many animal tissues e.g., liver, kidney.	Important in the synthesis of purines, pyrimidines and amino acids.	Anaemia

Vitamin C

HISTORY. As early as the thirteenth century scurvy was described as a disease common among the crusaders, but it became a much more conspicuous disease in the days of the long sea voyages of the sixteenth century. It is recorded that in a voyage to Newfoundland in 1535, 100 of the 103 men in the crew suffered from scurvy. In 1593 Admiral Sir Richard Hawkins noted that during his career 10,000 seamen had died of scurvy and adds that he found the most effective treatment for this was 'sower oranges and lemons'. The classic work is that of James Lind, chief physician of the Naval Hospital in Portsmouth in 1755 who stated that the disease could be eliminated from the British navy by supplying lemon juice to the seamen. He quotes the case of a seaman marooned to die because of scurvy who was almost paralysed with the disease and could only crawl feebly on the ground; he started to eat the vegetation on the salt marsh where he was marooned and in a short time he made a miraculous recovery. The vegetation that he ate was found to consist of Cochlearia officinalis, later called scurvy grass.

Lind proved at sea that simple things such as mustard and cress, oranges and lemons would cure the disease and remarked that the remedy was so simple that the Lords of the Admiralty disregarded his advice and he added that if he had called the cure 'an antiscorbutic golden elixir' they would have taken more notice of his advice. Captain Cook was made a Fellow of the Royal Society and received a gold medal for his successful treatment of scurvy rather than for his travels to the South pole and round the world. He had no scurvy on his ship, the 'Resolution', because he took Lind's advice and carried fresh fruit on the voyage. It was not until 1794, the year that Lind died, that the Admiralty first equipped its ships with lemon juice for long voyages. In 1804 regulations were issued enforcing daily rations of lemon juice. Scurvy was eradicated then from the British navy and 60 years later the Board of Trade applied the same treatment to the Mercantile Marine. It was then that as the sailors docked in America the Americans called them 'limeys'; the Americans thought that the British sailors were taking lime juice instead of lemon juice. It should be noted that lime juice is a relatively poor source of vitamin C and this brought discredit to Lind's ideas on the treatment of scurvy.

In 1932 King and Waugh succeeded in isolating a crystalline compound from concentrates of lemon juice which was potent in the treatment of scurvy; it was identified chemically as hexuronic acid. The substance was then synthesized, and because of its property in the prevention of scurvy it was called ascorbic acid.

CHEMISTRY. Ascorbic acid has the following structural formula:

$$CH_2OH \quad C - C - C = C - C \begin{array}{c} OH \ H \ OH \ OH \ O \end{array}$$

NATURAL SOURCES. Fresh fruit, especially lemons, black currants, oranges, rose hips, tomatoes, potatoes, green vegetables.

PHYSIOLOGICAL FUNCTION. The exact way in which ascorbic acid plays its role in the body is not yet understood. Ascorbic acid is readily oxidized in the body to form dehydroascorbic acid and this change is a reversible one; it is therefore thought that ascorbic acid plays an important role in biological oxidations and reduction.

There are a few examples of specific metabolic disturbances associated with deficiency of vitamin C. Firstly vitamin C is necessary for the conversion of folic acid to folinic acid (see p. 414). Secondly the metabolism of aromatic amino acids, particularly tyrosine and phenylalanine is disturbed in vitamin C deficiency, and this leads to a disturbance in melanin and pigment metabolism.

In the body vitamin C is found in high concentration in both the cortex and medulla of the adrenal gland; its exact role here is not understood, but because the vitamin disappears from the adrenal gland during stress situations or after a large dose of the pituitary hormone A.C.T.H. (see p. 255) it has been suggested that it may be necessary for the formation of the cortical steroid hormones.

DEFICIENCY DISEASE. A severe deficiency in vitamin C intake, produces the disease scurvy. In this condition there is a widespread disorder of the connective tissues and small blood vessels which produces the various symptoms and signs. In adults the condition is preceded by muscle and joint pains, weakness and loss of weight. Later bleeding gums appear and there may also be bleeding in a variety of sites including the skin, muscles, nose and even in the nervous system, gastrointestinal tract and urogenital tract. In infants the disease usually appears between the ages of six and eighteen months. Milk is a notoriously poor source of vitamin C and if vitamin supplements are not given (e.g. in the form of orange juice) or if weaning onto solid foods is greatly postponed there is the danger of the appearance of scurvy. The infant becomes irritable, loses appetite and weight; the irritability on handling of the infant is due to the presence of haemorrhages under the periosteum of the long bones of the limbs, and there may be also bleeding into the muscles or joints.

Vitamin D

HISTORY. Rickets is a disease with a long history, being first described by Whistler in 1645. Before the discovery of vitamin D about 80% of urban children suffered, to a lesser or greater extent, from rickets.

Mellanby in 1919 discovered that the disease in experimental animals could be prevented by cod liver oil and in 1919 Huldschinsky showed that sunlight could prevent rickets. Later it was found that foods after exposure to sunlight had a greater power to prevent the disease.

CHEMISTRY. Experiments have shown that there are many different substances which when exposed to sunlight have vitamin D activity. There are at least 10 different substances which possess vitamin D activity, and all these substances have the same fundamental chemical structure. The two most important substances are calciferol or vitamin D_2, and vitamin D_3.

Vitamin D_3 is obtained from animal sources e.g. fish liver oils. It is also formed when a substance called 7-dehydrocholesterol is exposed to ultraviolet light. Vitamin D₃ can be manufactured by mammals by the effect of sunlight upon 7-dehydrocholesterol in the skin. This latter substance is made by the dehydrogenation of cholesterol in the wall of the small intestine.

Calciferol can be produced by the effect of ultraviolet light on a substance called ergosterol; this substance is found in plants, including yeast, but not in animals:

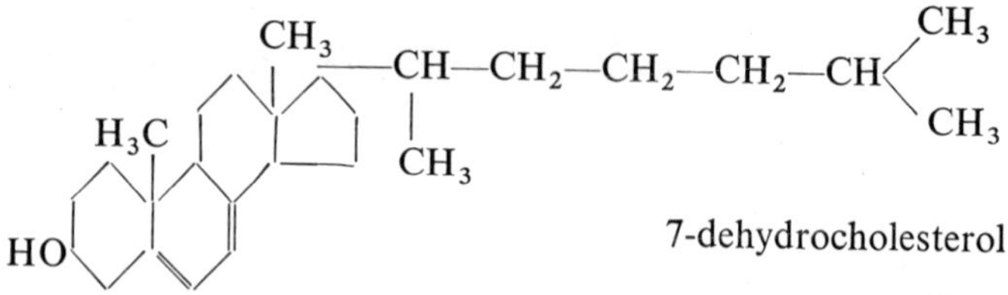

7-dehydrocholesterol

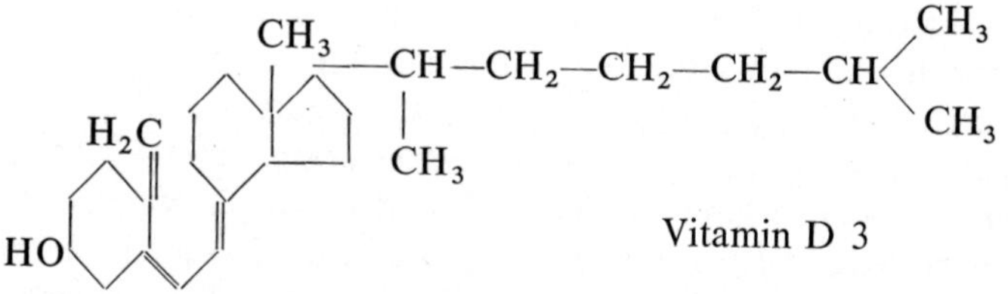

Vitamin D 3

SOURCES OF VITAMIN D. Liver oils, particularly fish (halibut, cod), milk, butter, eggs. Production within the skin by the effect of ultraviolet light.

PHYSIOLOGICAL FUNCTIONS. Vitamin D plays a vital role in the metabolism of calcium and phosphorus but the exact way in which it performs its functions is not known. The physiological function of the vitamin is to promote the absorption of calcium from the intestine and promote the deposition of calcium in the bones. In the absence of adequate amounts of vitamin D there is deficient absorption of calcium and there is a fall in the blood level of calcium. In response to this change in the blood calcium the parathyroid glands produce increasing amounts of the hormone parathormone, the effect of which is to encourage the release of calcium from the bones (see p. 642); this maintains the level of blood calcium at the expense of producing a demineralization of the bones. A deficiency of the vitamin is more acutely felt in the young growing mammal where growth of bone and deposition of calcium are taking place at a rapid rate.

Man and other animals obtain the vitamin from the effect of sunlight on 7-dehydrocholesterol in the skin, but in temperate latitudes, particularly in industrial areas where the smoky atmosphere filters out much of the available ultraviolet light, food becomes a more important source of the vitamin. Hence the vital role played by additions of the vitamin to dried milk powders and cod liver oil in the diet of infants.

DEFICIENCY DISEASE. A deficiency of vitamin D in childhood leads to the disease called rickets (fig. 185) in which there are widespread changes in the skeletal system because of the disturbance in normal bone development. There are enlargements of the ends of the long bones which may be curved in various directions depending upon the age at which the disease develops; after walking has started there is bowing of the legs and a deformity of the pelvic bones. There may be enlargement at the junction of bone and cartilage in the ribs giving rise to a vertical row of swellings on each side of the chest, sometimes called the ricketic rosary. Changes in the skull also develop, early a softening of the bones and later an overgrowth of the vault giving rise to 'bossing' of the head.

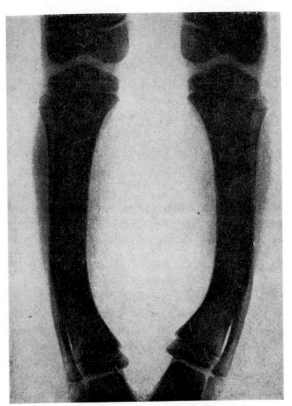

Fig. 185. *X-ray of the lower leg of a child with rickets showing bowing of the tibia. (By courtesy of Dr. F. Wilson Harlow, from Modern Surgery for Nurses', Heinemann Medical Books Ltd.)*

In the adult, when growth of bone has ceased, a deficiency in vitamin D intake leads to a generalized de-mineralization of the bones called osteomalacia; if this is severe it may lead to the appearance of spontaneous fractures in the bones.

Vitamin K

HISTORY. In 1929 it was observed by Dam that chickens fed on inadequate diets developed a deficiency disease in which there was spontaneous bleeding, apparently due to a deficiency of prothrombin in the blood. This condition was not cured by giving any of the then known vitamins. They found that it was cured by giving an unidentified fat-soluble substance to which Dam gave the name Vitamin K.

The pure vitamin was later isolated and its chemical structure discovered.

CHEMICAL STRUCTURE.

Vitamin K, (methyl-phytyl-naphthoquinone).

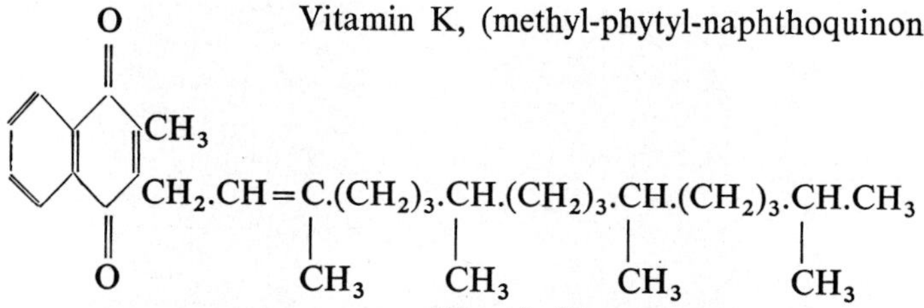

$$\text{CH}_2.\text{CH} = \text{C}.(\text{CH}_2)_3.\text{CH}.(\text{CH}_2)_3.\text{CH}.(\text{CH}_2)_3.\text{CH}.\text{CH}_3$$

Vitamin K_1

NATURAL SOURCES. Green plants and some animal tissues. It is synthesized by bacteria and the bacteria in the bowel form a source of the vitamin.

PHYSIOLOGICAL FUNCTION AND DEFICIENCY STATE. The physiological role of vitamin K in the formation of prothrombin by the liver is discussed on page 215, and the ways in which deficiency states arise is discussed.

Vitamin E

In spite of extensive studies the role of this vitamin in man is not understood. In experimental animals such as the rat a deficiency of vitamin E may lead to sterility or abortion in the pregnant female. In addition there are degenerative changes in voluntary and cardiac muscle. But there is no evidence that the vitamin is required in the diet of man.

It is a fat soluble substance and one of the best sources of the vitamin is in wheat germ oil. Other sources of the vitamin include egg, milk, butter, all seed embryos.

Chapter Nine   Tissue Respiration or Internal Respiration

Sources of energy

The continual supply of energy which is a pre-requisite for all life processes arises ultimately from the sun. The energy output of solar radiations is found in wavelengths from cosmic to infra-red rays. This energy of the sun arises from the continual formation of helium from hydrogen. During this process small amounts of mass which are 'left over' in the conversion of hydrogen to helium are converted into enormous amounts of free energy. Some of this solar energy can be used directly by animals insofar as it raises body temperature and

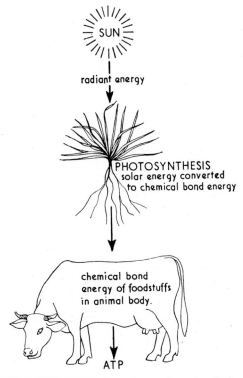

Fig. 186. *Sketch illustrating the transformation of solar energy.*

421

accelerates metabolic processes, but only to this very limited extent can animals directly utilize solar energy. Radiant solar energy cannot be utilized by animals for the synthesis of body materials, muscular activity, secretion etc. Plants however can utilize solar energy directly in the process of photosynthesis where solar energy is converted into chemical bond energy, originally in carbohydrate. This chemical bond energy is now available for the animals who eat the plants.

The unit of energy exchange in cells. ATP

We have seen that the free energy of solar radiation is trapped by plants during the process of photosynthesis as chemical bond energy. The cells of animals contain highly complex and imperfectly understood

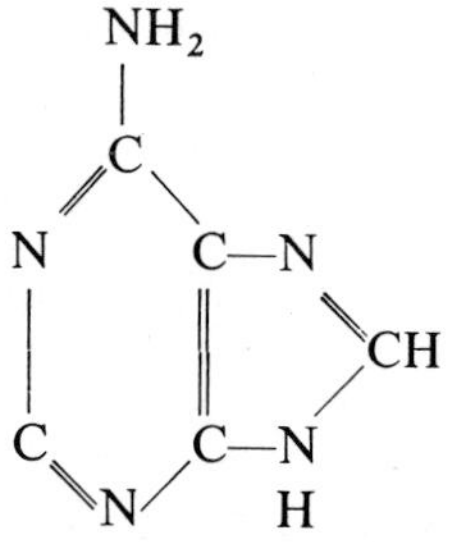

Adenine:

Adenosine (Adenine + ribose sugar)

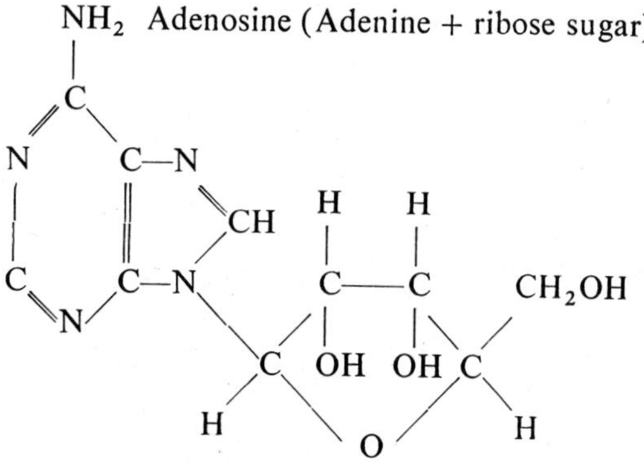

Adenosine phosphate:

Adenosine monophosphate.

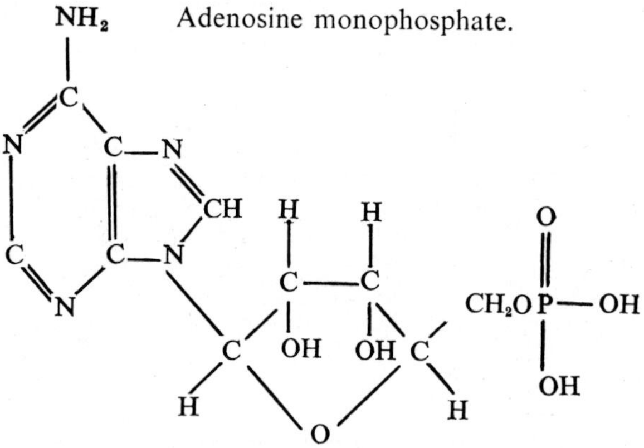

Fig. 187. *The structure of adenine, adenosine and adenosine monophosphate.*

mechanisms by which this chemical bond energy of foodstuffs can be made available to supply the energy which is needed for growth, maintenance, movement, nervous activity and so on. All these life processes utilize energy in the form of a common basic unit or currency. This energy currency is an energy-rich chemical bond in a substance called adenosine triphosphate (ATP) (fig. 186).

The structure of ATP

ATP is a compound consisting of a nitrogen containing base called adenine and a ribose sugar (figure 187). One of the —OH groups of the sugar can be phosphorylated, i.e. the H group replaced by H_2PO_3, producing adenosine phosphate. The adenosine phosphate can then be further phosphorylated to yield adenosine diphosphate (ADP) and eventually adenosine triphosphate.

Energy-rich bonds in ATP

Organophosphate compounds differ very markedly in their stability and in the amount of heat which is liberated during their hydrolysis. When phosphate esters such as adenosine monophosphate are hydrolysed approximately 3000 cals/mole are liberated. The phosphate is said to be attached by an energy-poor bond. In ADP the distribution of energy within the molecule is different and the second phosphate is attached by an energy-rich bond which releases 10,000 cals/mole when it is hydrolysed. In ATP there is one phosphate attached by an energy-poor bond and two by energy-rich bonds. The energy-poor bonds are represented by —P and the energy-rich bonds are represented by ~P. We may represent the three molecules of AMP, ADP and ATP as follows:

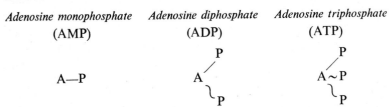

Adenosine monophosphate   *Adenosine diphosphate*   *Adenosine triphosphate*
    (AMP)   (ADP)   (ATP)

A—P   A   A~P

Coupling

This energy-rich phosphate bond is of supreme importance in bodily activities. It can be transferred without loss of energy to other substances such as creatine phosphate (page 621), phosphoenolpyruvate (page 450) or acetyl phosphate. The energy of the bond can also be utilized in the performance of muscular contraction (page 616), generation of the nerve impulse (page 284), secretion of ions (page 284) etc. The process of transfer of the energy of the $\sim$P to other substances or to provide energy for various activities is known as energetic coupling. Examples of this coupling or transfer of energy from the basic unit of energy currency is illustrated in figure 188.

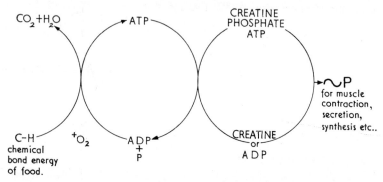

Fig. 188. *An example of energetic coupling.*

Oxidation of foodstuffs

We can now examine the way in which the chemical bond energy present in foodstuffs is transformed into the basic unit of energy exchange of the cell in the form of the energy-rich phosphate bonds of ATP. The oxidation of a foodstuff such as glucose takes place in many stages during which the glucose and its breakdown products are phosphorylated. There is a rearrangement of energy in the molecules so that eventually $\sim$P are formed. These substances with $\sim$P bonds are then allowed to react with ADP which is always present in small amounts in cells. The ADP is thus converted into ATP which becomes a temporary store for the small packets of useful energy which are released from the molecules undergoing oxidation. ATP may exist for only a minute fraction of a second before it transfers a $\sim$P to some more permanent store (e.g. to creatine phosphate in muscle) or supplies energy to some other metabolic process. The ADP which is now released becomes available to react with more $\sim$P from the respiratory substrate.

Biological oxidations

It seems that life began on this planet in the absence of atmospheric oxygen and endured this condition until plants had put oxygen into the atmosphere by the process of photosynthesis. It may be for this reason that biological oxidations are usually anaerobic, i.e. they do not involve the addition of oxygen to the compound being oxidized. This may appear confusing to the student who is beginning a study of biological oxidation and it is important at this stage to digress on what we mean by the term oxidation.

In fundamental terms oxidation is the removal of electrons from an atom or molecule. We regard atoms as composed of a positively charged nucleus and up to about 100 negatively charged electrons around it. In a neutral atom the total charge on the electrons exactly balances the charge on the nucleus. If there are fewer electrons the atom has a positive charge and if there are more electrons it has a negative charge. These charged atoms are called ions.

Originally the term oxidation meant the combination of oxygen with a metal to form a metal oxide, e.g. the oxidation of metallic copper to black copper oxide according to the equation:

$$2\,Cu + O_2 = 2\,Cuo$$

The reverse process, which involves the removal of oxygen, e.g. by carbon or hydrogen was called reduction of the oxide:

$$Cuo + H_2 = Cu + H_2O$$

in this reaction we can consider that hydrogen has been oxidized to water in the process if we replace the word *metal* in our original definition by the word *element*. If we regard the copper and hydrogen as competing for the oxygen in the copper it is obvious that the hydrogen has a greater avidity for oxygen, that is it is a stronger reducing agent.

The original concept of oxidation and reduction has been considerably extended. Compounds, as well as elements, may be oxidized. Further since hydrogen can often be used to remove oxygen from oxides, addition of hydrogen to an element or compound is considered to be the equivalent of reducing it. Conversely the removal of hydrogen from a compound is regarded as oxidation. Thus alcohol is oxidized in the presence of copper as a catalyst at 300°C according to the following equation:

$$C_2H_5OH = CH_3CHO + H_2$$

The terms oxidation and reduction are no longer restricted to reactions involving exchange of oxygen or hydrogen. There are two forms of iron oxide and salts which differ in the valency of the iron. Ferric iron is trivalent whereas ferrous iron is divalent. In the reaction:

$$2\,FeCl_2 + Cl_2 = 2\,FeCl_3$$

ferrous chloride   ferric chloride

ferrous chloride is regarded as being oxidized to ferric chloride. Similarly in the reaction:

$$2 \, FeCl_3 + SnCl_2 = 2 \, FeCl_2 + SnCl_4$$
$$\text{stannous} \qquad\qquad \text{stannic}$$

ferric chloride is reduced to ferrous chloride by stannous chloride which is oxidized in the process to stannic chloride. We can thus further redefine oxidation as the addition of oxygen, chlorine or a similar non-metallic constituent and reduction as the removal of oxygen and similar non-metallic components *or* the addition of hydrogen or a metallic constituent.

Our definition can be further modified by considering that most inorganic compounds form ions in aqueous solutions. Thus ferrous and ferric chlorides are present in solution as $Fe^{++} + Cl^-$ and Fe^{+++} and Cl^- respectively. The oxidation of ferrous to ferric iron may thus be represented simply by

$$Fe^{++} = Fe^{+++} + \varepsilon^-$$
$$\text{electron}$$

When chlorine is the oxidizing agent the complementary reduction of chlorine is represented by

$$\tfrac{1}{2}Cl_2 + \varepsilon^- = Cl^-$$
$$\text{electron}$$

Thus ferrous iron is oxidized to ferric iron by losing an electron. Conversely ferric iron can be reduced to ferrous iron by gaining an electron. Not all oxidation—reduction processes involve only the transfer of electrons. They may also involve the transfer of atoms. Thus when a solution of potassium permanganate in sulphuric acid is reduced to a salt of the Mn^{++} ion the manganese-oxygen bonds of the permanganate must be broken.

$$MnO_4^- + 8H^+ + 5e = Mn^{2+} + 4H_2O$$

Reduction of MnO_4^- to Mn^{2+} is achieved by the gain of electrons and by removal of oxygen by hydrogen. In this reaction the electrons are supplied by ferrous ions which are oxidized to ferric ions in the process. Since one permanganate ion can utilize five electrons and one ferrous ion can donate one electron when it is oxidized to ferric ion the overall reaction between ferrous irons and acidified potassium permanganate solution is represented as follows:

$$MnO_4^- + 8H^+ + 5Fe^{++} = Mn^{++} + 5Fe^{+++} + 4H_2O$$

We can thus regard oxidation and reduction as always involving the transfer of electrons but they may also involve the transfer of atoms. We have used the example of iron as a donator or acceptor of electrons because iron forms an important component of electron transfer systems in the mechanism of biological oxidation.

By passing known currents through electrolytes and weighing the electrodes it has been shown that in a particular electrolyte the weight of substance liberated or dissolved at a particular electrode is proportional to the amount of electricity which passes. The quantity of electricity needed to liberate (or dissolve) one gram atom of any element from an electrolyte is a small whole number multiple of the quantity of electricity needed to deposit one gram atom of silver. The amount of electricity needed to deposit one gram atom of silver is called 1 Faraday. Thus:

to deposit 1 gram atom of silver takes   1 Faraday

,,   ,,   1   ,,   ,,   ,, lead   ,,   2   ,,

,,   ,,   1   ,,   ,,   ,, ferric iron takes 3   ,,

If a substance is deposited at the negative electrode (cathode) it is said to be a positive substance (or cation) and conversely those liberated (or deposited) at the positive electrodes (anode) are called anions. Thus we designate ions as +ve or −ve and we can suggest the amount of charge needed to produce them thus Cl^-, Na^+, Ca^{++}, O^{--}, Fe^{+++}.

The 'oxidation state' is really a description of the number of electrons lost or gained. In the case of ions the oxidation state is numerically equal to the number of Faradays associated with one gram-ion.

+4							
+3							Fe^{+++}
+2		Cu^{++}					Fe^{++}
+1	Ag^+		Cu^+	H^+			
0	Ag	Cu	Cu	H	Cl	S	Fe
−1					Cl^-		
−2						S^{--}	

We have seen that the oxidation of foodstuffs by living things was achieved by anaerobic means. This is a primitive ability. As soon as the process of photosynthesis had yielded oxygen to the atmosphere it seems that organisms already capable of performing anaerobic oxidation by withdrawal of electrons or hydrogen from the foodstuffs developed another ability—the 'combustion' of hydrogen with oxygen. In present day organisms the earliest stages of oxidation of foodstuffs is invariably anaerobic, oxygen being involved only in the terminal phase of 'combustion' of hydrogen.

Details of oxidation of foodstuffs

The oxidation of a foodstuff, glucose, can be expressed simply in the following equation:

$$C_6H_{12}O_6 + 6O_2 \rightarrow 6CO_2 + 6H_2O + E \text{ (690 K.cals)}$$

This equation indicates only the raw materials and the end products of the reaction. The reaction in fact proceeds in a large number of intermediate stages, some of which result in the formation of ATP. Oxygen is involved only in the terminal stages of the reactions. When glucose burns in oxygen the energy is dissipated completely and rapidly and if such a process occurred in a living cell the cell could not trap the energy as $\sim P$. It is necessary therefore to liberate the energy slowly and in small amounts which can be captured as $\sim P$ and therefore oxidation of carbohydrate in cells is a long sequence of reactions.

The amount of energy in a molecule depends on its size and on its degree of oxidation. Obviously the bigger the molecule the greater we would expect its energy content to be. An oxidized substance has less energy than a reduced one, thus CO_2 is used for extinguishing fires; it is highly oxidized and has low energy whereas CH_4 is highly reduced and is the explosive gas of coal mines called methane. Fats have more energy than carbohydrates and have more hydrogen in their molecule, i.e. they are more highly reduced:

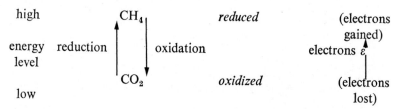

In the oxidation of carbohydrate we can recognize three main stages: glycolysis, the Krebs' cycle and the electron transfer system.

1. Glycolysis.

In this process the molecules of glucose (6 carbon atoms) undergo progressive chemical changes, including anaerobic oxidation, resulting in the formation of pyruvic acid (3 carbon atoms). During this process some of the chemical bond energy appear in the form of ∼P of ATP.

2. *Krebs' cycle or citric acid cycle.*

The fragmented glucose molecules (pyruvic acid) are now subjected to a cyclic sequence of chemical reactions. Carbon dioxide is removed from the molecules and anaerobic oxidation also occurs at various stages of the cycle. Oxidation occurs by means of enzymes which become reduced in the process. The re-oxidation of these enzymes depends ultimately on a supply of oxygen.

3. *The electron transfer chain—the transportation of hydrogen and its ultimate oxidation by oxygen.*

We have seen that the oxidation of glucose in the process of glycolysis and Krebs' cycle is achieved by the transfer of hydrogen from the foodstuff to various other substances. The substances which accept hydrogen from the foodstuff may be protein molecules which are called enzymes and are distinguished by the suffix -ase, e.g. various dehydrogenases. Other non-protein compounds act as co-enzymes for enzymes. One such co-enzyme is co-enzyme I or diphosphopyridine nucleotide (DPN) now more properly called NAD (nicotinamide–adenine dinucleotide). Another is co-enzyme II, now more properly known as NADP (nicotinamide–adenine–dinucleotide phosphate). Note that both of these co-enzymes contain a member of the vitamin B complex in their chemical make up—i.e. nicotinic acid.

As these co-enzymes become reduced during the oxidation of the foodstuffs they come to contain much of the potential energy of the foodstuffs. Each reduced co-enzyme is eventually oxidized by molecular oxygen to water and the oxidized co-enzyme is returned to Krebs' cycle for recharging. During the process of the 'combustion' of the hydrogen transported from the foodstuff by the co-enzymes further energy becomes available to the cell in the form of ATP. In fact for every reduced co-enzyme which is oxidized by oxygen three ∼P are formed.

We can now examine in more detail these three stages in the breakdown of glucose to supply energy for the cell.

GLYCOLYSIS

This is the first stage in the breakdown of the glucose molecule. It is an anaerobic process and it occurs in the general cytoplasmic matrix of the cell, i.e. in the water soluble fraction of the cell. The formation of pyruvic acid marks the completion of this stage.

When glucose arrives inside a cell from the interstitial fluid surrounding the cell it becomes converted into glucose-6-phosphate under the influence of an enzyme called hexokinase. The energy for this conversion is provided by ATP.

$$\text{Glucose} + \text{A.T.P} \xrightarrow{\text{Hexokinase}} \text{Glucose-6-phosphate} + \text{A.D.P.}$$
$$\text{G-6-P}$$

The glucose in its phosphorylated form can now enter various metabolic pathways in the cell. Here we are concerned with the glycolytic pathway. The next stage in this process is the conversion of G-6-P a six membered ring, to the five membered ring of fructose-6-phosphate. This compound is phosphorylated further by ATP to yield fructose 1:6 diphosphate.

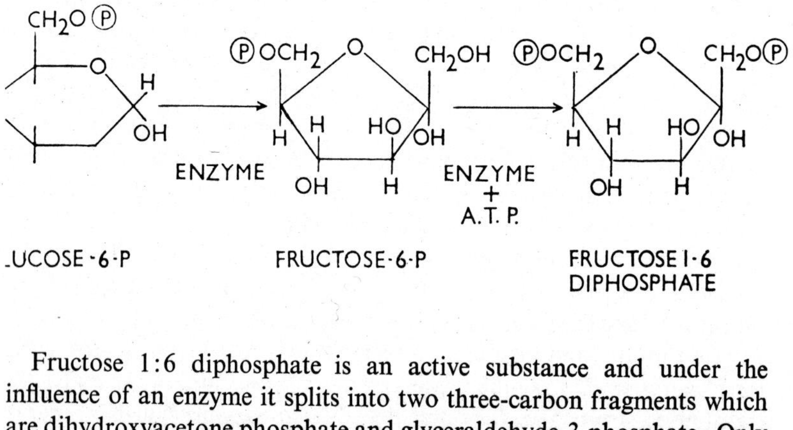

GLUCOSE-6-P   FRUCTOSE-6-P   FRUCTOSE 1-6 DIPHOSPHATE

Fructose 1:6 diphosphate is an active substance and under the influence of an enzyme it splits into two three-carbon fragments which are dihydroxyacetone phosphate and glyceraldehyde-3-phosphate. Only the latter can be used as a source of energy but dihydroxyacetone phosphate can be converted into glyceraldehyde-3-phosphate by means of an enzyme–phosphotriose isomerase.

Each molecule of glucose entering the glycolytic pathway has, under the influence of various enzymes and a source of energy (ATP) been converted to a 3-carbon compound, glyceraldehyde-3-phosphate. This compound now becomes oxidized by the co-enzyme NAD (nicotineadenine dinucleotide) in the presence of phosphoric acid. In this process di-phosphoglyceric acid is formed. During this oxidation process there

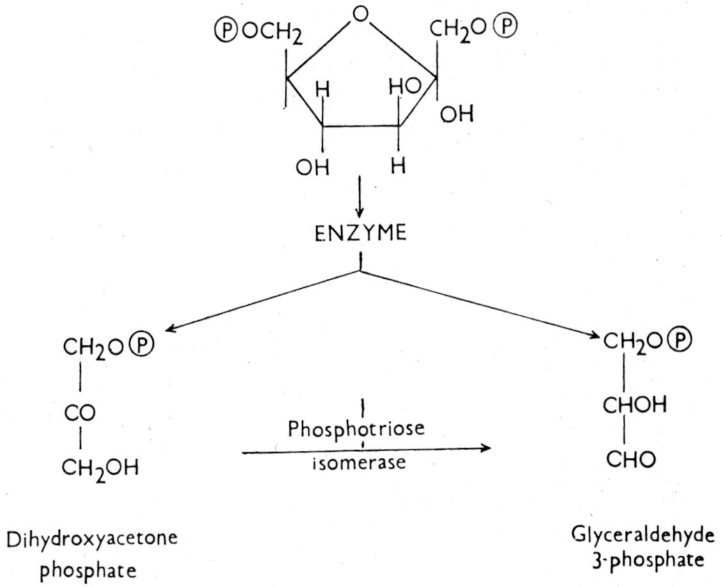

$$
\text{(P)OCH}_2 \qquad \text{CH}_2\text{O (P)}
$$

ENZYME

Dihydroxyacetone
phosphate

Glyceraldehyde
3-phosphate

$$
\begin{array}{c}
\text{CH}_2\text{O (P)} \\
| \\
\text{CO} \\
| \\
\text{CH}_2\text{OH}
\end{array}
\qquad
\xrightarrow[\text{isomerase}]{\text{Phosphotriose}}
\qquad
\begin{array}{c}
\text{CH}_2\text{O (P)} \\
| \\
\text{CHOH} \\
| \\
\text{CHO}
\end{array}
$$

is a redistribution of energy within the molecule resulting in the formation of a high-energy phosphate bond (carboxy-phosphate linkage). This high energy phosphate bond of di-phosphoglyceric acid can now be transferred to ADP forming ATP and phosphoglyceric acid.

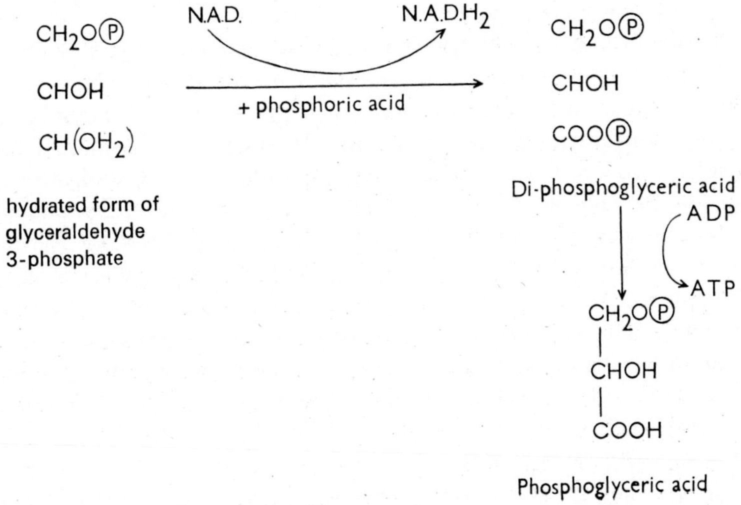

$$
\begin{array}{c}
\text{CH}_2\text{O (P)} \\
\text{CHOH} \\
\text{CH (OH}_2)
\end{array}
\xrightarrow[\text{+ phosphoric acid}]{\text{N.A.D.} \quad \text{N.A.D.H}_2}
\begin{array}{c}
\text{CH}_2\text{O (P)} \\
\text{CHOH} \\
\text{COO (P)}
\end{array}
$$

hydrated form of
glyceraldehyde
3-phosphate

Di-phosphoglyceric acid

ADP

ATP

$$
\begin{array}{c}
\text{CH}_2\text{O (P)} \\
| \\
\text{CHOH} \\
| \\
\text{COOH}
\end{array}
$$

Phosphoglyceric acid

There is now a further internal molecular arrangement of the phosphoglyceric acid producing a substance called phosphoenol pyruvic acid containing a $\sim P$. The $\sim P$ of enol-pyruvic acid is now transferred to ADP yielding ATP and enol-pyruvic acid. Enol-pyruvic acid then passes over into the more stable form of keto-pyruvic acid.

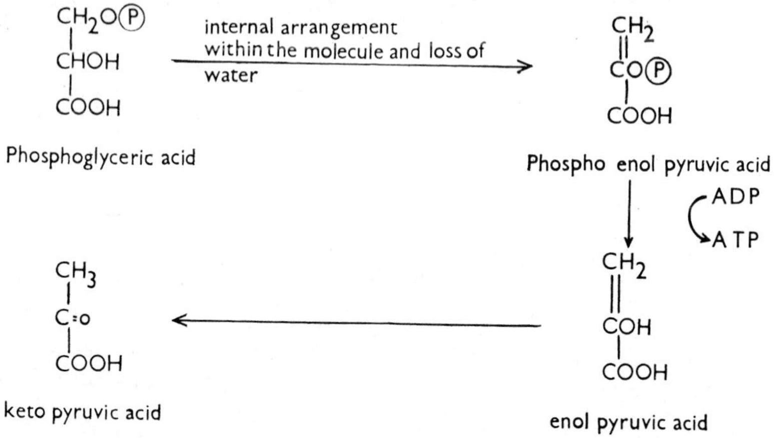

Phosphoglyceric acid

Phospho enol pyruvic acid

keto pyruvic acid

enol pyruvic acid

Since fructose 1:6 diphosphate gave rise to two 3-carbon fragments it can be seen that four molecules of ATP are produced in the breakdown of glucose to pyruvic acid. But since two molecules of ATP are utilized in the formation of fructose 1:6 diphosphate the net yield is only two molecules of ATP for each molecule of glucose which is broken down to pyruvic acid.

This is in fact a small yield but it can be achieved in the absence of oxygen. Furthermore, still further yields of $\sim P$ can be obtained by further oxidation of pyruvic acid and by oxidation of the reduced NAD by the electron transfer system which will be described in more detail in a later section. In the absence of oxygen for 'combustion' of hydrogen carried by reduced NAD the reduced co-enzyme must accumulate within the cell. Now there is only a limited amount of NAD in the cell and in order that the oxidation of glucose should continue it is necessary for reduced NAD to pass on its hydrogen to some other hydrogen acceptor. In mammalian muscle tissue in the absence of oxygen reduced NAD can pass its hydrogen to pyruvic acid itself and so produce lactic acid. Thus muscle tissue can temporarily obtain the ATP necessary for work in the absence of oxygen (see also page 622).

Yeast cells are also able to obtain their energy in the absence of oxygen. They are able to do this because the pyruvic acid obtained in the anaerobic oxidation of glucose can undergo the process of decarboxylation to produce acetaldehyde. Acetaldehyde under the influence of an enzyme alcohol dehydrogenase can accept hydrogen from

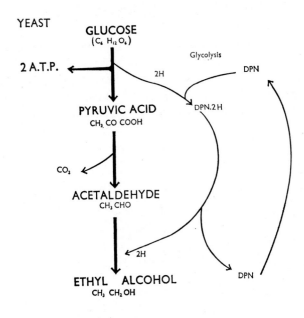

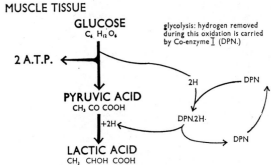

Fig. 189. *Diagram to show how A.T.P. is produced during the temporary absence of oxygen in muscle tissue and yeast.*

reduced NAD to produce ethyl alcohol. This process can go on until sufficient alcohol accumulates in the yeast culture to stop the growth of the culture and eventually kill it (figure 189).

THE FATE OF PYRUVIC ACID

Pyruvic acid is oxidized fully to carbon dioxide and water by most tissues if adequate amounts of oxygen are available.

$$CH_3 \ CO \ COOH + 5O \rightarrow 3 \ CO_2 + 2H_2O$$

Muscle has the greatest demand for carbohydrate and it is not surprising that the early work on aerobic respiration was done on minced muscle—which was obtained from pigeon breast. Szent Gyorgyi found that the respiratory rate of muscle could be maintained by adding catalytic amounts of succinic or fumaric acids. Later he found that malic and

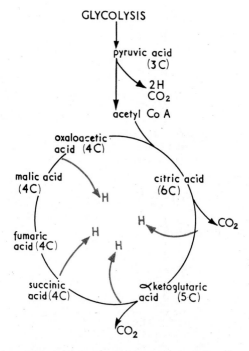

Fig. 190. *Diagram showing some of the major inter-mediate compounds in Krebs' citric acid cycle.* Also shown are the sites of CO_2 production and the sites at which oxidation (by removal of H) occur. Energy becomes available to the cell in the form of ATP during the transport of the hydrogen along the chain of respiratory enzymes to react eventually with molecular oxygen.

oxaloacetic acid had the same effect. Soon afterwards Hans Krebs found that α-ketoglutaric and citric acid catalyzed the breakdown of pyruvic acid to carbon dioxide and water. Krebs suggested that in tissues pyruvic acid (C_3) and oxaloacetic acid (C_4) join to make a C_7 substance from which other substances arose in the sequence of citric acid (C_6), α-ketoglutaric acid (C_5), succinic acid (C_4), fumaric acid (C_4) by the removal of carbon atoms (as CO_2).

The oxaloacetic acid feeds back into the system by reacting with pyruvic acid. As long as there is pyruvic acid present and oxygen to remove hydrogen from the reduced enzymes which are produced in the process then the cyclic reaction will go on. This cycle of activity has come to be known as Krebs' citric acid cycle (or the tricarboxylic acid cycle). So far every step of the cycle has been demonstrated with the exception of the C_7 compound. It is now suggested that the C_4 oxaloacetic acid reacts with a C_2 substance rather than with C_3 pyruvic acid. It is known that pyruvic acid is changed to a C_2 substance called acetyl Co A by process of oxidative decarboxylation (the co-carboxylase in this reaction is vitamin B_1).

$$CH_3 CO COOH + H_2O + DPN \rightarrow CH_3 COOH + CO_2 + DPNH_2$$
Pyruvic acid　　　　　　　　　　Acetic acid

This is a complex process involving the presence of magnesium ions phosphate, thiamine phosphate (vitamin B_1), lipoic acid, ADP DPN and co-enzyme A. Free acetic acid is not formed but instead the derivative of the acid, the very active acetyl Co A. Figure 190 summarizes the present-day information about Krebs' cycle.

KREBS' CYCLE AS A SOURCE OF ENERGY

When glucose is completely oxidized to carbon dioxide and water there is a liberation of free energy amounting to about 686,000 cals/g.mol. We have seen that the breakdown of 1 g.mol. of glucose under anaerobic conditions to pyruvic acid produces an energy yield of 2 ~P (i.e. about 20,000 cals). In the presence of oxygen however the reduced NAD which is produced during the anaerobic oxidation can transfer hydrogen to the electron transfer system.

The 'combustion' of this hydrogen results in a further yield of 3 ~P/ molecule of $NADH_2$. Since two molecules of triosephosphate arise from one molecule of glucose and reduce two molecules of the co-enzyme there is a net yield of 6 ~P. Thus the total yield of ~P in the aerobic conversion of one molecule of glucose into two molecules of pyruvate is 8 $\sim$P. i.e.

$$C_6H_{12}O_6 + O_2 \rightarrow CH_3COOH + 2H_2O + 8 \sim P$$

As hydrogen is progressively removed from molecules of the substances lying on the path of Krebs' cycle, further $\sim$P are produced by the transport of this hydrogen along the electron transport system. The stages at which oxidation (by removal of hydrogen) occurs in Krebs' cycle are shown below, together with the oxidizing agent.

Reaction	Oxidizing agent	Yield of $\sim$P
Pyruvate $\rightarrow$ Acetyl CoA $\rightarrow$ citrate	NAD	3
Isocitrate $\rightarrow$ oxalosuccinate	NADP	3
α ketoglutarate $\rightarrow$ succinate	NAD	4
Succinate $\rightarrow$ fumarate	cytochrome	2
Malate $\rightarrow$ oxaloacetate	NAD	3
	Total	15

Since two molecules of pyruvate are derived from one molecule of glucose the total yield of $\sim$P will be 30. Taking into account the energy yield of aerobic glycolysis this gives a grand total of 38 $\sim$P. The formation of 38 $\sim$P means that 38 × 11,500 calories are trapped in the form of $\sim$P from a total possible yield of 686,000 cals (i.e. the energy loss in the complete oxidation of glucose). This gives a figure for the efficiency of energy trapping of about 64%, a remarkable efficiency of living systems. The efficiency of a steam turbine or diesel engine is in fact lower than this, amounting to 35% and 40% respectively. In living cells the remaining 36% of the energy content of the food which is not trapped as $\sim$P appears mainly in the form of heat energy.

THE ELECTRON TRANSFER SYSTEM

We have seen that the anaerobic oxidation of the various respiratory substrates involves the removal of hydrogen from the molecules by various oxidizing agents or hydrogen carriers. The potential energy of the food becomes mainly located in these reduced carriers. The hydrogen is not directly oxidized to water by oxygen but instead it is transported along a bucket-brigade of hydrogen acceptors. Each member of the bucket-brigade is a more powerful oxidizing agent than the one before it. Only in the last component of the chain is the hydrogen reduced to water by oxygen. This process of the biological 'combustion' of hydrogen is thus split into many stages, each one involving small potential steps so that the energy of the oxidation is released in small manageable units of $\sim$P. For every reduced co-enzyme which is oxidized by this mechanism three $\sim$P are formed.

We have called this system one of electron transport. The substances are able to accept hydrogen atoms or electrons from a donor and pass them on to the next link in the carrier system. This question of transfer of hydrogen atoms or electrons may create some confusion in the mind of the reader. Remember that the hydrogen atom consists of a positively charged proton and a single negatively charged electron. We have seen that oxidation is the removal of an electron or hydrogen atom from a substance. The transfer of electrons or hydrogen atoms along the chain of carriers results in a progressive reduction of the line, one member becoming oxidized as it passes along one electron or hydrogen atoms to the adjacent more electronegative member of the series. It does not matter that some components of the system transfer electrons and some hydrogen atoms (i.e. proton and electron). The net result is the same. When only the electron is transferred the remaining component of the hydrogen atom, i.e. the positively charged hydrogen ion, remains close at hand and is available again to form the hydrogen atom at the appropriate time.

The components of the electron transfer system are located in the inner membrane of the mitochondrion. It is here that electrons are transferred from reduced co-enzymes to oxygen to form water and it is here that energy is trapped as $\sim$P during the stepwise transfer of electrons. Various chemical substances form part of the chain including NADP, NAD and a substance called FAD or flavine adenosine dinucleotide. These are succeeded in the chain by a series of respiratory pigments called cytochromes. These are conjugated proteins carrying an iron-containing prosthetic group which is very similar to the haem of haemoglobin. The iron component of cytochrome is vital for the function of electron transfer. In fact, the only observable difference between the oxidized and reduced form of cytochrome C is in the valency of

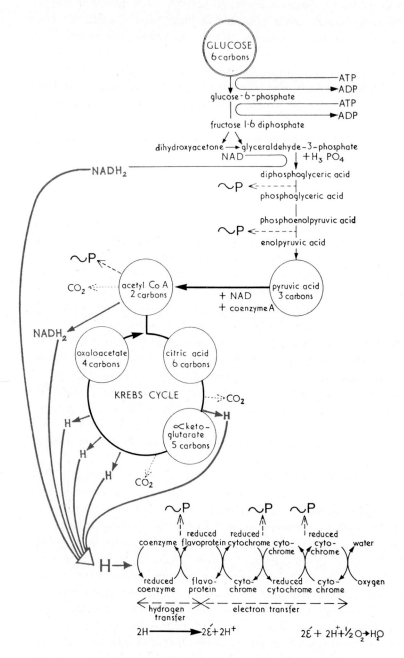

Fig. 191. *Summary of the process of tissue respiration.*

the iron atom in it. Figure 191 shows the change in valency of the iron of cytochromes as they pass an electron along the chain. The last member of the chain is oxidized by oxygen in the presence of an enzyme called cytochrome oxidase.

We still do not know the exact nature of the electron transfer system. One view is that the members of the chain are fixed as a kind of mosaic on the inner surface of the mitochondrial membrane. According to this view as each component accepts an electron it rotates on its axis so as to transfer the electron to the next component of the system. Of the several components of the chain only three are regarded as generating a high enough 'energy state' for the formation of energy-rich bonds.

Another view of the organization of the transfer system is based partly on the observed structure of the mitochondrion. During oxidation phosphorylation mitochondria undergo gross changes in shape and form. These have been regarded as reflecting changes in the shape of the sub-units of the mitochondrial membrane. The inner mitochondrial membrane appears to be composed of fused units of lipo-protein each of which has a 'stalk' and 'head piece' (fig. 192). It has been suggested that each of these units consists of a complex of electron carriers. The passage of electrons through a unit is presumed to cause a distortion of the complex, first at the base where the energy of electron transfer is 'stored' as in a wound-up watch spring. A wave of distortion then moves up the complex as the electron is transferred and the discharge of this distortion and return to the resting shape is presumed to be coupled to ATP formation. It is assumed that the sub-units of the membrane operate in triplets and the discharge of electrons across three sub-units results in the formation of three $\sim$P.

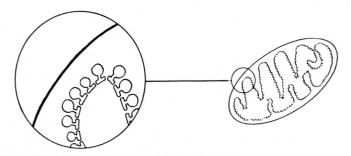

Fig. 192. *Mitochondrion* (Right) *and a H.P. sketch of the structure of the inner mitochondrial membrane* (Left).

This hypothesis may at first sight seem somewhat bizarre. But remember that the reverse sequence of events is known to happen in muscle where the new energy associated with the fission of ATP results in changes in the molecular configuration of the proteins actin and myosin which produce contraction of the muscle.

Regulation of respiration

We have described the process of respiration as a complex series of reactions the different stages of which may be localized in different parts of the cell structure. Individual reactions are influenced by specific enzymes. The rate at which these reactions proceed is highly variable in time and space. Different cells may have different respiratory rates depending upon their metabolic activity. Individual cells may also show variations in the rate of oxidation of substrate. These variations may be very marked as in the case of a muscle cell which can be rapidly transformed from a resting cell to one which contracts and performs work.

One basic mechanism of regulation of the rate of respiration seems to depend on the amount of ADP within the cell. If a cell is performing minimal amounts of work, e.g. in terms of synthesis of protein or fat, transfer of ions, contraction, secretion and the like, then ATP will accumulate in the cell and ADP disappear. There is no longer any acceptor for $\sim$P and the rate of production of $\sim$P (i.e. respiration) now declines.

Conversely if ATP disappears from the cell because it is being utilized at a fast rate then ADP accumulates and its presence appears to stimulate the process of respiration.

Certain drugs and hormones can strongly influence the process by which the energy of foodstuffs appears in the form of $\sim$P. These substances can reduce the transfer of energy to $\sim$P so that ATP production declines and ADP accumulates in cells. Now in the presence of free ADP respiration proceeds at a faster rate. Because the energy of the food is no longer appearing as $\sim$P it is dissipated as heat. We describe this process as an uncoupling of oxidation, i.e. a break in the links which transfer the energy of foodstuffs into $\sim$P. The hormone thyroxine is thought to act on cells in such a fashion. Under the influence of thyroxine cells produce less ATP, their rate of oxidation increases and this means that more oxygen is utilized and more heat generated. Hormones which can produce this effect by reducing the efficiency of energy transfer can markedly increase the heat production of the body and this forms an important mechanism of regulation of body temperature during exposure to cold.

THE RELATIONSHIP BETWEEN CARBOHYDRATE METABOLISM AND PROTEIN. Protein has been described as a body-builder, that is it becomes incorporated into the protoplasmic framework of the tissues of the body. But protein in excess of these anabolic needs can be used as a source of energy. In order to act as a source of energy the amino-acids must be broken down to yield substances which lie on the chain of compounds in Krebs' cycle. The first step in breakdown of an amino-acid is the removal of the amino group $-NH_2$. The amino group may become incorporated into the ornithine-arginine cycle (see p. 516) to produce urea, leaving a keto acid residue, or it may be passed onto other compounds and so form amino acids (this process is called transamination). In the following example the amino-acid glutamic acid is reacting with pyruvic acid in a transamination reaction to produce α keto-glutaric acid and the amino-acid alanine.

Glutamic acid   Pyruvic acid   α ketoglutaric   alanine
                                             acid

Both pyruvic acid and α ketoglutaric acid are involved in Krebs' cycle, and glutamic acid and alanine are common amino acids. Thus amino acids can be linked to the respiratory cycle. The prosthetic group of the transaminases in all organisms is pyridoxal phosphate, which is a member of the vitamin B complex.

After deamination or transamination of the amino acid the fatty acid so produced can be converted to acetyl CoA, but the ways in which this occurs are too complex for study here. An often used path is to convert the fatty acid to oxaloacetic acid and thence to pyruvic acid. The pyruvic acid is then converted into acetyl CoA which passes into Krebs' cycle. By a reversal of the mechanisms described in the Embden Meyerhof route pyruvic acid can be converted into glucose.

THE METABOLISM OF PROTEIN AND THE RELATIONSHIP BETWEEN PROTEIN
AND CARBOHYDRATE METABOLISM.

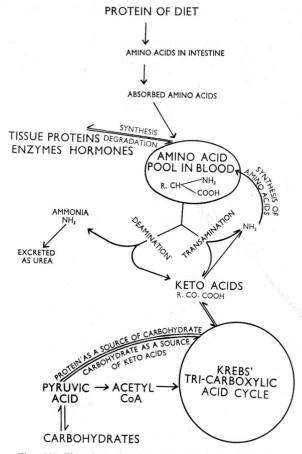

Fig. 193. *The relationships between protein and carbohydrate
metabolism.*

RELEASE OF ENERGY FROM FAT. In the body, fats have two important
functions. Firstly they enter into the structure of protoplasm and cell
membranes. Secondly they act as stores of energy. In addition certain
fatty acids may have special metabolic functions within the organism;
if particular fatty acids are removed from the diet of rats they become
diseased.

Most cells contain lipolytic enzymes and it appears that the first stage
in the breakdown of fats is the hydrolysis to glycerol and fatty acids.
Glycerol can pass into the pathway of carbohydrate metabolism by

becoming phosphorylated at the expense of A.T.P. to produce glycerol phosphate which is then converted to triose-phosphate. Once it has entered the Embden Meyerhof pathway of carbohydrate metabolism the glycerol can be broken down to pyruvic acid, yielding energy, or it can take the reverse process and be converted into glucose and eventually into glycogen.

The fatty acids produced from the hydrolysis of the fat are broken down into two carbon fragments by a process known as β oxidation. These two carbon fragments are acetyl CoA molecules which can enter the Krebs' cycle. In addition to this source of fatty acids from the hydrolysis of fat, fatty acids are also derived from the bacterial digestion of cellulose within the alimentary tract of herbivores. In these animals the fatty acids from bacterial digestion of cellulose form the major energy source (see also p. 399). These fatty acids are activated by A.T.P. and are broken down into C_2 fragments of acetyl CoA and enter the Krebs' cycle.

The net energy yield from a 6 carbon fatty acid has been estimated at 44 molecules of A.T.P. compared with 38 molecules of A.T.P. from a 6 carbon sugar such as glucose. We see therefore that, carbon for carbon, fat has more biologically useful energy in its molecule than has carbohydrate.

THE INTERRELATIONSHIP BETWEEN FAT AND CARBOHYDRATE METABOLISM. When there is an excess intake of carbohydrate into the body it is converted into fat and stored within the various fat depots of the body. When it is desired to fatten farm animals their carbohydrate intake is stepped up. The reactions involved probably include the conversion of carbohydrate into pyruvic acid which then by a process of oxidative decarboxylation is converted into acetyl CoA, which can be used in the synthesis of fats. The reverse reaction does not seem to occur for acetyl CoA is not converted back into pyruvic acid, for the balance of the reaction is too heavily in favour of the formation of acetyl CoA. A mechanism for the conversion of fat into carbohydrate by an alternative route has been suggested, called the glyoxylic acid cycle. In this cycle acetyl CoA combines with oxaloacetic acid to form iso-citric acid which breaks down into succinic acid and another acid called glyoxylic acid. The latter combines with more acetyl CoA and eventually gives rise to more oxaloacetic which is made to react with acetyl CoA to start the cycle again. The succinic acid lies on Krebs' cycle, which by operating in reverse yields pyruvic acid. The pyruvic acid can then be converted into glucose by reversing the Embden Meyerhof path. This glyoxylic acid cycle has not yet been proved to exist in mammals.

One of the problems of fat metabolism is to discover the source of ketone compounds which are produced when animals are metabolizing a lot of fat. In humans with diabetes mellitus glucose cannot be readily oxidized because of the lack of insulin. The hormone insulin controls the uptake of glucose by muscle and fat cells (p. 266). Acetyl CoA must accumulate because of the emphasis on fat as a source of energy instead of carbohydrate. Krebs' cycle will not work efficiently in untreated diabetics since the various compounds in this cyclic chain of reactions are derived from carbohydrate breakdown; thus acetyl CoA is not rapidly incorporated into the cycle. It is thought that the acetone bodies e.g. aceto-acetic acid are derived from the accumulating amounts of acetyl CoA.

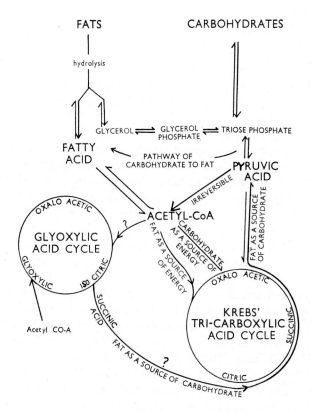

Fig. 194. *The relationships between fat and carbohydrate metabolism.*

Chapter Ten

Energy Exchange and its Regulation

Sources, forms and transformation of energy

The term energy exchange includes all the various ways in which the animal body gains or loses energy in its relations with the external environment. In animals we can consider four types of energy— chemical, mechanical, thermal and electrical. These various kinds of energy arise in the body from one source, which is chemical bond energy in the various foodstuffs, where it occurs mainly in the form of bonds uniting carbon and hydrogen. In order to be available as an energy source these C—H bonds must be present in a molecule having a similar structure to those of the body i.e. in the form of lipids, carbohydrates or proteins. Mineral oils, although they contain the appropriate chemical bonds, cannot serve as a source of energy for the body. The energy which is liberated during the oxidation of these carbon containing compounds becomes available for performing chemical work (e.g. synthesis of molecules), mechanical work (e.g. pumping of blood, lifting of objects, locomotion) or electrical work. The mechanisms of energy transfer in the body e.g. the conversion of chemical bond energy into mechanical energy, is relatively inefficient. Usually less than 30% of the energy of foodstuffs is transferred for the performance of work. The remainder appears as heat energy. Indeed in the resting, fasting adult all the energy of metabolic processes (i.e. the sum total of chemical reactions necessary for life), appear as heat. Thus although the heart may perform work in discharging blood into the aorta the kinetic energy of the moving column of blood ultimately appears as heat energy as it overcomes the frictional resistance of the smaller blood vessels. Similarly the energy involved in the chemical work performed when hydrogen ions are secreted into the gastric juice ultimately appears as the heat of neutralization as acid contents enter the duodenum and meet the alkaline secretion of the pancreas.

The maintenance of living things involves the expenditure of energy in the performance of various kinds of work. We have already seen the relative inefficiency of these transformations of energy. Mechanical work is performed when an object is lifted or given acceleration. This kind of work is measured in terms of the product of force × distance moved along the line of the force, and is expressed in various units e.g. 1 joule = 1 Kg. moved 1 metre by one newton. 1 erg = 1 g moved a cm by 1 dyne. Since work measures the transfer of energy from one form to another, it should not surprise us that the units used for work also serve for the measurement of energy. Chemical work is carried out during the synthesis of a great variety of compounds. Secretory work is involved in the transfer of substances, often against a concentration gradient. Such is the case when cells 'pump' sodium ions from the cell interior into the interstitial fluid bathing the cells, and in so doing generate concentration and electrical potential gradients.

Basal metabolic rate

The energy exchanges of the organism can be described by the following simple relationship:

Food intake = work + heat energy + energy storage.

The significance of energy storage depends upon the sex, age and metabolic state. In the growing animal, energy is being stored in the form of the chemical materials of cells. Similarly in the adult animal eating more food than is necessary to supply current energy needs, energy is stored in the form of chemical bond energy of fats in adipose tissue.

If we consider an adult man at complete rest and in a fasting state then we can in the above relationship ignore food and work leaving heat as the only route of energy dissipation, and energy storage (fat, glycogen and protein) as the only source of energy. If we measure the heat production of an individual under these conditions it will give us an estimate of the energy exchange involved in maintaining the life processes—the energy requirements for breathing movements, circulation of the blood, kidney function etc. This quantity is called the basal metabolic rate. This is not the lowest possible rate of energy exchange for in sleep the energy output falls even lower. This difference is no doubt partly due to a greater relaxation of muscles during sleep. Thus the term basal metabolic rate refers to measurements of energy output when the individual is awake, but resting, fasting and in an equable temperature.

Measurement of Basal Metabolic Rate

1. DIRECT CALORIMETRY

The heat output of an individual can be measured directly by recording the rise in temperature of a known weight of water used to absorb the heat. In order to make this measurement, complex apparatus is required. The subject is placed in a thermally insulated box (fig. 195) through which water is circulated. The temperature of the water is checked entering and leaving the chamber. In addition air must be circulated through the chamber and its temperature change recorded. Further the water content of the air must be determined because heat is utilized in the evaporation of water from the skin and respiratory passages. In short all avenues of heat loss from the individual must be accounted for. Using this technique Harris and Benedict estimated that a fasting resting male subject had a heat output of 1600 K cal/day.

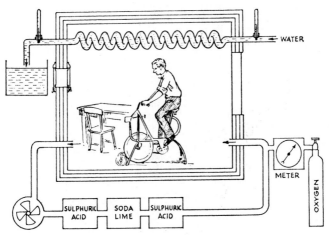

Fig. 195. *The Atwater human calorimeter.*

(*By courtesy of Dr. J. V. G. A. Durnin and Dr. R. Passmore, from Energy, Work and Leisure,' Heinemann Educational Books, Ltd.*)

2. INDIRECT CALORIMETRY

The technical problems of direct calorimetry make it unsuitable for routine use. The energy of basal metabolism can be estimated indirectly by studying the rate of production of the end products of metabolism—CO_2, H_2O and N-containing compounds (mainly urea) arising from protein.

Estimation of protein metabolism

Since almost all of the nitrogen derived from the oxidation of protein is excreted in the urine, the nitrogen output can be readily obtained from analysis of urine. Study of proteins show that they contain on average 16% of nitrogen. Thus the urinary nitrogen content × 6·25 = weight of protein used in metabolism during the period of urine collection. The calorific value of protein can be determined by burning protein in oxygen and measuring the heat produced. This is performed in a special instrument called the bomb calorimeter (fig. 196). These studies give a value of 4·3 K cal/G protein. We now

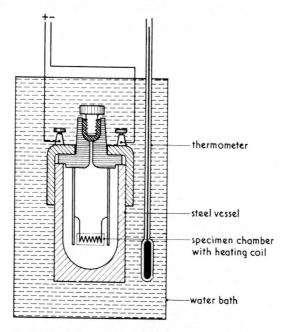

Fig. 196. *The principles of the bomb calorimeter.* The direct measurement of the calorific value of food is simple in principle but complex in practice. The figure shows only the principles of the technique. A weighed amount of food is placed in the specimen chamber. The platinum lined steel vessel is filled with oxygen at 20–25 atmosphere pressure. An electric current is passed through the platinum heating coil in the specimen chamber causing complete oxidation of the food, producing heat, H_2O, CO_2 and oxides of N, S + P from protein. The heat of combustion is calculated from the change in temperature of the weight of water in the bath, after appropriate correction for heat storage in the apparatus. Some typical results in K.cal/G food: glucose 3·74, glycogen 4·19, fats—palmetic acid 9·28, stearic acid 9·55, glycerol 4·31, animal proteins about 5·6 K.cal.

Note that the value for proteins is higher than that for oxidation in the body (4·3 K.cal/G) because they are not completely oxidised in the body.

have an estimate of the amount of protein oxidized during the period of measurement of the BMR and an estimate of the calories yielded by this oxidation.

Estimation of oxidation of carbohydrate and fat

The amounts of carbohydrate and fat oxidized during the period of measurement of the BMR can be estimated from measurements of oxygen consumption and carbon dioxide production after these measurements have been corrected for oxygen consumption and carbon dioxide production by the oxidation of protein. We have already discussed how it is possible to estimate the amount of protein oxidized.

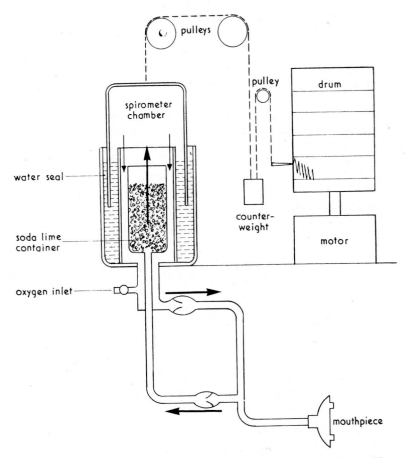

Fig. 197. *Recording spirometer for direct measurement of oxygen consumption.* The apparatus consists of an oxygen filled spirometer chamber, water sealed and connected to the subject by a mouth piece and breathing tubes. Inspired air is drawn directly from the chamber—but expired air is diverted through a soda-lime container —for CO_2 absorption—on its return to the spirometer. As oxygen is absorbed in the lungs of the subject the volume of gas in the spirometer chamber decreases and the spirometer chamber sinks into the water seal. The descent of the chamber is recorded on the revolving drum which is calibrated to show the volume of oxygen consumed over a given time.

The gas exchange during the oxidation of this amount of protein is calculated from the combining weights, using the empirical formula of a typical protein in the following equation:

$$\text{Protein} + O_2 \rightarrow CO_2 + H_2O + UREA$$

Carbon dioxide production can be estimated in various ways. Thus expired gases can be drawn through tubes containing soda lime which absorbs CO_2. The amount of CO_2 absorbed can be determined from the gain in weight of the soda lime after correction for water exchange in the reaction. Alternatively the gas expired during the period of measurement of the BMR can be collected in a bag (Douglas bag). A sample of air in the bag is removed for estimation of CO_2 and O_2 content. The volume of air in the bag can be determined by connecting the bag to a gas meter and emptying the bag through the meter. Knowing the volume of air collected and its composition one can readily calculate the CO_2 production and O_2 utilization. The oxygen consumption can also be measured directly by rebreathing from a spirometer (fig. 197) filled with oxygen. CO_2 is removed from the expired air and the descent of the spirometer in the water chamber is recorded directly on the rotating smoked drum.

Respiratory quotient

The respiratory quotient describes the ratio of the volumes of CO_2 production and O_2 utilization.

$$\frac{\text{vol. of } CO_2}{\text{vol. of } O_2} = RQ$$

Each class of foodstuffs—protein, carbohydrates, fat, yield a different RQ value when the food is burned in oxygen.

Carbohydrate

The oxidation of glucose can be represented as follows:

$$C_6H_{12}O_6 + 6O_2 \rightarrow 6CO_2 + 6H_2O$$
$$180\,G \qquad 192\,G \qquad 264\,G \qquad 108\,G$$

The volumes of oxygen consumed and carbon dioxide produced are approximately $22.4 \times 6 = 134.4$ litres. Thus 1 G of glucose reacts with $\frac{134.4}{180} = 0.75$ litres of oxygen and produces the same volume of carbon dioxide.

$$\frac{0.75}{0.75} = RQ\ 1$$

For glucose then the RQ is unity. The oxidation of 1 G of glucose in the bomb calorimeter liberates 3·74 K cal of heat. Knowing this we can calculate the energy release during the oxidation of glucose by measuring the oxygen consumption.

$$1 \text{ litre oxygen} = \frac{\text{calorific value I G glucose}}{\text{amount of oxygen utilized in oxidation of 1 G glucose}} = \frac{3 \cdot 74}{0 \cdot 75} = 5 \text{ K cal.}$$

Similar calculations can be made for fat and protein. For fat the RQ is approximately 0·7 and the calorific value of oxygen when fat is being oxidized is 4·7 cal. For protein the RQ is 0·8 with a calorific value of oxygen when protein is being oxidized of 4·5 K cal.

We have seen how it is possible to determine the amount of protein oxidized, its energy yield and the carbon dioxide production and oxygen utilization attributable to the oxidation of protein. These values for gas exchange attributable to oxidation of protein can now be subtracted from the total gas exchange during the period of measurement of the BMR. Fig. 198 shows a linear relationship between the non-protein RQ and the oxygen used in the oxidation of carbohydrate. Using this graph it is thus possible to assess from the RQ the amount of oxygen used in the oxidation of carbohydrate (or fat). The heat produced by the oxidation of carbohydrate or fat can now be determined

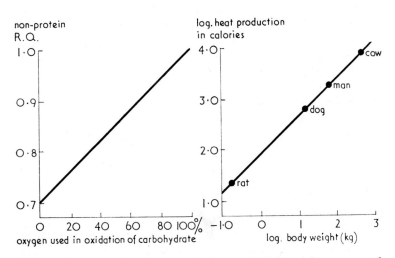

Fig. 198. *Left: the relationship between the non-protein R.Q. and the percentage of the non-protein oxygen consumption which is attributable to the oxidation of carbohydrate. Right: plot of log. of metabolism/log. of body weight.*

by multiplying the volume of oxygen by the calorific value of oxygen for that type of compound. The sum of the various quantities of heat (from oxidation of fat, carbohydrate and protein) gives an estimate of the heat output of basal metabolism.

Summary of the process of indirect calorimetry

1. Estimation of urinary nitrogen. Calculation of amount of protein oxidized. Calculation of heat output from the oxidation of this amount of protein from experiments with the bomb calorimeter. Calculation of oxygen utilization and carbon dioxide production from oxidation of this amount of protein in order to estimate the non-protein RQ.

2. Measurement of oxygen consumption and carbon dioxide production of whole organism. Subtraction of gas exchange attributable to the oxidation of protein (from 1) and calculation of the non-protein RQ.

3. Having determined the non-protein RQ it is possible to estimate the amounts of carbohydrate and fat in the metabolic mixture by referring to tables. The nearer the RQ is to unity the more carbohydrate is being oxidized. The nearer the RQ is to 0·7 the more fat is being oxidized.

Knowing the composition of the metabolic mixture (i.e. the amounts of fat, carbohydrate and protein oxidized) it is now possible to calculate the heat production in the oxidation of each class of compound.

It is important to stress that the RQ is determined solely by the rate of respiratory exchange of carbon dioxide and oxygen and that a variety of factors other than the composition of the metabolic mixture can influence this exchange. Thus changes in ventilation rate due to acid-base disturbances of blood can profoundly affect the RQ (see page 513).

Simplified indirect calorimetry

The method of indirect calorimetry as described above was of great value in showing that living systems obey the laws of thermodynamics. It has however been shown that measurements of energy output of comparable accuracy can be made in a much simpler fashion. Modern studies no longer involve an estimation of the composition of the metabolic mixture of substrates. It has been found that under the standard conditions of BMR measurement—i.e. resting and fasting— the RQ is approximately 0·82 with a calorific value of oxygen of 4·8 K cal/litre. These values are now assumed and all that is necessary to determine basal metabolism is to measure the volume of oxygen consumed.

Energy (K cal/min) = calorific value oxygen × litres of oxygen
consumed.

$$= 4.8 \times \frac{\text{vol expired air L/min.}}{100} \quad Oi - Oe$$

where   Oi = oxygen content of inspired air
Oe = oxygen content expired air.

Factors affecting BMR

1. BODY SIZE

Large individuals have greater energy exchanges than small individuals—they eat more and can perform more work. Even under resting conditions their heat output is greater than that of small individuals. However there is no direct correlation between body size and metabolic rate. By plotting log. weight/log. heat production a straight line relationship can be obtained (fig. 198).

In medical work it has become traditional to express BMR in terms of surface area. This is based on the conception that since all mammals have a similar body temperature and since losses of heat occur mainly at the body surface then mammals must produce heat in proportion to their surface area. The following table shows that if basal metabolism is expressed in terms of surface area (K cal/M^2) then very similar values are obtained from animals differing in size as much as a cow and a mouse. If basal metabolism is expressed in terms of body weight then marked differences are apparent between animals of different sizes.

THE RELATION BETWEEN SURFACE AREA AND BASAL METABOLISM
IN VARIOUS SPECIES

Species	K.cal/day	Basal metabolism	
		K.cal/kg	K.cal/m^2
Mouse	5·2	180	739
Dog	542	36	831
Man	1666	27	910
Cow	5678	15	1245

There are considerable difficulties in the measurement of surface area of the body. This was originally determined by covering the surface with pieces of paper and then measuring the surface area of the pieces with a planimeter. Dubois and Dubois derived an equation for the determination of surface area in sq. metres from readily obtainable measurement of weight (Kgs) and height (cms).

Surface area = weight 0·425 × height 0·725 × 0·007184

Expression of basal metabolism in terms of surface area has its limitations. The surface of heat loss is not identical with the surface area of the skin—because of the effect of posture, body coverings etc. Perhaps a more accurate relationship is that between lean body mass and basal metabolism.

2. AGE AND GROWTH

During periods of growth e.g. adolescence there is a higher basal metabolic rate. This is because growth—synthesis of body constituents —is an energy consuming process. When growth ceases basal metabolism levels out and begins to decline with increasing age (fig. 199). Pregnancy and lactation have a similar effect upon basal metabolism.

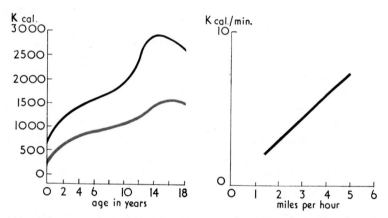

Fig. 199. *Left: the effect of age on total energy exchange (black) and basal metabolism (red) in girls. Right: the energy expenditure of walking on the level at different speeds.*

3. FOOD

We have seen that food provides the raw material for the synthesis of body structure and also the energy for the performance of this and other kinds of work. Food also has another effect on the organism. The intake of food by a resting fasting individual results in an increased rate of heat production. This means that the energy of foodstuffs is released less efficiently than from the body stores of carbohydrate, protein and fat. Of all foods taken protein has the greatest effect on heat output. This effect is called the specific dynamic action of food. This extra heat does not appear to arise in the alimentary tract during the processes of digestion and absorption of the food but probably appears in the liver during metabolism of the products of digestion, e.g. the deamination of amino acids in the process of urea formation.

4. TEMPERATURE OF THE ENVIRONMENT

Extremes of temperature can affect BMR in various ways. Thus during exposure to cold various mechanisms are activated which promote increases in tissue metabolism and thus heat production. These mechanisms are discussed in more detail on page 468. Here we can say that heat production increases due to shivering, increased activity and an increased output of hormones such as adrenaline and thyroxine which accelerate cellular metabolism in various tissues. For these reasons measurements of basal metabolic rate are made under conditions of equable environmental temperature.

Energy expenditure in activity

So far we have dealt with the energy expenditure in the resting fasting state—i.e. in basal metabolism. We can now consider the energy expenditure which is involved in various physical activities.

METHODS OF MEASUREMENT OF ENERGY EXPENDITURE

The methods of measurement of energy expenditure whilst various activities are carried out are basically the same as those used in the measurement of basal metabolic rate and involve direct or indirect calorimetry.

DIRECT CALORIMETRY

The principles of direct calorimetry have already been described. Atwater was the first physiologist to construct a human calorimeter. The principles of the technique are shown in fig. 195. The chamber consisted of a room in which a man could live for several days, sleeping, lying down awake, sitting or exercising on a bicycle ergometer (a bicycle with a fixed front wheel and a mounted mobile rear wheel to which various frictional loads could be applied). The walls of the room were insulated to prevent heat loss. A hatch in the wall provided a means of supplying food and water to the subject and removal of excreted material. Heat output from the subject was measured by a change in temperature of water circulating through pipes in the room. Air was pumped through a circuit so that air from the chamber passed through absorbants for water (sulphuric acid) and carbon dioxide (soda lime). Changes in weight of the absorbants provided the information needed for calculation of $H_2O + CO_2$ production. Oxygen was added to the air in the circuit to maintain a normal concentration of oxygen in the chamber. Oxygen utilization during the period of measurement could be determined from loss of weight of the oxygen cylinder. As already described this kind of experiment presented many great practical problems. These problems were indeed overcome by Atwater and his co-workers and they were able to show that over a period of several days the energy of food exactly equals the energy loss as heat. In spite of refinements of the technique of direct calorimetry the indirect measurements are far simpler and convenient for the measurement of energy exchange during various physical activities. Many measurements could not, in fact, be made within the confines of the calorimeter chamber—e.g. the energy cost of a sprint.

INDIRECT CALORIMETRY

The measurement of energy exchange by indirect calorimetry has already been described. Using the simplified technique already described (page 452) the only measurement needed is the rate of oxygen consumption. In order to be able to measure the rate of oxygen consumption in an exercising individual some form of mobile equipment is necessary and some examples of these will be described below. By breathing in and out of a mouth gag or mask which contains valves it is possible to separate the inspired and expired air. The composition of inspired air is known, or it can readily be determined. What is necessary is to measure the volume of expired air and determine its oxygen content (and CO_2 production if calculations of RQ are needed).

COLLECTION OF EXPIRED AIR

Douglas bag: The simplest method of measuring the volume and composition of expired air is to direct expired gases into a leakproof rubberized bag—the Douglas bag. The volume of gas in the bag is determined by forcing the contents of the bag through a gas meter. A sample is removed for analysis of O_2 and CO_2 composition. This method although suitable for laboratory studies or for static exercise, e.g. on a bicycle ergometer is cumbersome and unsuitable for use in larger surveys under active conditions.

The Max Planck respirometer

During the second world war the Max Planck respirometer was developed by German physiologists who used it to study energy expenditure in various industries to provide information for a fair system of food rationing. The respirometer consists of a gas meter, weighing about 6 lbs. (3 kg) and worn on the back, held there by shoulder straps like a haversack. The nose of the subject is clipped and he breaths through a mouthpiece connected to valves which permit him to breath directly from the air but which directs expired gases by way of a corrugated rubber tube to the gas meter on the back. The gas meter contains a system which diverts a small percentage of the expired air (less than 1%) into a rubber bag. This is later analysed for O_2 and CO_2 content. Figure 200 shows a man wearing the Max Planck respirometer whilst playing squash. The meter was constructed to accurately record rates and flow of expired air ranging from 15–50 L/min.

Fig. 200. *Man wearing a Max Planck respirometer when playing squash.*
(*By courtesy of Dr. J. V. G. A. Durnin and Dr. R. Passmore, from 'Energy, Work and Leisure,' Heinemann Educational Books Ltd.*)

Problems arise at rates below and above these. First the meter tends to under record the volume of expired air and at higher ventilation rates the resistance to flow of air in the valves and meter restricts the breathing of the subject. More elegant systems have been devised for measurement of expired air when ventilation rates are high.

MEASUREMENT OF THE COMPOSITION OF EXPIRED AIR

A variety of methods are available for the measurement of oxygen and carbon dioxide content of expired air. Which of them one uses depends on requirements of accuracy and speed of measurement and capital available. Classical chemical methods are still widely used and can be extremely accurate. In fact even in laboratories which use more elegant and expensive physical methods for measurement of CO_2 and O_2 the apparatus needs to be calibrated and checked by the older chemical methods. In Haldane's classical chemical method a sample of expired air is taken into a burette where its volume is recorded. The air is then exposed to potassium hydroxide solution to absorb carbon dioxide and then to alkaline pyrogallol to absorb oxygen. Changes in the volume of the sample of expired air during these procedures gives the appropriate information for calculation of the composition of the air.

ENERGY EXPENDITURE DURING ACTIVITY

Using the techniques we have already described it is possible to measure the energy expenditure during particular activities or even throughout the period of a day or indeed several days. In order to be able to estimate energy exchange over long periods of time it is necessary to make accurate measurements of the time spent doing particular activities as well as the oxygen utilization during these particular activities.

Daily energy expenditure can then be calculated as follows: daily energy expenditure = time spent in each activity (min.) + metabolic rate of activity (k cal/min).

UNITS OF MEASUREMENT

In examples given below the units of energy expenditure are expressed in k cal/min. This value is gross or total energy expenditure and thus includes the amount of energy of basal metabolism. The figures are corrected for body weight. For men the figures refer to k cal/min/65 kg. For women the values are k cal/min/55 kg. (1 k cal = 4·18 kJ.)

Example of energy exchange

At rest average men have a resting energy expenditure of 1·14 k cal/min (equivalent to an oxygen utilization 238 ml/min). Fat or frankly obese men of the same weight, however, have a lower energy exchange (1·04 k cal/min).

This is accounted for by a smaller mass of metabolically active tissue. The average woman has a somewhat lower resting metabolic rate than average men, amounting to an energy exchange of 0·91 k cal/min/55 kg. Even women as heavy as average men (65 kg) have a lower energy expenditure of 1·05 k cal/min. This is no doubt accounted for by the greater amounts of adipose tissue in the female. With increasing fat content the average woman shows even lower rates of energy exchange.

Given these resting values we can now look briefly at the effect of increasing physical activity on energy expenditure. On changing from the lying to sitting position increasing amounts of energy are needed by the postural muscles. This amount increases on standing.

	Average man (k cal/min/65 kg)	Average woman (k cal/min/55 kg)
Lying down	1·14	0·91
Sitting	1·38	1·1
Standing	1·72	1·3

On walking increasing amounts of energy are needed, the amounts depending on rate of walking, gradient, load carried (including body weight).

Figure 199 shows the relationship between speed of walking and energy expenditure. The relationship between speed and energy expenditure is linear except for speeds above and below the normal rates of walking because of the inefficiency of walking at these speeds. These values were obtained for walking on a smooth level surface but energy expenditure increases if the surface is uneven, if there is a gradient or if a load is carried. Steep gradients, e.g. those found on climbing stairs, may involve much energy expenditure. The combined exercise of going up and down stairs may have an energy expenditure of up to 12 k cal/min.

The following table shows the energy expenditure in various occupations and leisure activities:

	k cal/min/65 kg man
Watch repairer	1·6
Piloting an aircraft	1·7
Draughtsman	1·9
Wall papering	3·1
Garage—general repairs	4·1
Stoker of locomotive	4·8
Cabinet maker	5·6
Ploughing	6·0
Coal mining—drilling coal or rock ..	6·1
Road work—digging with pick and shovel	7·1

					k.cal/min/65 kg man
Coal mining—pushing tubs	..	..			8·0
Felling trees	..	..	..	..	8·0
Card games ..	..	..	..	..	Up to 2·5
Billiards				
Bowls				
Golf	..	..	..	..	2·5–5·0
Sailing				
Table tennis				
Dancing				
Badminton				
Tennis	..	..	..	..	5·0–7·5
Horse-riding				
Hockey				
Football				
Squash	..	..	..	..	7·5
Rowing				

There are obviously great variations in energy expenditure depending upon the kinds of occupation we follow and the nature of our leisure activities.   These variations are reflected in the calorific requirements provided by food intake. The mechanisms which determine this balance of food intake and energy expenditure will be dealt with in a separate section.

The regulation of energy exchange

When men and women become adults—that is when sexual maturity is achieved and growth stops—body weight remains relatively constant. Some individuals maintain a relatively constant body weight over a period of many years—during which time they may consume tons of food. This stabilization of body weight indicates that there is a regulation of energy exchange of the individual so that equal amounts of energy are taken in and given out. We have already indicated that the energy exchange of the organism can be described by the following relationship:

Food intake = work energy + heat energy + energy storage

Of these four components of energy exchange animals achieve long term energy balance by regulating food intake. Although over short periods of time our food intake may vary considerably from day to day—we may for example eat more at weekends when leisure and good food may be available—over the course of weeks there is a remarkable matching of food intake and energy output. The way in which this balance is achieved is far from completely understood. We will look here in some detail at the factors we suppose are involved in this regulation. And in a later section we will consider another variation in energy exchange—heat production and loss—which is modulated to achieve a relatively constant body temperature in mammals.

The regulation of food intake

NERVOUS CENTRES

Although the search for food and the act of feeding and digestion involves a large number of body structures and their nervous connections—eyes, nose, mouth, pharynx, oesophagus, stomach, arms, hands, legs etc.—there are certain parts of the brain which are specially associated with appetite and feeding. There are various complex neural systems occurring at several regions of the brain-stem which regulate feeding behaviour. Of these the most significant one in the regulation of caloric balance is the hypothalamus. This small area of the brain—weighing only about 4 grams of a total human brain weight of about 1200 G—forms the floor and part of the lateral walls of the third ventricle of the brain. It is concerned in the regulation of a large number of vital body functions—in water balance, temperature regulation, sexual differentiation and behaviour, control of anterior pituitary gland function, sleeping and waking behaviour etc. The importance of the hypothalamus for feeding behaviour is indicated by the term 'appestat' which has been applied to it.

Within the hypothalamus there is an organization of centres concerned with different aspects of feeding behaviour. The role of different areas of the hypothalamus has been studied by various techniques which include electrical stimulation by means of electrodes inserted through the skull, or by destruction of localized areas by passing heating currents by way of needles inserted into the brain. If the ventromedial area of the hypothalamus is destroyed by such a technique then marked changes in feeding behaviour occurs. Animals with these lesions rapidly devour food and within a space of minutes they will devour a day's food ration. If food is freely available they continue to eat and eventually become grossly obese as the excess energy intake is stored as fat. These animals do not, as it were, know when they have had enough food. The ventro-medial region of the hypothalamus is called thus a 'satiety centre'. After these animals have become grossly obese they do achieve a new balance in which energy input matches energy output. It is as though some information from the fat depots modifies the disturbance of appetite to return energy intake and output to normal levels.

Quite different changes in appetite result from destruction of more lateral regions of the hypothalamus. Here there is a refusal of food and animals will starve to death even when abundant food is available. This part of the hypothalamus is thus called the 'feeding centre'.

Factors influencing the hypothalamus feeding and satiety centres

We have seen the key role played by the hypothalamus in determining energy balance. In order to perform this function the hypothalamus must be supplied with information of the level of caloric balance— e.g. the rate of energy expenditure, body temperature, and energy stores (principally). We are by no means certain what kind of

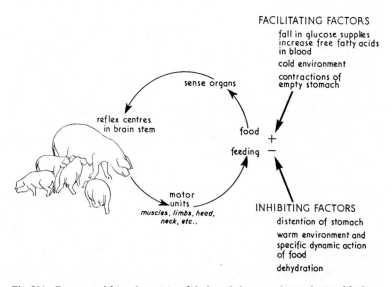

Fig. 201. *Factors modifying the activity of the hypothalamus in the regulation of feeding.*

signals carry this information. Some of the factors which are thought to modify the role of the hypothalamus in determining energy balance are shown in figure 201. Some of these are labelled facilitating factors indicating that they promote appetite, others as inhibitory meaning that they suppress appetite. We can now consider briefly some of these factors.

Factors influencing the nerve control centres for appetite

BLOOD GLUCOSE

Blood glucose concentration tends to fall during fasting and it has been suggested that in the hypothalamus there are gluco-receptors which are sensitive to the concentration of glucose in blood or to the degree that they can utilize blood glucose. The electrical activity of nerve cells in the feeding and satiety centres of the hypothalamus can be recorded by means of electrodes inserted into these areas of the hypothalamus. One worker found that a rise in blood glucose concentration increased the electrical activity of the satiety centre whilst a fall in blood glucose concentration reduced the activity of this centre. The activity of the feeding centre however changed in opposite directions.

Environmental and body temperature

It is well known that changes in body temperature or environmental temperature can influence appetite. In cold environments heat production by the body is high—due to increased oxidation of foodstuffs. This is associated with increased appetite to supply the necessary increase in energy intake. Hot environments on the other hand tend to reduce appetite. There are indeed temperature sensitive cells in the hypothalamus which are involved in mechanisms which regulate body temperature. The significance of these sensitive cells for appetite is not clear although some experiments have shown that direct cooling or heating of the hypothalamus can modify feeding behaviour.

Fat stores and blood content of fatty acids

During fasting individuals obtain energy supplies by breakdown of fat stores in adipose tissues. These energy supplies are transported from adipose tissue in the form of fatty acids (non-esterified fatty acids or NEFA in blood plasma). During fasting blood levels of NEFA are high but fall after feeding. It has been suggested that NEFA may be a chemical stimulus to the hypothalamus promoting appetite. Fat stores may yield information to the hypothalamus in other ways. Adipose tissue is fairly well innervated and it is conceivable that the degree of distention of fat cells might provide information to the hypothalamus on the size of energy stores.

Role of the gastro-intestinal tract

It was once thought that the periodic contractions of the empty stomach were an important stimulus to feeding. However, there is no known correlation between food intake and these gastric hunger contractions. Complete denervation of the stomach or removal of most of the stomach does not significantly reduce appetite or food intake. There is, however, clear evidence of the role of the stomach in the cessation of feeding, i.e. in satiation mechanisms. The role of stomach distention in satiation mechanisms is shown by sham feeding experiments. Sham feeding is carried out by bringing the oesophagus in the neck to the body surface so that food which is swallowed passes through the oesophagal opening in the neck and does not reach the stomach. In these animals feeding time is considerably prolonged but can be shortened by placing food directly into the stomach (e.g. via a gastric fistula) or by dilating the stomach with a balloon. However since extensive denervation of the stomach or surgical removal of most of the organ does not affect intake or the regulation of energy exchange the information supplied by the stomach seems to be dispensable for regulation of energy exchange.

Long and short term regulation of feeding

It is important to distinguish between short and long term regulators of feeding. Most of the factors already mentioned—blood glucose, blood NEFA, body and environmental temperatures act as short term regulators—that is they determine the size of the next meal. We have already seen that energy exchange is achieved over a period of days or weeks. Over shorter periods of time there may be marked inbalance of energy intake and expenditure. These short term factors operate over the scale of minutes. Some long term regulators must be operative to prevent the accumulation of errors of these fast acting measures. To do this it would seem to be necessary to monitor total energy store. We do not know how this is achieved. The situation in man is even more complicated by the acquisition of regular feeding habits which may lead to feeding without real hunger. For western societies at least it is important for men to learn how to prevent eating habits which lead to an energy intake which is greater than energy expenditure—which over the long term eventually leads to obesity.

The regulation of body temperature

We have already discussed (page 238) the significance of a constant body temperature in the evolution and life of mammals. A constant body temperature implies that heat production and heat loss must be matched and we shall examine how the mammal can modulate both heat production and heat loss to achieve homoiothermy. We must remember that the term 'constant body temperature' is a relative one.

Body temperature in mammals does vary both in space and time. If we apply thermometers or temperature sensitive thermocouples to different regions of the body, e.g. mouth, skin, rectum, eardrum, we can readily show that there may be marked differences in temperature of these different parts. The temperature of the outerlayers of the body particularly may vary since they are influenced by such factors as movement, exposure, sweating, blood flow etc. The aim of temperature regulation devices is to maintain a near constant *deep* body temperature. This is achieved by varying the temperature gradient between the internal and external environment, e.g. by increasing or decreasing the heat loss from the body surface.

We can now look at mechanisms of heat loss and heat production.

Heat loss

PHYSICAL MECHANISMS

Heat is lost from the body in three physical ways—by radiation and conduction of heat and by the latent heat of the vaporization of water. The different amounts of heat which are lost in these ways can vary greatly according to the circumstances. Heat is lost by radiation from the body to objects in the environment without physical contact with them. The amount of heat lost by the body by radiation to surrounding objects depends on the temperature gradient between them. Of course the body can also gain heat by radiation from the external environment if there is an appropriate temperature gradient. Conduction of heat energy to objects in contact with the body is a less important avenue of heat loss. The vaporization of water from the skin and respiratory membranes also removes heat from the body. Each gram of water takes up about 0·6 K cal from its surroundings when it passes from a liquid to a vapour. Thus heat is continually lost from the skin and respiratory membranes in amounts which depend upon the saturation of the surrounding air with water vapour. Under hot dry conditions, e.g. deserts, large amounts of water and thus heat may be lost from the body by this means.

THE REGULATION OF PHYSICAL EXCHANGE OF HEAT WITH THE ENVIRONMENT

The amount of heat lost by the body by these three mechanisms varies and is to some extent capable of regulation by the body. The amount of heat lost depends upon a variety of factors including:

1. The temperature gradient between the body and the environment.
2. The surface area available for heat exchange.

3. Changes in the amount of fluid available for evaporation i.e. secretion of sweat.

The surface area available for heat exchange varies according to the size of animal but it also can be modified by changes in posture and body covering—hair, clothes etc. Mammals can also modify the temperature gradient between the body and the environment. Since the skin is the main avenue of heat exchange it is by changing the rate of heat supply to the skin and rate of heat loss from the skin that the temperature gradient between body and environment can be modified. The heat supply to the skin is by way of the blood stream. Thus appropriate modification of the blood supply to the skin—by dilatation or contraction of arterioles—can either increase or decrease the supply of heat to the skin which is available for exchange with the environment. The heat supplied to the skin by the blood can also be conserved or dissipated by changes in the insulating properties of hair—by seasonal changes in the quality of fur and hair or by minute to minute changes in the position of the hairs which is determined by the activity of those tiny muscles, the arrectores pilorum attached to the hair follicles. We shall see later how the blood supply and activity of the arrectores pilorum are regulated by the nervous system. The situation of heat exchange in man is complicated by his control of the environment, by the wearing of clothes, living in buildings, heat, air-conditioning and the like. Heat loss by the evaporation of fluid from the skin and respiratory membranes can also be regulated. There is always some loss of fluid and thus heat from the body surfaces. In man this amounts to about 50 mls of water per hour and is called the insensible loss to distinguish it from sweating or sensible loss. Heat loss by these processes amounts to about 24% of metabolic heat under resting conditions. The insensible loss of water is inevitable and is not under the control of the body. However, loss of water and thus of heat can be increased in ways which vary from one species of mammal to another.

Mammals which do not sweat can increase heat lost in evaporation of water by increasing the flow of air over the moist membranes of the upper and lower respiratory tract, i.e. by panting. However this is not as an efficient way of increasing heat loss as sweating because of the increased heat production in the muscular activity of panting. Some non-sweating animals can further increase heat loss due to evaporation of water by licking copious saliva over the body. This is seen in animals such as cats and opossums and greatly increases the ability of these animals to withstand the stresses of exposure to heat. But it is sweating which is the most efficient way to increase heat loss due to evaporation of water. Large amounts of water can be lost in this way—up to 1·6 litres/hour working in a hot dry environment with an equivalent heat loss of 870 K cal/hour. Of course sweat is not pure water and considerable amounts of mineral salts, particlarly sodium, can be lost in this way. This means that in hot dry environments both water and salt intake should be adequate to cover skin losses if danger of heat cramps and collapse are to be avoided. Fortunately continued exposure to hot environments is followed by some physiological adaptation and the sodium content of both sweat and urine declines which assists in the balance of sodium intake and losses. The secretory activity of sweat glands is regulated by both nervous and humoral influences.

THE ROLE OF THE NERVOUS SYSTEMS AND HORMONES IN TEMPERATURE REGULATION

The importance of the brain in temperature regulation was appreciated as long ago as 1884 when two workers, Aronsohn and Sachs damaged an area of the brain in the midline and noted that this was followed by a marked rise in body temperature. We now know that the area of the brain they damaged was the hypothalamus. The control of body temperature implies that there must be temperature sensitive cells which provide a 'set point' for body temperature. These cells lie in the hypothalamus. When the anterior part of the hypothalamus was heated in cats by means of high frequency currents, typical heat loss responses appeared including panting. If the anterior part of the hypothalamus is damaged, e.g. by electrocautery then fever appears and if the animal survives there is a permanent disorder of regulation of body temperature in hot environments. This part of the hypothalamus which protects the animal from hyperthermia is called the 'heat loss centre'. The sensitivity of the hypothalamus to changes in deep body temperature is exquisite and even changes as small as 0·01°C produce measurable responses in temperature regulation. The posterior region of the hypothalamus protects the animal from cooling. If this area of the hypothalamus is destroyed there is a failure in adaptation to cold environments and large lesions result in failure of shivering or erection of hairs.

The hypothalamus is not only an exquisite thermostat sensitive to minute changes in deep body temperature but it also initiates activities, e.g. shivering, panting, sweating and erection of hairs. The hypothalamus also receives information of temperature of other parts of the body, particularly from the skin. In man at least information from the skin rises to consciousness and initiates other adaptive measures in temperature regulation such as changing of amount of clothing, seeking a cooler or warmer environment and so on.

It is important to stress that temperature regulating mechanisms do not function independently of other body functions. It may be necessary to accept some deviation of body temperature in some circumstances. Thus during severe muscular exercise the body may have to endure a period of hyperthermia. This is because the pattern of blood flow cannot be determined solely by the needs of temperature regulation, and the blood flow to the lungs, heart and muscles increases at the expense of the skin blood flow and hence at the expense of heat loss mechanisms.

Hormones and temperature regulation

Hormones are involved in thermoregulation mainly by modifying the mechanisms of heat production. They act by carrying a general or localized increase in the metabolic rate of tissues. They may also increase the rate of heat production by reducing the efficiency of metabolism so that less of the energy contained in the substrate appears in metabolically useful forms and more appears as heat. A variety of hormones are involved in thermoregulation and there is a complex interplay of one endocrine upon another which is incompletely understood.

NORADRENALINE AND THYROXINE IN THERMOREGULATION

When adult mammals are rapidly moved from a warm to a cold environment various mechanisms are activated which conserve body heat and increase heat production.

Heat is conserved by a reduction in skin blood flow and by erection of hairs which increases the thickness of the heat insulating barrier. Heat production increases at first by a general increase in muscular activity and by shivering—rapid involuntary muscle tremors. As other forms of heat production develop then shivering declines. Both noradrenaline and thyroxine secretion increases during exposure to cold. Noradrenaline increases the metabolism of both fat and muscle tissues. Fatty acids are liberated from adipose tissue which may act as a source of fuel for muscle tissue. Thyroxine has widespread effects on tissue metabolism increasing oxygen consumption and heat production.

Chapter Eleven

Breathing

An outline of the process of respiration. Living organisms need a constant supply of energy in order to maintain themselves and carry out their activities. Internal respiration, or tissue-respiration, is the name given to the processes which release energy from food substances; this process of tissue respiration is carried out in every living cell. Many complex cyclical reactions are involved and some idea of the nature of this chemical process has been given in Chapter VIII. The glucose, which is being used as a source of energy within the cell, is oxidized by removing hydrogen from it; this hydrogen is transferred along a bucket-brigade of hydrogen accepting enzymes until finally it is handed to a substance called cytochrome oxidase. Unless the cytochrome oxidase can hand on its hydrogen to oxygen, and so form water, the whole process of tissue respiration will cease; this is in fact what happens during poisoning by hydrocyanic acid (prussic acid), when cytochrome oxidase accepts a cyanide molecule in the place where it normally holds oxygen—the cyanide blocks the acceptance of oxygen and so tissue respiration stops and death occurs in minutes. Therefore, although in the process of the tissue respiration the *release* of energy is an anaerobic process (i.e. occurs in the absence of oxygen), it has to be supported by the intake of oxygen, if the hydrogen produced is to be removed. Carbon dioxide is also a product of tissue respiration and this gas must be constantly removed from the body.

Breathing. The physical process of obtaining oxygen from the air, and returning carbon dioxide to the air is called breathing, or external respiration, to distinguish it from the chemical reactions which occur inside the cells (called internal or tissue respiration). Historically the term respiration was first used to describe the process of breathing only, but as knowledge has accumulated it now embraces the following:

1. Breathing; the exchange of gases in the lungs, external respiration
2. The transport of gases in the blood
3. Complex energy releasing chemical reactions inside the cells—tissue respiration.

The present chapter is concerned only with the first section, the exchange of gases within the lungs. The transport of gases between the lungs and tissues is dealt with in Chapter V, and the process of tissue respiration is dealt with in Chapter VIII.

External Respiration in Mammals. In small animals such as protozoa, which are less than 0·25 mm. in diameter, the surface/volume ratio is sufficiently large for the surface of the organism to supply the oxygen needs. During evolution there has been an increase in size and complexity until at the mammalian level—and indeed long before—special cells have been set aside for the absorption of oxygen, and the blood transport system conveys the oxygen from these special structures to the tissues requiring oxygen.

Conservation of water is a great problem for all animals living in air and mammals have overcome this problem by covering the outside of the body with a horny keratin which is relatively impervious to water. Such a layer is obviously impermeable to oxygen. The requirements of an organ of respiratory exchange are first, that the membrane through which the gases diffuse must be very thin, and secondly the membrane must be kept moist, so that the oxygen dissolves in the moisture and diffuses in the aqueous phase into the capillaries of the respiratory organ. If the respiratory membrane must be moist it is obvious that it cannot be situated on the surface of the body like the external gills of a tadpole, or they would dry up and become inefficient, or lose too much body water. The solution of this problem was the infolding of the respiratory membrane inside the body. This infolded respiratory surface is so constructed that it has an enormous surface area, amounting to about 93 square metres in man; this is achieved by division of the respiratory membrane into very many closely packed pouches called alveoli—such is the fundamental structure of the lung.

The structure of the respiratory system in man.

THE UPPER RESPIRATORY TRACT (fig. 202). Air enters the body via the nostrils, or the mouth. After passing through the nostrils the air enters the nasal chambers which are separated by the nasal septum. The surface area of the nasal chambers is increased by outfoldings of mucous membrane, supported by delicate bones, the turbinals or scroll bones. The mucous membrane contains glandular elements whose secretions keep it moist, and tracts of ciliated epithelium keep these secretions moving posteriorly where they mingle with the secretions of the pharynx and buccal cavity and are swallowed. In the roof of the nasal chambers lie the sensory olfactory epithelial cells, connected by nerves to the olfactory areas of the brain, and responsible for the sense of smell. The functions of the nasal chambers include warming and moistening the inspired air; the larger particles of dust are deposited on the moist membrane and are carried backwards in the flow of secretions and are swallowed instead of being inhaled, thus avoiding the risk of damaging the delicate respiratory epithelium of the lungs.

Leaving the nasal chambers the air passes through the pharynx, passing en route the openings of the eustachian tubes, air filled canals which connect the middle ear chamber with the pharynx. The aperture of the tubes is opened during swallowing so that there is a free connection of air in the middle ear with that in the upper respiratory system, enabling the pressure in the middle ear to be kept at atmospheric levels.

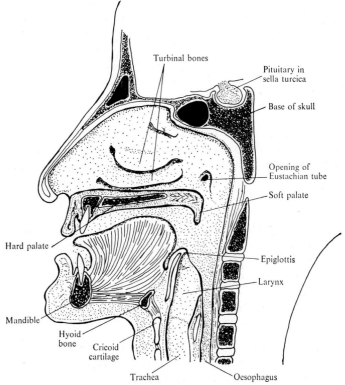

Fig. 202. *Diagram of a section through a human head showing the upper respiratory tract.*

Thus the effect of repeated acts of swallowing during rapid ascent or descent in aircraft is to prevent those unpleasant sensations within the ear which are the result of differences in pressure of the air in the external auditory meatus and the middle ear.

Air enters the trachea from the pharynx through a structure at the upper end of the trachea called the larynx, or voice-box. The opening into the upper end of the larynx is a slit like aperture in a fold of membrane, called the glottis. The size of the glottis varies and is determined by the action of muscles of the larynx. Below the level of the glottis in the larynx there is a fold of membrane on the two side walls of the larynx, the vocal cords. The pharynx is the common channel for both air and food, and at its lower end it opens into the larynx and trachea anteriorly and the oesophagus posteriorly. Food is prevented from entering the larynx during swallowing by several mechanisms. During the act of swallowing inspiration is inhibited, the glottis is closed, and the whole larynx is lifted up behind the base of the tongue. Further, there is a fold of membrane, supported by cartilage, called the epiglottis which is situated in front of the larynx, and when the whole larynx is lifted upwards, the epiglottis is directed backwards to partly shield the opening into the glottis. If by chance these barriers are penetrated and some food does enter the larynx then it meets an extremely sensitive eptihelium, which sets off violent reflex coughing in order to expel the foreign body from the larynx.

THE TRACHEA, or windpipe, is a tube which is supported by C-shaped rings of cartilage; the rings are deficient posteriorly where the trachea abuts onto the oesophagus, so that a bolus of food can pass down the oesophagus without friction against cartilaginous rings. In man the trachea is about 11 cms. long. It is lined by a mucous membrane which has a ciliated epithelium containing numerous goblet cells secreting mucus. Because of the action of the cilia there is a constant stream of mucus, containing dust particles from the inspired air, upwards to the larynx; by the action of intermittent coughs collections of mucus are passed into the pharynx and are swallowed with saliva. The importance of this cough mechanism is seen in patients with paralyzed respiratory muscles; because of their inability to cough, large collections of secretions develop in the trachea needing suction to remove them.

THE BRONCHI. The trachea bifurcates at its lower end into the left and right bronchus, which pass to the corresponding lung.

Like the trachea the lumen of each bronchus is kept open by means of C-shaped rings of cartilage. As the bronchus reaches the lung it begins to give off branches which supply various parts of the lung. These branches further subdivide until the ultimate branches are only about 1 mm. diameter. These small divisions of the bronchial tree are called bronchioles; their wall is lined by mucous membrane (containing no goblet cells). As soon as the bronchi enter the lung the ring shaped pieces of cartilage are replaced by irregular plates of cartilage, and these disappear when the smaller divisions of 1 mm. diameter are reached.

ALVEOLI. The terminal bronchioles are called respiratory bronchioles because they have a few small sacs, called alveoli, arising from their walls. These respiratory bronchioles branch into a varying number of ducts called alveolar ducts, from which arise large numbers of thin walled sacs, the alveoli (fig. 203). The lining of the alveolus is very thin

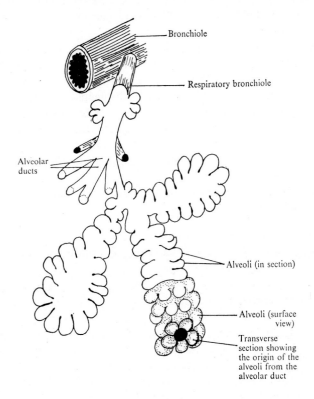

Fig. 203. *Diagram to show the terminal divisions of the lower respiratory tract.*

(about 0·1 μ) and its nature is not well understood; some workers claiming that there are holes in it where the walls of the capillary blood vessels surrounding the alveolus are exposed to the air in the alveolus. The important fact is that the air in the alveolus is separated only by a very thin layer from the blood in the capillaries of the pulmonary blood vessels. It is estimated that there are about 7,000,000 alveoli in man, with a surface area of about 93m². It is here at the alvaolar surface that oxygen can diffuse into the blood and carbon dioxide can leave the blood. The alveoli provide a very thin moist membrane, a large area for diffusion and suitable concentration gradients for the exchange of oxygen and carbon-dioxide.

Diffusion is the movement of molecules from an area of high concentration to an area of low concentration (p. 19). The blood returning from the body tissues, via the heart to the lungs, has lost much of its oxygen and has gained carbon dioxide; the partial pressure of carbon dioxide in venous blood is about 45 mm.Hg whereas it is only 40 mm.Hg in alveolar air.

Carbon dioxide passes from the high pressure region of carbon-dioxide in the blood vessels into the alveolar air. Blood reaching the alveoli from the tissues of the body contains oxygen at about 40 mm. pressure of mercury, whilst in the alveoli the partial pressure of oxygen is 101 mms. of mercury; thus oxygen diffuses from the alveoli into the blood vessels. Blood draining from the alveoli has oxygen at 100 mm. partial pressure, and carbon-dioxide at 40 mm. pressure; thus blood reaching the lungs has rapidly come into equilibrium with the alveolar air.

We must now consider the way in which fresh supplies of air are delivered to the alveoli.

Mechanism of breathing. The lungs are passive elastic structures and the movement of air through them is caused by movements of the surrounding structures. The lungs are surrounded closely by a thin layer of epithelium which is reflected back onto the chest wall at the root of the lung, forming another continuous layer of epithelium covering the inner layer of the chest wall (fig. 204). These two layers of epithelium, one covering the lung and one lining the chest wall, are called the pleura. The space between the outer (or parietal) layer and the inner (or visceral) layer is only a potential one and contains a thin film of fluid. There is a negative pressure amounting to −4 mm. of mercury in the intrapleural space; this negative pressure gradually develops from birth and appears to be due to the unequal growth of the chest wall and the lung, the chest wall growing at a greater rate than the lungs. It is this negative intrapleural pressure which is responsible for keeping the elastic lungs expanded against the chest wall and if for any reason this negative pressure is obliterated e.g. by a wound penetrating the chest wall and pleura, so allowing the intrapleural pressure to rise to atmospheric pressure, then the lung collapses. Air may be introduced into the pleural space through a needle, producing collapse of a lung and this is used in medicine in order to rest a lung; this is called a pneumothorax.

The air in the alveoli is in direct communication with the atmosphere external to the body by way of the upper respiratory passages, trachea, bronchi and respiratory bronchioles, and is at atmospheric pressure. In order to draw air into the lungs the pressure of air within the alveoli must be reduced below that of atmospheric pressure so that air will flow into the lungs along a pressure gradient; since the pressure of a gas is inversely proportional to its volume (Boyle's law) the volume of the thorax must be increased in order to draw air into the lungs. Air is exhaled from the lungs by the reverse process, a reduction in volume of the thorax leads to a consequent rise in pressure of the alveolar air.

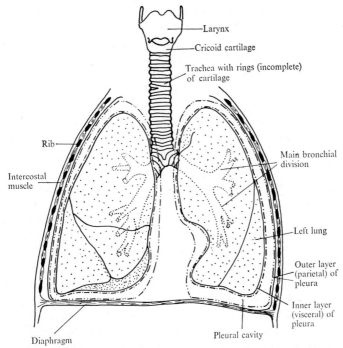

Fig. 204. *Diagram of the human chest showing the respiratory organs. The two layers of pleura are shown more widely separated than they occur in life.*

There are two mechanisms for altering the volume of the thorax, by movements of the ribs and by movements of the diaphragm; these mechanisms are mainly concerned in increasing the volume of the thorax. Reduction in the volume of the thorax at expiration is due mainly to passive elastic recoil of the chest wall.

The wall of the thorax is bounded laterally by the twelve pairs of ribs, ventrally by the sternum, and dorsally by the vertebral column. Each rib is jointed dorsally to the vertebral column and ventrally to the sternum. Since the ribs are curved they move outwards when they are raised, rather like the handle of a bucket does when it is picked up from a position resting on the side of the bucket. Thus when the ribs are raised by the effect of contraction of the intercostal muscles which pass from one rib to another, the diameter of the chest is increased, the volume of the thorax is increased and air passes into the lung so that its internal pressure comes into equilibrium with the atmosphere. When the muscles which move the ribs relax then the chest returns to the resting position because of the elastic recoil of the chest wall.

The diaphragm is a thin sheet of muscle which separates the contents of the thorax from those of the abdomen, but it is pierced by the blood vessels, nerves and lymphatics which pass from one cavity to the other. The margins of this sheet of muscle are attached to the lumbar vertebrae behind, to the ribs laterally and to the sternum anteriorly. The central area of the diaphragm is tendinous. In its resting position it is dome shaped, the left and right sides of the diaphragm being lifted above the central tendinous area by means of the pressure of the abdominal contents. When the diaphragm contracts it becomes flattened and descends into the abdomen, the abdominal wall relaxing at the same time to accommodate the descent of the abdominal organs which occurs. With the descent of the diaphragm the volume of the thorax is increased and inspiration occurs. When the muscle relaxes again the reverse process takes place.

The extent to which the movements of the chest wall and diaphragm play a part in inspiration varies from animal to animal, and from year to year in the same animal. In quadrupeds the thorax takes its part in weight bearing and thus is relatively immobile, leaving to the diaphragm the major role in inspiration. When the thorax is relieved of weight bearing in the bipedal and arboreal animals, movements of the chest wall play a more important part in inspiration of air. In man movement of the thoracic wall plays a varying part in inspiration; in the infant the ribs are nearly at right angles to the vertebral column, so that movement in any direction would tend to *reduce* the volume of the thorax, and diaphragmatic movements are more important: with the development of the downward slant of the ribs of the adult, movements of the chest become more important.

At the end of a normal passive expiration the lungs still contain large amounts of air, amounting in man to about 2·5 litres. During periods of greater activity the lungs can be further emptied of air by bringing into play certain accessory muscles; these include the abdominal muscles the contraction of which forces the abdominal viscera up into the relaxed diaphragm and further reduces the volume of the thorax. Other muscles situated around the thorax e.g. upper limb girdle muscles, can compress the thorax and reduce its volume during expiration.

Because the air passages are never emptied of air, the air in the lungs is not completely changed during an expiratory and inspiratory movement—it is merely 'freshened up'. In fact the air within the alveoli is of constant composition and its oxygen and carbon dioxide content does not fluctuate during the respiratory movements. Of course if the breath is held then the alveolar air gradually changes its composition, its oxygen content falling and its carbon dioxide content rising. Its constant composition is maintained by regulating the rate and depth of breathing movements as will be explained in the section on the control of breathing.

A man's lungs hold about 5000 ccs. of air and during an expiratory movement about 500 ccs. of air pass from the alveoli into the larger air tubes, which themselves hold about 150 ccs. of air. The volume of air inhaled and exhaled in a single breath (500 ccs.) is called the tidal air.

The composition of inspired and alveolar air is given in the following table:

*Composition of inspired and expired air in volumes per cent.**

	inspired	expired	alveolar
oxygen	20·71	14·6	13·2
carbon dioxide	0·04	3·8	5·0
water vapour	1·25	6·2	6·2
nitrogen	78·00	75·4	75·6

The control of breathing movements. The process of breathing is a complex one involving many series of muscles, which include the inter-costal muscles, muscles surrounding the thorax, the diaphragm and abdominal musculature, and the muscles of the larynx; thus during inspiration the vocal cords open, the process of swallowing is inhibited, the intercostal muscles and diaphragm contract and the abdominal muscles relax, so that the increased volume of the abdomen can be accommodated. The integration of these various muscles depends upon a complex series of reflexes maintained by the action of the nervous system. Further this series of reflexes must be adapted to meet the changing needs of the body for oxygen.

NERVOUS MECHANISMS IN THE CONTROL OF BREATHING. The basic centre for the control of breathing movements is situated in the medulla

*From Winton & Bayliss. Human Physiology. Churchill.

oblongata and this centre consists of two parts, an inspiratory centre and an expiratory centre; when one centre is active the other one is inhibited. The cells of the respiratory centre are connected by nerves, both sensory and motor, to the breathing apparatus. Further the cells themselves are sensitive to changes in the chemical composition of the blood, particularly to the level of carbon dioxide and the hydrogen ion concentration.

The Hering-Breuer reflex. There are sense endings in the walls of the respiratory bronchioles, air sacs and alveoli, which are stimulated by the change in tension in the walls of these air passages. At the height of inspiration these receptors fire off stimuli to the respiratory centre, via

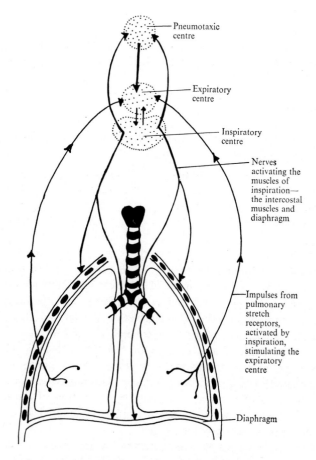

Fig. 205. *Mechanisms maintaining the breathing rhythm. The arrows indicate the direction of the flow of nervous impulses.*

the vagus nerve, which inhibit any further inspiratory movement. Further there are deflation receptors which are stimulated by deflation of the lungs; these fire off nerve impulses which stimulate the succeeding inspiration.

Under normal conditions this periodic inhibition of inspiration by afferent impulses in the vagal nerves, from receptors stimulated by inflation of the lungs is largely responsible for the rhythm of breathing. But there is another inhibitory mechanism of inspiration in the form of an inhibitory nerve centre in the medulla connected to the inspiratory and expiratory divisions of the respiratory centre. During the phase of inspiration the inspiratory division of the respiratory centre discharges impulses to the inhibitory or pneumotaxic centre as it is called. The pneumotaxic centre is then activated and sends impulses to the expiratory centre. When this barrage of impulses becomes sufficiently intense the expiratory centre is activated and the inspiratory centre is reflexly inhibited.

These two mechanisms for maintaining rhythmic respiration, the Hering-Breuer reflex and the pneumotaxic centre are illustrated in fig. 205.

CHEMICAL CONTROL OF BREATHING. In addition to the nervous mechanism which maintains the rhythmic quality of breathing movements, the ventilation of the lungs must be capable of modification to suit the metabolic needs of the organism. This is achieved by chemical mechanisms. There are two ways in which the activity of the cells of the respiratory centre are modified to suit the metabolic needs of the mammal. Firstly the neurones of the respiratory centre themselves are sensitive to changes in the chemistry of the blood, and secondly there are special receptors in the carotid and aortic bodies (see also p. 513) which are similarly sensitive to changes in the chemistry of the blood; these chemoreceptors are connected to the respiratory centre of the medulla oblongata by way of the 9th and 10th cranial nerves.

The neurones of the respiratory centre are sensitive to changes in the oxygen and hydrogen ion concentration of the blood. When there is a rise in the carbon dioxide or hydrogen ion concentration of the blood the neurones of the respiratory centre are stimulated, and initiate an increased ventilation of the lungs, which corrects the change in blood chemistry by removing excess carbon dioxide from the lungs. This is the most important cause for the increased ventilation of the lungs during exercise. If there is a fall of carbon dioxide content of the blood or a fall in the hydrogen ion concentration then ventilation of the lungs is depressed until sufficient carbon dioxide accumulates to rectify the change in blood chemistry. The sensitivity of the neurones of the respiratory centre is very great and very small changes in the carbon dioxide concentration of the blood will produce changes in the ventilation of the lungs.

If you take four successive rapid but very deep breaths forcing air out of the lungs at the end of each breath and then stop you will notice that there is a pause before normal spontaneous breathing movements return. You have over ventilated your lungs and reduced the amount of carbon dioxide in the alveoli and the blood and therefore the respiratory

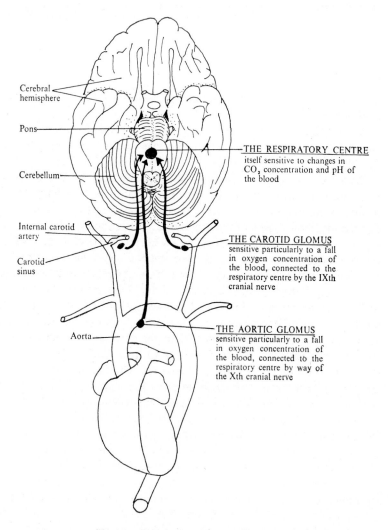

Cerebral hemisphere

Pons

Cerebellum

Internal carotid artery

Carotid sinus

Aorta

THE RESPIRATORY CENTRE
itself sensitive to changes in CO_2 concentration and pH of the blood

THE CAROTID GLOMUS
sensitive particularly to a fall in oxygen concentration of the blood, connected to the respiratory centre by the IXth cranial nerve

THE AORTIC GLOMUS
sensitive particularly to a fall in oxygen concentration of the blood, connected to the respiratory centre by way of the Xth cranial nerve

Fig. 206. *The chemical control of breathing.*

centre in the medulla is not getting adequate stimulation. The normal spontaneous breathing only returns when sufficient carbon dioxide has accumulated in the blood. (This experiment should only be tried by the healthy subject.)

Since the concentration of carbon dioxide in the alveolar air does not fluctuate throughout the normal expiratory-inspiratory cycle only changes in the metabolic activity will produce changes in the activity of the respiratory centre, by way of changes in the carbon dioxide and hydrogen ion concentration of the blood.

On page 232 we have described receptors situated at the bifurcation of the carotid arteries and on the arch of the aorta, which are sensitive to changes in the pressure of the blood; these pressure receptors are important in the reflex control of the circulatory system. There is a further set of receptors which are chemoreceptors and sensitive to a decrease in oxygen pressure, a rise in carbon dioxide pressure and an increase in the acidity of the blood. These chemoreceptors are located in small epithelial bodies, the aortic and carotid glomi (see fig. 206). The carotid glomi are situated on the external carotid artery, and the aortic glomus is within the concavity of the aortic arch. The pressure receptors on the other hand are situated within the wall of the blood vessels, in the carotid sinus (the swelling at the base of the internal carotid artery) and the wall of the aortic arch and innominate artery.

The main role of the chemoreceptors in the carotid and aortic glomi is to stimulate breathing when there is a fall in oxygen pressure in the blood. Under conditions of oxygen lack most of the cells of the body, particularly those of the nervous system e.g. the respiratory centre, are depressed; under these conditions the chemoreceptors, by means of their nervous connections with the brain, act as an important drive to respiration.

Mountain sickness and adaptation. When man climbs to a height of 3000 m. he enters a rarified atmosphere in which there is a much smaller amount of oxygen and he experiences a state which approximates to drunkenness. He can no longer calculate accurately, and he may experience a sense of elation or giddiness or sickness. It is this sort of sensation that made the early balloonists wish to go higher and higher once they got to high altitudes. If man ascends to 6000 m. he may do permanent damage to his nervous system through oxygen lack.

If the ascent is made slowly and the climbers have undergone a period of training at high altitudes it is possible to climb to altitudes higher than would be expected. On the Everest expedition of 1924 a height of 8500 m. was reached without any oxygen apparatus. The climbers are reported to have taken ten breaths for each pace forward. That training does lead to adaptation is shown by the fact that at 5000 m. one would expect, from the normal dissociation curve of oxyhaemoglobin, only 38 mm. pressure of oxygen in the alveoli; but experiments on trained climbers adapted to high altitudes showed an alveolar oxygen pressure of 52 mm. Adaptation to altitudes involves several factors including an increased number of red blood cells in the blood, increased ventilation of the lungs and an increased output of blood from the heart.

Flying in aircraft. Although the composition by volume of air does not change much from ground level to 21 000 m. the actual amounts, and therefore the pressure, of the gases gets less as one moves away from the earth's surface. Variations in temperature and pressure cause mass movements of air which result in the weather. Above 10 000 m. the temperature is constant at $-55°C$; pressure changes also get less as one gets higher. Modern jet engines are more economical on fuel at higher altitudes and since pressure and temperature changes are less at these altitudes one can fly as it were above the weather. High flying may also be necessary for tactical reasons in military aircraft.

There are physiological difficulties to be overcome in flying at these altitudes. There is a lack of oxygen and a danger of decompression sickness. Because the partial pressure of nitrogen falls as one ascends there is a danger of nitrogen coming out of solution into the body fluids and accumulating as gas bubbles (see Caisson sickness).

Royal Air Force regulations for the use of oxygen say that air crew flying at cabin altitudes above 3000 m. will always use oxygen. Oxygen should be used for passengers at 3700 m. Above 10 000 m. the efficiency of airmen using oxygen equipment falls and at 12 000 m. breathing 100% oxygen, man is in much the same state as regards oxygen supply as he is at 3000 m. breathing air. The maximum height of flight in unpressurized aircraft is therefore at 12 000 m. It is possible to pressurize the breathing equipment, or indeed the whole cabin, and this makes stratosphere flying above 12 000 m. possible.

It would be ideal physiologically to have the cabins of aircraft full of air at atmospheric pressure (about 101 325 Pa) but there are other considerations to be made e.g. weight of pressure equipment, and in military aircraft there is the danger of losing pressure rapidly from perforation of the cabin by missiles. Thus cabins are not so highly pressurized. V-Bombers are pressurized at 60 800 Pa and this gives conditions in the cabin similar to those at 2500 m. when the aircraft is actually flying at 14 000 m. Since one does not need oxygen at 2500 m. then there is no danger of oxygen lack, or of decompression sickness, even when flying at 14 000 m. In civilian airlines high pressurization is also used e.g. 50 662 Pa giving an effective 2500 m. equivalent at 10 500 m.

Caisson sickness. In engineering work under water, a diving bell or caisson is used. The water is pumped out of the bell by compressed air and then the divers are allowed to enter the bell and descend to the depth of the water where they are working. The pressure in the bell may be about 4 atmospheres and men can work safely at this pressure. The danger occurs when the men come to the surface again. At four atmospheres pressure the nitrogen in the air dissolves to a greater extent in the body fluids; but when the men return to conditions at atmospheric pressure the nitrogen comes out of solution in the body fluids. If the change from four atmospheres pressure to atmospheric pressure is sudden then the nitrogen comes out of solution in the form of gas bubbles which may obstruct the smaller blood vessels or cause damage to delicate tissues such as the brain. The condition produced is called 'the bends' and there may be severe abdominal pain, paralysis, sickness or even collapse and death. In order to avoid Caisson sickness the period of exposure to high pressures is restricted so that large amounts of nitrogen do not go into solution in the body fluids, and the change of pressure on returning to the surface is brought about gradually in special decompression chambers.

Diving mammals and respiration. Whales are very highly adapted to their existence in water, so much so that they are often mistaken for fish. They are of course mammals and not fish. Their streamlined shape, absence of hind limbs, conversion of fore-limbs into flippers, and the development of horizontal tail flukes, their loss of hair and development of blubber under the smooth skin are obvious adaptations. The adaptations which they have for diving are very exciting.

Although many whales such as the whalebone whales (which filter off the plankton for food) do not dive to great depths, there are others such as the sperm whales (which possess teeth and hunt squids in deep water) which need to dive very deep for periods up to half an hour. The problem is that the whale is a mammal and must breathe using lungs and cannot use gills like a fish; it has a breathing apparatus for use in air but it lives its life diving deep in the sea. The large 21 m. fin whale can take down about 3350 litres of oxygen which is enough to last it about 16 minutes if it moved at about 5 knots. But these whales move faster than this underwater and have been known to stay down for half an hour. It seems therefore that they must use oxygen at a very low rate whilst they are submerged. The muscles work anaerobically and are capable of building up a very large oxygen debt, which is paid off when the whale comes to the surface (for further discussion of oxygen debt see p. 624). The nostril is single and easily pokes out of the water because of its position on top of the head; the whale then takes very deep breaths which ventilate the lungs rapidly. In expiration the breath which is very heavily laden with moisture is blown out and much of the moisture condenses in the atmosphere. This process is called spouting and it used to be thought that the whale was blowing out a jet of water which it had drunk under water. The direction of the spouts and the number (for some whales have two nostrils) enable the whaler to recognize the whales.

We have seen that muscle tissue in whales can stand a very large oxygen debt. Whale meat has a very deep red colour because it contains large amounts of the respiratory pigment myoglobin in the muscle cells. Muscle haemoglobin or myoglobin has a greater affinity for oxygen than has blood haemoglobin and does not give up its oxygen until the partial pressure of oxygen in the surroundings is very low. This myoglobin is responsible for much of the oxygen which the whale takes down with it. In the fin whale of the 3350 litres of oxygen taken down 9% is in the lungs, 48% in the blood as oxyhaemoglobin, 42% in the myoglobin of the muscles and 7% dissolved in the tissue fluids.

Although muscle can live without oxygen for about half an hour because of its myoglobin and its ability to withstand oxygen debt, the brain cells cannot tolerate this deprivation of oxygen; the brain and spinal cord must have a good supply of oxygenated blood at all times. Diving mammals have networks of fine blood vessels around the brain and spinal cord called retia mirabilia; when the animal dives the circulation to the rest of the body, that is other than to the central nervous system, is reduced, and the retia mirabilia serve to take up the increased amount of blood which has been diverted from the rest of the body.

The rate of the heart beat of seals has been shown to fall from 180/minute to 35/minute during a dive; in spite of the resulting fall in the output of blood by the heart, the arterial blood pressure is maintained by means of vasoconstriction in the large capillary beds in the seals' skin. It is obviously difficult to record the heart beat of a 21 m. whale during a dive and there is no reliable information on this point.

Thus we have seen that whales have special mechanisms to provide the brain with oxygen-rich blood whilst the rest of the body relies for survival on oxygen stored as oxy-myoglobin and upon the ability to develop considerable oxygen debts. This oxygen debt is paid off during the efficient ventilation which occurs during spouting. There are many problems still unsolved. For example, how does a whale hold its breath for periods up to half an hour? We have seen that in most mammals breathing is regulated by the amount of carbon dioxide in the blood. The respiratory centre of whales seems to be indifferent to the amount of carbon dioxide in the blood. And how does a whale avoid decompression sickness? The blood of a whale will absorb nitrogen just as easily as human blood; if a man stayed underwater breathing air and surfaced as rapidly as a whale he would get the 'bends'. The whale only takes down the air in its lungs, and the larger deep diving whales have proportionately smaller lungs. When the whale dives the pressure of water acting upon the abdomen forces the abdominal contents forwards so that they compress the lungs; this is possible because the diaphragm is very obliquely inclined, being attached to the dorsal body wall at a rather posterior position. Thus the air in the alveoli is compressed and forced into the dead space of the air passages. Now here little gaseous exchange occurs so that little or no nitrogen finds its way into the blood. Thus the whale is free from the dangers of decompression sickness.

Chapter Twelve

Excretion

Excretion

Definition. Excretion is the process whereby the waste products of the body's metabolism and the substances which are in excess of requirements are removed from the body. The waste products of metabolism include carbon dioxide, produced during the oxidation of carbon containing compounds in tissue respiration, and nitrogenous waste products produced from protein metabolism. Urea is the commonest nitrogenous waste product in man. It is manufactured in the liver and then passes into the blood stream from which it is excreted by the kidneys. The lungs and the kidneys are the main excretory organs, the lungs excreting carbon dioxide, the kidneys excreting nitrogenous waste.

The definition of excretion includes the removal of substances which are present in the body in excess of requirements. The cells of the mammal are highly specialized and can only function properly in a restricted range of pH and osmotic pressure. The kidney, by excreting excess mineral salts and water is able to exercise control of the chemistry of the body fluids. Thus if we drink too many fluids the body fluids become in danger of dilution and the excess water must be excreted by the kidneys. Even an invaluable substance such as water is a danger when in excess, and this excess must be excreted.

The main excretory products may be grouped into four categories.

1. Carbon dioxide—derived from tissue respiration.
2. Nitrogenous waste products—from protein metabolism.
3. Water in excess of requirements.
4. Mineral salts in excess of requirements.

The excretion of carbon dioxide. This takes place mainly from the lungs. The carbon dioxide present in the capillary blood in the lungs diffuses into alveoli through the thin moist membrane of the alveoli and is removed from the body in the process of breathing (Chapter IX).

Some carbon dioxide leaves the body in the form of bicarbonate through the kidneys, and changes in the bicarbonate excretion by the kidneys plays an important role in the maintenance of the acid-base balance of the body.

Excretion by the kidney. The kidney is the organ which produces urine, containing all the important excretory products, except gaseous carbon dioxide. By alterations in the amount of water excreted and changes in the pattern of salt excretion the kidney plays a vital role in the maintenance of the constancy of the internal chemical environment of the body. The present chapter will be concerned with the kidney and its functions.

The kidney

The kidneys are paired organs which lie closely adpressed to the muscles of the back, and bound to the dorsal wall of the abdomen by a layer of peritoneum, so that they do not move about as the animal moves. There are often large collections of fatty tissue around the kidneys. They are oval structures, shaped like a broad bean in most mammals although in some mammals they may have a different shape. In aquatic mammals such as whales, and in ungulates, they have a lobulated surface, but in the typical mammalian kidney the surface is smooth.

The edge of the kidney facing the midline is concave; this concave border is called the hilum. From the hilum arises a thin muscular tube,

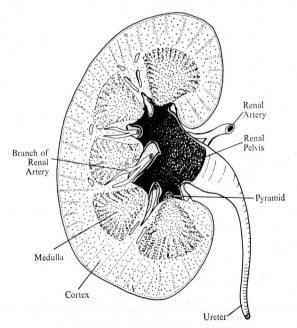

Fig. 207. *Vertical section of a human kidney.*

the ureter, which passes down along the dorsal wall of the abdomen and pelvis to enter the bladder near its base, where there are two separate ureteric openings. Peristaltic waves pass down the ureters, and urine drains continually from the kidneys and ureters into the bladder, a pear shaped muscular sac which opens to the exterior by way of a single duct, the urethra. The neck of the bladder is kept closed by a ring of muscle, the bladder sphincter, which relaxes to allow the escape of urine when the bladder is full. The relaxation of this bladder sphincter is a reflex action which is initiated when the pressure receptors in the wall of the bladder record that the bladder is full; the control of the sphincter is, of course, under voluntary control, although this has to be learned by the human baby. The wall of the bladder is made up of smooth elastic muscle so that the bladder expands as it fills with urine, and the pressure on the wall does not rise until the bladder is almost full, when one becomes conscious of a desire to micturate.

The section of the kidney (fig. 207) shows two well marked zones, an outer cortex and an inner medulla, which has well marked projections called the pyramids. Urine drains from the tips of the pyramids into a small cavity within the kidney, called the renal pelvis, which opens into the ureter.

The nephron. Apart from blood vessels, and small amounts of connective tissue, the mammalian kidney consists of masses of tubules called nephrons; there are about a million nephrons in each human kidney. Each nephron is a hollow tube which commences in the cortex as a blind ended thin walled sac called Bowman's capsule, into which is invaginated a knot of blood vessels called the glomerulus; Bowman's capsule together with its glomerulus is called a Malpighian corpuscle.

Proceeding from Bowman's capsule the next part of the nephron consists of a series of coiled loops situated in the renal cortex and called the proximal convoluted tubule. Following this the nephron loops down into the medulla as a thin walled tube called the loop of Henle which then returns to the cortex again to end in another series of coils called the distal convoluted tubule. The distal convoluted tubules open into a series of collecting ducts which drain away the urine from the nephrons, and discharge it into the renal pelvis at the pyramids (see fig. 208A).

BLOOD SUPPLY OF THE NEPHRON. The blood supply to the kidney is such that most of the blood passes into the glomeruli before reaching the rest of the kidney. The blood in the glomerular vessels is at relatively high pressure because of the rapid way in which the renal artery divides into arterioles. After passing through the glomerular vessels the blood is collected into an efferent arteriole which then supplies the tubules with blood: the medulla of the kidney is supplied by long loops of blood vessels (fig. 208B) called vasa rectae which arise from the efferent glomerular arterioles of those glomeruli which lie adjacent to the cortico-medullary junction. These juxta-medullary glomeruli are rather different from the glomeruli situated further out in the cortex

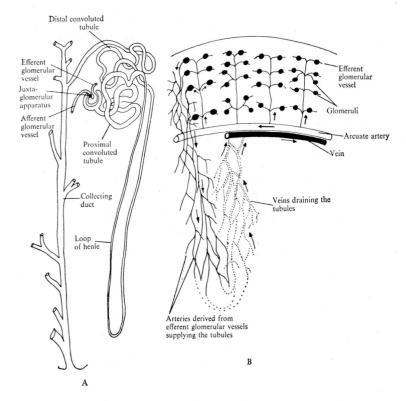

Fig. 208. A. *Diagram of a nephron and its relationship to a collecting duct.* B. *Representation of the blood supply to the cortex and medulla. The glomeruli are supplied directly from the main divisions of the renal artery. The efferent glomerular vessels pass inwards to the medulla to supply the tubules.*

in that their nephrons have very long loops of Henlé which reach the tips of the pyramids. This population of nephrons with their long loops of Henlé and their associated vasa rectae are especially concerned in the mechanism of concentration of urine. The blood entering the medulla by way of the vasa rectae is drained away into the arcuate veins by vessels called venae rectae.

Thus the characteristic features of the blood supply of the kidney are:

1. the immediate supply of arterial blood at relatively high pressure to Bowman's capsule,

2. the provision of two capillary beds in series, one in the Bowman's capsule, the second around the tubular parts of the nephron. (See fig. 208B).

The formation of urine

Bowman's capsule, as described above, is the swollen thin walled blind end of the nephron; into it protrudes the glomerulus, which thus converts the capsule into a hollow cup-shaped structure (see fig. 209). The inner layer of the cup is closely applied to the glomerulus.

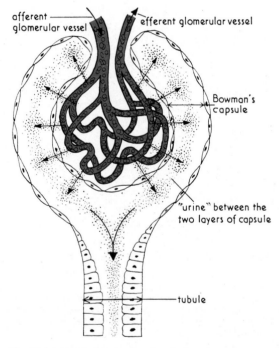

Fig. 209. *The Malpighian body, showing that fluid and solutes pass from the glomerular blood vessels into the Bowman's capsule from which it drains into the tubule.*

The layers which separate the plasma in the glomerular capillaries from the cavity of Bowman's capsule include the capillary endothelium, a basement membrane on which the endothelium rests, and the visceral layer of Bowman's capsule. Of these three layers only the basement membrane acts as a barrier to the free diffusion of substances. The cytoplasm of the endothelial cells is 'fenestrated' so that over certain areas of the endothelial cell the blood plasma is in direct contact with the basement membrane. The cells of the visceral layer of Bowman's

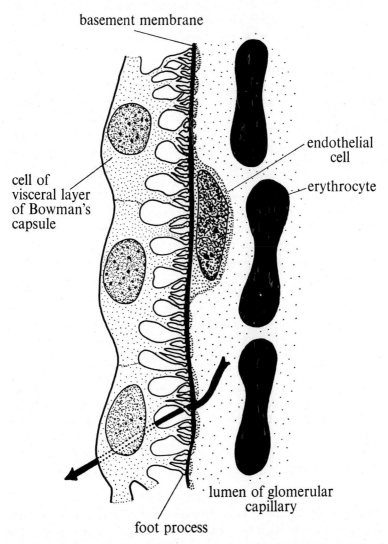

basement membrane

endothelial cell

cell of visceral layer of Bowman's capsule

erythrocyte

lumen of glomerular capillary

foot process

Fig. 210. *Diagram of the structure of the layers separating blood in the glomerular capillaries from the cavity of Bowman's capsule (made from electron micrographs). The arrow indicates the movement of water and solutes from the plasma in the glomerular capillary through fenestrations in the endothelial cells, through the basement membrane, between the foot processes and between the cells of the visceral layer into the cavity of Bowman's capsule,*

capsule bear many branched projections, the various branches having somewhat swollen ends which abut directly onto the basement membrane. Between these swollen terminals or foot processes, are gaps through which water and solutes can diffuse from the capillary lumen, through the basement membrane, between the cells of the visceral layer of the capsule, into the cavity of Bowman's capsule without having to traverse the cytoplasm of either the endothelial cells or the visceral layer of the capsule (fig. 210).

Just as tissue is formed by water and solutes passing from the arterial end of the capillaries into the tissue spaces, so urine is formed initially. Water and solutes pass from the high pressure glomerular blood vessels, through the pores in the endothelial cells, through the basement membrane and gain access to the cavity of Bowman's capsule through the slit pores between the foot processes of the visceral layer cells. Because the blood in the glomerular capillaries is at high pressure throughout the length of the capillary and so much higher than the osmotic pressure of the plasma proteins there is no tendency for fluids and solutes to return to the glomerular blood vessel. This is unlike the condition at the venous end of the capillaries in other tissues where the hydrostatic pressure is lower and where there is a return of fluid into the capillary. As in the formation of tissue fluid, cellular elements of the blood and the substances of higher molecular weight i.e. proteins, are retained within the capillary network and only water and substances of lower molecular weight viz. mineral salts, glucose, amino acids, urea etc., pass out of the capillary. This process is essentially one of ultrafiltration, the solutes of lower molecular weight being filtered, with water, through the basement membrane under the force of the blood pressure. If there is a fall in blood pressure below a critical level e.g. as in haemorrhage, then the filtration process stops and no urine is formed. It is possible to obtain samples of this glomerular filtrate by piercing the parietal layer of Bowman's capsule with a micro-pipette. Analysis of this fluid confirms that it is an ultrafiltrate of plasma containing water, salts, glucose, amino acids and excretory products, but none of the cellular elements of blood or the proteins of higher molecular weight.

The role of the tubules in urine formation and in homeostasis. A large volume of fluid containing solutes is filtered into the nephrons. The two million or so nephrons of both kidneys in man receive about 120 ml./min. of fluid filtered from all the glomeruli, i.e. over 7 litres an hour. If this fluid were excreted unchanged then the body would lose its entire water and salt content in less than half a day. In one day 600G of sodium, 35G potassium, 200G of glucose, 65G of amino acid and 180 litres of water are filtered into the nephrons. But in one day only 1 to 2 litres of urine containing approximately 6G sodium, 2G potassium, no glucose and only 2G of amino acids are excreted. The prime role of the tubules is thus the conservation of water and solutes, and there are great changes in the volume and constitution of the fluid filtered off in the Bowman's capsules as it trickles down the nephrons. By varying the amount of water and solutes which are excreted the tubules play a very important role in the maintenance of the chemical composition of the internal environment. Thus the kidney is not only concerned with the excretion of nitrogenous waste products, e.g. urea and uric acid, but is vitally concerned in regulating the composition of the rest of the body.

The prime role of the tubules is thus the reabsorption of large volumes of fluids and large quantities of solute. The tubules can, however, actively secrete some substances into the urine. This role of the tubules is seen, par excellance, in those bony fishes e.g. the sea horse, Hippocampus, which have no Bowman's capsules and glomeruli in their kidneys, and which depend entirely on the secretory activity of the tubule for the excretion of waste products. In mammals, only a few minor organic constituents of the urine are excreted in this way e.g. hippuric acid, creatine and urobilinogen. However certain important inorganic constitutents are actively secreted into the urine, including potassium and hydrogen ions.

Methods of investigation of renal function. In recent years it has become increasingly possible to investigate the way in which the kidney deals with particular substances and to localize certain activities in particular regions of the nephron. Indirect methods of studying kidney function include the measurement of the volume of urine and the concentration of its solutes over fixed periods of time. From this data one can express the rate in mg. per min. at which the kidney excretes a substance. Knowing the amount of substance which is delivered to the nephrons in unit time (calculated from a knowledge of the plasma concentration of the substance concerned and the rate of glomerular filtration) one can then assess the role of the nephrons in the 'renal handling' of this substance. Thus if the urine output was 1ml./min. and contained 120 mg./ml. of a substance x, the plasma concentration of x was 1 mg./ml. and the rate of plasma filtration from all the glomeruli was 120 ml./min. then 120 mg./min. would be delivered to the nephrons and 120 mg./min. would be excreted in the urine. This would imply that the tubules neither reabsorbed nor secreted the substance x. Some substances are excreted in this way e.g. the polysaccharides inulin and mannitol. For these substances the following relationship holds:

plasma concentration × rate of plasma filtration = amount excreted in urine
      (mg./ml.)   (ml./min.)   (in mg./min.)

Knowing the plasma concentration and the rate of excretion in the urine one can calculate the rate of plasma filtration using these particular substances. Data from this type of experiment indicates that in man about 120 ml. of plasma is filtered off collectively by all the glomeruli of both kidneys in each minute. Provided that a substance is free to be filtered off in the glomeruli, that is it is not bound to plasma proteins or cellular elements, knowing the plasma concentration of the substance and the rate of plasma filtration one can calculate the rate in mg./min. at which the substance is delivered to the nephrons. An analysis of urine will also provide information about the rate of excretion of the substance. If the amount excreted in the urine is less than the amount of substance delivered to the nephrons one can safely assume that the renal tubules are absorbing some of this substance. Conversely if more of the substance appears in the urine than was filtered off by the glomeruli then the tubules must have secreted some of the substance into the urine.

More direct methods of study are also available including the puncture of the renal tubules with micropipettes at various accessible sites. Analysis of the small amounts of fluids so obtained will reveal changes in the concentration of substances along the nephron. If one uses a double-barrelled micropipette one can introduce, under direct microscopic vision, a drop of coloured oil into a nephron, followed by a drop of a test solution, followed by a second droplet of coloured oil. This gives a column of fluid bounded in front and behind by coloured oil so that one can watch its progress along the nephron and note any changes in volume. Another direct method is to place a very fine polythene tube into one of the larger collecting ducts and allow the urine to drain through it. Then the tube is clamped so that urine cannot escape. The nephrons draining into that collecting duct become filled with a column of static fluid and each part of the tubule can more completely alter the fluid with which it is in contact. Then the clamp is released and small specimens of urine are collected in quick succession. The first sample obtained is fluid derived from the collecting ducts, the second sample is from the distal convoluted tubules, and so on. Analysis of these samples will reflect the activities of different parts of the tubules.

The reabsorption of solutes. Much of the solute filtered off in the glomerular filtrate is reabsorbed in the proximal convoluted tubules.

This reabsorption of solute is accompanied by the passive reabsorption of water so that of the 120 ml./min. of fluid delivered to the proximal tubules about 100 ml./min. is reabsorbed. There is no glucose or amino acid and a greatly reduced potassium content in the fluid which is passed on from the proximal convoluted tubules into the more distal parts of the system. The reabsorption of these solutes is against a concentration gradient since the solute concentration of the glomerular filtrate soon falls below that of the blood plasma as the solutes are reabsorbed. This reabsorption against a diffusion gradient is an active process involving the supply of energy by the cells of the proximal tubule. If the energy liberating processes of the tubule cells is hampered by cooling

luminal surface

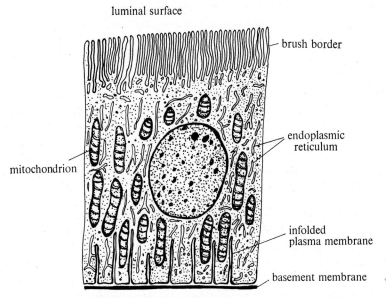

brush border

endoplasmic reticulum

mitochondrion

infolded plasma membrane

basement membrane

Fig. 211. *Diagram of the highly organized structure of a proximal tubule cell made from electron micrographs.*

of the kidney or by metabolic poisons such as cyanide then the proximal tubules lose the ability to reabsorb solute and the urine soon resembles scarcely modified glomerular filtrate. The high absorptive capacity of the cells of the proximal tubule is reflected in their structure (figs. 211, 212). The luminal surface area of the cell is vastly increased by the presence of many finger-like processes, the brush border, and the basal plasma membrane of the cell is infolded. The cytoplasm of the cell contains large numbers of mitochondria, the sites of energy release by oxidative metabolism, and there is a rich endoplasmic reticulum.

(*a*) *Glucose.* In normal conditions all the glucose which is present in the glomerular filtrate is reabsorbed in the proximal tubules. There is, however a maximal capacity of 230 mg. per minute for glucose reabsorption. If more than this quantity of glucose/min. is delivered to the proximal tubules then the amount in excess of 230 mg./min. is excreted in the urine. Amounts in excess of this maximal tubular capacity for glucose reabsorption are reached when the glucose concentration of the plasma is abnormally high, a state of affairs which occurs in diabetes mellitus.

(*b*) *Amino Acids.* In addition to glucose almost all of the filtered amino acids are reabsorbed in the proximal tubules and appreciable quantities of amino acids do not appear in the urine, unless there is damage to the proximal tubule cells or there is an abnormally high concentration of amino acid in the blood, when the filtered load of amino acids exceeds the reabsorptive capacities of the tubule cells.

(*c*) *Sodium.* Much of the filtered load of sodium is also reabsorbed by active processes in the proximal tubule. Negatively charged anions such as chloride move with the positively charged sodium ions and electrochemical neutrality is maintained. Water passively accompanies the movement of sodium and chloride ions to maintain osmotic equilibrium. Since both water and solute are absorbed by the proximal tubule cells the fluid remaining in the tubular lumen stays isotonic with plasma, although its volume is considerably reduced.

(*d*) *Bicarbonate.* Bicarbonate ions are also conserved by the proximal tubule, although in a rather indirect fashion. The tubule cells are capable of producing amounts of carbonic acid from the carbon dioxide generated in their metabolism because they are rich in the enzyme carbonic anhydrase. This carbonic acid dissociates into hydrogen ions and bicarbonate ions. The hydrogen ions are actively secreted *into* the tubular fluid and the bicarbonate ions pass out into the blood surrounding the tubule. When the hydrogen ions reach the tubular fluid they react with bicarbonate ions which have been filtered off from plasma in the glomerular filtrate to form water and carbon dioxide. Thus although the bicarbonate ions of the glomerular filtrate are destroyed by the secretion of hydrogen ions by the tubule cells, for each bicarbonate ion destroyed one new bicarbonate ion is added to the plasma by the tubule cell. If the enzyme carbonic anhydrase is inactivated by giving an animal an inhibitor such as acetazoleamide then the tubule cells can no longer secrete hydrogen ions into the urine at the normal rate, bicarbonate cannot be conserved and up to as much as half of the filtered load of bicarbonate appears in the urine, making the urine alkaline.

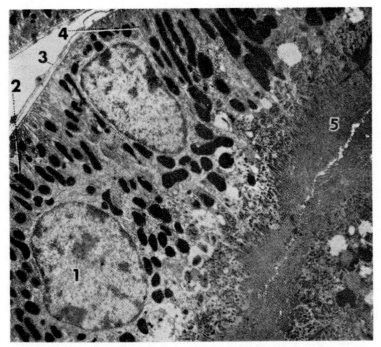

Fig. 212. *Electron micrograph of a section of a cell of the proximal convoluted tubule.* 1. nucleus.   2. mitochondrion.   3. basement membrane.   4. infolded plasma membrane.   5. brush border.

(*By courtesy of Dr. S. Bradbury.*)

The regulation of body water. The kidneys, by altering the amount of water excreted, play a vital role in controlling body water, in the face of large variations in water intake. When mammals are deprived of water the urine produced is scanty and concentrated and when excess water is taken in the urine becomes copious and dilute. Of the 120 ml./min. filtered into the nephrons in man, about 100 ml. is always reabsorbed in the proximal tubules. This is a passive reabsorption of water which moves along an osmotic gradient as solutes are actively reabsorbed. The remaining 20 ml. or so of water is passed on to the more distal parts of the nephron and it is this fraction of the filtered water which the kidney modifies according to the state of body hydration. If the body is dehydrated then most of this 20 ml. is reabsorbed and only about 0·5 ml./min. may appear as urine. If the body is over hydrated then almost all of this 20 ml./min. can be excreted in the urine. There

are two questions which need answering about this facultative reabsorption of water, as it is called. First, how does the kidney absorb
this water and second how is the kidneys' activity regulated in relation
to the state of body hydration?

The mechanism of concentration of urine. The ability of mammals to
conserve body water by producing a highly concentrated urine varies
greatly between species. Desert living mammals, such as the kangaroo
rat, can produce a urine which is ten times more concentrated than the
urine of an animal such as the beaver. If one looks at the kidneys of
these animals (fig. 213) one sees that the kidney of the beaver has a
shallow medulla and all the nephrons have short loops of Henlé whereas
the kidney of desert living mammals has a deep medulla and many
nephrons with long loops of Henlé which reach to the tips of the
elongated pyramids. Kidneys of amphibia have no loops of Henlé in
the nephrons and these animals are unable to produce a concentrated
urine.

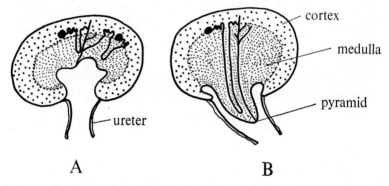

Fig. 213. *Diagram of vertical section of a beaver kidney* (A) *and the kidney of a desert
rodent* (B). *Note the shallow medulla and short loops of Henlé in the beaver kidney. The
kidney of the desert rodent has a deep medulla with elongated pyramids and the loops
of Henlé reach to the apex of the pyramid.*

The basis of the renal mechanism for the concentration of urine is
a 'sodium pump' located in the ascending limb of the loop of Henlé
which extrudes sodium ions from the tubular fluid into the tissue fluid
between the tubules, so producing a hypertonic tissue fluid in the
region of the medulla. The collecting ducts draining all nephrons have
to pass through this medullary region of hypertonic tissue fluid in order
to empty the urine into the pelvis of the kidney, and water diffuses out
of the urine in the collecting ducts along the osmotic gradient. In order

that the ascending limb of Henlé can perform this function of producing a hypertonic interstitial fluid it must be itself impermeable to water otherwise water would diffuse out with the sodium ions and maintain osmotic equilibrium. And in order that the urine concentrating ability of the kidney can be varied to meet the requirements of the body's state of hydration, the permeability of the distal convoluted tubule and collecting ducts to water must also be capable of regulation. If the body is dehydrated then the distal tubules and collecting ducts remain permeable to water so that urine diffuses out from the lumen of the tubule into the hypertonic interstitial fluid of the medulla, water is thereby conserved and a scanty concentrated urine is produced. If the body is overhydrated then these distal parts of the tubules become impermeable to water so that water cannot move along the osmotic gradient and a copious dilute urine is produced (fig. 214). The arrangement of

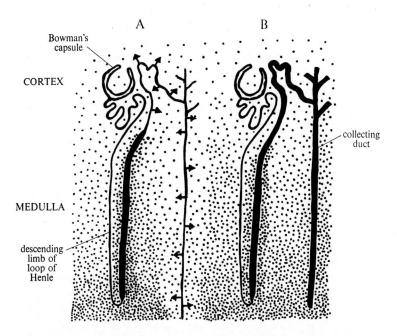

Fig. 214. *Diagram of two nephrons, one* (A) *antidiuresis, the other* (B) *in diuresis. The concentration of dots represents the tonicity of the interstitial fluid, the more dots the greater the tonicity. The tonicity of the interstitial fluid increased from the cortex to the tip of the medulla. The thickened regions of the nephrons represent parts impermeable to water. In A, in the presence of antidiuretic hormone, the distal convoluted tubules and collecting ducts are permeable to water, and this diffuses out of the tubule along an osmotic gradient, into the hypertonic interstitial fluid, so diluting this fluid. This movement of water is indicated by the arrows. B represents a nephron in diuresis, in the absence of antidiuretic hormone. The distal convoluted tubule and collecting duct is now impermeable to water and water cannot leave the lumen of the tubule along the osmotic gradient. A copious dilute urine is thus formed.*

the loop of Henlé as two closely aligned tubes in one of which is located a sodium extrusion mechanism imparts a special character to the mechanism of urine concentration. As the glomerular filtrate trickles down the descending limb of the loop, sodium ions diffuse into the limb from the surrounding hypertonic interstitial fluid and adjacent ascending limb. Thus the isotonic fluid delivered to the loop by the proximal convoluted tubule becomes increasingly hypertonic as it gains sodium ions. The ascending limb of the loop thus receives a fluid rich in sodium ions and by the expenditure of a small amount of energy the cells are able to extrude sodium ions into the interstitial fluid, so maintaining the hypertonicity of the interstitial fluid in the region of the medulla. If the tubular fluid of the descending limb of Henlé did not equilibrate with the interstitial fluid then a fluid isotonic with plasma would be delivered to the ascending limb of the loop and much more energy would be required in order to extrude sodium ions from the isotonic tubular fluid into the hypertonic interstitial fluid, because of the net tendency of sodium ions to diffuse into the ascending limb along a diffusion gradient. The final concentration of sodium ions in the interstitial fluid would be much smaller for an equivalent energy expenditure of the cells of the ascending limb.

This system is called a countercurrent multiplication system. 'Countercurrent' describes the organization of flow in which the incoming fluid (in the descending limb) equilibrates with the outgoing fluid (in the ascending limb). Because of this countercurrent exchange the effect of work done by the sodium pump in producing a hypertonic interstitial fluid is multiplied. The longer the tubes in which countercurrent exchange occurs the greater will be the effect, because more time is given for diffusion to reach equilibrium; thus the long loops of the nephrons of desert rodents enable a much more hypertonic interstitial fluid to be developed in the medulla of these kidneys than in, for example, the kidney of the beaver which possesses a shallow medulla and nephrons all of which have short loops of Henlé. The concentration of sodium ions in the tubular fluid is greatest at the apex of the countercurrent exchange system, and a gradient of increasing tonicity is built up, increasing from the cortico-medullary junction to the tips of the pyramids.

The more hypertonic the interstitial fluid of the medulla, the greater will be the diffusion gradient of water from the distal convoluted tubules and collecting ducts to the interstitial fluid, and the more concentrated the ultimate urine will be.

The ascending limb of Henle delivers a hypotonic fluid to the distal convoluted tubules. This is because sodium ions have been extruded from the tubular fluid and water has remained in the tubular fluid. Thus if the walls of the distal convoluted tubules are permeable to water then water will diffuse out of the tubules into the isotonic interstitial fluid of the cortex. This renders the fluid in the distal convoluted tubules isotonic with blood plasma. This isotonic fluid which is now delivered to the collecting ducts can now lose increasing amounts of water to the hypertonic interstitial fluid of the medulla, provided that the tubules remain permeable to water. Thus in passing down the length of the nephron various changes in tonicity occur. The isotonic fluid delivered to the proximal tubules remains isotonic because both water and solute are reabsorbed. In the descending limb of Henlé the fluid becomes hypertonic because water diffuses out into the hypertonic interstitial fluid of the medulla and sodium ions move in along their diffusion gradient. The hypertonic fluid delivered to the ascending limb from the descending limb gradually becomes hypotonic as sodium ions, but not water, are extruded. In the distal convoluted tubules this hypotonic fluid becomes isotonic as it loses water, and in the collecting ducts the fluid becomes increasingly hypertonic to plasma as water is lost to the hypertonic medullary interstitial fluid.

The concentration of urine then depends upon the ability of the kidney to develop and maintain a hypertonic interstitial fluid in the medullary region. The flow of blood through this region however threatens the maintenance of this mechanism because of the net tendency of sodium ions to be lost by diffusion from the medullary interstitial fluid into the blood plasma. This threat is reduced to a minimum by two mechanisms. First, there is an absolutely low rate of blood flow through the medulla and the cells of the medullary region rely heavily on anaerobic metabolism for their supplies of energy. Second, the arrangement of the blood vessels themselves in longitudinally aligned vessels, the vasa rectae and venae rectae, form a countercurrent exchange system. This means that as blood which has collected sodium ions leaves the medulla it circulates past blood entering the kidney, and the concentration of sodium ions in the blood entering the medulla increases by exchange from blood leaving the kidney. This quantity of sodium derived from the outflowing blood effectively reduces the otherwise steep diffusion gradient for sodium ions between the hypertonic medullary interstitial fluid and the blood plasma. This countercurrent exchange of sodium ions means that for a given rate of blood flow the amount of sodium lost from the medulla is reduced.

THE REGULATION OF THE PERMEABILITY OF THE COLLECTING DUCTS BY
ANTI-DIURETIC HORMONE.

The mechanism for water reabsorption by the kidney is the development and maintenance of a hypertonic interstitial fluid in the medullary
region into which water diffuses from the collecting ducts along an
osmotic gradient. However, the kidney must be able to alter the amount
of water reabsorbed depending upon the state of body hydration. This
is achieved by altering the permeability of the distal convoluted tubules
and collecting ducts to water. The permeability of the distal parts of
the nephron to water is determined by the presence of a hormone,
called anti-diuretic hormone, liberated from the posterior nervous lobe
of the pituitary gland. The name of the hormone describes its action
on the kidney. The term diuresis describes the condition in which large
volumes of dilute urine are passed. In anti-diuresis, water is reabsorbed
from the urine and the urine passed is reduced in volume and is concentrated. The hormone increases the permeability of the distal parts of the
nephron to water, permitting water to diffuse out of the nephron along
the osmotic gradient, and the urine becomes concentrated. In the
absence of the hormone these parts of the nephron become impermeable to water and no concentration of the urine occurs. There is a
condition called diabetes insipidus in which the posterior pituitary
gland does not liberate the hormone, or insufficient amounts of it, and
so conservation of body water by concentration of the urine is markedly
deficient. These patients pass large quantities of dilute urine, and need
to drink large amounts of water to replace the losses of water in the
urine.

The water content of the body is reflected in the osmotic pressure of
the circulating blood. When the body is deprived of water, losses of
water from the skin, respiratory tract and in the faeces continue.
Further, production of urine still continues because of the need to
excrete nitrogenous waste products, although the urine passed is
maximally concentrated. As water is lost from the body the osmotic
pressure of the blood rises. Conversely when large volumes of water
are drunk there is a dilution of the plasma and extracellular fluids and
there is a fall in osmotic pressure of the blood. The osmotic pressure
of the blood is being constantly monitored by nerve cells in the floor
of the third ventricle of the brain, in the hypothalamus. These nerve
cells, called osmoreceptors have long axons which travel downwards to
the posterior nervous lobe of the pituitary gland where they terminate
around blood vessels. These neurones secrete the antidiuretic hormone,
which passes down the axons, in the form of protein bound granules,

towards the posterior lobe of the pituitary. A rise in osmotic pressure of the blood, indicating a deficit of body water, promotes a release of ADH from the neurones and this gains access to the circulating blood which transports it to the kidney where the hormone promotes an increased permeability of the distal part of the nephron to water. The urine becomes more concentrated and body water is conserved. Conversely if the osmotic pressure of the blood falls, indicating an excess of body water then the secretion of antidiuretic hormone ceases, the distal parts of the nephron become impermeable to water, and the excess water is excreted in the urine. The osmoreceptors of the hypothalamus are a very sensitive mechanism and small amounts of either hypotonic or hypertonic solutions injected into the hypothalamic area of the brain produce marked changes in the secretion of antidiuretic hormone and thus the renal handling of water, in spite of the fact that these minute local injections do not significantly alter total body water content or its osmotic pressure.

The osmotic pressure of body fluids obviously cannot be the only factor which determines the secretion of ADH from the posterior pituitary gland. If this were so then drinking a salt solution isosmotic with body fluids would produce progressive expansion of the volume of body fluids. This does not happen. The reason for this is that the secretion of ADH is also influenced by changes in the volume of fluid in the body and in particular the volume of blood in the vessels. This type of reflex response to changes in the volume of fluid within the blood vessels requires the presence of pressure-sensing elements (baroreceptors) in the vascular system. There are indeed baroreceptors in various parts of the vascular system—in the great veins, aorta carotid sinus and within the heart itself. Many of these baroreceptors have been found to influence the secretion of ADH. When there is a loss of body fluids e.g. because of haemorrhage, the fall in blood pressure stimulates the secretion of ADH by altering the discharge of impulses from the baroreceptors. The ADH which is released causes an increased retention of water by the kidney, thus helping to compensate for the loss of blood. The fall in blood pressure also stimulates the chemoreceptors of the carotid body; this stimulation is caused by anoxia of certain sensitive cells i.e. a deficiency of oxygen due to reduced blood flow. Sense cells within the carotid body increase their rate of 'firing' of nerve impulses to the central nervous system, which results in an increased rate of secretion of ADH.

The various factors which influence the secretion of ADH from the posterior pituitary gland are shown in fig. 215.

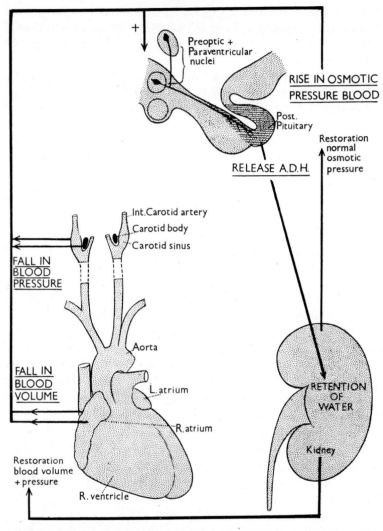

Fig. 215. *Diagram showing the factors which determine the release of ADH from the posterior pituitary gland. (From Clegg and Clegg, 'Hormones, Cells and Organisms', Heinemann Educational Books Ltd.)*

ALDOSTERONE AND SODIUM AND POTASSIUM REGULATION

In addition to controlling body water the kidney regulates the plasma concentration of the important electrolytes sodium and potassium. The osmotic pressure of the plasma is determined mainly by the sodium ion (with the associated anions e.g. Cl). If the body loses sodium ions, then water is also lost because of the overriding necessity to maintain a normal osmotic pressure of body fluids. Conversely if sodium is retained by the body then water is also retained. Of the filtered load of sodium in the kidney, the bulk (90 %) is reabsorbed in the proximal tubules. This active reabsorption of sodium is also responsible for absorption of water which occurs in this region, water diffusing passively with the sodium ions to maintain osmotic equilibrium.

Only 10 % of the filtered load of sodium is passed on to the more distal parts of the nephron. In the loop of Henlé sodium accumulates in the descending limb as the tubular fluid equilibrates with the hypertonic interstitial fluid. However in the ascending limb the accumulated sodium is extruded from the nephron by the 'sodium pump' mechanism in the tubule cells. Of the filtered load only a small fraction (up to 10 %) eventually reaches the distal tubule. In the distal convoluted tubule some of the sodium ions are reabsorbed in exchange for hydrogen ions and potassium ions which are being secreted into the tubule in this region (p. 515). All of the filtered load of potassium is reabsorbed in the proximal convoluted tubules and the potassium which appears in the urine is derived by active secretion of the ion into the fluid of the distal convoluted tubule in exchange for sodium ions. It is this small fraction of the filtered load of sodium, which is absorbed in exchange for secreted potassium ions, that the body regulates to achieve sodium balance. The exchange process which occurs in the distal tubule is stimulated by the presence in the blood of the hormone aldosterone secreted by the adrenal cortex. An excess of the hormone stimulates this exchange process so that there is a loss of potassium from the body and a retention of sodium. A deficiency in the supply of the hormone as in Addison's disease (p. 258) causes a net loss of sodium from the body and a retention of potassium. In Addison's disease there is thus a continued loss of sodium ions in the urine and with them water is also lost. This leads to a progressive shrinking of the volume of circulating plasma and extracellular fluid and eventually leads to a circulatory collapse. When a normal animal is deprived of sodium in the diet there is an increased outpouring of the hormone aldosterone from the adrenal cortex, which results in a conservation of sodium by the exchange mechanism of the distal convoluted tubule mechanisms regulating the secretion of aldosterone.

In order that the adrenal cortex can regulate the sodium balance of the body by way of the effect of the hormone aldosterone on the kidney, it must be provided with information about the state of the body's sodium economy. One possibility is that the adrenal cortex 'senses' the concentration of sodium in the blood. This possibility was examined by means of an ingenious technique which made the adrenal gland readily accessible for experimental study. An adrenal gland of a sheep was removed from the abdomen and transplanted to the neck of the sheep within a skin flap, the adrenal artery and vein being joined to blood vessels in the neck. With the adrenal gland in this position it was possible, in the conscious animal, to study the effect of changes in the sodium content of blood on the secretion of aldosterone. A needle could be inserted into the adrenal artery to perfuse the gland with blood of known sodium content and another needle in the adrenal vein could provide specimens of blood for analysis of aldosterone content. Using this technique it proved possible to show that the production of aldosterone by the adrenal gland does not increase when the sodium content of the blood entering the gland was lowered. Further in sheep which were already responding to a deficiency of body sodium by increased secretion of aldosterone the perfusion of the transplanted adrenal with sodium-enriched blood did not reduce the amount of aldosterone which was being secreted.

It seems then that the adrenal gland does not directly 'sense' the sodium content of the blood. Some other part of the body must perform this function and then relay the information to the adrenal gland. This information appears to be in the form of one or more hormones. This was shown by depriving a sheep of sodium until its own adrenal glands responded by increased secretion of aldosterone. A small amount of blood removed from this sheep and injected into the adrenal gland in a neck pouch of a sheep in normal sodium balance caused this animal to produce increased amounts of aldosterone.

We still do not have a complete story of the way in which the adrenal cortex, receives information about sodium balance. However, studies which have been made since the last edition of this book was published indicate that the kidney itself produces one of the hormones which regulate the secretion of aldosterone. The kidney, like the adrenal cortex, does not appear to directly 'sense' the concentration in the blood. Rather does it 'sense' the effect of sodium upon blood volume. We have seen that sodium and water tend to be handled together by the body. When sodium is lost from the body water is also lost and when sodium is gained water is also gained; these reciprocal fluxes of sodium and water being determined by the need to maintain a normal osmotic pressure of body fluids. Thus changes in body sodium content are reflected in water content—i.e. the volume of body fluids. It is the change in blood volume (and blood pressure) which the body senses. The sense organ in the kidney i.e. the baroreceptor—is in the afferent arteriole of the glomerulus, in the so-called juxta-glomerular apparatus.

As the afferent arteriole approaches the glomerulus the smooth muscle cells in the wall of the arteriole become replaced by large epithelial type cells containing secretory granules. This tissue forms a cuff in the afferent arteriole which butts against the wall of Bowman's capsule. There is a localized region of the distal convoluted tubule of the nephron containing enlarged cells which is associated with the polar cuff (fig. 216). This region of the distal convoluted tubule is called the

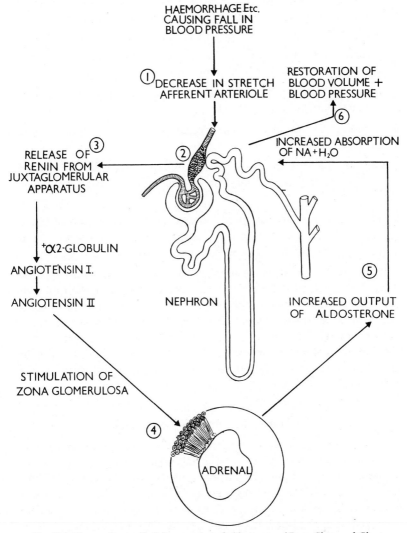

Fig. 216. *The renal control of the secretion of aldosterone.* (*From Clegg and Clegg,*
'*Hormones, Cells and Organisms, Heinemann Educational Books Ltd.*)

macula densa. Together with the polar cuff the macula densa form the juxta-glomerular apparatus.

We do not yet know the function of the macula densa but the polar cuff appears to produce a hormone in response to changes in blood pressure in the afferent arteriole of the glomerulus. Fig. 226 illustrates the way in which we imagine the juxta-glomerular apparatus regulates the secretion of aldosterone by the adrenal cortex.

1. The renal afferent arteriole forms the 'volume receptor', the signal to the receptor being the degree of stretch of the arteriole.

2. When there is a decrease in stretch of the afferent arteriole caused by a decrease in blood volume (= sodium deficiency) the juxta-glomerular apparatus releases the hormone renin into the blood stream.

3. Renin reacts with a member of the plasma proteins to form another active substance called angiotensin.

4. Angiotensin stimulates the adrenal cortex to produce more aldosterone.

5. Aldosterone is transported in the blood stream to the kidney where it stimulates the reabsorption of sodium in the distal convoluted tubules.

6. The retention of sodium (and water) causes an expansion of blood volume, a rise in blood pressure, an increased renal blood flow . . . and a decline in the production and release of renin.

Returning to our original example—the loss of blood—it can be seen the fall in blood pressure which results stimulates the reabsorption of both salt and water by the kidney. The secretion of ADH (producing retention of water) is brought about by the effect of a fall of blood pressure on baroreceptors in the neck and chest. The secretion of aldosterone (causing retention of sodium by the kidney) is brought about by the effect of a fall in blood pressure on baroreceptors in the kidney itself.

Acid-base balance, and the kidney

There is an ever present tendency for the body fluids to become increasingly acid in reaction. In one day, in man, metabolic processes liberate 13 000 m. moles of carbon dioxide, which if converted into carbonic acid would form an amount of acid equivalent to thirteen litres of N hydrochloric acid. Various mechanisms exist which regulate the reaction of the body fluids, and among these the kidneys have an important role to play.

Acidity, alkalinity and neutrality. A solution is defined as acid in reaction when there is a net excess of hydrogen ions (H^+) over hydroxyl ions (OH^-). A solution is alkaline when there is an excess of hydroxyl over hydrogen ions, and neutral when there are equal concentrations of the two ions.

EXPRESSION OF DEGREES OF ALKALINITY AND ACIDITY

At a particular temperature the product of the concentration of hydrogen ions [H^+], and the concentration of hydroxyl ions [OH^-], in a particular solution is constant. This means that one can express either acidity or alkalinity in terms of the concentration of one of these ions alone, and the concentration of hydrogen ions is used for this purpose. In any but the most acid solutions the concentration of free H^+ is so small that one would have to use inconvenient numbers to express them. So instead one uses the logarithm of the reciprocal of concentration (i.e. dilution). This expression, the log. of the reciprocal of hydrogen ions concentration, is called the pH value.

$$pH = \log \frac{1}{[H]^+}$$

In the pH scale, the neutral point is 7·0. Acid solutions have a pH value less than 7, the smaller the pH value the more acid the solution. Alkaline solutions have a pH value above 7, the higher the pH value the more alkaline the solution. The pH of body fluids is kept fairly constant around 7·4, although the range of survival covers 0·4 of a pH unit on either side of this point. The pH scale is a logarithmic one, a change of one pH unit implying a ten-fold change in concentration, so that this range of tolerance of tissues from pH 7·0–7·8 is a fairly wide one.

ACIDS AND ALKALIS

An acid is a substance which tends to give up a proton i.e. the hydrogen atom after losing its single electron. The more readily the acid gives off a proton the stronger it is. One can characterize this in the following way.

$$HB \rightleftharpoons H^+ + B^-$$
acid   proton base

A base is a substance which tends to accept a proton, the greater the avidity for protons the stronger the base. In the above formulation of the reversible dissociation of an acid it can be seen that a base is left after an acid has given off its proton. In this reversible reaction the stronger the tendency for the reaction to proceed to the right i.e. the stronger the acid, then the weaker is the base, and vice versa. The base left after dissociation of the acid is called the conjugate base of the acid in question and the two, acid and base, form a conjugate pair. Thus hydrochloric acid is a strong acid because of its marked tendency to dissociate into free hydrogen ions and the weak conjugate base (chloride ions).

$$HCl \rightleftharpoons H^+ + Cl^-$$
hydrochloric acid $\quad$ proton $\quad$ chloride ion.

A variety of acids are present in body fluids including,

1. *Carbonic acid.*

$$H_2CO_3 \rightleftharpoons H^+ + HCO_3^-$$
carbonic acid $\qquad$ bicarbonate ion (base).

2. *Primary phosphate ion.*

$$H_2PO_4^- \rightleftharpoons H^+ + HPO_4^{--}$$
secondary phosphate ion (base).

3. *Ammonium ion.*

$$NH_4^+ \rightleftharpoons H^+ + NH_3$$
ammonia (base).

The reaction of the body fluids (pH 7·4) is alkaline and this is determined by the presence of various bases. The chloride ion (a base) is the conjugate base of such a strong acid (hydrochloric acid) that is, it has very little tendency to accept protons. But in addition to chloride ions there are smaller amounts of three stronger bases which contribute to maintaining the alkaline reaction of the body fluids. These bases include bicarbonate ions, secondary phosphate ions and proteins (which exist in plasma as negatively charged anions capable of accepting protons). These three bases are the conjugate bases of very weak acids which are important because changes in pH around the neutral point can alter equilibrium point of dissociation of the weak acid into the proton and the stronger base.

Weak acid $\rightleftharpoons$ proton + strong base
$$HB \rightleftharpoons H^+ + B^-$$

The equilibrium is shifted to the right if the pH of the solution rises (i.e. the concentration of free hydrogen ions falls) so that more hydrogen ions are formed. The equilibrium is shifted to the left if the pH of the solution falls (i.e. the concentration of free hydrogen ions increases). Because of this tendency to release or take up hydrogen ions depending on the pH of the solution these acids and bases tend to stabilize the pH of a solution containing them. For this reason they are called buffers.

pH of buffer solutions. One can calculate the pH of a solution containing a buffer by means of the modified law of mass action called the Henderson–Hasselbach equation.

$$pH = pK + \log \frac{\text{concentration of buffer base}}{\text{concentration of buffer acid}}$$

The pK value is a constant, under certain specified conditions, for a particular buffer system.

Three important buffer systems exist in blood to stabilize the pH.

1. *Carbonic acid-bicarbonate buffer system.*

$$pH = 6{\cdot}2 + \log \frac{[HCO_3^-]}{[H_2CO_3]}$$

$$= 6{\cdot}2 + \log \frac{20}{1}$$

In this buffer system if there is a rise in the concentration of free H^+ ions then these react with the buffer base to form carbonic acid. e.g.

$$HCl + HCO_3^- \rightleftharpoons H_2CO_3 + Cl^-$$

$$\underset{\text{acid}}{\text{hydrochloric}} + \underset{\text{ion}}{\text{bicarbonate}} \rightleftharpoons \underset{\text{acid}}{\text{carbonic}} + \underset{\text{ion}}{\text{chloride}}$$

In the above example electrochemical neutrality is maintained in buffering by the loss of one negative charge on the bicarbonate ion which is replaced by a negative charge on the chloride ion. The strong acid HCl is replaced by the much weaker carbonic acid, H_2CO_3 (i.e. there is much less tendency for this acid to dissociate to give rise to free hydrogen ions).

The concentration of bicarbonate ions, in the process of buffering, falls since the pH of the solution is determined by the ratio of HCO_3^-/H_2CO_3.

However, the pH falls much less than if free hydrochloric acid remained in the solution and was unbuffered. If an alkali is added to this system e.g. sodium hydroxide, then this would also be buffered:

$$NaOH + H_2CO_3 \rightleftharpoons NaHCO_3 + H_2O$$
$$\Updownarrow$$
$$Na + HCO_3^-$$

In this reaction the strong base sodium hydroxide is replaced by the much weaker base HCO_3^-. As the numerator of the equation

$$pH = 6\cdot2 + \log \frac{[HCO_3^-]}{[H_2CO_3]}$$

increases, the pH of the solution rises but much less than if the sodium hydroxide had not been buffered.

2. *The phosphate buffer system.*

$$pH = 6\cdot8 + \log \frac{[B_2HPO_4]}{[BH_2PO_4]} \quad \text{where B = Na or K}$$

Buffering of acids by this system occurs in the following way

$$B_2HPO_4 + H^+ \rightleftharpoons BH_2PO_4 + B^+$$

Electrochemical neutrality is maintained by the freeing on one cation (either a sodium or potassium ion). Buffering occurs because free hydrogen ions are replaced by the weak acid BH_2PO_4 which has little tendency to give up a proton.

3. *Proteins.* One cannot give a single pK value for proteins because of the large number of ionisable groups on the protein. At the pH of plasma, proteins exists as colloidal anions which can give up protons becoming more negatively charged or take up protons becoming less negatively charged.

$$\text{Alkali}$$
Buffering of alkali: $HProt^{(n-1)-} + OH^- \longrightarrow Prot^{n-} + H_2O.$

$$\text{Acid}$$
Buffering of acid: $Prot^{n-} + H^+ \longrightarrow HProt^{(n-1)-}$

The proteins form a very important system, both intra and extra-cellularly. In the blood the protein haemoglobin is a potent buffer which enables most of the carbon dioxide generated in metabolism to be transported from the tissues to the lungs without change in the pH of the blood (page 220).

Sequence of events on the introduction of hydrogen ions into the blood. When hydrogen ions are introduced into the circulation, either artificially or from acids generated in metabolism, they are first diluted in the large volume of body water and then they are buffered. Buffering of hydrogen ions involves a reduction in the amount of body bases (bicarbonate ions, secondary phosphate ions, and a fall in the buffering capacity of proteins). Considering the bicarbonate-carbonic acid buffer system one might have a shift of HCO_3^-/H_2CO_3 from the normal ratio of 20/1 to 10/1. The condition of acidosis is characterised by a fall in the blood concentration of buffer base. This bicarbonate buffer system is of great importance in the body because each component of the system is under independent physiological control. The concentration of carbonic acid in blood is under the control of the respiratory centre of the medulla and the lungs (p. 479) and the bicarbonate concentration of blood is under renal control. Carbon dioxide dissolved in blood is a potent regulator of respiratory rate through the direct action on the respiratory centre of the medulla and the indirect action through the chemoreceptors of the aortic and carotid bodies. Under normal circumstances breathing is controlled by these means so as to maintain a partial pressure of CO_2 in the alveoli and arterial blood of 40 mm. Hg. An excess of CO_2 in the blood stimulates a rise in ventilation of the lungs to 'flush out' the excess CO_2 from the blood. A fall in pp. CO_2 in the arterial blood depresses ventilation so that less of the CO_2 generated in metabolism is excreted from the lungs and it accumulates in the blood until the normal concentration is achieved. Thus the denominator of the Henderson–Hasselbach equation, H_2CO_3, is maintained constant under normal conditions.

The chemoreceptors of the aortic and carotid glomi are also sensitive to the pH of the blood circulating through them. A fall in pH acting through them stimulates an increase in ventilation rate, and an increase in pH of the blood depresses ventilation. A change of 0·1 of a pH unit of the blood alters the ventilation two fold. Thus, after buffering of acids by the blood buffers and a fall in blood pH, modifications in ventilation occur which tend to restore the pH of the blood towards normal values. In the hypothetical example considered above, buffering of acid was associated with a change in the ratio of HCO_3^-/H_2CO_3 from 20/1 to 10/1. This is associated with a fall in the pH of the blood. Acting via the chemoreceptors this stimulates ventilation so that more

CO_2 and thus more H_2CO_3 is excreted from the blood, producing a further change in the ratio of buffers e.g. from 10/1 to 10/0·5.

This establishes a normal ratio of the two buffers and the pH of the blood returns to normal levels. Although the pH of the blood is quickly restored to normal by the change in ventilation of the lungs, the buffering *capacity* of the blood, which is determined by the total amount of substances rather than their ratio, has been greatly reduced—in the above example buffering capacity has been halved.

The restoration of buffer capacity is carried out by the kidney which restores bicarbonate ions to the plasma. This, however, is a slower process, taking hours and days contrasting with the immediate ventilatory response which may be accomplished within minutes.

By means of the process already described on p. 496 the kidney tubule cells generate new hydrogen ions equivalent to those which have been buffered and these are excreted in the urine, and for each hydrogen ion excreted in the urine one new bicarbonate ion is returned to the plasma to restore the buffer capacity. The concentration of bicarbonate ions in the plasma gradually rises and the process of buffering is reversed.

1. The concentration of bicarbonate ions in the bicarbonate–carbonic acid system is returned to normal.

2. Primary phosphate is reconverted to the more basic secondary phosphate.

$$H_2PO_4^- + HCO_3^- \longrightarrow HPO_4^{--} + H_2CO_3$$
$$\downarrow$$
$$H_2O + CO_2 \longrightarrow \text{excreted via lungs.}$$

3. Hydrogen ions taken up by proteins are now given up.

$$HProt^{(n-1)-} + HCO_3^- \longrightarrow Prot^{n-} + H_2CO_3$$
$$\downarrow$$
$$H_2O = CO_2 \longrightarrow \text{excreted by lungs.}$$

On p. 496 the reabsorption of bicarbonate ions from the glomerular filtrate was described. By the end of the proximal tubules of all the filtered bicarbonate ions have been destroyed by the tubular secretion of hydrogen ions, but each destroyed bicarbonate ion has been replaced by a new bicarbonate ion generated in the tubule cells. After the bicarbonate has been removed from the filtrate by this mechanism the secretion of hydrogen ions continues in the distal convoluted tubules and the bicarbonate ions, which pass into the blood from the tubule cells, are now in excess of the bicarbonate ions filtered off in the glomerular filtrate. It is this new additional bicarbonate which restores the buffer capacity of the blood.

Free hydrogen ions cannot be excreted into the urine for obvious reasons. The hydrogen ions secreted into the tubules are buffered by secondary phosphate ions which have been filtered off in the glomerular filtrate.

$$HPO_4^{--} + H^+ \longrightarrow H_2PO_4^-$$

When all the phosphate has been used up in this way the pH of the urine is at its lowest limit of pH 4·3. However, further hydrogen ions can be excreted in the urine, and thus additional bicarbonate is restored to the plasma without further lowering the pH of the urine. This is because the kidney cells can produce the base ammonia, either from the amino acid glutamine or from the oxidative deamination of other amino acids. This ammonia is secreted into the tubular lumen with the hydrogen ions, as the very weak acid the ammonium ion.

$$\underset{\text{ammonia}}{NH_3} + H^+ \longrightarrow \underset{\text{weak acid.}}{NH_4^+}$$

The amount of ammonium ions excreted by the kidney varies according to the state of the reaction of the body fluids, and large amounts can be excreted in acidosis.

The most important factor which determines how much of the bicarbonate base is restored to the blood passing through the kidney is the amount of bicarbonate which is filtered off in the glomerular filtrate which in turn depends upon the concentration of bicarbonate in the plasma. If this is low, e.g. because of buffering of acids by the blood, then less bicarbonate is filtered off in the kidneys and fewer of the hydrogen ions secreted by the tubule cells are used in 'reabsorbing' filtered bicarbonate ions and a greater quantity of 'new' bicarbonate ions are restored to the blood. If the concentration of bicarbonate ions in the plasma is high (alkalosis) more bicarbonate ions are filtered off in the kidneys, more hydrogen ions are used in 'reabsorbing' this bicarbonate and a reduced or zero quantity of 'new' bicarbonate ions is returned to the plasma. In fact all the hydrogen ions secreted by the tubule cells may be utilised in the 'reabsorption' of filtered bicarbonate ions, and these now begin to appear in the urine, making the urine alkaline.

Constituents of Urine and their origin

COMPOSITION OF HUMAN URINE in grammes per 24 hours.

Urea	12–35 G
Ammonia	0·6–1·2 G
Creatinine	0·8–2·0 G
Uric acid	0·3–0·8 G
Hippuric acid	0·7 G
NaCl	10–15 G
P	1·2 G
Na	2·5 G
S	1·2 G
K	1·5–2·0 G.

UREA (CARBAMIDE) $(NH_2)_2CO$. This is the main nitrogenous excretory product found in mammals. It is very soluble in water and fairly toxic. Human blood has normally between 20–38 mg. urea/100 mls. of blood. Urea is manufactured in the liver in mammals in a cyclic chemical reaction called the ornithine cycle (see fig. 217). It was once thought to be produced directly from ammonia but it is now known to be produced from ammonia by way of the amino acid arginine. Under the influence of an enzyme arginase the arginine breaks down to urea and the amino acid ornithine. The ornithine is reconverted into arginine by taking up further ammonia derived from the breakdown of waste and excess amino-acids.

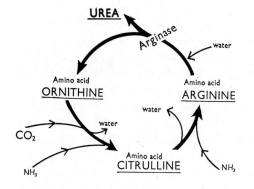

Fig. 217. *The Ornithine cycle for the production of urea.*

AMMONIA. The ammonia found in the urine is formed in the kidney, as described on page 515. It is produced by the deamination of the amino acid glutamine in the cells of the renal tubules.

The process of deamination is an important one; it occurs also in the mammalian liver where the ammonia, however, is converted into urea by way of the ornithine cycle. Chemically, deamination is usually an oxidative process:

$$
\begin{array}{cc}
\text{H} & \\
| & \\
\text{R.C—NH}_2 + \tfrac{1}{2}\text{O}_2 \rightarrow & \text{R.C}{=}\text{O} + \text{NH}_3 \\
| & | \\
\text{COOH} & \text{COOH} \\
\text{amino acid} & \text{keto-acid} \quad + \text{ammonia}
\end{array}
$$

The keto-acids produced are respired to produce energy and in the liver ammonia is incorporated into the amino acid arginine. Ammonia is a very toxic substance and as little as 5 mg./100 ml. of blood will kill a rabbit. There is normally only about 0·001 mg./100 ml. in mammalian blood.

Since ammonia is so poisonous it is not surprising to find that it is an important excretory product only in those animals that can quickly expel it from their bodies. In fresh water bony fishes where water is passing through the body in large quantities and there is a copious dilute urine, ammonia accounts for over 50% of the nitrogenous excretory products.

URIC ACID. Uric acid is almost insoluble in water and therefore does not enter into the chemical reactions of the body. It is found only in small amounts (about 0·5 G/day) in man, but in reptiles and birds it forms the major excretory product. It is essential for an animal which lives for a long embryonic period in the egg to be able to produce an excretory product which will not poison the embryo, for there is no means of removing the excretory product when a shelled egg is laid on land. Reptiles and birds, which have large shelled eggs on land, have the ability to excrete most of their waste nitrogen as uric acid. In passing we may note that although the young mammal has a long embryonic period it can survive in spite of a relatively toxic soluble excretory product (urea) because the excretory waste is removed via the placenta and the kidney of the mother.

The small amount of uric acid produced in the urine of man comes from oxidative deamination of the purines adenine and guanine, which are produced when nucleic acids break down.

$$\text{Nucleic acids} \begin{array}{l} \nearrow \text{adenine} \rightarrow \text{hypoxanthine} \\ \qquad\qquad\qquad\qquad \downarrow \\ \searrow \text{guanine} \rightarrow \text{xanthine} \end{array}$$

$$\downarrow \text{xanthine oxidase}$$

$$\text{uric acid}$$

An abnormality in the metabolism of uric acid in man may produce the disease gout, in which deposits of uric acid crystals may accumulate in certain joints and soft tissues; a diet excluding foods rich in purines is advised.

PHOSPHATE AND SULPHATE. These are derived mainly from the break-down of proteins containing phosphorus and sulphur, in addition to which there is a variable intake of sulphates and phosphates in food-stuffs.

CHLORIDES. The chlorides present in the urine come from the dietary salt, which is a varying quantity.

Chapter Thirteen

Reproduction

THE STRUCTURE OF THE REPRODUCTIVE SYSTEM

The Male

The primary male organs consist of paired glands called the testes, and these produce the male gametes or spermatozoa. The testes develop in the abdominal cavity adjacent to the adrenal glands but in most mammals they migrate downwards through the body cavity into a special fold of skin called the scrotum, or scrotal sacs. The reason for this is that spermatogenesis will not take place at body temperature, and within the thin walled scrotum the temperature is several degrees below that of the body cavity. If for some reason the descent does not occur then the animal is infertile. In amphibia, reptiles and in some mammals e.g. the elephant, the testes remain within the abdominal cavity. The testis is guided in its descent by a long cord, the gubernaculum, which extends from the lower pole of the testis to the scrotum, and the gubernaculum anchors the mature testis in the scrotum.

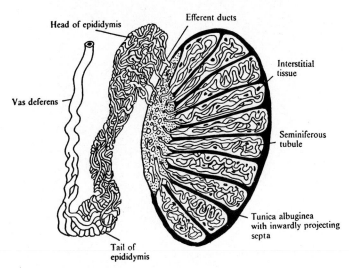

Head of epididymis

Efferent ducts

Interstitial tissue

Vas deferens

Seminiferous tubule

Tunica albuginea with inwardly projecting septa

Tail of epididymis

Fig. 218. *Diagram of a vertical section of the testis.*

519

The secondary sexual organs include a variety of ducts and glands which convey the spermatozoa in special secretions to the exterior of the body, so that they can be deposited within the female; the secondary sex organs also include those characters of the male which distinguish it from the female e.g. the long mane of the male lion. These secondary sex organs are developed and maintained by means of sex hormones produced within the testis itself. Male sex hormones are called androgens.

Structure of the testis (figs. 218–222). The testis is a tubular gland surrounded by a fibrous capsule the tunica albuginea. It is divided into several hundred compartments by means of fibrous tissue septa and each compartment contains several tubules, called seminiferous tubules. Each tubule is about 50 cms. long in man and is coiled upon itself, hence the name convoluted seminiferous tubules. All the tubules drain into one

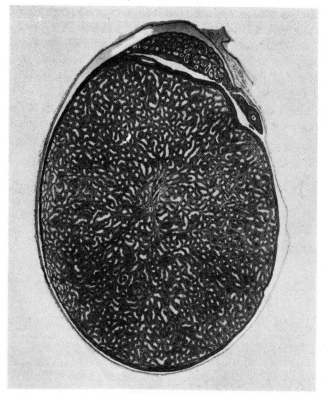

Fig. 219. *L.P. photomicrograph of a section of human testis and epididymis* (R). Note the fibrous capsule surrounding the organ which is composed of a large number of tubules in section. (*By courtesy of Dr. S. Bradbury.*)

border of the testis, into larger collecting tubules which are coiled together in a mass called the epididymis which is applied to the surface of the testis. Between the seminiferous tubules inside the testis, there is

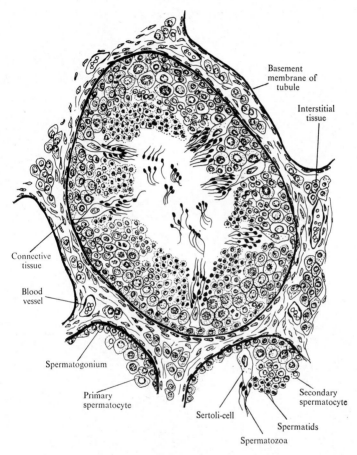

Fig. 220. *Transverse section of a seminiferous tubule seen at high magnification showing the seminiferous epithelium and the stages of spermatogenesis.*

connective tissue containing blood vessels and the glandular cells which are called interstitial cells or Leydig's cells and are responsible for the production of the male sex hormone.

STRUCTURE OF THE TUBULE. In the adult the seminiferous tubule consists of a basement membrane lined by the seminiferous epithelium; this seminiferous epithelium (figs. 220–222) consists of two types of cell.

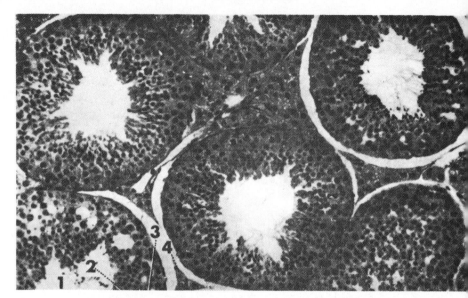

Fig. 221. *L.P. photomicrograph of a section of testis showing seminiferous tubules and interstitial tissue.* 1. lumen of seminiferous tubule. 2. spermatids. 3. spermatocytes lying on basement membrane. 4. interstitial tissue. (*By courtesy of Dr. S. Bradbury.*)

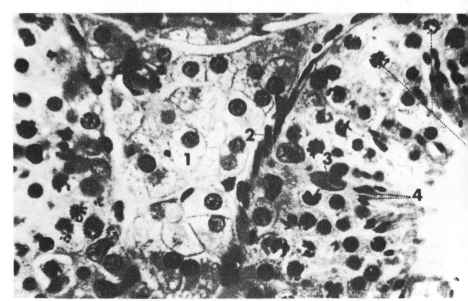

Fig. 222. *H.P. photomicrograph of a section of testis showing interstitial cells and various components of the seminiferous epithelium.* 1. Interstitial cells. 2. Basement membrane of tubule. 3. nucleus of sertoli cell. 4. spermatozoa attached to sertoli cell. 5. spermatocyte undergoing cell division. (*By courtesy of Dr. S. Bradbury.*)

First the germ cells themselves, secondly the cells of Sertoli which support and nourish the germ cells. The youngest germ cells are those lying close to the wall of the tubule and these divide to produce cells which pass nearer to the lumen of the tubule where the mature spermatozoa occur. The details of this transformation are described in more detail on p. 530. The Sertoli cells are slender pillar like cells attached at their base to the basement membrane. At a certain stage of spermatogenesis the germ cells become closely attached to these Sertoli cells. The mature spermatozoa pass from the lumen of the seminiferous tubules into the larger collecting tubules of the epididymis, whose walls bear a ciliated epithelium which moves the spermatozoa into the vas deferens.

Secondary sex organs (fig. 223). The spermatozoa are conveyed by the vasa deferentia to the base of the bladder where they can pass into the urethra. In structure the vas deferens is a hollow muscular tube, which, by means of muscular contraction, rapidly transports the spermatozoa before they are deposited in the female. Opening into the vas deferens

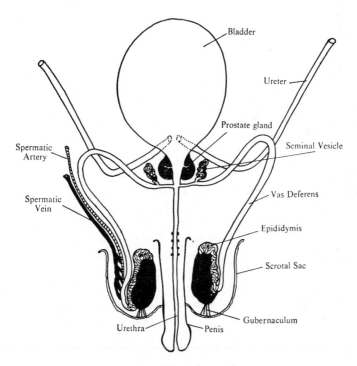

Fig. 223. *Diagram of the male reproductive organs.*

before it joins the bladder neck is the duct of the seminal vesicle; the seminal vesicles were once thought to store sperm but their function is to produce a thick secretion to provide the bulk of the fluid in which the sperms are transported. Surrounding the base of the bladder where the vasa deferentia open into the urethra is another gland, the prostate gland, whose secretions are discharged into the urethra together with the spermatozoa and the secretions from the seminal vesicles. The prostate gland produces a thin alkaline secretion; this helps to neutralize any acid urine remaining in the urethra and also to neutralize some of the acid secretions of the female vagina after the sperms have been placed into the female.

These various secretions, spermatozoa and the products of the seminal vesicles and prostate gland are together called semen; the semen is discharged through the urethra which passes through the penis. The penis, through which both urine and semen pass, contains in its walls, sponge-like systems of blood spaces which can become filled with blood so making the penis a more rigid organ so that the semen can be deposited within the female.

The female

The primary sex organs in the female consist of paired ovaries situated within the abdominal cavity. The germ cells are liberated from the surface of the ovary into the peritoneal cavity from whence they pass into the secondary sex organs—the Fallopian tubes, the uterus and vagina, leading to the exterior of the body. Paired secretory organs, the mammary glands, are included in the secondary sexual organs of the female, and serve to nourish the newly-born mammal.

Structure of the ovary (fig. 224). The ovary is attached to the wall of the body cavity by a fold of peritoneum. The free surface of the ovary bulges into the peritoneal cavity into which the germ cells are liberated. The ovary is studded with follicles (see p. 527) in various stages of development, containing the germ cells. When the follicles are ripe they come to the surface of the ovary where they rupture.

The free surface of the ovary is covered by a thin layer of germinal epithelium from which the germ cells arise in the embryonic period. In some species of mammal it seems that even in the adult the germinal epithelium can give rise to successive crops of new germ cells which pass inwards to mature in the tissues of the ovary. The timing of the successive crops of new germ cells coincides with the rupture of mature Graafian follicles in which follicular fluid rich in the hormone oestradiol pours over the surface of the ovary. This hormone has been called a 'mitogenic' hormone because of its effect on the germinal epithelium in stimulating cell division and the formation of new crops of germ cells. Beneath the germinal epithelium is a layer of dense connective tissue, the tunica abluginea. Beneath the tunica albuginea the thicker outer part of the ovary, or cortex, contains the follicles in various stages of development. The central part of the ovary or medulla contains a loose connective tissue containing masses of blood vessels.

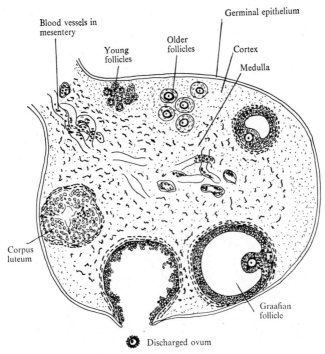

Fig. 224. *Schematic representation of a section through the ovary showing the stages of follicle development in sequence.*

The interstitial connective tissue of the ovarian cortex consists of connective tissue fibres and various types of cells. Some of these cells, large polyhedral 'epithelioid' cells are given the name interstitial cells. The number of these cells varies throughout the life of the female mammal. In some mammals with large litters e.g. rodents, there may be enormous numbers of these cells. They may arise from the walls of degenerating Graafian follicles.

Not all of the follicles in the ovary undergo the course of development into mature Graafian follicles as described on page 527. Very many of them undergo a degenerative change called atresia in which there is a hypertrophy of the cells forming the wall of the follicle together with a degeneration of the ovum. The interstitial cells and the cells of the atretic follicles are considered to have an endocrine function.

The oviduct or Fallopian tube. The oviducts are muscular tubes which serve to convey the germ cells from the ovaries to the uterus. The outer end of the tube, nearest to the ovary is expanded, and its edge is split up into fringes, the fibriae, which are closely applied to the surface of the ovary. The lumen of the oviduct is lined by a secretory mucous membrane, in which there are many ciliated epithelial cells. The germ cells are conveyed down the tube to the uterus by means of peristaltic movements of the tube itself and by the effect of the ciliated epithelium.

The uterus. The uterus is a thick-walled muscular structure within which the embryo develops. Its wall has three layers, an outer serous coat, a thick middle coat consisting of interlaced smooth muscle fibres (the myometrium) and an inner vascular mucous layer, the endometrium.

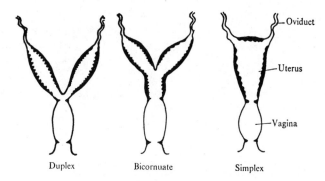

Duplex   Bicornuate   Simplex

Fig. 225. *Types of mammalian uteri.*

In primitive mammals there are two uteri, each opening into the vagina and this is called the duplex condition and is found in marsupials, many rodents (e.g. rats, mice, rabbit) and bats. In most mammals the distal end of the two uteri is fused to give a bicornuate uterus. In higher primates, including man, the two uteri are completely fused together to give a single organ, the uterus simplex (fig. 225).

The vagina is a distensible tube lined by squamous epithelial cells which connects the uterus to the outside world.

GAMETOGENESIS

The process of formation of gametes is called gametogenesis; the formation of eggs is called oogenesis and the formation of sperms is called spermatogenesis. Gametogenesis may be conveniently divided into three stages. The first stage is one in which the cells of the germinal epithelium divide and is called the stage of multiplication. The second stage is one of growth when each of the tiny cells produced in the multiplication stage grows to a larger size. The cell at the end of this stage is called the primary oocyte in the case of the female, and the primary spermatocyte in the case of the male. The third stage in gametogenesis is a period of maturation; during this period very important changes occur in the nucleus with the result that the chromosome number is reduced from the diploid to the haploid number. This reduction in chromosome number takes place in the so-called reduction division of a special kind of cell division called meiosis. When the primary oocyte divides by the reduction division it does so unequally; the nucleus divides into two equal parts, but almost all of the cytoplasm goes with one half of the nucleus, whereas the remaining half of the nucleus has very little cytoplasm. The latter is called the first polar body. The nucleus with most of the cytoplasm is now called the secondary oocyte and it contains only the haploid number of chromosomes. In mammals it is at this stage that the female gamete is released from the ovary; and before this gamete can be considered as fully mature another division of the nucleus has to take place, and this again is an unequal division resulting in the production of a second polar body. The production of this second polar body takes place, in a mammal, when the egg is fertilized by the sperm. We have seen that the development of the female gamete involves three stages, the first of multiplication, the second of growth ending in the formation of the primary oocyte, and the third of maturation involving the production of the secondary oocyte with polar bodies.

The first stage of gametogenesis is complete in the female embryo by the end of intra-uterine life, and she is born with all the oogonia already formed within the ovary. The second stage of growth of the oogonia continues throughout the life of the mammal and we will now look at this growth phase in more detail.

Development of the Graafian follicle (fig. 226–27). In the cortex of the ovary of the mature mammal are many small clusters of cells called the primary follicles, which have been formed during the embryonic period from invaginations of the germinal epithelium, called sex cords. The primary follicles are very small and there are about 400,000 in the human female at maturity. At birth the first phase of oogenesis, the phase of multiplication has already started producing small collections of germinal cells, called primary follicles. One of the cells in the primary follicle is larger than the rest and is the oogonium, whilst the smaller surrounding cells are called follicular cells. In the second phase of oogenesis, the growth phase, the primary follicle develops and changes occur, in the oogonium as it becomes the primary oocyte, in the follicular cells and also in the connective tissue which surrounds the follicles. The oogonium enlarges, its nucleus gets bigger, and a few yolk granules begin to appear in its cytoplasm. At this stage a well defined shining layer appears around the surface of the oogonium called the zona pellucida. In the primary follicle the oogonium was surrounded by a simple columnar epithelium but as the follicle grows the follicular cells multiply to produce an epithelium which is several layers in thickness. The cells of this epithelium secrete a follicular fluid which accumulates in spaces which begin to appear between the cells.

The follicle by this stage in the human female is about 2 mm. in diameter and is now called the Graafian follicle. The follicle increases in size with the accumulation of more fluid within it and the oogonium is pushed to one side of the follicle where it is attached to the wall of the follicle in a group of columnar cells called the discus proligerus. The cavity of the follicle is lined by a few layers of columnar cells called the membrana granulosa. The connective tissue surrounding the Graafian follicle has become organized into a membrane called the theca (consisting of two layers, the theca interna and externa). Eventually the Graafian follicle may reach a size of 10 mm. in diameter in the human female and bulges from the surface of the ovary; by this time the oogonium has grown to its full extent and is called the primary oocyte. The fluid within the follicle is formed at a faster rate than the follicle wall grows and the follicle eventually ruptures.

By the time the follicle has ruptured the primary oocyte has undergone the meiotic division producing the first polar body and is now called the secondary oocyte. When the secondary oocyte is released it is surrounded by a few columnar cells which form the corona radiata, which may have a nutritive function similar to that of Sertoli cells in the male. When the Graafian follicle has ruptured and liberated the oocyte it collapses and the hole left by the departing oocyte becomes plugged with a blood clot. There is now a multiplication of the remaining cells of the follicle, the granulosa and theca cells. The cells enlarge and develop deposits of a yellow pigment called lutein. The

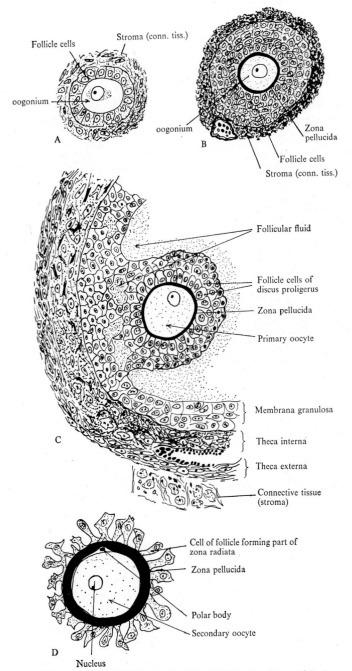

Fig. 226. *Development of the Graafian follicle. A. Young follicle.*
B. Older follicle. C. Part of a mature Graafian follicle. D. The maturing
ovum liberated by rupture of the Graafian follicle.

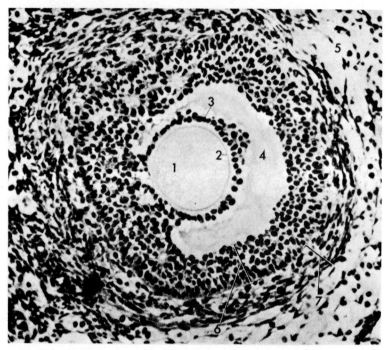

Fig. 227. *H.P. photomicrograph of a section of ovary showing the structure of a Graafian follicle.* 1. oocyte. 2. zona pellucida. 3. discus proligerus. 4. follicular fluid. 5. connective tissue stroma of ovary. 6. membrana granulosa. 7. theca (interna and externa). (*By courtesy of Dr. S. Bradbury.*)

whole structure so produced is called the corpus luteum, a solid ball of yellow pigment cells, which produces hormones which prepare the uterus to receive the fertilized oocyte.

Spermatogenesis, the formation of spermatozoa (fig. 228). Unlike the female, where the germinal epithelium forms the outermost layer of the ovary, in the male the germinal epithelium lines the walls of the semi-niferous tubules. The cells nearest to the wall of the tubule, the sperma-togonia, are the most primitive, undifferentiated cells of the tubule and they give rise to the other cells by mitotic division. As in oogenesis there are three stages in spermatogenesis.

The first stage of spermatogenesis is that in which the spermatogonia multiply by mitotic division. Some of the products of these divisions pass inwards nearer the lumen of the tubule where they enter the

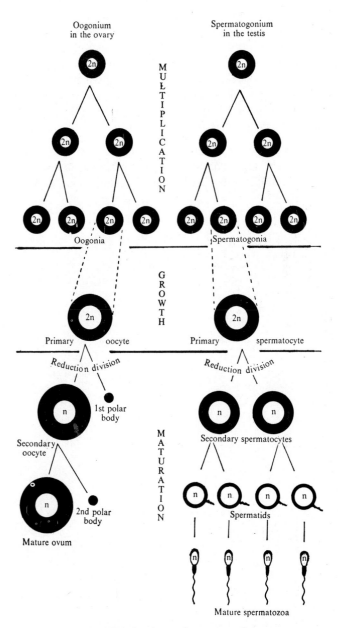

Fig. 228. *The phases of gametogenesis.*

second phase of spermatogenesis, the growth phase. During this phase the spermatogonia become larger, producing the primary spermatocytes. In the third phase each primary spermatocyte undergoes a reduction division producing two secondary spermatocytes containing the haploid number of chromosomes. Each secondary spermatocyte then divides to produce two spermatids. The spermatids do not undergo division but a series of changes occur which transform the spermatid into the mature sperm. During this maturation of the spermatids they are attached to the cells of Sertoli.

THE MATURE SPERM (human). The mature sperm consists of a head, middle piece and tail. The head consists of the condensed nucleus of the spermatid and is a flattened ovoid structure about 5 microns long.

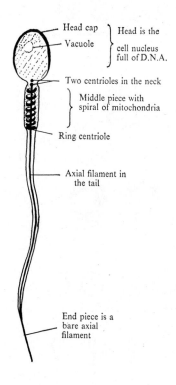

— Head cap ⎫ Head is the
— Vacuole   ⎬ cell nucleus
            ⎭ full of D.N.A.

— Two centrioles in the neck

⎫ Middle piece with
⎬ spiral of mitochondria
⎭

— Ring centriole

Fig. 229. *Diagram of the structure of a human sperm.*

— Axial filament in the tail

End piece is a bare axial filament

The head is capped by a sheath of material called the head cap. In the middle piece of the sperm there is a centriole from which arises a long axial filament which passes through the middle piece and the tail.

Surrounding the axial filament in the middle piece is wound a sheath of mitochondrial material, the mitochondrial sheath, which is probably concerned in the respiration of the sperm. In the tail the axial filament is covered by a sheath (see fig. 229).

The spermatozoa remain inactive until they pass from the testis. During their passage from the testis to the penis they are activated by the secretions of the accessory glands. The sperm is capable then of active swimming during which S-shaped waves pass along the tail. The energy for this is derived from the anaerobic breakdown of fructose which is present in the prostate secretions.

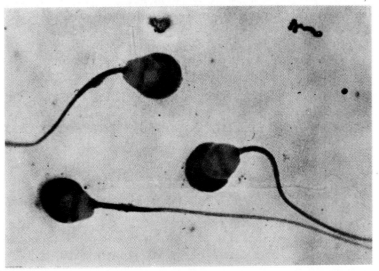

Fig. 230. *H.P. photomicrograph of guinea-pig spermatozoa. (By courtesy of Dr. S. Bradbury.)*

THE PHYSIOLOGY OF REPRODUCTION

Hormones and reproduction in the male

In the sexually immature mammal the primary and secondary sex organs are small and undeveloped. The growth and development of the sex organs is dependent upon the activity of the pituitary gland. The pituitary exerts its effect by the production of hormones called gonadotrophic hormones because of their growth effects upon the gonads. The gonadotrophic hormones are complex protein substances which have been isolated in a relatively pure state; their chemical structure is not yet elucidated. There are almost certainly two gonadotrophic hormones. One of these hormones is called the follicle stimulating hormone (or F.S.H.) because of its effect in the female in stimulating the growth of the follicles in the ovary. F.S.H. stimulates the growth of the seminiferous tubules of the testis and stimulates the activity of the germinal epithelium. The other pituitary hormone is called the luteinizing hormone (L.H.) or interstitial cell stimulating hormone. L.H. stimulates the growth and secretory activity of the interstitial or Leydig cells of the testis. These cells produce certain steroid hormones called sex hormones, because of their effect on the sex organs.

The most important sex hormone in the male is called testosterone. However, there are also female sex hormones or oestrogens produced in the male. The chemical structure of testosterone is shown in fig. 231.

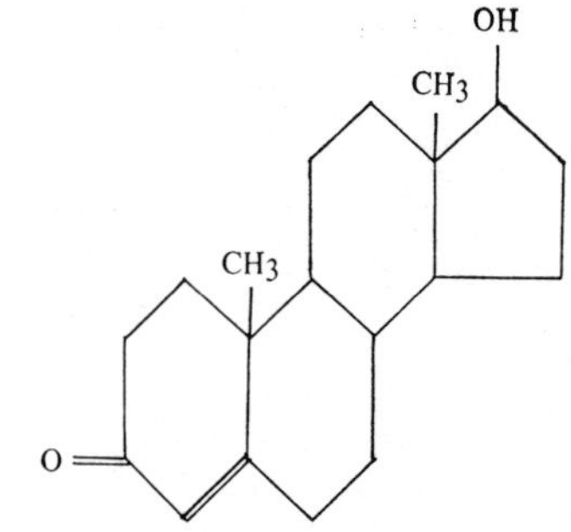

Fig. 231. *Formula of testosterone.*

The effect of testosterone is to promote the growth of the various sex organs—the vas deferens, seminal vesicles, prostate gland, penis etc. It also promotes the development of those other secondary sex characters which vary from one mammalian species to another; in man these include the growth of the beard, enlargement of the larynx with the development of a deeper voice, the male distribution of hair on the body, and the greater development of muscle. The effect on muscle growth is due to the fact that testosterone is what is called a protein anabolic hormone, that is it promotes the retention and incorporation of protein in the tissues.

In the immature animal testosterone will promote the precocious development of the sex organs, and in those species of mammals in which the testes do not descend into the scrotum until sexual maturity it stimulates the descent of the testes. In animals which have been castrated the sex organs gradually atrophy and the administration of testosterone can reverse these changes.

Testosterone has also a direct effect on the pituitary gland and a certain level of testosterone in the blood will inhibit the pituitary from producing gonadotrophic hormones; when the pituitary production of luteinizing hormone falls then the Leydig cells of the testis stop producing testosterone and the blood level of testosterone falls. With the falling production of testosterone by the Leydig cells the pituitary gland is now released from the inhibitory effect of testosterone and it begins to produce the gonadotrophic hormones again. By this feed-back mechanism (see also Chapter VI) the secretion of testosterone is controlled.

This effect of testosterone upon the pituitary gland also explains the varying results that experimenters have had in the administration of

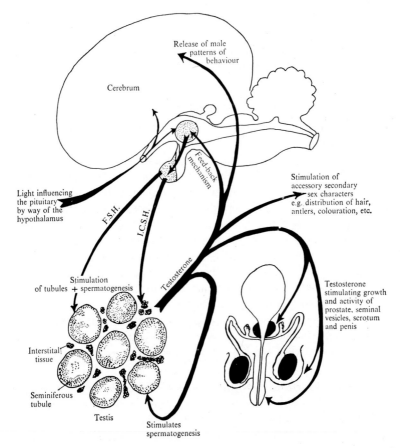

Fig. 232. *Diagrammatic summary of hormonal control of the sex organs in the male mammal.*

testosterone to animals. Some workers have found that in some animals testosterone will cause a stimulation of the germinal epithelium of the testis and the production of spermatozoa; because of the appearance of large numbers of dividing cells in the germinal epithelium the hormone has been called 'mitogenic', that is one which stimulates mitotic division within the cells. In large doses however testosterone can depress the growth of the testis, presumably because it inhibits the anterior pituitary gland from producing the gonadotrophic hormones, by the feed-back mechanism.

Testosterone also appears to be responsible for behaviour changes in animals, and administration of the hormone to sexually immature males results in the appearance of male breeding behaviour. In some species even the female will show a masculine pattern of breeding or sexual behaviour when given testosterone. The aggressiveness of many male mammals can be promoted by giving testosterone. In higher primates, including man, sexual behaviour cannot be controlled so simply, and cultural factors play a much more important role in the determination of the direction of sexual impulses.

Hormones and reproduction in the female

In the mature female mammal sexual activity tends to be an intermittent phenomenon during the breeding season, which is the period when mating can occur. The breeding season may consist of several weeks or months of the year, and there may be more than one breeding season in the year. During the breeding season itself sexual activity is a cyclic phenomenon, with periods of sexual activity or oestrus ('heat') alternating with periods of sexual inactivity. These cycles of activity stop of course as soon as a pregnancy is started. In some animals, including primates and rodents, these oestrous cycles are continuous throughout the sexual life of the animal and are not restricted to breeding seasons; these animals are said to experience poly-oestrus. In man, because of the unusual feature of menstruation the oestrous cycles are known as menstrual cycles; but the menstrual cycle is fundamentally similar to the oestrous cycle of other mammals. The variations in breeding activity are best illustrated by referring to particular examples:

Horse. The mating season of the horse is between March and August, although some breeds will mate in the autumn and winter in England. During the mating season, if pregnancy does not occur then there is a regular occurrence of oestrus, each period of oestrus lasting about 20 days. The horse is said to be seasonally poly-oestrous.

Dog. The domestic dog has two breeding seasons in the year, in late Winter and early Spring and in the Autumn. During each season there is only one period of oestrous, and the dog is said to be monoestrous.

Golden Hamster. This rodent is polyoestrous, coming into heat at all times of the year, the oestrous cycles recurring every ten days, each cycle lasting about four days.

Roe Deer. Like the dog the Roe Deer is monoestrous. The breeding season is in July and August during which there is only one period of oestrous.

The oestrous cycle (fig. 233). During the oestrous cycle there are widespread changes in the structure and behaviour of the female; all the changes that occur are under the control of hormones, produced mainly by the anterior pituitary gland and the ovary. The aim of these changes is to mature an ovum and prepare the uterus to receive and nurture the ovum if it is fertilized by a sperm.

In the early phase of an oestrous cycle the ovary is activated by the secretion of follicle-stimulating hormone (F.S.H.) from the anterior pituitary gland. The ovary responds by the progressive growth of one or more Graafian follicles, depending upon the species of mammal. Small amounts of luteinizing hormone (L.H.) are produced by the anterior pituitary gland at this time and in some way it assists the response of the ovary to F.S.H.; L.H. is said to have a synergistic effect. The ripening Graafian follicles secrete a steroid sex hormone called oestradiol which has widespread effects upon the secondary sex organs, and also has an effect on the secretory activity of the pituitary gland. Oestradiol is only one of a number of substances isolated from the ovary, blood and urine of the female mammal, all of which have some effect on the sex organs; the name oestrogens has been used to describe this class of substances, although oestradiol is the most potent naturally occurring oestrogen.

Oestradiol stimulates the growth of the uterus; the myometrium increases in thickness because of growth of its individual cells, its vascularity increases and the endometrium thickens, becoming more vascular as its secretory glands grow in length. The Fallopian tubes and vagina are also stimulated. In the sexually inactive phase or anoestrus, the epithelial lining of the vagina is thin, only one or two cells thick, but after stimulation by oestrogen the epithelium thickens and cornifies, and flattened cornified squames appear in the vaginal secretions. The mammary glands also increase in size under the influence of oestrogen, which stimulates the growth of the duct system.

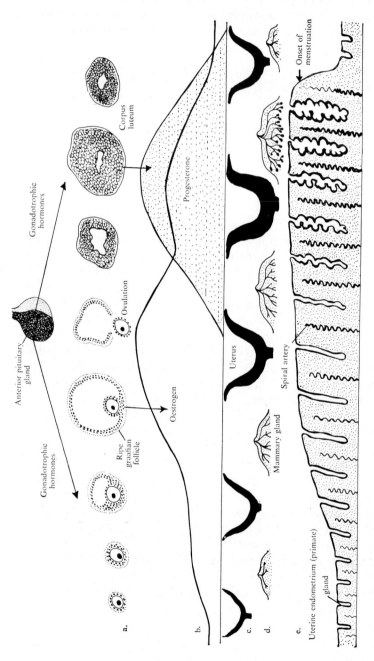

Fig. 233. *Diagram illustrating the changes in various structures throughout the oestrous (or menstrual cycle).* a. *Shows the development of the Graafian follicle, ovulation and the corpus luteum.* b. *Shows the blood levels of oestrogen and progesterone.* c. *Shows the progressive development of the uterus.* d. *Shows the development of milk forming tissue in the mammary glands. Early there is a growth of the duct system, and later in the progesterone phase the development of glandular elements.* e. *Shows the development of the primate endometrium with its peculiar feature of menstruation.*

The rising level of oestradiol in the blood has important effects upon the pituitary gland. It inhibits the formation of F.S.H. and at the same time stimulates the production of further amounts of L.H. which brings about ovulation and the formation of the corpus luteum. This first phase of the oestrous cycle, as described above, is called the *follicular phase*, because of the growth of the Graafian follicles, or the oestrogen phase, because of the importance of this hormone in this part of the cycle. In the uterus the follicular phase is associated with growth, both of the myometrium and endometrium, and this phase of uterine change is called the proliferative phase. The follicular phase finishes at ovulation when the egg escapes from the ovary and the remains of the Graafian follicles grow to produce special glandular structures under the influence of the pituitary luteinizing hormone (L.H.), called corpora lutea.

We now enter the *luteal phase* of the oestrous cycle. This is often called the progesterone phase, because of the importance of this hormone at this time. The corpora lutea are large yellow pigmented bodies studded in the ovarian cortex. They are formed by the proliferation and growth of cells in the wall of the ruptured Graafian follicle. Under the influence of another pituitary hormone called luteotrophin or lactogenic hormone the corpora lutea secrete a hormone

Fig. 234. *L.P. photomicrograph of secretion of human ovary showing corpus luteum.*
(By courtesy of Dr. S. Bradbury.)

called progesterone, in addition to small amounts of oestradiol. Progesterone produces further changes in the endometrium of the uterus; the increase in thickness and vascularity of the uterus progresses and the glands of the endometrium become tortuous and begin to pour secretions into the cavity of the uterus. Because of these glandular changes this phase of uterine activity is called the *secretory phase*.

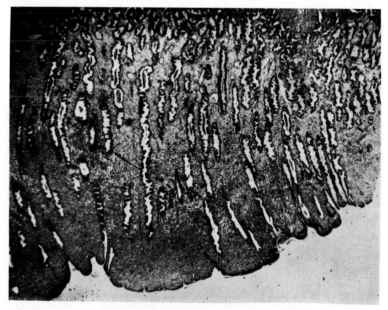

Fig. 235. *Section of human endometrium showing tortuous glands opening onto the surface (i.e. into the lumen of the uterus). (By courtesy of Dr. S. Bradbury.)*

Progesterone also has effects on the mammary glands which have already been primed by the effects of oestradiol; now, glandular elements begin to appear around the ends of the duct systems of the mammary glands.

We have seen that the time of maximal oestradiol activity is at the time of ovulation and after this time the level of oestrogen production gradually falls, as progesterone comes to play a more important part in the cycle. It is at the time of maximal oestradiol production that the female mammal is most willing to receive the male and this is the true period of 'heat'. Mating and fertilization usually occur about this time. If fertilization of an ovum does not occur then corpora lutea gradually disintegrate and retrogressive changes occur in the secondary sex organs. The falling level of blood oestrogen in the luteal phase of the cycle, together with some inhibitory effect of progesterone on the anterior pituitary gland, are responsible for a gradual decline in the production of L.H. and therefore the corpora lutea degenerate.

Menstrual cycle. In the human female when the corpus luteum disintegrates at the end of the luteal phase of the cycle, there is a complete breakdown of the hypertrophied endometrium; blood and broken-down tissues are discharged from the vagina and this constitutes menstruation. In the human the menstrual cycle lasts about 28 days. In the first half of the cycle there is the follicular phase, culminating in ovulation at about the fourteenth day. In the second half of the cycle, the luteal phase, the corpus luteum is formed and the endometrium enters the secretory phase, and in the absence of fertilization the luteal phase is ended by the appearance of the menstrual flow (see fig. 233). Following menstruation only fragments of endometrium are remaining and these lie in crypts in the myometrium. From these fragments the entire endometrium is reformed during the proliferative phase of the next cycle. The primate endometrium undergoes this almost complete breakdown because of a peculiarity in its blood supply. When the supply of progesterone is waning as the corpus luteum degenerates at the end of the luteal phase, certain spiral arteries of the endometrium go into such intense spasm that the tissues supplied by them die and undergo degenerative changes. The whole of the dead endometrium is then sloughed off from the uterine wall together with blood.

The oestrous cycle and pregnancy. If during the luteal phase of the oestrous cycle an ovum is fertilized and settles in the uterine cavity then the retrogressive changes in the secondary sex organs do not occur, nor does the corpus luteum degenerate. We have seen that in the absence of pregnancy the corpus luteum degenerates; it does this because of the decline in the production of pituitary gonadotrophic hormones. As soon as there is a union established between the fertilized ovum and the uterine wall increasing amounts of gonadotrophins appear in the maternal blood and this maintains the structure and function of the corpus luteum. The gonadotrophic hormone is produced by the placenta, the organ which unites the mother and foetus, and through which it receives its nourishment. The placenta also produces large amounts of oestrogen and progesterone and gradually replaces the corpus luteum as a source of these hormones, so that later in the pregnancy one can remove both ovaries from the female mammal without disturbing the pregnancy. The function of the large amounts of sex hormones produced by the placenta is to promote further growth of the uterus, vagina and mammary glands.

During pregnancy, growth of further Graafian follicles and ovulation is prevented because the large amounts of oestrogen produced by the placenta inhibit the anterior pituitary gland from producing follicle-stimulating hormone.

Reproduction and the environment

The reasons why animals tend to breed at relatively restricted times of the year have been studied and classified under two headings, internal physiological mechanisms on the one hand, and external environmental factors on the other. An internal physiological 'clock' cannot be the sole factor in determining periodic breeding; seasonal breeding has an adaptive significance in that young are produced at favourable times of the year, and obviously a rigid internal mechanism would, through the course of time, fail to adapt the animal to changing climatic conditions.

It appears that animals have become adapted to respond to certain environmental factors which herald the oncoming favourable season. In temperate latitudes many animals have as it were harnessed their breeding behaviour to the length of day and respond to an increase in day length by development of the sex organs, so that the young will develop in a favourable season with warmth and an adequate food supply. The first study of the effect of light on sexual cycles was made on the Canadian bunting, a bird which normally breeds in Spring. By giving these birds extra periods of light in the Autumn it was possible to cause development of the testes, even at a time when the temperature was below freezing point—such is the potency of extra light. In many animals, reptiles, fish, birds and mammals it has been possible to cause the development of the sex organs out of the breeding season. The light exerts its effect by way of the eyes and connections with the hypothalamus and anterior pituitary gland. Not only is the increase in day length of importance but in some autumn breeding animals the shortening days may also act as the stimulus to sexual development.

When some animals with fixed breeding seasons in temperate latitudes are transferred from the Southern to the Northern Hemisphere, after a certain time the breeding season becomes adapted to the new conditions. However animals imported from tropical countries tend to continue with their former breeding habits, since they do not respond to those differences in daylight to which the species inhabiting temperate zones respond. This may be due to the fact that owing to the comparative uniformity of conditions in their own countries they have never acquired the capacity to respond to variations in light intensity or duration, characteristic of many animals living under seasonally changing conditions. Thus Java deer when imported to England continue to produce young in late Autumn, as they are believed to do in Java, a condition which is abnormal for any deer inhabiting temperate countries. But even in tropical climates animals may have special breeding seasons; if reproduction were continued throughout the year the competition for food for the young might be too intense for survival of the species. A further advantage of seasonal breeding may be the synchronization of the male and female sexual cycles by the fact that they are both adapted to the same environmental factor. What these environmental factors are in tropical climates is uncertain. That breeding cycles in some tropical animals are harnessed to some environmental factor and not dependent upon an internal rhythm seems certain; a tropical insectivorous bat lives throughout the period of daylight until about ten minutes before sunset in dark and almost thermostatic caves and yet it was found that in one year of observation no pregnancies occurred until a few days at the beginning of September. It seems impossible that such synchronization of the sexual cycles of such a group of bats could be achieved by an internal physiological mechanism.

The homiothermic vertebrates (warm blooded) have become almost independent of the temperature of their surroundings and are able to carry out breeding at any time of the year. They do, however, tend to breed in Spring in temperate latitudes to ensure a favourable environment with adequate food for the young. But even in homiothermic animals temperature may play some role in the timing of breeding behaviour. If Autumn weather is warm and food supply is good, sexual behaviour in the robin and other birds may be pronounced and may even lead to reproduction in a few birds, although they normally only breed in Spring. In this Autumn breeding phase sexual activity is developing when daylight is decreasing.

We have now seen something of the way in which sexual cycles are controlled by the endocrine organs, and the way in which the endocrine organs, by way of the hypothalamus of the brain, are synchronized with external environmental factors so that the young are cared for under favourable conditions.

THE PLACENTA

The uterus has been preparing to receive a fertilized ovum throughout the oestrous cycle. In the follicular phase of the cycle there is growth both of the myometrium and endometrium and an increase in the blood supply of the uterus. The endometrium is thickened, new blood vessels grow and its glands increase in length. In the luteal phase of the cycle these changes continue and the endometrium takes up a secretory character. The glands become tortuous and their epithelium becomes active, secretions passing into the uterine cavity.

The early development of the fertilized egg occurs during its passage through the Fallopian tube, since fertilization is usually achieved high up in the Fallopian tube. The fertilized egg or zygote undergoes divisions to produce a ball of cells called the morula. The morula then differentiates into an outer layer and an inner cell mass producing the blastula (see fig. 236). The cells of the outer layer form the trophoblast or trophoblastic ectoderm. This enters into the formation of the chorion, the outermost covering of the developing zygote. This outer layer is called trophoblast because it enters into the formation of the placenta, the organ concerned in nourishing the foetus. The inner cell mass of the blastula contains the cells from which the embryo will develop.

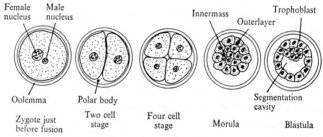

Fig. 236. *Diagrams showing the early development of the fertilized egg of the mammal up to the blastula stage.*

During its passage down the Fallopian tube the nutrition of the zygote is dependent on the small amounts of food stored in the protoplasm. When the morula reaches the uterine cavity it is bathed in the secretions of the uterine glands, and these secretions probably have some nutritive function. In some species the blastocyst lies in the cavity of the uterus, and in contact, by means of its trophoblast, with the endometrium all over its surface. In some other species the blastocyst becomes attached to the endometrium on one surface, and projects freely into the uterine cavity on its other surface. In other types the blastocyst sinks into the endometrium and becomes surrounded completely by maternal tissues; this type of implantation of the blastocyst is found in man and is called the interstitial type. In order to understand the varied types of placenta found in mammals it is necessary to examine in some detail the formation of the various foetal membranes which take part in the formation of the placenta. There are four foetal membranes concerned in the adaptation of the foetus to life in the uterus, the amnion and chorion, formed from the original embryonic body wall, and the yolk sac and allantois, parts of the original gut of the embryo.

We had left the development of the embryo at the blastula stage, consisting of an inner cell mass and an outer layer, the trophoblast; a two layered vesicle is produced by a growth of endoderm cells around the inner layer of the trophoblast (enclosing the yolk of the egg, when this is present) (fig. 237). Only a part of the wall of the blastula is destined to form the embryo and this is called the embryonic area. This embryonic area gradually sinks into the centre of the blastula as it becomes covered over by the amniotic folds. In man the embryo is not covered by the amniotic folds and there is merely a hollowing out of the cells of the embryonic area to produce a cavity, the amnion. The amniotic folds meet and fuse above the embryo and because of the double walled nature of the folds the embryo becomes to be surrounded by two membranes, an outer chorion and an inner amnion (fig. 238). The chorion thus becomes the membrane which is in contact with the endometrium of the uterus, and projections called chorionic villi grow out from its surface to make a more intimate contact with the maternal tissues. The amniotic membrane surrounds a fluid filled cavity, the amniotic cavity, in which the embryo floats.

We have seen above how the endoderm grows round the inner surface of the trophoblast to produce a two layered vesicle. In the heavily yolked eggs of birds and reptiles this layer encloses the yolk of the egg and is called the yolk sac. The primitive gut of the embryo, a groove on the under surface of the embryonic area, is open into this yolk sac but later as the gut is folded off it is separated from the yolk sac except for a narrow passage, the yolk sac stalk. Later mesoderm grows out from the embryo between ectoderm and endoderm; this process provides a double layer of mesoderm in the amniotic folds (see fig. 238) and a layer of mesoderm between the ectoderm and endoderm of the yolk sac. It is in this layer of mesoderm that the embryonic blood vessels develop. In the heavily yolked eggs of birds and reptiles the inner layer of the yolk sac becomes vascular in order to absorb the nourishment from within the yolk sac. But in mammals there is very little nourishment within the yolk sac and it is from outside that it has to look for nourishment. Thus the outer layer of the trophoblast becomes highly developed and supplied with blood vessels from the mesoderm. In primate embryos, including man, the yolk sac is a rudimentary structure only. In Marsupial mammals which have a less developed placenta the yolk sac is a more important organ, and is the major organ

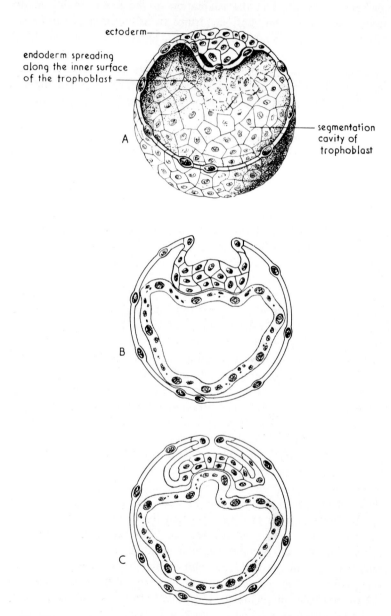

ectoderm

endoderm spreading
along the inner surface
of the trophoblast

segmentation
cavity of
trophoblast

A

B

C

Fig. 237. *Development of the late blastula.* Note the embryonic area gradually sinking into the blastula in A, B and C, gradually becoming covered by the amniotic folds. The endoderm is shown progressively spreading over the inner surface of the trophoblast.

for nourishment of the embryo. In rodents the yolk sac becomes partly vascularized; the non vascularized area becomes eroded away so that the cavity of the yolk sac opens into the uterine cavity. The inner wall of the yolk sac is applied directly to the endometrial lining of the uterus, producing what is known as an inverted yolk sac placenta.

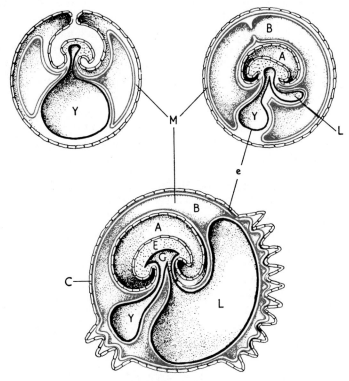

Fig. 238. *Diagram of the stages in development of the embryo showing the formation of foetal membranes.* Mesodermal tissue is shown in red. A—amniotic cavity limited by the amniotic membrane.   B—extra-embryonic coelom enclosed by layers of mesoderm.   C—chorion.   e—endoderm.   G—primitive gut.   L—allantois.   T—yolk sac.   The pro-projections of ectoderm and mesoderm on the right hand side are the villi of the placenta (allanto-chorionic villi).

In most mammals the chief absorbing surface occurs on a structure called the allantois. This is a blind ended outgrowth of the hind gut carrying mesoderm on its surface as it grows into the extra embryonic coelom. In the eggs of reptiles and birds the allantois functions as a bladder for the storage of excretory products, but it also serves to transport oxygen which diffuses through the shell, through the blood vessels of the allantois to the developing embryo. In mammals the allantois has become progressively more important in the nourishment of the foetus, and blood vessels of the allantois pass into the villi of the chorion which make connection with the lining of the uterus.

The processes which grow out from the trophoblast (i.e. extra-embryonic ectoderm of the chorion) are called trophoblastic or chorionic villi. The villi are connected to the embryonic blood vessels through the yolk sac or allantois, depending upon the species of mammal. Not all of the surface of the trophoblast is covered by villi and the arrangement of the villi varies from species to species. In the horse and the pig the villi are diffusely arranged over the surface of the trophoblast producing a diffuse placenta. In the sheep and cow the villi are localized in patches called cotyledons producing the cotyledonary placenta. In carnivores the villi are restricted to a band encircling the embryo producing a zonary placenta. In man (also rodents and insectivores) the villi are at first scattered over the whole trophoblast and later become limited to a disc shaped area producing a discoidal placenta.

Types of placental union with the uterine wall. There is a great variation in the intimacy of contact between the foetal and maternal tissues in the placenta of mammalian species. On the foetal side of the placenta there are three layers of tissue, the chorion, the mesenchymal tissues and the endothelium of the capillary vessels. On the maternal side there are uterine secretions, the endometrial epithelium, connective tissues and the endothelium of the blood vessels. There are thus at least six possible layers of tissue separating the foetal and maternal blood. In the diffuse placenta found in the pig and horse these six layers which separate the maternal and foetal blood streams persist, producing what is called an epithelio-chorial placenta (p. 551 and fig. 239).

This description of the placental structure of the pig and the horse give a rather exaggerated picture of the separation of maternal and foetal blood. The foetal cells of the chorion are low cuboidal in shape and foetal capillaries ramify close to the chorion and may actually penetrate between the chorionic cells to form intra-epithelial plexuses. The capillaries at places displace the cuboidal cells and compress their cytoplasm into thinned out plates. Thus over a large part of the placental surface only a very thin plate of epithelial cytoplasm separates the foetal and maternal blood. In other mammals the barrier between maternal and foetal blood is reduced by the erosion of the layers of the uterine wall by the trophoblastic villi. In the syndesmochorial placenta of cattle and sheep the endometrial epithelium is eroded and the epithelium of the trophoblastic villi is in contact with the uterine connective tissue (p. 551 and fig. 240) and in the endotheliochorial placenta of the cat and dog the uterine tissues are further eroded and the epithelium of the trophoblastic villi is separated from the maternal blood only by the endothelium of the maternal capillaries. In insectivores, rodents and man the maternal capillaries are eroded so that the trophoblastic villi are bathed in lakes of maternal blood (p. 551 and fig. 240). The

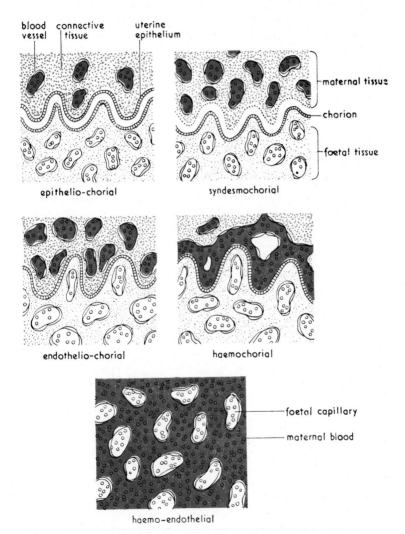

Fig. 239. *Diagrammatic representation of different placental types. Maternal blood is shown in red.*

epithelial cells of the chorion are provided with many minute projections called microvilli which interdigitate with maternal tissue and provide a very intimate form of contact between the two. The long processes of foetal tissue, called trophoblastic villi, run vertically from the surface of the placenta into maternal tissues. Between these trophoblastic plates are the maternal capillaries with a thick endothelium. These capillary endothelial cells show, in electron micrographs, an interesting structure which may well be related to the way in which some substances are transferred by the placenta between maternal and foetal blood. The cytoplasm of the maternal capillary endothelium is spongy, containing many vacuoles. In the cytoplasm between these vacuoles are mitochondria and Golgi material. The vacuoles are sometimes seen to connect directly with the lumen of the capillary and it seems probable that the vacuoles are formed by invagination of the surface membrane of the cell. Thus substances included in the vacuoles may be transferred by these means from maternal blood to foetal tissues without having to pass through the 'cytoplasm' of the endothelial cells. The basement membrane of these endothelial cells varies a great deal in thickness, being absent in some places so that the plasma membrane of the maternal endothelial cells may be in direct contact with the cells of the foetal trophoblastic villi. The irregular basement membrane may represent erosion by the trophoblastic villi.

In the human placenta (fig. 240) the trophoblast consists of an inner layer of cells (Langhans cells) lying next to the foetal connective tissue and blood vessels and an outer syncitial layer which is bathed by the maternal blood. The Langhans cells become flattened and many disappear as pregnancy advances. The cytoplasm of the syncytium is foamy and many ovoid nuclei are scattered through it. The foaminess of the cytoplasm is due to the presence of many vacuoles, each with a granular outline, the granules probably being of R.N.A. Synthesis of foetal protein may occur at these sites until the foetal liver takes over this role.

In addition to these smaller vesicles there are larger vacuoles, usually near to the outer surface of the syncytium under the microvilli which occur here. These vacuoles may arise by invaginations of the surface membrane of the cell and may be concerned with the 'absorption' of maternal plasma into the cell by the process of pinocytosis. The large number of microvilli increases the surface area of the cell for absorption or secretion.

In the rabbit, guinea pig and rat there is even more intimate contact since the epithelium and mesenchymal tissues of the chorionic villi disappear leaving only the endothelium of the trophoblastic capillaries separating maternal and foetal blood, producing the haemo-endothelial placenta.

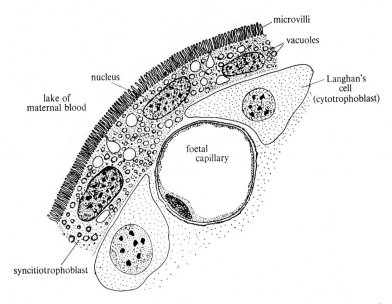

Fig. 240. *Diagram of a small area of human placenta made from electron micrographs. Note the microvilli of the syncytio-trophoblast projecting into a pool of maternal blood. Some of the vacuoles of the syncitio-trophoblast are shown opening between the bases of the microvilli.*

Summary of types of placenta

Placental type	Epithelio-chorial	Syndesmo-chorial	Endothelio-chorial	Haemo-chorial	Haemo-Endothelial
Maternal tissues					
Endothelium . .	+	+	+	−	−
Connective tissue .	+	+	−	−	−
Epithelium . .	+	−	−	−	−
Foetal tissues					
Chorionic epithelium	+	+	+	+	−
Mesenchyme . .	+	+	+	+	−
Endothelium . .	+	+	+	+	+
Examples . .	Horse Pig	Cattle Sheep	Dog Cat	Insectivores Man, Lower Rodents	Rat Rabbit

+ indicates presence
− indicates absence

Type of placenta, efficiency and length of gestation period. In the ungulates with epithelio-chorial or syndesmochorial placentae the barrier between maternal and foetal circulations is much greater than in the other placental types. Nutrition of the foetus in ungulates is assisted by secretions of the uterine glands called uterine milk. It is difficult to correlate the efficiency of the placenta with the type of placenta; even in one species the placental structure is not constant throughout development, and in some species e.g. rat, there are additional structures to the chorio-allantoic placenta in the form of the yolk sac placenta.

However there is a general tendency for animals possessing a more highly developed placenta (i.e. fewer layers) to have a shorter gestation period, although the young are born in a helpless condition. This feature is illustrated in the following list of gestation periods:

Mouse. gestation period 19 days. Haemoendothelial placenta. Young very immature at birth.

Rat. gestation period 21 days. Haemoendothelial placenta. Young helpless and blind.

Dog. gestation period 63 days. Endothelio-chorial placenta. Young are helpless and blind.

Cat. gestation period 65 days. Endothelio-chorial placenta. Young are helpless and blind.

In the above examples the young are in a very similar condition at birth, although the gestation period of the cat and dog is three times as long as that of the mouse and rat. This difference is explained on the grounds of a more efficient placenta in rats and mice.

Guinea Pig. gestation period approx. 68 days. The young are born active with eyes open. There is an accessory yolk sac placenta throughout the gestation period. The additional period of gestation compared to the mouse and rat is used to further the development of the young.

Pig. gestation period 119 days. Epitheliochorial placenta. This is a primitive type of placenta and a long gestation period is necessary.

Thus it is seen that the more primitive type of placenta is associated with a longer gestation period, the guinea pig being an exception to this general rule perhaps because the young are so well developed by the time they are born.

PHYSIOLOGY OF THE FOETUS

The placenta and metabolism of the foetus

Through the placenta the foetus obtains its nourishment; water, carbohydrates, proteins, fats, mineral salts, vitamins and oxygen, and through the placenta pass the waste products of the foetus, including carbon-dioxide and urea, to be excreted eventually by the lungs and

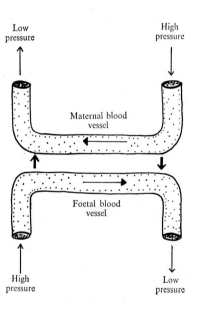

Fig. 241. *Diagram to illustrate the principle of the arrangement of blood vessels in the placenta. The two short thick arrows indicate the direction of the exchange of materials.*

kidneys of the mother. The blood vessels of the foetus and mother do not join one another and all these substances must diffuse across the barrier between the two sets of blood vessels. There are three devices in the placenta which promote the passage of materials across the barrier between the two sets of vessels. We have already seen two of these devices: first the large surface area of the placenta provided by the branching villi and secondly the reduction in the thickness of the placental barrier which varies from species to species. The third device consists in the arrangement of maternal and foetal blood vessels so that the maternal blood vessels containing blood at relatively high pressure are first adjacent to foetal vessels containing blood at low pressures, whilst foetal blood vessels containing blood at relatively high pressure lie adjacent to maternal vessels containing blood at low pressure; this arrangement encourages an efficient transfer of substances between the two series of blood vessels see fig. 241.

Respiration of the foetus. The need for an increasing supply of oxygen by the foetus is somewhat anticipated by the maternal part of the placenta, in that the maternal vascular bed in the uterus grows more rapidly than the foetal contribution to the blood vessels of the placenta. Thus at an early stage in gestation the maternal blood leaving the placenta still contains large amounts of oxygen. As the gestation period progresses the maternal blood leaving the placenta becomes progressively more de-oxygenated; the placenta reaches its maximum size at a time when the foetus is still growing rapidly, and at the end of the gestation period the supply of oxygen is critical.

The foetus is adapted in two main ways to obtain oxygen efficiently from the placenta. First there is the special arrangement of foetal blood vessels which has already been described above. Secondly, in many species foetal haemoglobin has somewhat different properties from adult haemoglobin, in that it takes up oxygen with a greater avidity and gives it up less readily. Foetal haemoglobin can take up oxygen at partial pressures at which maternal haemoglobin would give up oxygen. The foetus is a relatively inactive creature and can exist with haemoglobin which retains its oxygen more avidly than does adult haemoglobin. But once the animal is born it may need to use oxygen as rapidly as does an active adult, particularly in those species such as the horse where the young are very active at birth. It has been found that in some animals, e.g. the goat, there is a gradual alteration of the properties of foetal haemoglobin produced towards the end of the gestation period to approximately those of the adult, so that the animal will be better adapted to an active life breathing air.

Carbohydrate metabolism. As in the adult, the foetus uses glucose as an important source of energy. Glucose can pass across the placenta from mother to foetus in all animals. There may be persistent differences between the blood sugar level of the foetus and mother which may imply that the transfer of glucose across the placental membranes is not merely one of diffusion. In most mammals the foetal blood sugar tends to be lower than that of the mother but in some epithelio-chorial placentae, e.g. pig, the blood sugar is higher on the foetal side.

In early foetal life the maternal side of the placenta acts as a temporary liver for the foetus in that it holds stores of glycogen at a time when the foetal liver contains little or no glycogen. Later in the gestation period when the foetal liver has developed stores of glycogen, less is stored in the placenta. From experiments it has been found that the foetus exercises a strict glycogen economy, drawing upon glycogen stores only in emergencies.

Lipid metabolism. The foetus usually has good stores of fat. This may come from several sources. It may be able to synthesize fat from carbohydrates and amino acids. Some lipids pass across the placenta and become available for the synthesis of fat, but the mechanism of transfer across the placental membranes is not understood. Most of the fat fed to the mother and stored in her tissues does not pass unchanged across the placental barrier. If the food fat is stained with the dye Sudan III then the mother's stores of fat are intensely stained red, although none appears in the foetus. Further evidence which indicates that fat is not transferred directly across the placental membranes is the different chemical constitution of foetal and maternal fat.

Protein metabolism. Foetal protein may be derived directly by transfer of intact protein molecules from the mother or by transfer of amino acids. Antibody proteins (gamma globulins) are known to be transferred intact from mother to foetus in some species (e.g. rabbit) and this confers on the foetus a passive type of immunity. In other species much or all of the antibody proteins are received by way of the colostrum, the first pale secretion of the mammary glands of the mother. Amino acids probably do not pass across the placenta by simple diffusion and active transport mechanism may be involved.

Endocrine function of the placenta

In addition to serving the nutrition of the foetus the placenta is an endocrine organ producing several hormones in large quantities. Gonadotrophin is produced from the chorion and is called chorionic gonadotrophin (see also p. 541). In the human this hormone is produced in such large quantities by the end of the second month of pregnancy that a specimen of urine, in which the hormone is excreted, will cause the growth of the ovaries and ovulation when injected into a sexually immature mouse. Modifications of this effect are used in the early diagnosis of pregnancy. When the urine is injected into a female toad called Xenopus, eggs are shed into the surrounding water within 24 hours. Even mature male amphibians have been used in pregnancy tests and injections of chorionic gonadotrophin are followed by a release of sperms, which have to be identified in the urine microscopically. In addition to gonadotrophin the placenta also produces large amounts of oestrogen and progesterone. The exact significance of these various hormones is not well understood but they undoubtedly play their part in the development and maintenance of the sex organs during the pregnancy and prepare the mammary glands for their function after the birth of the young. The high local concentration of progesterone at the placental site, by increasing the membrane potential of the uterine smooth muscle cells (p. 620) renders them less excitable. Thus the uterus is mechanically inactive and the foetus is safeguarded.

The circulation of the foetus and modifications at birth

The circulation is adapted to life in utero, in which the placenta is the organ of nutrition and respiration; the foetal lungs and gastro-intestinal tract do not function as the respiratory and nutritive organs in utero and their blood supply is correspondingly small. But immediately at birth the placenta ceases to have these functions and there must be a rapid adjustment of the organism to an independent life. The lungs are rapidly converted from semi-solid organs with a small blood supply into air filled organs with a large blood supply and are responsible for the gaseous exchange of the whole organism. The gastro-intestinal tract takes over the nutritive functions of the placenta as the young animal begins to feed. In order to make these transformations possible there must be special devices within the cardio-vascular system of the foetus so that blood which once went to the placenta can now be diverted to the lungs and gastro-intestinal tract.

The general pattern of the circulation of the foetus is shown in fig. 242A. Foetal oxygenated blood is returned from the placenta by way of the umbilical vein which passes to the liver where it joins the hepatic portal vein. Some of the oxygenated blood goes to the liver but most of it is shunted away through a connection of the umbilical vein with the inferior vena cava called the ductus venosus, a feature only of the foetal circulation. In the inferior vena cava the oxygenated blood mixes with venous blood draining from the lower part of the body. When this blood reaches the heart most of it is shunted from the right auricle through an opening in the septum between the two auricles, into the left auricle. The opening in the septum, called the foramen ovale, is a special feature of the foetal heart and it ensures that the oxygenated blood returning from the placenta is diverted away from the right ventricle and lungs to supply the head (i.e. central nervous system) and upper limbs of the foetus by way of the left ventricle and aorta. This flow of blood has been confirmed using X-ray studies of the foetus after radio-opaque material has been injected into the circulation. The blood is diverted from the right auricle into the left auricle by a valvular arrangement in the wall of the right auricle.

The venous blood which drains into the right auricle from the head and neck of the foetus by way of the superior vena cava passes through the right auricle into the right ventricle, from which it passes out into the pulmonary artery. But since the foetal lungs are not functioning as organs of gaseous exchange, only a small amount of blood is needed, sufficient to meet the metabolic needs of the tissues in the lung and much of the blood in the pulmonary artery passes directly into the aorta by way of a connection called the ductus arteriosus. This blood supplies the lower part of the body, and much of it passes into the umbilical arteries (branches of the internal iliac arteries) supplying the placenta with deoxygenated blood.

There are thus four special features of the foetal circulation:

1. The placental circulation.

2. The ductus venosus which shunts blood from the umbilical vein away from the liver into the inferior vena cava.

3. The foramen ovale through which oxygenated blood returning to to the heart via the inferior vena cava passes into the left auricle and so to supply the upper part of the body.

4. The ductus arteriosus which shunts blood from the pulmonary arch into the aorta, so by-passing the lungs.

It will be seen that the right ventricle of the foetus, pumping blood to the lungs and to the lower part of the body and placenta, performs more work than the left ventricle. This position is reversed after birth.

At birth there are dramatic changes in this circulation. The umbilical vessels become increasingly irritable towards the end of the gestation period and with the physical stimuli of birth these vessels go into spasm, thus excluding the placental circulation from the foetus. This means that when the mother bites through the umbilical cord to free the young animal it does not bleed to death. The ductus venosus also goes into spasm so that blood in the hepatic portal vein now has to pass through the tissues of the liver and cannot be directly diverted into the inferior vena cava. With the expansion of the lungs at birth larger amounts of blood pass into the lungs from the pulmonary arch, and the wall of the ductus arteriosus contracts so that blood in the pulmonary arch can no longer be diverted into the aorta. With the increased supply of blood to the lungs there is an increasing amount of blood returning to the left auricle and the pressure of blood in the left auricle rises. This rise in pressure pushes a loose flap of tissue against the foramen ovale, closing the connection between the left and right auricles. The adult type of circulation is now achieved. The ductus arteriosus, ductus venosus and umbilical vessels, initially closed by muscular spasm are gradually permanently obliterated as fibrous tissue grows in the

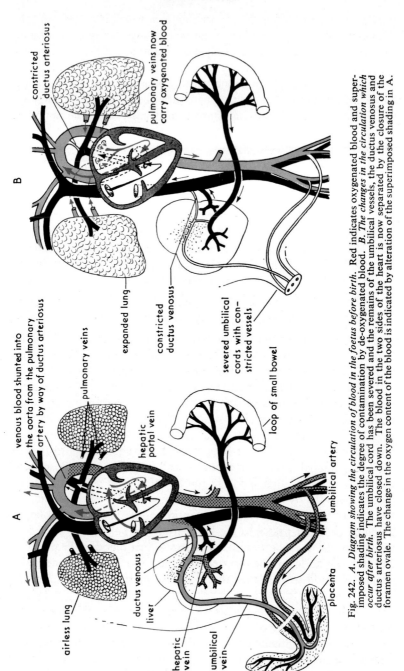

Fig. 242. *A. Diagram showing the circulation of blood in the foetus before birth.* Red indicates oxygenated blood and super-imposed shading indicates the degree of contamination by de-oxygenated blood. *B. The changes in the circulation which occur after birth.* The umbilical cord has been severed and the remains of the umbilical vessels, the ductus venosus and ductus arteriosus have closed down. The blood in the two sides of the heart is now separated by the closure of the foramen ovale. The change in the oxygen content of the blood is indicated by alteration of the superimposed shading in A.

lumen of the vessel, and the flap of tissue which is closing the foramen ovale gradually fuses with the septum (fig. 241B).

Parturition. At the end of the gestation period parturition occurs when the young are expelled from the uterus into the outside world by means of powerful intermittent contractions of the myometrium. What initiates the process of parturition is not known but it has been suggested that the declining production of the hormone progesterone, which has an inhibitory effect on uterine motility, sensitizes the uterus to the posterior pituitary hormone oxytocin, which stimulates intermittent powerful contractions of the uterus. In cases of slow labour associated with weak uterine contractions, injections of oxytocin are used to produce more powerful uterine contractions.

In those species in which there is an intimate mingling of foetal and maternal tissues in the placenta the uterine contractions not only propel the young through the birth canal but they serve to separate the placenta from the uterine wall, and after birth of the young the persistent contraction of the uterine muscles serve to close the blood vessels which have been torn open during the separation of the placenta.

Lactation. The mammary glands have been prepared for their function during the pregnancy by the effect of the sex hormone produced by the placenta. As mentioned previously, oestrogen causes a growth of the duct system of the glands and progesterone initiates the development of the glandular elements around the ducts. But the glands are unable to produce milk during pregnancy because the large amounts of oestrogen inhibit the pituitary gland from producing a hormone vital for lactation, the lactogenic hormone. Towards the end of pregnancy the oestrogen production by the placenta gradually falls, and after parturition, when the placenta is expelled, the level of oestrogen in the body falls still further. Lactogenic hormone is now produced by the anterior pituitary gland and lactation starts soon after parturition. The first pale secretions called colostrum are rich in proteins, and in some species they contain antibodies, which serve to protect the young for some months until they have been broken down by the body (see p. 581).

During the lactation period milk is secreted by the glandular cells of the mammary gland continuously. Milk, however, only passes along the larger ducts that open onto the nipple during suckling. When the infant mammal grasps the nipple in its mouth stimulation of sense organs in the nipple cause a reflex liberation of the hormone oxytocin from the posterior pituitary gland. This hormone reaches the mammary gland by way of the blood stream. On reaching the mammary gland the hormone stimulates certain contractile cells surrounding the glandular cells causing an emptying of milk into the larger ducts which open onto the nipple. Milk now appears from pores on the nipple and the young, by compressing the dilated ducts which underly the nipple, increase the flow of milk.

Chapter Fourteen

Immunity

Introduction. The resistance of the body to invasion by disease-causing organisms is a highly complex system. Indeed, it is composed of a whole hierarchy of systems ranging from primitive reactions that are non-specific and are mobilized on any 'irritation' of tissues (be it physical, chemical or biological), to highly specific reactions which operate against particular chemical substances of micro-organisms. A decription of these specific reactions is complicated by the fact that the same reactions are involved in body processes quite unrelated to defence against infection. They are concerned with what has come to be called 'maintaining the integrity of the body', with the 'weeding out' of abnormal cell types that continually arise in the highly complex body of the mammal. We thus find that the same mechanisms that are involved in resistance to infection by micro-organisms also become mobilized when 'cancer' cells arise in tissues or when surgeons transplant into the body of one individual an organ or tissue removed from the body of another individual. These specific defence systems have enormous implications for medicine and surgery, and research is progressing at a fantastic pace to probe their mechanisms and to develop new ways of *promoting* the defences and so increase resistance to infection or to the abnormal cell types of a malignant growth (cancer), and new ways of *blocking* these defences so as to permit organ or tissue transplants. In this field of study there has been a virtual explosion of new information in recent years, so much so that in a book of this sort it is impossible to review the subject comprehensively.

External defence mechanisms

We will first look at the barrier that surrounds the body and then at the reactions that occur when this barrier becomes breached by the action of physical agents (cuts, heat and cold injury etc.), chemicals, or by invading micro-organisms.

The organs of the body are surrounded by various kinds of tissue such as skin or mucous membranes that act not only as a limiting surface but also as a barrier against various agents in the external world, be they animate or inanimate. Let us first look at skin which surrounds the entire body. An important property of skin is its mechanical strength conferred by many layers of epithelial cells and a tough outer layer of dead, keratinized (horny) cells. Intact skin is not only virtually impervious to penetration by micro-organisms but it also has its own mechanisms for decontamination. These mechanisms are both biological (the effect of an established flora of harmless commensal bacteria) and chemical (various chemicals in secretions of the skin).

The respiratory tract is continuously exposed to large amounts of foreign material—gases (fumes from cigarettes, cars, open fires, industrial fumes etc.), dust, and micro-organisms occurring in dust or in minute fluid droplets. The nose is the first line of defence and it may trap as much as 90 per cent of inhaled particles. Most of the particles are trapped by hairs in the nose and by a moist blanket of mucus which covers all the internal surfaces of the nose and contains a high concentration of the antibacterial enzyme called lysozyme, as well as other antibacterial substances. This blanket of mucus is in continuous movement, propelled by the action of minute cilia on the cells of the nasal epithelium. The cilia drive the mucus blanket in two directions, downwards to the external openings of the nose and backwards to the pharynx, where the mucus is swallowed. There is a similar mucus blanket in the trachea and bronchi; here the cilia transport the mucus and trapped particles to the larynx where it accumulates until it is coughed into the pharynx and swallowed. Only the smallest of particles reach the end structures of the air passages—the alveoli. Here they meet a further defence mechanism in the form of cells called alveolar macrophages that engulf the particles and kill them if they are bacteria.

The mouth, pharynx, alimentary tract, vagina, and the outer surface of the eye have their own defence mechanisms. Here we can mention the cleansing action of saliva, the effect of stomach acid on swallowed micro-organisms, antibodies secreted into the intestine, the cleansing action of tears which also destroy or inhibit bacteria by means of chemical substances such as the enzyme lysozyme and antibodies, and the acid conditions of the adult vagina which inhibits the growth of many pathogens.

Internal defence mechanisms

Inflammation

The ability of living tissue to react to injury is a very fundamental property. In mammals this is a complex reaction which may begin as a local reaction to damage and then extend to involve many systems of the body. Let us look first at the local reaction of a tissue to injury. This reaction is called inflammation and it continues as long as tissue damage continues. When the local injury stops then the debris of the inflammatory reaction is removed by scavenger cells and, unless there has been much loss of tissue, the tissue returns to normal. We are all familiar with the appearance of inflamed tissue, for our skin from time to time becomes cut, scratched, burned, over-exposed to sunlight or is infected by pathogenic bacteria. Whatever the cause of the damage to the skin, the reaction is similar; the skin becomes reddened, warm and painful, and if the inflammation is more severe then swelling appears. A very basic change in inflamed tissues occurs in the blood vessels. You can readily observe the effect of these changes in blood vessels in your own skin. Expose the delicate skin of the inner side of your fore-arm and then take a fine blunt instrument, such as the head of a pin or needle or even a finger nail, and draw a line a couple of inches in length on the skin. In making this first line use little pressure. You will notice a white reaction along the line of pressure. This white line will increase in intensity for a few seconds and then fade over the next few minutes; it is due to constriction of the capillaries that follows direct stimulation by the instrument. This reaction is not very relevant to the process of inflammation, so now apply your instrument more firmly to the skin. After this strong stimulation you may or may not see a white line, but if it does occur then it is rapidly replaced by a red line. The intensity and duration of this red line depends on the amount of pressure that you applied to the skin. The reaction is due to dilatation —opening up—of the capillaries, so that there is more blood in the area. If you now apply an even stronger pressure or repeat the last pressure often enough, then a bright red flush will spread outwards from the red line. This red flare is due to dilatation of arterioles. If the stimulus is even stronger, the red line will become pale and raised above the surface of the surrounding skin. This pale swollen area of skin is called a wheal, and it is caused by the escape of fluid from the capillaries into the tissue spaces around the capillaries. The accumulation of fluid in tissues is known as oedema. After a while the wheal loses its sharpness by becoming wider and less raised, and finally it disappears altogether as the fluid passes back into the capillaries or into the lymph channels of the tissue.

This simple procedure illustrates two important changes in inflamed tissues—changes in blood vessels and the accumulation of fluid in the tissue spaces. Let us now look at these changes in a little more detail, beginning with the vascular changes. The dilatation of blood vessels that follows the initial constriction persists for the duration of the inflammation, and it affects all of the small blood vessels in the tissue, arterioles, capillaries and venules. The dilatation of the arterioles which supply the tissue with blood results in an increased flow of blood to the tissue and a rise in temperature (this rise in temperature is a feature of inflammation near the cooler body surface and is not a feature of internal inflammation). In spite of an increased supply of blood to inflamed tissue, the dilatation of capillaries results in a slowing of the flow of blood. This produces various effects that include an increase in the 'leakiness' of the capillary wall so that the fluid part of blood (plasma) oozes out to accumulate in the spaces in the tissue around the capillaries. There are also changes in the cells that line the capillary —the endothelial cells; these cells become swollen, and white blood cells and platelets stick to the swollen cells. White blood cells now begin to migrate through the capillary wall, squeezing their way between the endothelial cells, to accumulate in the tissues.

The tissue spaces of an inflamed tissue thus come to be filled with a fluid similar to blood plasma and packed with white blood cells of various sorts, particularly neutrophil polymorphonuclear leukocytes— we will call these polymorphs for brevity—and monocytes. This cell-loaded fluid, the exudate, fulfils a variety of functions. It carries with it any antibacterial substances that are present in the blood, either naturally occurring ones such as specific antibodies or any drugs or antibiotics that have been given to the individual. The fluid of the exudate also dilutes any irritant that has been introduced into the tissue, whatever its nature. The exudate also contains the blood protein called fibrinogen, which when converted to fibrin becomes the basis of a clot of blood. This fibrin is often transformed into a clot of fibrin in the tissue spaces. The presence of fibrin clots in the tissue spaces help to unite severed tissues, as in a cut, and they may temporarily act as a barrier against invasion by bacteria. Fibrin clots may also form in the lymph channels of the inflamed tissue; there they obstruct the flow of fluid from the tissues into the lymph stream and so help to prevent the distribution of pathogenic micro-organisms from an infected inflamed tissue.

The accumulation of white blood cells of various sorts in an inflamed area is an important component of the defence system. One important action of the white blood cells is the engulfing of bacteria, cell débris and the like into the cytoplasm, a process called phagocytosis; once inside the cytoplasm the material may be digested and destroyed by means of enzymes. The polymorphs are among the first cells to begin this scavenging activity. Later the monocytes that enter the inflamed area become transformed into cells that can carry out phago-cytosis. The scavenger cells that can engulf material into their cytoplasm are called macrophages.

The polymorphs are attracted to any area where there is cell damage, whatever its cause. There are about 25 billion of these cells in the blood of a normal adult, and about a similar number are attached to the lining of blood vessels in the process of moving out into the tissue spaces, where they are eventually destroyed by other macrophages which are permanent residents in connective tissues. For every polymorph in the circulating blood there are about a hundred in the bone marrow so that there are obviously enormous reserves of this defensive cell. The ability of polymorphs to ingest micro-organisms is greatly enhanced by the presence in the plasma of certain antibodies which become bound to the surface of the organisms. The kind of antibodies that have this effect are given the special name of opsonins.

If the polymorphs cannot ingest and kill the pathogenic micro-organisms in an inflamed area, then a second line of phagocytic defence is provided by the other macrophages which can carry on a prolonged battle. If a local invasion of pathogens cannot be arrested by these mechanisms then the general defence mechanisms of the body are activated; but more of these later. If a local invasion has been arrested by the inflammatory process, the rate at which the tissue returns to normal depends upon the degree of tissue damage. When tissue damage has been slight, the exudate is absorbed back into the blood stream and any fibrin clots are broken down. The blood flow to the area decreases and the tissue resumes its normal appearance. If there has been much tissue damage, this gradual return to normal is not possible. The dead tissues are softened by means of proteolytic (i.e. protein-digesting) enzymes which are released from the dying bodies of polymorphs that have accumulated in the region. The fluid that results from this action is called pus and is contained within a cavity called an abscess. As pus accumulates, the pressure inside the abscess rises and the pus then begins to track along a line of least resistance until a free surface (external or internal) is reached. Now the abscess bursts and discharges its fluid-contents—with a dramatic relief of pain—unless this has been anticipated by a surgeon and the abscess drained by cutting into its wall. Sometimes an abscess does not rupture but remains embedded in the tissue; its contents thicken to a porridge-like consistency as fluid is absorbed, and the walls of the abscess becomes thickened with fibrous tissue.

The cause of inflammation

We have seen that there are dramatic changes in the blood supply and in the character of the blood vessels in an inflamed area. How are these brought about? It is thought that various chemical substances appear in the damaged tissue to trigger off the changes in the blood vessels by means of a direct action on the vessels. Many substances have been isolated from inflamed tissues that have effects on blood vessels, but there is still no general agreement as to which of these substances are normally responsible for triggering off the vascular changes. Some of the substances such as histamine and 5-hydroxytryptamine are pre-formed materials that are stored within cells that occur in most tissues. These chemicals can be released by rupture of the storage cells (see mast cells in fig. 243) when the tissue is damaged. Other chemicals are not stored in tissues but circulate in the blood stream in an inactive form. They can be converted into their active form very readily by many factors that can disturb the normal equilibrium of plasma, and it is thought that these active substances appear in inflamed tissues. These chemicals are called the plasma kinins and

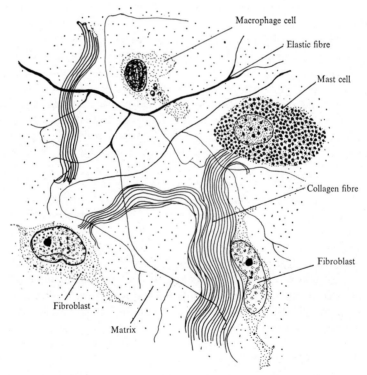

Fig. 243. *Some of the components of loose (areolar) connective tissue as seen at high magnification. The mast-cells are pharmacological 'time-bombs'; when they rupture they release powerful chemicals that can affect various tissues nearby, such as blood vessels or smooth muscle.*

extraordinarily small amounts of them can produce effects on blood vessels.

Antigens

The defence mechanisms that we have so far considered—i.e. external 'barriers' and inflammation—are rather non-specific mechanisms in that they can cope with a variety of damaging agents, physical, chemical or biological (i.e. micro-organisms). These are, in terms of evolution, rather primitive mechanisms. We now come to much more sophisticated systems that have become perfected in the evolution of mammals, systems that may have functions much wider than defence against invasion by pathogenic micro-organisms. Animals low down in the evolutionary scale—say insects or crustacea—can tolerate exchanges of tissue between one individual and another. The ability of an organism to *recognize* a transplanted tissue as being different from its own tissues and then to take steps to *reject* this foreign material from its body, appears higher in the evolutionary scale; it comes with the evolution of a cell type called the lymphocyte.

An important feature of these defence mechanisms is that they are specific. This means that the body can develop a mechanism which can protect it from a particular foreign material, be it a particular strain of bacteria, viruses, fungi or the cells of a particular species of animal. Thus the defence mechanism which can protect the body against the smallpox virus are quite ineffective against any other virus, or indeed any other micro-organism. We give the name *antigen* to any substance that can alter the properties of certain body cells so that whenever the antigen is re-introduced into the body these cells can in one way or another 'neutralize' the antigen. We can now look at the nature of antigens, the kind of body cells that respond to them, and at the nature of their response.

The nature of antigens

Antigens are large organic molecules. It was once thought that only proteins were antigenic. Certainly proteins make good antigens, i.e. they provoke marked responses in an animal, but we now know that some carbohydrates are also antigenic. In order that a substance can act as an antigen in an animal, it must be sufficiently different from the body's own chemical substances, i.e. it must be foreign to the animal. If we take a sample of blood from an animal and extract a protein, say plasma albumen, and inject it back into the animal, there will be no disturbance, no response by the immune systems of the body. Similarly if we remove a piece of skin from an animal and sew it back in an abnormal position, the graft will 'take' and the immune systems of the body will not be disturbed. If we repeat these studies using plasma albumen or a skin graft from another animal of a different species then the immune systems are activated and various changes occur in the circulating blood and in the properties of certain cells. The kind of response of the immune system is rather different in the two examples. In the case of the experiment with plasma albumen certain proteins called antibodies appear in the circulating blood, antibodies that unite specifically with the foreign albumen. In the case of the skin graft, lymphocytes with special properties accumulate in and around the graft of 'foreign' skin and ultimately lead to the death of the graft, i.e. its rejection from the body. We thus have two distinct mechanisms in immunity, one in the form of specific antibodies in the circulating blood, the other in the form of various populations of immune lymphocytes, each population of which is generated by the introduction into the body of a particular antigen; both antibodies and immune lymphocytes are capable of reacting with the antigens that caused their production.

The reaction of antigen with cells

There are two main groups of body cells that are involved in the immune responses described above. First we have cells that have the ability to 'recognize' antigens. These cells are called antigen-sensitive or immunocompetent cells. When they come into contact with antigen, one of two reactions occurs. First, the cells can multiply and produce cells that mature into special 'factories' for the production of specific antibodies. These antibodies are liberated into the circulating blood and appear in that fraction of the plasma proteins that are called globulins. A second reaction is that the cells multiply to produce lymphocytes that have special properties; these cells carry the 'memory' of previous contact with the antigen and can react with it whenever they meet it. Thus we have two types of immune responses:

1. humoral—mediated by circulating antibodies, produced by plaxma cells.
2. cellular—mediated by 'sensitized' lymphocytes.

There are various other groups of cells that play an important accessory role in immunity. These cells are various kinds of macrophages that can engulf foreign material (antigens). They may travel far in the body, 'processing' antigen and presenting it to the antigen-sensitive cells.

The thymus and immunity

Until a few years ago the function of the thymus, a two lobed structure that lies in the upper part of the chest overlying the heart and great vessels, was a complete mystery. This gland is very prominent at birth but it begins to shrink in size in the early years of life so that only traces remain in the adult (see fig. 244). For years the organ had been removed from experimental animals and from human beings during the

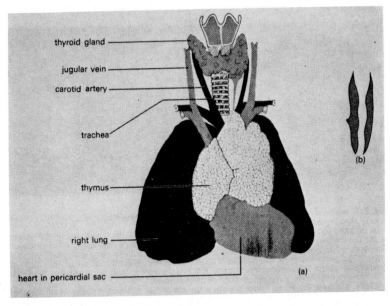

thyroid gland

jugular vein

carotid artery

trachea

thymus

right lung

heart in pericardial sac

(a)

(b)

Fig. 244. (a) *Shows the position and size of the thymus in an infant.* (b) *Shows the relative size of the gland taken from an adult aged 30 years.* (*From 'Man Against Disease', Clegg and Clegg, Heinemann Educational Books.*)

course of operations of the chest, without causing any ill effects. The reason for our long-continued ignorance of the functions of the thymus is that the gland begins its functions during the embryonic period and these functions are completed fairly soon after birth. Only when in 1961 the thymus was sucked out from the minute bodies of anaesthetized newborn mice did the vital function of this organ become apparent. Although the mice developed normally for a few months, they eventually died. These mice had a deficiency of lymphocytes in the circulating blood and also a deficiency of lymphocytes in the lymphatic tissues of the body, such as the lymph modes, spleen and bowel wall. There were also defects in the immune responses of the mice, the most dramatic of

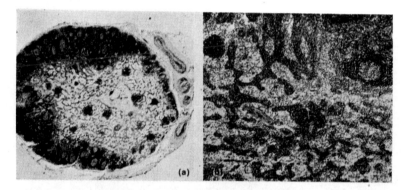

Fig. 245. (a) *Low power section of a lymph node.* (b) *High power view of a part of a lymph node showing a few cortical nodules with germinal centres.*

In (a) note the dense outer rim of the node, the cortex, and the loosely arranged tissues of the core of the gland, the medulla. In the cortex are round structures called nodules. The dark outer rim of each nodule consists of masses of densely arranged small lymphocytes. The paler centre of the nodule is called the germinal centre and it is here that the production of new lymphocytes occurs. Each germinal centre is perhaps a clone of cells developing from a single stimulated immunocyte.

There is a large traffic of lymphocytes, into and out of the lymph node, small lymphocytes entering the gland from the blood stream, and other lymphocytes leaving the node via the lymph vessels. Antigens also reach the lymph nodes and are taken up by phagocytic cells lining the various lymph spaces within the node. Here they stimulate some lymphocytes to multiply and transform into plasma cells which manufacture antibody; the antibody then leaves the node to eventually reach the circulating blood. It is in the loose tissues of the medulla that the plasma cells tend to be found. (*Photographs by Brian Bracegirdle, B.Sc., M.I.Biol., F.R.P.S. (From 'Man Against Disease', Clegg and Clegg, Heinemann Educational Books.)*

which was the inability of the mice to reject transplants of the skin taken from other mice, or even from rats! These mice were thus incapable of being able to recognize rat skin as being foreign.

These experiments were the fore-runners of very many studies aimed at elucidating the role of the thymus in immunity. It seems that during embryonic life, and for a period after birth, the thymus becomes populated with what are called stem-cells; these come from the bone marrow. The stem-cells are undifferentiated in form or function but under the influence of the thymus the cells become transformed into immunocytes, a term that we use to describe any cell that is involved in the immune response. Many of the cells mature into lymphocytes, some of which leave the thymus to seed the lymphatic organs of the body—the lymph nodes (fig. 245), spleen and intestinal wall. In these situations the lymphocytes continue to divide. In addition to seeding the lymphoid tissues, the thymus produces some chemical substance that enters the circulating blood to stimulate the multiplication of lymphocytes. This hormone also alters the properties of lymphocytes so that they are able to react with antigens.

The thymus is thus a vital structure in the production of antigen-sensitive cells. An important property of these cells is that they do not react with the body's own proteins and other large molecules that would certainly act as antigens if they were injected into another person. In the language of modern immunology, this is the ability of the body to distinguish 'self' from 'non-self'. In these processes the thymus plays a vital role. In order to understand the role of the thymus in preventing lymphocytes from responding to the proteins of the body as though they were antigens—a response that would eventually lead to self-destruction—we have to take a brief look at theories of immunity.

A classical view is called the 'instructive theory' which assumes that an immunocyte is incapable of producing antibody until it has been in contact with antigen. According to this view, when an antigen is taken up by an immunocyte the antigen acts as a framework on which the antibody is built. Thus the antigen and antibody unite like a lock and key, and a particular antibody can only get into close contact with one antigen (see fig. 246). The instructive theory of immunity described above is no longer acceptable to many immunologists who favour Burnet's theory of 'clonal-selection' or a modification of it. According to the instructive theory the immunocytes of the young mammal are unable to produce particular antibodies until the cells have taken up the antigen and, as it were, 'learned' to manufacture antibody using the antigen as a template on which antibody is produced. The clonal-selection theory takes an entirely opposite view and regards the body

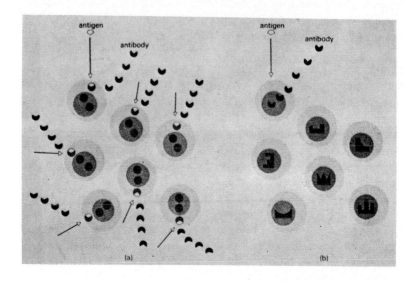

Fig. 246. *The two theories of immunity.*
(a) shows the classical instructive theory of immunity. Here all of the immunocytes are capable of response to contact with the antigen. Antigen enters all cells and acts as a template on which an antibody with a complementary structure is manufactured.
(b) shows the 'clonal selection' theory in which only one immunocyte is able to respond to the antigen; this ability is genetically determined. The remaining immunocytes are capable of producing other forms of antibody but they do not do so until they have been in contact with the appropriate antigen. (*From 'Man Against Disease', Clegg and Clegg, Heinemann Educational Books.*)

as being capable of producing antibodies against a vast variety of potential antigens, i.e. the ability to produce particular antibodies is genetically determined. Large numbers of small clones, or populations, of cells exist, each of which is capable of producing a particular kind of antibody. When an antigen is introduced into the body it reacts with a particular clone of cells, causing the cells to mulitply and transform themselves into the cells that manufacture antibody (plasma cells) or into cells that are the vehicle of cellular immunity. There is considerable support for this view, both experimental and theoretical.

A great advantage of this theory is that it goes a long way to explain the mechanism of self-recognition, as follows. Assume that in the body's population of lymphocytes there exists a vast number of small groups, or clones, of cells, each one capable of reacting with a particular antigen, including the body's own antigens. We know that the reaction of a lymphocyte to antigen depends greatly on the 'dose' of antigen. *Small* amounts of antigen—protein, nucleic acid etc.—such as would normally be liberated by bacteria or viruses during the course of infection, stimulate a clone of cells to proliferate to produce large numbers of cells each of which is either capable of reacting with the particular antigen which initiated the process, or capable of producing specific antibody which can combine with the antigen. We also know that a *large* dose of an antigen can have a damaging effect on the immune mechanisms, so much so that the body may tolerate the antigen and come to treat it as a 'self component. Let us look at one experimental example of this kind of effect using as antigen a pure protein, albumen, prepared from the blood of a cow (bovine serum albumen). No animal can produce antibody until it is a few weeks old, but after an interval the ability to produce antibody rapidly builds up. If we inject a rabbit at a few months old with a small dose of albumen, there is a fairly rapid production of antibody to the albumen. The antibody binds to the albumen and the latter begins to disappear rapidly from the blood of the rabbit. If we repeat this injection after a few months, the production of antibody is even more rapid and intense and the albumen rapidly disappears from the circulating blood. This is a fairly general reaction to antigens; a first contact with antigen primes the body's defences and a second contact with antigen produces a rapid and vigorous antibody response; this behaviour is made use of in the 'booster' doses of vaccines used to protect individuals against infectious diseases.

Now let us look at the effect of *large* doses of albumen on the rabbit. Let us give the first large dose of albumen just after birth when the animal has not yet fully developed the ability to distinguish 'self' from 'non-self'. The albumen produces no antibody response and the albumen is only slowly eliminated from the blood of the rabbit. We can repeat these injections of large amounts of albumen at short intervals of a few weeks; the rabbit continues to fail to respond by the production of specific antibody. However, if a long period, say six months, passes without an injection of albumen and then a dose of the antigen is administered, the animal will respond to the antigen by the production of specific antibody. Thus it seems that continuous and heavy contact with antigen can prevent the body from reacting to it. We have this same state of affairs with the body's own tissue antigens, i.e. they are present in continuous and heavy contact with immunocompetent cells.

What has all this to do with the thymus? It is thought that it is in the thymus that the clones of lymphocytes which are capable of reacting with the boyd's own proteins and nucleic acid are eliminated. In the environment of the thymus we see the expression of this heavy and continuous contact of body components with lymphocytes. A large number of lymphocytes are destroyed in the young thymus gland where the scavenging macrophages are constantly loaded with the dying and dead remains of lymphocytes. We presume that these lymphocytes are the ones that were capable of reacting with 'self'.

We can now summarize the functions of the thymus; this gland receives stem cells from the bone marrow and transforms them into lymphocytes. These lymphocytes form a very mixed population, each

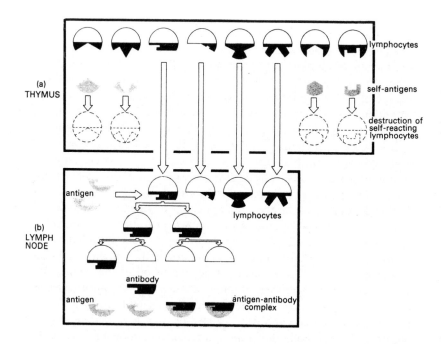

Fig. 247. (a) *A view of the function of the thymus. Large numbers of different populations of cells are present in the thymus (only eight of which are represented in the drawing) each of which is capable of reacting with a particular antigen. In the thymus these cells are exposed to 'self-components'. Those cells which are capable of reacting with these 'self-components', which are present in high concentration, are destroyed. The remaining cells leave the thymus to populate the lymphoid tissues.*

(b) *In the lymphoid tissues these cells are ready to meet and react with any non-self component (antigen) that reaches them. Shown here is a particular antigen stimulating a cell to produce antibody which is specific to that antigen. The cell multiplies and the offspring mature inot plasma cells that liberate antibody. The antibody unites with the antigen to form a complex, i.e. the antigen is neutralized. (From 'Man Against Disease', Clegg and Clegg, Heinemann Educational Books.)*

small group of which is capable of reacting only with one particular antigen. Within the thymus the lymphocytes are exposed to 'self components' and any of the cells that react with these components are destroyed, thus protecting the body from self-destruction. Only the 'safe' lymphocytes leave the thymus; they go and populate the various lymphoid organs of the body where they multiply and, under the influence of 'thymic hormone' in the blood, become transformed into populations of cells that carry out the immune reactions. The theory that accounts for the role of the thymus in eliminating 'self-reacting' components is illustrated diagrammatically in fig. 247.

Nature of immune reactions

Having looked at the origin of immunocompetent cells, we can now look in a little detail at the nature of the immune reactions of the body. We have already seen that these reactions are basically of two kinds, humoral (i.e. circulating antibodies) and cellular, and we can perhaps best begin by looking at what happens in individuals who lack one component of this two-pronged attack on antigens.

Agammaglobulinaemia

Agammaglobulinaemia is a rare disorder, some cases of which are inherited as a sex-linked recessive condition (the gene is carried on the X chromosome). The basic defect in this disease is an inability of lymphocytes to transform themselves into plasma cells, the cells which manufacture antibodies. Antibodies are mainly found in the fraction of plasma proteins called globulins—hence the name of the disease, agammaglobulinaemia. In these individuals there is no antibody production in response to the introduction of any antigen into the body. This defect shows itself by repeated attacks of infection, particularly in the respiratory tract—bronchitis, pneumonia etc. Before the intro- duction of antibiotics and other drugs to control infections, sufferers from agammaglobulinaemia used to die early in life, and the disease illustrates the importance of circulating antibodies in resistance to infection. On the other hand the thymus is normal in these individuals and they reject transplants of tissues from other individuals as well as a normal person. Moreover, when they suffer from an attack of measles the illness runs its normal course and they develop an immunity to a further attack of the disease. Obviously there are immune mechanisms other than circulating antibodies. These mechanisms are mediated by cells, principally lymphocytes.

Di-George's syndrome

This disease is the reverse of agammaglobulinaemia. These individuals have normal antibody responses to infections and other antigenic stimulation. The thymus is, however, absent and they show deficiencies in that part of the immune system which is carried out by means of cells. We can now look at these two mechanisms of immunity, circulating antibody and cell-mediated responses.

Antibody production

When an antigen (bacteria, virus, fungus, foreign proteins in sera, vaccines etc.) appears in a tissue, much of it is destroyed by the various phagocytic cells that appear during the inflammatory reaction. Some of the antigen, however, persists in macrophages, remaining there for weeks or months. This antigen is carried away from the scene of the invasion when the macrophages migrate to other tissues, including the spleen, lymph nodes and other lymphoid tissues. Here the antigen seems to persist even though the cells that engulfed it die. It is passed on to yet other macrophages. The persistence of antigen seems to be due to its combination with ribonucleic acid in the cell, a component which protects the antigen from destruction by the enzymes inside the cell. These macrophages are, however, not the cells that manufacture antibody. It seems that the mere contact of the antigen-loaded cells with lymphocytes triggers off multiplication of some of the lymphocytes and their transformation into plasma cells which manufacture antibody. There is no trace of antigen in these plasma cells, which is evidence against the instructive theory of immunity that we have already discussed (page 571). Unlike the lymphocytes with their thin peripheral rim of cytoplasm, the plasma cells carry all the cell machinery which is indicative of an ability for extensive protein synthesis, features such as a rich endoplasmic reticulum bearing ribosomes (fig. 248).

The following account of a fairly recent experiment will illustrate how the function of plasma cells can be demonstrated in a very direct fashion. The first step is to choose an antigen which is to be injected into an animal to stimulate antibody production. The antigen chosen is the protein extracted from the flagellae of Salmonella bacteria. The choice of this antigen is very deliberate, for the movements of the flagellae enable movement of the bacteria, and antibody to the protein of these structures will paralyse the flagellae and stop the movement of the bacteria. The immobilization of Salmonella bacteria can thus be used as a very sensitive test for the presence of specific antibody against them. The purified flagellar proteins are now injected into the skin of a rat. After sufficient time has elapsed for the animal to react to antigen (a matter of days), the lymph nodes that receive lymph draining from the skin injected with antigen are removed. The lymph nodes are teased

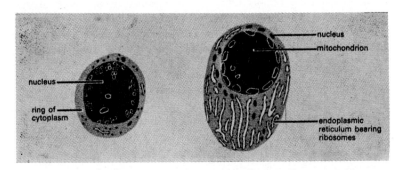

Fig. 248. (a) *a lymphocyte and* (b) *a plasma cell.* (*From 'Man Against Disease', Clegg and Clegg, Heinemann Educational Books.*)

out into individual cells using very fine instruments, the whole process being viewed under a powerful microscope. Using this process it is possible to isolate individual living cells, lymphocytes, macrophages and plasma cells. A single cell can be taken up into a micro-pipette and dropped into a tiny drop of culture fluid hanging onto the under-side of a thin strip of glass which is supported on a glass slide. The edges of the slip are now sealed with an oil to prevent the drop of fluid from drying out. The slide bearing its single cell floating in a drop of culture medium can now be incubated at 37°C and left for a time to allow it to manufacture antibody, if it is capable of this synthesis. The presence of antibody released into the culture medium can now be tested by injecting a few motile Salmonella bacteria into the drop and observing their behaviour. It is found that only slides containing plasma cells are capable of immobilizing the bacteria, and then only the exact strain of Salmonella that was injected into the rat at the beginning of the experiment. Cultures of lymphocytes or macrophages have no ability to immobilize the bacteria (see fig. 249).

This sort of experiment demonstrates in a very elegant fashion that antibodies are manufactured by plasma cells. The plasma cells are derived from certain lymphocytes that have reacted to contact with the particular antigen. We have seen one way of detecting antibodies—the immobilization of bacteria. In some cases cells immobilized in such a way stick together; the antibodies that produce such an effect are called agglutinins, i.e. they agglutinate (stick together) particles carrying the corresponding antigen. There are many such names that are given to antibodies according to the kind of visible reaction that they have with antigens, or to their functions in the body. Often these observable results of antigen-antibody union are determined by the physical state of the antigen—i.e. whether it exists on the surface of cells, particles, as solutions etc.—rather than on the nature of the antibody itself. Some antibodies are called precipitins because a precipitate of antigen-antibody

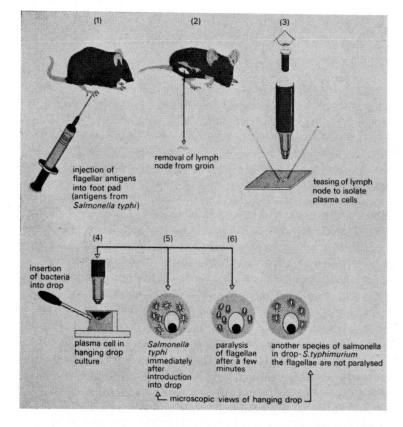

Fig. 249. *An experiment illustrating the function of plasma cells (see text). (From 'Man Against Disease', Clegg and Clegg, Heinemann Educational Books.)*

complex may appear when antigen in the form of a colloidal solution is mixed with antibody under suitable conditions. Some antibodies can be seen to produce effects by stimulating macrophages to engulf micro-organisms or particles that carry the antigen; these antibodies are called opsonins. Antibodies that cause micro-organisms to disintegrate are called lysins. *Neutralizing* antibodies are those that either unite with a toxin and inactivate it or combine with a virus to neutralize its ability to penetrate living cells. All these descriptions tell us little of the nature of the antibodies themselves, they merely tell us that antibody has united with antigen to produce some visible or measurable effect. Before we look at the nature of antibodies we might mention that these observable effects of antigen-antibody union do have great practical significance. One use is the identification of strains of a species of bacteria. Thus there are very many different strains of Salmonella bacteria which have an identical appearance under the microscope. Antibodies can be prepared which are specific to each strain by inject-ing purified proteins from each strain into animals; specific antibodies can later be harvested from the blood of the animals. An unknown strain of Salmonella can be identified by adding some of the bacteria to a series of tubes each containing a different specific antibody. The bacteria react only with the antibody which is specific to its own strain.

The nature of antibodies

As long ago as the 1930s it was known that antibody was a protein in blood plasma, a protein that belonged to the group called the gamma globulins. For many years antibody was known as gamma globulin. More recent studies of plasma proteins has shown that there are several kinds of globulin that can contain antibodies, and it is now conven-tional to call these immunoglobulins. The immunoglobulins form a *vast* family of proteins, each member of the family presumably being produced in response to stimulation by one particular antigen. This family of proteins can be subdivided into various groups on the basis of such differences as molecular weight and biological properties. These major sub-groups are called IgG, IgM and IgA (Ig = immuno-globlulin). Because of the enormous number of different antibodies, the chemical studies of antibody structure have been concentrated on certain kinds that can be obtained in relatively large amounts that can be easily purified. Fig. 250 shows the structure of one kind of antibody, IgG. The whole molecule of the antibody can be broken down into four parts, four chains that come in two pairs, a pair of heavy chains and a pair of lighter chains. In any particular IgG antibody the two heavy

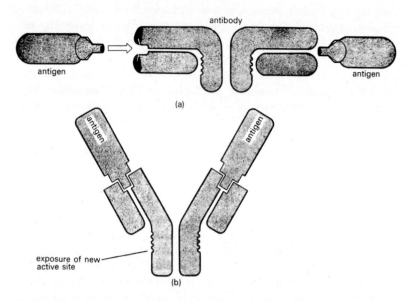

Fig. 250. *The structure of* (a) *immunoglobulin G and* (b) *a possible effect of union of antigen with antibody*. (*From 'Man Against Disease', Clegg and Clegg, Heinemann Educational Books*.)

chains are identical with one another, as are the two light chains. Antibodies have two kinds of function to perform—first they have to recognize a specific antigen: second, they have to do something to the antigen or trigger off some cellular defence response. One suggestion is that the combining site with antigen is shared by the heavy and the light chains (fig. 250A). IgG antibody is a mixture of very many different antibodies, each specific to a particular antigen, and presumably this specificity is reflected in the structure of the combining site. Fig. 250B illustrates what might conceivably happen when antigen combines with antibody. The molecule becomes transformed, revealing other 'active sites' which may be responsible for triggering off other effects, such as the stimulus which makes lymphocytes multiply and transform into plasma cells that manufacture more of the same antibody.

Biological properties of antibodies

We have so far been mainly concerned with the role of antibody in the circulating blood, where it does not come into contact with antigen unless the antigen is artificially injected or reaches the tissues in the form of invading pathogenic micro-organisms. Antibodies are, however, much more widely distributed than this. They are, for example, poured out onto the mucous surfaces of the respiratory and alimentary tract, and are present in tears; in these places they act as parts of the first line of defence of the body. Some antibodies cross the placenta to reach the blood of the foetus so that the newly born infant comes to carry a sample of its mother's antibodies. The nature of these antibodies will depend, of course, on the kind of micro-organisms that have infected the mother or the kind of vaccinations she has received. These antibodies are not permanent features of the infant. They are gradually broken down and excreted, as are any proteins of the body. They do, however, give protection against some infections until the infant begins to produce its own antibodies against the micro-organisms that it meets. This kind of temporary protection is called passive immunity. In this case it is conferred by antibodies that have passed across the placenta from the blood stream of the mother. In some mammals passive immunization of the offspring is transferred by way of the first secretion of the mammary glands that we cali colostrum. These kinds of passive immunity are given the name 'natural' to distinguish them from the artificial passive immunity that results when we try to protect an individual from developing an infection by injecting immunoglobulins that have been extracted from the blood of another individual, often from an individual who is convalescing from an attack of the infection. These kinds of passave immunity must be clearly distinguished from the immunity that results from an infection or from the administration of a vaccine. In the latter case the individual produces his own antibodies from his own plasma cells. This is *active* immunity and differs from passive immunity in another important respect, for the body's ability to produce antibody persists for long periods, often for a lifetime. After an initial contact with the antigen there is a moblization of the immunological defence mechanisms in which a particular set of lymphocytes is stimulated to divide and transform into plasma cells that manufacture specific antibody. The antibody produced in this first reaction to the antigen performs its function, i.e. the control of the infection, and then the level of antibody in the blood gradually falls over months or years. However, the memory of this first contact persists, perhaps in the form of a small group of lymphocytes that are, as it were, committed to producing this particular antibody. If there is a second contact with the same antigen then these lymphocytes multiply rapidly and transform into plasma cells; the production of antibody is more rapid and more intense than in the first contact with the antigen.

Cell mediated immunity

We now come to the second type of immune response that is carried out by cells, a response that can occur in the absence of any circulating antibody. The study of this kind of immunity is still in its infancy but without any doubt this will become a very important branch of immunology, for it is the kind of immune response that is involved in the destruction of an organ or tissue grafted from one individual to another, and is probably also involved in the body's defences against internal hazards in the form of cancer cells.

The kind of cell that determines this form of immunity is the lymphocyte. If we inject some foreign cells into an animal we can later remove lymphocytes from the animal which, when added to the same foreign cells growing in artificial culture in the laboratory, will 'recognize' these cells, more closely to them and cause their destruction.

Although the discovery of the wide significance of cellular immunity is a recent event, the first hint that cells had a direct role to play in immunity came from studies made as long ago as 1880. There are some infectious diseases, such as tuberculosis, in which an infection may produce a long-lasting immunity without any protective antibodies appearing in the blood stream. Clues as to the nature of this immunity came when Koch discovered an unusual reaction in guinea-pigs suffering from tuberculosis. He found that if he injected living tubercle bacilli into the skin of animals suffering from tuberculosis, then a severe localized inflammation occurred at the point of injection and this damage progressed to produce an ulcer. This reaction was very different from that in uninfected animals in which the inoculation site healed rapidly, perhaps to develop some ten to fourteen days later a small nodule. In tuberuclous guinea-pigs the same severe local inflammation occurred even after injection of killed tubercle bacilli or a cell-free extract of the bacteria. This reaction was later used as a test for the presence of tuberculosis in man and animals and is called the Mantoux test, after the man who first used it for this purpose.

Because of the delay of hours or days in the development of the response described above, the reaction has come to be known as 'delayed hypersensitivity'. Delayed hypersensitivity has been found to occur with antigens other than those of the tubercle bacillus—antigens from other bacteria and from fungi and viruses. It is possible to transfer the ability of tissues to react in this way to an antigen by transferring lymphocytes (but not plasma) from one animal to another. Thus a guinea-pig without tuberculosis can be made to respond to an injection of tubercle bacilli with a vigorous but delayed local inflammation by previously injecting into the skin a suspension of lymphocytes taken from another guinea-pig suffering from tuberculosis (fig. 251).

The real significance of this particular reaction is unknown. Animals and man can become immune to tuberculosis and yet the blood contains no antibodies against the tubercle bacillus. In animals with experimental infections of tubercle bacilli, the unrestricted multiplication of the bacilli stops when delayed hypersensitivity appears in the tissues of the animal. Thus in tuberculosis delayed hypersensitivity develops in parallel with immunity, and it is tempting to think that the

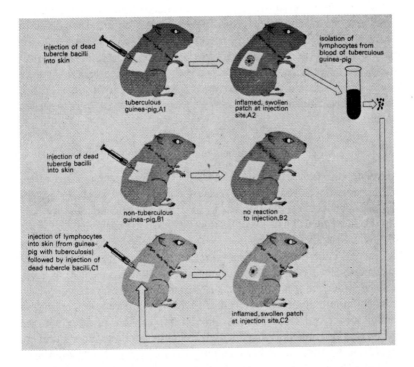

Fig. 251. *The delayed hypersensitivity reaction.* In A1 a guinea-pig with tuberculosis receives an injection of dead tubercle bacilli into the skin. After a few days an area of inflammation appears at the site of injection, A2. A guinea-pig not suffering from tuberculosis does not react in this way (B1 and B2) but it can be made to do so by injecting into the skin lymphocytes that have been isolated from the blood of an animal suffering from tuberculosis (C1 and C2). (*From 'Man Against Disease', Clegg and Clegg, Heinemann Educational Books.*)

one is the cause of the other. At the site of a delayed hypersensitivity response there is a striking accumulation of macrophages. This accumulation is no doubt due to the presence of 'sensitized' lymphocytes which, in contact with tubercle bacilli, release a substance that immoblilizes macrophages. In the laboratory a single sensitized lymphocyte will, when in contact with the products of tubercle bacilli, release enough material to immobilize one hundred macrophages.

We have already hinted at the importance of lymphocytes in the destruction of transplanted foreign tissue and of cancer cells. It is impossible here to consider even the basic evidence for the role of lymphocytes in these reactions.

Auto-immune diseases

There is increasing evidence that a variety of human diseases are the result of a failure of the immune mechanisms to eliminate lymphocytes that have the ability to react with the antigens found in the tissues of the body. In these diseases, appropriately called auto-immune diseases, one or more of the body tissues such as thyroid gland, adrenal cortex, joints etc., become progressively damaged by the appearance of specific antibodies or sensitized lymphocytes that react with the tissue. Part of the treatment of these diseases is the use of drugs that suppress immune mechanisms.

Vaccination (Immunization)

Introduction. We have looked at the body's defence mechanisms against pathogenic micro-organisms. We saw that after the successful control of an infection caused by a particular micro-organism, the defences may be so primed that a second infection does not occur in the individual's life-time, at least rarely so. We say that the individual is immune to the particular disease. This long-lasting immunity is particularly striking after diseases such as smallpox, diphtheria, measles, mumps, chicken pox, and poliomyelitis, although for certain reasons a permanent immunity does not usually follow influenza or infections by staphylococci or streptococci. The term vaccination describes the technique of artificially introducing the antigens of pathogenic micro-organisms so that the body's defences are stimulated and immunity achieved without the penalty of having to suffer the effects of infection by the living virulent pathogens. This technique of artificially and safely producing immunity against certain infections is called vaccination for historical reasons. A more modern and perhaps more rational name is immunization, which describes both the aim and the result of the procedure. These two names, vaccination and immunization, are interchangeable. Here we will be looking at the history of this technique, which is possibly the greatest achievement of modern medicine.

Variolation

The first way in which immunity was artificially produced was by the technique called variolation. A person was inoculated with material (containing living virus) from the skin rash of a patient suffering from a mild form of smallpox. The inoculated person developed smallpox, usually, but not invariably, in a mild form; following this mild infection the person was immune to even the most virulent and dangerous form of smallpox. This technique goes back to ancient times and it was practised in India and in China. It was introduced into Europe in the early part of the eighteenth century by a Lady Mary Wortley Montagu who was wife of the British Ambassador in Turkey. She wrote from Turkey to her friends in England telling them how, in Turkey, groups of people collected together to be visited by an old woman who brought with her a nutshell full of the matter taken from the skin of the best form of smallpox. She used a needle to make a scratch in the skin and then introduced a small amount of the pox matter into the scratch. The aim of this procedure was to produce a mild form of smallpox and so develop an immunity which would protect the individual against the severe forms of the disease. The procedure was, in fact, far from safe; some people became seriously ill and perhaps three in a hundred died. However, in these times naturally occurring smallpox was a common illness having a mortality of some 20–30 per cent; obviously variolation was an improvement on this, and was probably worthwhile until something better was introduced. Variolation became so popular in England that even George I was persuaded to have some of the royal family treated. Inoculation centres were set up in various parts of the country for the purpose of variolation. Of course, opposition developed when some variolated person died or when smallpox in a variolated person spread to other people in whom it sometimes produced a more severe form of the disease.

Vaccination

Variolation was gradually replaced by the much safer technique of vaccination, introduced mainly due to the efforts of Edward Jenner of Berkeley, Gloucestershire. In vaccination the skin was inoculated not with material from a human case of smallpox, but with material from the udder of a cow suffering from a related disease called cow pox.

Inoculation with this material produced only a small localized pustule on human skin, but afterwards the individual had a high degree of immunity to human smallpox. Jenner cannot be credited with the discovery that infection with cow pox produces immunity to smallpox, but his scientific approach to the problem did a great deal to popularize the procedure and establish it as a safe way of preventing smallpox. In country districts the 'superstition' was often held that those who contracted cow pox by milking cows infected with the disease were subsequently immune to the ravages of human smallpox. When a local milkmaid—Sarah Nelmes—developed a typical sore of cow pox on her hand (fig. 252) Jenner took some of the material from the sore and scratched it into the skin on the arm of a healthy boy, James Phipps.

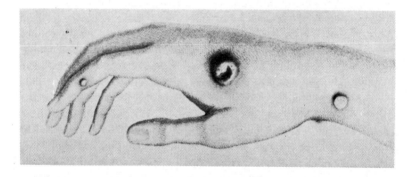

Fig. 252. (*The hand of Sarah Nelmes*. (*From Jenner* 'Inquiry,' 1798, plate 1. By courtesy of the Wellcome Trustees. (*From* '*Man Against Disease*', *Clegg and Clegg, Heinemann Educational Books*.)

On the boy's arm a sore developed where the material had been intro-duced into the skin (fig. 253), but otherwise the boy remained well. Then six weeks later Jenner put the superstition to the critical test. He inoculated into Phipps' arm material from the skin of a case of smallpox. Nothing happened to Phipps! Again, months later, Jenner repeated the test, but the boy remained immune to smallpox. Jenner then repeated these tests on a number of other subjects. The results were the same. Infection by cow pox produced immunity to smallpox. Jenner published his findings privately (his paper on the subject was rejected by the Royal Society because of the 'lack of adequate proof') and in spite of various objections from doctors who were not entirely successful in repeating his work, the popularity of vaccination grew and grew. Special sessions were held for the purpose of vaccination

Fig. 253. *Jenner inoculating James Phipps.* (From a drawing by William Thompson, c. 1880. By courtesy of the Wellcome Trustees.)

in different parts of the country. In recognition of his services parliament granted him £10,000 in 1802. However, it was not until 1840 that the dangerous practice of variolation was made illegal in England.

Pasteur's contribution

Jenner's success, although founded on a scientific test of the validity of country 'folk lore', was not based on any understanding of microorganisms, and he did not know why vaccination prevented smallpox. The scientific basis of artificial immunization against infectious disease was established by Louis Pasteur who developed the 'germ theory' of disease. Pasteur's first studies of the ways in which animals and men could be protected from attack by microbes was in the field of infections in poultry. At that time fowl cholera was a serious threat to poultry in

France. Pasteur's discovery of a method of making poultry immune to cholera was in part due to an accident. He had previously been able to culture the cholera bacteria in his laboratory, and whenever he wished to produce cholera in chickens all that he had to do was inject a small number of the bacteria from his laboratory cultures. The bacteria, however, had to come from a fresh culture; old cultures failed to produce the typical disease in many of the chickens and most of them recovered. On one occasion when Pasteur wished to inject fresh cultures of virulent cholera bacteria into chickens, his technicians gave him birds that had survived after an injection from an old culture of cholera bacteria. He found that these birds were resistant to the virulent bacteria and after a mild illness they recovered. These birds had obviously been rendered immune to cholera by the previous injection of old bacteria. Soon, these suspensions of bacteria from old cultures were being used in France to protect poultry from cholera.

Later, Pasteur was able to produce a vaccine to protect animals from the killer disease anthrax. Because anthrax bacilli produce resistant spores, old cultures of the bacilli were useless as vaccines for they were as dangerous fresh cultures. He found, however, that if cultures of anthrax bacilli were kept at 42–43°C for about a week (instead of at 37°C, the temperature of the human body) the virulence of the bacteria was reduced and they lost their ability to produce resistant spores. The virulence of these modified (attenuated bacteria) was related to the period at which they had been held at 42°C, and the longer they had been kept at this temperature the less virulent they became. He was thus able to produce samples of anthrax bacilli of varying virulence. He immunized farm animals by first injecting them with a suspension of bacteria of low virulence, and following this twelve days later by an injection of bacteria of greater virulence. About a month after these injections, the animals—sheep, goats, cattle—could be given injections of anthrax bacilli of full virulence without them succumbing to anthrax. These living attentuated anthrax vaccines radically reduced the mortality of farm animals in France. In these early studies several accidents occurred; many animals died because anthrax bacilli were used which had not been sufficiently attenuated.

Several years later, in 1885, Pasteur produced what is perhaps his greatest achievement, a living attenuated vaccine for rabies (a virus infection). Pasteur was quite unable to see the minute rabies virus but he was able to establish that it lived and multiplied in the brain and spinal cord of infected animals. The vaccine he used to protect men and dogs from this terrible disease contained living virus that had been weakened (attenuated) by passage through the brains of a whole series of living rabbits. He injected rabies virus (from an infected dog) into the brain of a rabbit, then later he killed the animal and took a small amount of its brain substance for injection into the brain of another living rabbit, and so on until he produced a virus of reduced virulence for dogs and man. The virulence of the virus was further reduced by drying the spinal cord of the infected rabbits in sterile air over a desiccating (drying) agent at room temperature. The longer the spinal cord was dried, the less virulent was the virus it contained. After fourteen days of drying, the rabbit's spinal cord was not infective and was used as the first dose of a course of injections.

On the following days cord was used that had been dried for shorter periods, thirteen days, twelve days, eleven days and so on until the final dose, the fourteenth, was of fully virulent fresh spinal cord. Using this method Pasteur was able to save the lives of dogs that had been bitten by another animal infected with rabies. Rabies is a disease with a long incubation period, and provided that the course of injections were started soon after the bite then few animals died. In 1885 Pasteur tested this method of vaccination on a nine-year-old boy, Joseph Meister, who had been brought to him from Alsace for treatment. The boy had many wounds inflicted by a dog suffering from rabies and he would almost certainly have died. Pasteur gave this child a course of his vaccines and he survived. Further human cases received the same treatment, although not all were saved, either because the vaccination was started only after some days' delay or because the rabies virus had been introduced close to the brain from bites on the face. The death rate of those who received Pasteur's vaccine after the bite of a rabid animal was less than one per cent compared with the death rate of 15–20 per cent of those who received no treatment after the bite. Of course not everyone who is bitten by a rabid animal develops rabies, but all of those who do develop rabies die of the disease.

The use of dead micro-organisms and the contribution of Almroth Wright

The vaccines we have seen so far are composed of living micro-organisms. In the case of fowl cholera, anthrax or rabies the micro-organism has been attenuated by changes in the culture of the organism in the laboratory—ageing the organism, growing it at high temperatures, or repeatedly passing it through the brains of rabbits. Vaccines against tuberculosis, brucellosis, yellow fever, poliomyelitis, German measles (rubella) and measles also contain living micro-organisms, the virulence of which has been modified by various procedures (see Vaccine Table).

Types of vaccine Table

---

1. Living attenuated micro-organisms
    Rabies   Tuberculosis
    Brucellosis   German measles (rubella)
    Yellow fever   Measles
    Poliomyelitis

2. Dead micro-organisms
    Typhoid and paratyphoid fevers   Influenza
    Cholera   Whooping cough
    Plague   Leptospirosis

3. Toxoids
    Diphtheria   Tetanus

Injections of dead micro-organisms may also produce some immunity against a particular infectious disease, although the immunity may not be as complete or as long-lasting as that which follows the use of living, attenuated micro-organisms.

The first disease to be attacked by means of injections of dead organisms was typhoid fever, and the success of this venture was due to the work of a British bacteriologist, Sir Almroth Wright. Wright started his researches on vaccines in 1892 at a time when it was believed that any vaccine likely to be effective must contain living organisms. He began his studies of the effects of dead vaccines in the prevention of a prolonged weakening infectious disease that attacked the garrison troops in Malta, a disease appropriately called Malta fever (=brucellosis). This illness with its prolonged fever was often confused with other diseases such as typhoid fever or tuberculosis. After some preliminary research on monkeys, Wright demonstrated his faith in vaccines in a most realistic way by injecting himself with the dead bacteria that were the cause of Malta fever, followed at a later date by an injection of living bacteria. Unfortunately the injection of dead bacteria did not produce enough immunity and Wright succumbed to Malta fever, with a prolonged illness. This experience did not, however, deter him and he next made plans to prepare a dead vaccine for immunization against typhoid fever.

His first experiments with a vaccine prepared from dead typhoid organisms were made on himself and other volunteers at the Royal Victoria Hospital at Netley. Although the vaccine made some of the volunteers rather ill, soon there was evidence that the blood of Wright and his other volunteers had developed the ability to destroy typhoid bacteria. Because of the rather severe reactions to the vaccine, further experiments were needed to find out how small a dose of the vaccine could protect individuals from typhoid fever. The chance to make these experiments soon occurred when an outbreak of typhoid fever appeared in an asylum in Kent. Wright prepared a vaccine from cultures of typhoid bacteria in a broth. The bacteria were killed by heating to 53°C and by the addition of 0·4 per cent lysol. Although there were still local reactions at the site of injection of the vaccine (swelling and discomfort) and general reactions (fever etc.), the blood of those who received the vaccine developed a marked ability to destroy the typhoid bacteria.

At this time the Boer War (1899–1902) was in progress, a war in which there were 58 000 cases of typhoid (15 per cent of all troops) with about 9000 deaths, more deaths than were due to wounds (8000). Wright was unable to persuade the War Office to inoculate all troops embarking for the Boer War. However, compulsory vaccination was later introduced for troops leaving for India and other foreign countries. By 1909 nearly all the British soldiers in India had been inoculated with typhoid vaccine, and this produced a reduction in hospital admissions due to typhoid fever from 8·9 per thousand (with a death rate of 1·58 per cent) to 2·3 per thousand (with a death rate of 0·25 per cent) in 1913. After the introduction of Wright's vaccine no other British Army campaign suffered the misery and death due to typhoid fever that had harassed the soldiers in the Boer War. In the First World War (1914–18), when typhoid vaccination was compulsory, only 2·35 per thousand soldiers suffered from typhoid fever, compared to 10·5 per thousand in the Boer War. This reduction in typhoid fever was achieved in spite of the insanitary trench warfare of the First World War. After 1916 paratyphoid organisms were included in the typhoid vaccine, which came to be known as T.A.B. vaccine. After the First World War there were considerable improvements in this vaccine, particularly in the strains of bacteria that were used in its preparation.

Modern vaccines against typhoid fever reduce the risk of developing the disease by about 75 per cent. Since protection is not absolute, unnecessary risks from drinking water and food that are likely to be contaminated by typhoid organisms should still be avoided. Immunity is not permanent and booster doses need to be given from time to time. Because paratyphoid fever is usually less severe than typhoid fever, the paratyphoid organisms are now omitted from *some* modern vaccines. These vaccines also cause less severe local and general reactions than T.A.B.

Filtrates (toxins and toxoids) as vaccines

We have seen that early in the history of immunization, living attenuated organisms or dead organisms were used in the production of vaccines. A new milestone in preventive medicine appeared when it was discovered that whole bacteria, either living or dead, were not necessary for immunization against certain infectious diseases, such as diphtheria and tetanus. In these diseases, the bacteria produce their damaging effects by the liberation of toxins. These toxins were discovered early in the history of bacteriology. As long ago as 1888 two of Pasteur's followers, Roux and Yersin, produced a bacteria-free filtrate of a broth culture of the diphtheria bacillus which contained a chemical toxin that could produce effects in animals identical to those produced by living virulent bacteria. Similar powerful toxins were isolated from cultures of tetanus bacilli. It was found that when animals were injected with small doses of toxins, they developed powerful substances in the circulating blood which could neutralize the effects of toxins. These substances in the blood stream were called anti-toxins. They could be extracted from the blood of animals (mainly horses) to treat cases of diphtheria or tetanus (see fig. 254). Each anti-toxin is specific to the kind of bacterial toxin that stimulates its manufacture in the body. Injections of diphtheria toxin result in the appearance of anti-toxin which neutralizes only diphtheria toxin, and the same holds for tetanus toxin.

In addition to this use of anti-toxins in the treatment of established cases of disease, anti-toxins had another use. It was found that if toxin and anti-toxin were mixed together, the combination lost the damaging effect of the toxin but still retained its ability to stimulate the body to produce anti-toxin. Anti-toxins could then be produced by injecting horses with safe mixtures of toxin and anti-toxin; later blood was removed for isolation of serum (i.e. blood minus cells and fibrinogen) containing anti-toxin. For many years these 'sera' have been used in the treatment of various bacterial infections, mainly diphtheria and tetanus but also staphylococcal infections, gas-gangrene and botulism. For various reasons these sera have almost disappeared as methods of treatment. Diseases such as tetanus and diphtheria are more easily prevented than treated (see below for immunization) and the use of horse serum carries risks for some individuals who are allergic to horse serum and who may have alarming reactions after injection of the material. For other diseases, such as staphylococcal infections and gas-gangrene, much more effective methods of treatment are now available.

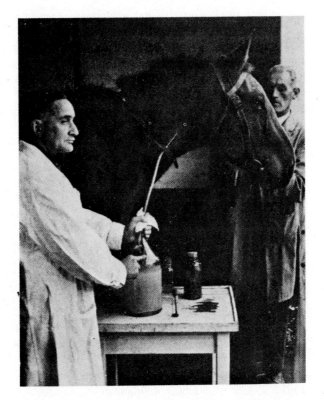

Fig. 254. *A horse being bled for the production of anti-toxins.* (*Photograph generously supplied by Prof. L. H. Collier of the Lister Institute of Preventitive Medicine.*)

Immunization against diphtheria and tetanus in man

The discovery that mixtures of toxin and antitoxin could safely be injected into animals and still stimulate the manufacture of anti-toxins had far reaching effects on the prospects of the eradication of diphtheria and tetanus from human populations. In 1901 Von Behring was awarded a Nobel prize in Physiology and Medicine for his work on immunization of animals by injection of toxin + anti-toxin mixtures. A few years later this work was applied to man and in 1913 he used a toxin + anti-toxin mixture which produced rapid, lasting and fairly safe immunity to diphtheria in man. An even safer method of inactivating toxin was discovered by Alexander Glenny. Glenny was puzzled when he found that a certain batch of diphtheria toxin had very little damaging effect in animals although it produced as strong an immunity against diphtheria as did highly toxic batches. His safe toxins had been stored in containers that had been sterilized with formalin. The formalin had in some way 'detoxicated' the toxin and yet left its ability to make the animals immune to diphtheria. These inactivated toxins are called toxoids and are used nowadays in vaccines for immunization against diphtheria and tetanus in man.

The use of a vaccine against diphtheria is a classic example of the dramatic possibilities for the prevention of infectious disease. We have already seen the early work on the preparation of a vaccine by modifying diphtheria toxin so that it could safely be injected into the body to produce a prolonged immunity to diphtheria. There were early trials, in the 1920s and 30s, of the use of diphtheria vaccines in various institutions and training centres in Britain. All of these various projects were successful in the prevention and control of diphtheria, but a mass immunization campaign was a late development in Britain. There were various mishaps in immunization campaigns in certain parts of the world which were very discouraging. In some of these accidents, laboratories carelessly issued toxin alone as a vaccine, omitting the anti-toxin which would have produced a safe mixture. Such an accident occurred at Baden, Austria where seven infants died, and this was followed by the forbidding of immunization in Austria. Other accidents were due to bacterial contamination of the vaccines which produced serious and sometimes fatal infections in the vaccinated. This sort of error led to the famous Bundaberg accident in Queensland, Australia, when the omission of a preservative from diphtheria vaccine led to a large multi-dose container of vaccine being contaminated by a virulent strain of *Staphylococcus pyogenes*. This error led to the death of twelve children. When we look at the development of vaccines against many infections, we will see that the early stages of the formulation of the vaccines has often led to various sorts of accidents that have hindered, at least for a while, the general acceptance of the vaccine. Partly as the result of the various accidents with the early diphtheria vaccines, the British Therapeutic Substances Act was passed in 1925; the purpose of this was to control the quality of all the vaccines and antisera prepared for use in this country. This Act of 1925 has been reviewed and modified continuously since it first became effective in 1931.

The British Ministry of Health at last supported large-scale vaccination against diphtheria in 1941, and for the first time the materials were issued free by the Government. Now the control of diphtheria changed from the treatment of established disease by antisera (which were sometimes dangerous in themselves and often ineffective against some

strains of diphtheria bacilli) to active immunization of a large proportion of the young population of the country with toxoid. Anti-toxin
itself did indeed save many lives, reducing the mortality from around
30 per cent to 8 per cent, but it did not reduce the incidence of the
disease; in the years before 1940 there was an average of 58 000 cases of
diphtheria a year with 2800 deaths per year. By the end of the 1939–
45 war, over 60 per cent of the child population had been actively
immunized with toxoid, producing a dramatic fall in the number of
cases of diphtheria (see Table).

Deaths and cases of diphtheria

Year	Number of cases per year	Deaths per year
1946	11 986	472
1960	49	5
1964	20	—

Immunization against diphtheria with toxoid is so successful that
there can be little doubt that if sufficient numbers of children received
the protection of toxoid, then the disease would disappear from our
country. That the disease has not disappeared is in no small measure
due to apathy or ignorance on the part of parents who fail to take their
children for immunization or fail to ensure that they complete the
course of injections. Unfortunately individuals who are immune to
the disease can still 'carry' virulent strains of diphtheria organisms in
the nose or throat and can thus infect other individuals who have not
had the benefit of immunization. For these reasons diphtheria has not
been completely conquered in this country. In 1970 there were twenty-
two cases of diphtheria with three deaths. This was an increase over
the previous two years in which there were seventeen cases in each year
with no deaths.

At the time of writing there was a recent outbreak of diphtheria
within Manchester in February 1971. There were four cases in children,
three of whom were admitted to hospital with a 'membranous ton-
sillitis'. Fortunately in none of the cases was the illness severe. Three
of the children had never been immunized and one child had received
only a primary course. Throat and nose swabs were taken from over
2000 children and family contacts of these cases. This resulted in the
detection of twenty-six carriers of a virulent strain of diphtheria; these
individuals were admitted to hospital for observation and treatment to

eradicate the organisms. As a result of this outbreak over 7000 children attending the local schools or who lived in the neighbourhood of the cases received immunization against diphtheria.

This great achievement, the virtual eradication of diphtheria from Great Britain, does not end our story of the conquest of infectious diseases by vaccines. Later years saw the beginning of the conquest of tuberculosis, measles, and German measles.

Schedules of immunization

The kinds of vaccine that a child may receive have changed considerably since the introduction of the first vaccine—that against smallpox. New vaccines against particular diseases have been introduced from time to time, together with improvements in the nature of particular vaccines. The vaccines in routine use today in Great Britain are:

1. Smallpox vaccine.
2. Vaccines against diphtheria, tetanus and whooping cough, usually combined in a single injection but sometimes used separately.
3. Live poliomyelitis vaccine, given orally.
4. Live measles vaccine.
5. Live tuberculosis vaccine—B.C.G.
6. Live German measles vaccine.

Additionally, vaccines such as those against yellow fever, typhoid fever, and cholera have to be given if an individual travels to an area where these diseases are common.

The effectiveness of a particular immunization programme in a child depends upon many factors. The age at which vaccination begins is important because antibodies transferred from the mother to the child by way of the placenta can neutralize the effect of vaccines. A period has to elapse before vaccination begins to allow for the disappearance of these maternal antibodies from the body of the child. Once immunization has begun many factors now operate to determine the effectiveness of the vaccination, factors such as the size of the dose, the interval between doses and the use of combined vaccines (e.g. diphtheria-tetanus-whooping cough) in which the balance of the different components is critically important. Not all children are automatically given a course of vaccination that includes all of the six vaccines listed. Before each child begins its vaccination programme, careful consideration has to be given to whether a particular vaccine will carry more risk for the child than the disease which he *may* develop if he is not protected by vaccine. There are special contra-indications for certain vaccines. Here we can mention the dangers of whooping cough vaccine in children who have a history of convulsion or the dangers of smallpox vaccination in individuals who suffer from eczema.

The following is a recommended schedule of immunization to be given where there are no special reasons for avoiding a particular vaccine.

Age	Immunization
4–6 months	Diphtheria-tetanus-pertussis (whooping cough) mixture plus oral polio. 1st dose.
6–8 months	Diphtheria-tetanus-pertussis mixture plus oral polio. 2nd dose.
12–16 months	Diphtheria-tetanus-pertussis mixture plus oral polio. 3rd dose.
16–24 months	Live measles followed 3–4 weeks later by smallpox.
5 years or at school entry	Diphtheria-tetanus-pertussis booster followed 3–4 weeks later by smallpox vaccination.
10–13 years	B.C.G. (against tuberculosis).
15–19 years	Diphtheria-tetanus plus oral polio booster doses, followed 3–4 weeks later by smallpox vaccination.

N.B.—This schedule does not include the recently introduced vaccine against German measles (rubella). The main aim of this vaccine is to produce immunity in *women*, thus preventing the possible damaging effect of rubella on the foetus. Immunization against rubella can be carried out at any age (preferably after the first year of life since below this age the presence of maternal antibodies in the blood can reduce the immune response) but pregnant women must not be vaccinated because there is a risk that the foetus could be infected and damaged by the attenuated virus. Women of child-bearing age must not become pregnant for at least two months after this vaccination.

In view of the proven value of the various vaccines in preventing serious infectious diseases, it is indeed surprising that all children do not benefit from this protection. A few children do not receive their vaccination schedules, or do not complete them once begun, because of special contra-indications. The vast bulk of omissions, however, is due to apathy or ignorance on the part of parents. The number of children immunized varies from one part of the country to another, ranging from as high as 95 per cent of all children to as low as 20 per cent, and for vaccination against smallpox the figures are much lower than these. The persistence of diphtheria in these islands, a disease that *can* be eradicated by 100 per cent vaccination, provides a cautionary tale to all parents and all those involved in health education.

Smallpox has now been eradicated from most parts of the world such that the risks of vaccination against smallpox are greater than that of contracting the disease itself. Smallpox vaccination is thus no longer recommended as a routine measure in Great Britain. Smallpox vaccination is only used for travellers to areas of the world where the disease is still prevalent.

Chapter Fifteen

Muscles and the Skeleton

MUSCLE ACTION

Introduction. The muscles of the mammal constitute an important tissue in that they form 40–50% of the body weight. We have already seen in Chapter IV that there are three basic types of muscles, voluntary or striated muscle, involuntary or unstriped muscle and cardiac muscle. Most of the voluntary muscles of the body are connected to a system of movable levers, the skeleton, and by changes in their length or tension the muscles are able to maintain the posture of the animal in addition to moving part or whole of the organism in response to internal or external needs. The voluntary muscles and skeletal system thus play an important part in the adaptation of the organism to changes in the environment by enabling appropriate action to be taken. The voluntary muscles have two insertions into the skeletal system: an *origin* from some relatively fixed part against which the muscle can act and an *insertion* into some more movable part.

The involuntary or unstriped muscles lack strict origins and insertions and they form layers of tissue around hollow structures such as the alimentary tract, ureters, bladder, uterus, vas deferens, and blood vessels. Many of them are concerned with the movement of substances within the hollow organs of which they are part. The properties of unstriped muscle differ from voluntary muscle in that whilst the latter are rapidly contracting (tetanic), unstriped muscle shows slow, rhythmic, sustained contractions. Most of the present chapter will not be concerned with these unstriped muscles.

The properties of cardiac muscle have already been described in Chapter V.

Muscle attachments. In order to act at some movable part of the skeleton a muscle must have a secure attachment at the opposite end, its origin. The way in which the muscles are attached to the skeleton is very variable, sometimes taking the form of a wide-spread fleshy attachment whilst in other cases the union is restricted to a narrow band of tendon, see figs. 256, 257. In those muscles attached to bone by way of a tendon there is a close relationship between the collagenous fibres of the tendon and the collagen fibres of the connective tissue which surrounds bundles of muscle fibres (the perimysium). There is also a fusion of the sarcolemma at the end of the muscle fibres with the collagen tissue of the tendon. There are prolongations of the tendon into the tissue of the muscle which provide a more intimate relationship between the tissues of muscle and tendon and permit the use of short muscle fibres instead of fibres which have to traverse the whole length of the muscle from origin to insertion (fig. 255) although the latter type of fibre does exist.

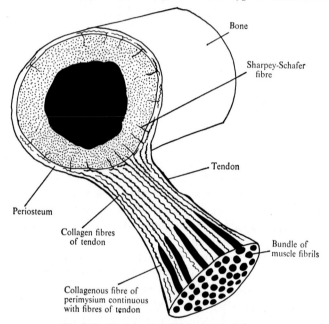

Fig. 255. *The insertion of a tendon into a bone.*

Types of muscle action. The fundamental property of muscle is that when it contracts it is able to exert tension between the two sites of its attachment to the skeleton, its origin and insertion. The contracting muscle may shorten and so move the point at which it is inserted. This type of contraction is called *isotonic* contraction. In this type of contraction the tension is utilised in performing work, in moving some part of the organism. But there is another type of contraction, called *isometric* contraction, in which no external work is done; the tension developed in the muscle is utilized in opposing some other force e.g. the force of gravity. The same muscle may at one moment contract isotonically and at another moment isometrically. In fact probably no muscle contracts purely isometrically or isotonically; there is seldom shortening without some rise in tension in the muscle and vice versa. We can now consider a simple example of these two different types of muscle action, which can be demonstrated on oneself.

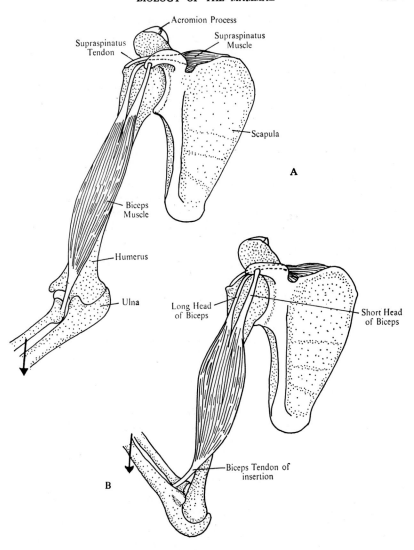

Fig. 256. *Drawings showing the biceps muscle during isometric* (A)
and isotonic contraction (B). *In A with the elbow held at about 90°
the hand is supporting a weight whose force acts downwards as
indicated by the arrow. The contracting biceps prevents the elbow
being straightened; the muscle does not shorten and is acting isometrically. In B the forearm is moved towards the shoulder, work
being done by the biceps shortening (isotonic action). The arrow
indicates the direction of the force caused by the weight of the
forearm. In the figures the supraspinatus muscle is shown, arising
from the upper surface of the scapula and inserted into the head of the
humerus. This muscle by holding the rounded head of the humerus
in the shallow joint surface of the scapula acts as a fixator of the
shoulder joint.*

Hold the arm with the elbow joint flexed at about 90° whilst holding a fairly heavy book in the hand. With the other hand feel the biceps muscle which is situated on the front of the upper arm. The muscle will be found to be tense and its tendon of insertion into the upper end of the radius will be felt projecting tautly in the bend of the arm. The biceps muscle has its origin from the shoulder girdle and passes over the anterior surface of the humerus, over the elbow joint to be inserted by a narrow tendon into the upper end of the radius. Contraction of the muscle with shortening produces bending (flexion) of the elbow joint. But when the hand is holding a heavy object with the elbow at 90° the muscle is active in preventing straightening (extension) of the elbow joint; in this case the muscle is acting isometrically and the muscle belly can be felt not to shorten appreciably (see fig. 256A). Now bend the elbow joint to bring the arm towards the chest, at the same time feeling the belly of the muscle which will be found to shorten; the muscle is now functioning isotonically and is performing work (see fig. 256B).

The different capacities of muscle action. A muscle may act in various capacities including:

1. Prime mover.
2. Antagonist.
3. Fixator.
4. Synergist.

When a muscle acts as a *prime mover* it is responsible for the actual movement which take place. Thus when the biceps muscle contracts isotonically and shortens, it flexes the forearm on the upper arm and is acting as a prime mover.

In a particular movement an *antagonist* is a muscle which by contracting is able to produce the opposite movement. Thus the triceps muscle originating from the posterior surface of the humerus and the pectoral girdle and inserted into the olecranon process of the ulna is the antagonist to the biceps muscle. When the triceps contracts it produces straightening of the elbow (extension).

When the biceps muscle is acting the triceps relaxes and vice versa. But the situation is not as simple as this because the antagonist may actually contract when the prime mover is operating, but it does so in a controlled manner, gradually 'paying out' to produce a smooth movement effected by the prime mover.

In addition to prime movers and antagonists other muscles may be acting during the movement of a part, in order to stabilize the origin

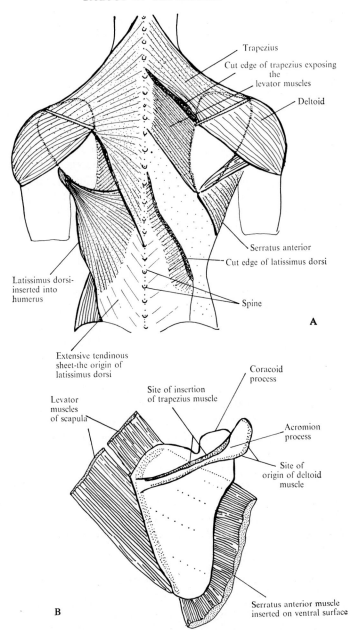

Fig. 257. *Fixator muscles of the scapula.* **A.** *Shows the back of a man. On the left side the large trapezius muscle is shown, with its wide origin from the back bone and its insertion into the spine of the scapula. On the right, part of the trapezius muscle has been removed to show the underlying mucles attached to the scapula. The outline of the scapula is indicated by the dotted line.* **B.** *The dorsal surface of the right scapula showing the insertion of muscles.*

of the prime mover to produce a firm basis, or in order to stabilize the joint which is moving. These muscles are then operating as *fixators*. The wide range of movements which is permitted at the shoulder joint in man is determined by the rounded head of the humerus fitting into a very shallow cup on the scapula; stability at the joint, which is not weight bearing, is sacrificed at the expense of mobility. In order that the head of the humerus is not dislocated from its shallow cup during movements of the arm, there are special muscles which originate from the scapula and are inserted into the head of the humerus, that serve to maintain the stability of the joint. One such muscle, the supraspinatus is shown in fig. 243. In order to produce mobility the whole shoulder girdle, embedded in muscle has only one attachment to the rest of the body skeleton, by way of the inner end of the clavicle which is articulated with the sternum. In addition to the stabilization of the shoulder joint by muscles, the scapula itself is stabilized by muscle action. Thus the trapezius muscle and serratus anterior, in addition to being prime movers of the scapula, also act as fixators (see fig. 257).

Synergists are special examples of fixators in that they control the position of intermediate joints in those cases where a prime mover passes over several joints on its way to its insertion. Thus the long flexor muscles of the fingers pass from their origin at the upper end of the medial side of the forearm across the wrist joint and carpal joints before their tendons are inserted into the fingers. If these long flexors contracted alone then in addition to causing flexion of the fingers they would produce flexion of the wrist. This is prevented normally by simultaneous contraction of the extensor muscles of the wrist which pass from the posterior surface of the forearm to the carpal bones on the dorsal side of the wrist (see fig. 258).

It is thus apparent that even in the simplest of movements there are several muscles involved, and to this concept is given the name of 'group action of muscles'. In the cerebral cortex of man individual muscles are not represented; we are not conscious of the action of individual muscles, but of movements in which several muscles are involved.

The innervation of muscles and the gradation of muscle activity. When a muscle is exposed in the body it may be made to contract by a variety of stimuli including mechanical, chemical and electrical changes; the muscle will, for example, contract if pinched with a pair of forceps. But muscles are made to contract by means of nervous stimuli, each motor neurone in the nerve supplying the muscle supplies several muscle fibres; as the neurone enters the muscle it divides into branches,

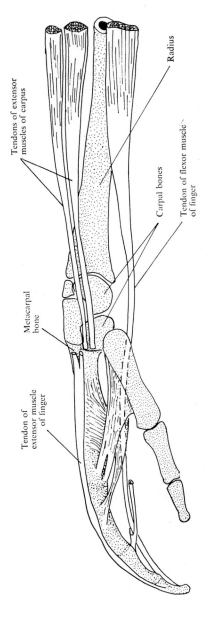

Tendons of extensor muscles of carpus

Radius

Carpal bones

Tendon of flexor muscle of finger

Metacarpal bone

Tendon of extensor muscle of finger

Fig. 258. *Flexors of the fingers and syngergist extensors of the wrist. The extensor muscles of the carpus hold the wrist firm during the contraction of the flexor muscles of the fingers.*

each division supplying a single muscle fibre. The relationship between the muscle and the termination of the neurone as a motor end plate has already been described on page 176. The motor neurone and the muscle fibres supplied by it are known as the *motor unit*. The division of the muscle into functional motor units provides the basis for one way in which the strength of muscle contraction can be graded, by increasing or decreasing the number of motor units active. It will be apparent that the fewer the number of muscle fibres in each motor unit the more delicate will be the control of muscle action. It is found that in those muscles performing delicate movements, such as the hand and extrinsic muscles of the eye, there are fewer muscle fibres in each motor unit than in other skeletal muscles. The other mechanism for the gradation of the strength of muscle contraction depends upon the property of summation of nerve impulses in muscle, and this mechanism will be described subsequently.

The mechanical response of muscle to nervous stimulation

The mechanical response of a muscle to nervous stimulation can be recorded by dissecting out the muscle, fixing its origin, and attaching

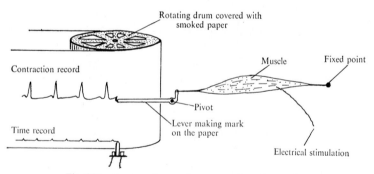

Fig. 259. *Kymograph recording contraction of a muscle.*

the tendon of insertion to a movable lever. When the muscle contracts (isotonically) the lever moves and this movement may be recorded on a moving smoked drum (kymograph) fig. 259.

The twitch contraction, summation and tetanus. The muscle responds to a single stimulus by a twitch contraction, the record of which is shown in fig. 260A. There is a short period between the electrical stimulation of the muscle and the mechanical response of shortening called the latent period. The upstroke of the curve in fig. 260A represents the

period of contraction, the downstroke of the curve the period of relaxation of the muscle. By increasing the strength of the stimulus to
the muscle, increasing numbers of motor units are activated until a
maximal response from the muscle is obtained. One can artificially
increase the strength of the stimulus under experimental conditions but
when a certain level of response has been reached further increases in
the strength of the stimulus bring about no increase in the response of

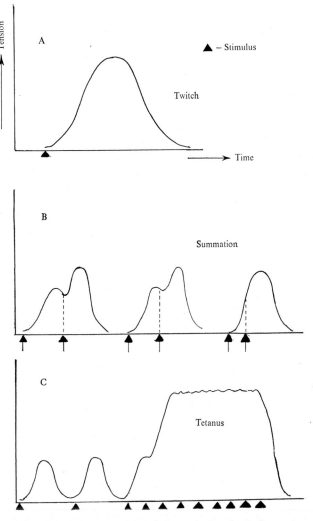

Fig. 260. *Records of muscle contraction.* A. *Twitch.* B. *Summation.*
C. *Tetanus.*

the muscle. However, if a second stimulus is applied to the muscle during the short period when the muscle is contracting in response to the first stimulus then the effect of the two stimuli is combined and there is a greater response by the muscle; there is a *summation* of the response to the two stimuli (see fig. 260B). If now a series of maximal stimuli is applied in rapid succession there is a complete fusion of the responses of the muscle to produce a smooth sustained contraction, a condition known as *tetanus* (see fig. 260C). Thus an increase in the power of muscle contraction can be obtained not only by increasing the number of motor units in action but also by increasing the frequency of stimuli to the muscle to produce a state of sustained contraction called tetanus. The tension which is developed in the state of tetanus is greater than can be explained on a simple additive effect of the stimuli; the effect of repeated stimuli somehow increases the ability of the muscle to develop tension.

These two methods of grading the power of muscular contraction are used in the normal physiological state in the mammal.

Neuromuscular transmission. The area of contact between axons and muscle in the motor end plate has already been described (p. 176). They are areas of contact between the axon and the sarcoplasm of the muscle. When a nerve impulse arrives at the motor end plate there is the liberation of minute amounts of a substance called acetyl choline. This substance alters the property of the adjacent muscle membrane so that it becomes permeable to ions. This altered permeability results in a flow of ions across the end plate membrane producing an electrical change in the end plate called the end plate potential. The end plate potential produces electrical changes in the surrounding muscle membrane which spreads and in some way initiates the contractile mechanism of the muscle fibre. The acetyl choline produced at the end plate is rapidly destroyed by an enzyme called cholinesterase. The excitability of the muscle to acetyl choline is restricted to the region of the motor end plate. Acetyl choline applied directly to the motor end plate will cause contraction of the fibre and if injected into the artery supplying the muscle it will produce a contraction of the whole muscle. It has been found that the arrow poison curare and the synthetic d-tubo-curarine which is now widely used in anaesthesia produce paralysis of the voluntary muscles by reducing the sensitivity of the end plate region to acetyl choline. Formerly, large doses of anaesthetics had to be used to produce the degree of muscular relaxation necessary for surgery, but using tubocurarine the muscles are completely relaxed (including the breathing muscles which means that the lungs of the patient have to be ventilated mechanically) and only sufficient anaesthetic to produce unconsciousness is needed. An antidote to the effect of curare is a drug called neostigmine which blocks the effect of the cholinesterase, so allowing acetyl choline to accumulate at the neuro-muscular junction in sufficient quantities to overcome the block induced by curare. This drug is given at the end of an anaesthetic in which tubocurarine is used.

Acetyl choline is a very important substance in the organism, not only in neuromuscular transmission. It is also the transmitting agent at the synapse between the pre- and post-ganglionic fibres of the autonomic nervous system and it is liberated at the termination of the post-ganglionic fibres of the parasympathetic division of the autonomic system. Thus acetyl choline is produced at the terminals of the vagus nerve (parasympathetic) in the heart and is the agent which causes the slowing of the heart which results from vagal stimulation (fig. 261).

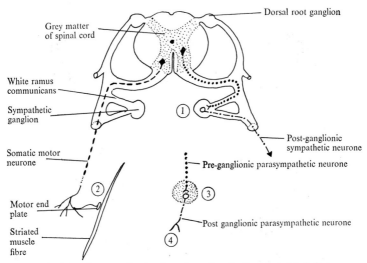

Fig. 261. *Diagram showing four sites of production of acetyl choline.*

Proprioreceptors and muscle reflexes

The stretch reflex. When a muscle is stretched certain sense organs in the muscle are stimulated to send out a volley of impulses via a sensory nerve to the spinal cord, resulting in a reflex contraction of the muscle to oppose the stretch. The knee jerk reflex described on p. 276 is a good example of a stretch reflex. The tap on the tendon of the quadriceps femoris muscle, whilst the knee is flexed, stretches the muscle, causing a sharp volley of impulses from the muscle sense organs which results in a brief reflex contraction of the quadriceps muscle causing extension of the knee joint. A similar reflex may be elicited by a sharp tap on the Achilles tendon in the ankle resulting in the ventro-flexion of the foot by the action of the gastrocnemius and soleus muscles of the calf of the leg.

All muscles have the ability to respond in some degree to the stimulus of stretch but this ability is particularly well developed in the extensor muscles, which by operating against the effects of the force of gravity maintain the posture of the body. The weight of the body tends to flex

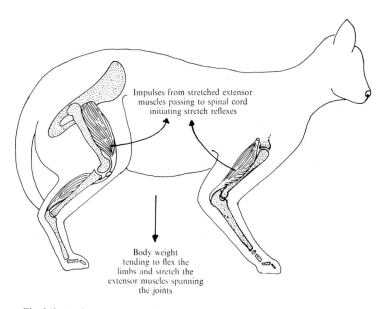

Impulses from stretched extensor muscles passing to spinal cord initiating stretch reflexes

Body weight tending to flex the limbs and stretch the extensor muscles spanning the joints

Fig. 262. *Reflex contraction of the extensor muscles of the limbs in opposing the effect of gravity.*

the limbs and produce a stretching of the extensor muscles which span the limb joints. Reflex contraction of the extensors straightens the limbs and opposes the effect of gravity (fig. 262).

When the tendon of the quadriceps is struck the muscle is stretched and all the muscle stretch receptors fire off a short burst of impulses at the same time. This results in a brief volley of impulses returning to the quadriceps causing a brief strong contraction of the muscle and the jerking of the knee. This is an artificial stimulation of the muscle sense

organs and is rather different from what happens under normal physiological conditions. Under normal conditions all the muscle stretch
receptors are not stimulated together but act asynchronously i.e. they

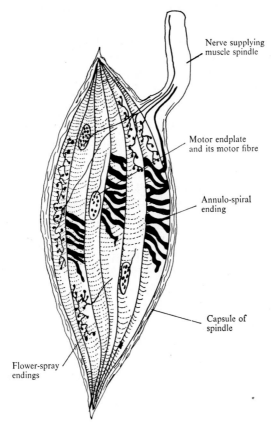

Nerve supplying
muscle spindle

Motor endplate
and its motor fibre

Annulo-spiral
ending

Capsule of
spindle

Flower-spray
endings

Fig. 263. *Diagram of a muscle spindle showing three
types of nerve endings.*

are acting in different parts of the muscle at different times. Thus
impulses come down the motor nerve to the muscle at irregular intervals
over long periods of time and the result is a varied but sustained contraction of the muscle, which is responsible for muscle tone. If the
tendon of a muscle is cut so that no tension can be applied to it then
the muscle becomes completely flaccid—it loses its tone.

In the knee jerk the whole of the quadriceps femoris muscle responds,
but the response can be fractionated so that only a certain segment

of the muscle responds. Thus, as the limbs shift, the pattern of weight bearing alters so that different segments of the muscle are stretched, and the response is limited to the segment stretched. This is an important feature in maintaining posture.

The stretch reflex is also important in locomotion. In progression the foot is first carried forward by a flexion of the limb, carrying the foot clear of the ground, then the limb is straightened bringing the foot in contact with the ground. In the flexion phase the extensor muscles are stretched but during this time whilst the flexor muscles are active the extensors are reflexly inhibited. When the flexor muscles are not operating, however, this inhibition of the extensors is released and the stretch reflex takes over resulting in contraction of the extensor muscles, which then extend the limb. As the limb straightens the stretch reflex dies away.

Proprioceptors—the muscle spindle (fig. 263). In the general sense, proprioceptors detect changes within the body e.g. the pressure in blood vessels, but the term is more often used in a narrower sense to signify those receptors which detect changes in the skeletal and muscular system. An important proprioceptor in muscle is a sense organ called the muscle spindle, which is responsible for initiating the muscle stretch reflex. Each spindle is formed from several modified muscle fibres called intrafusal fibres. These special muscle cells are separated from the surrounding normal striated muscle by a connective tissue sheath which is penetrated by nerves of three kinds.

1. *Small motor neurones* from the ventral root of the spinal cord, which end in typical motor end plates, bringing impulses to the fibres which cause them to contract.

2. *Sensory nerves.*

    (a) Some sensory neurones begin in fine branches (flower spray endings) which are closely applied to the sarcolemma of the intrafusal fibres.

    (b) Larger sensory neurones begin by wrapping spirally around the intrafusal fibres in annulo-spiral endings.

Both flower-spray and annulo-spiral endings are stimulated when the muscle and intrafusal fibres are stretched and this produces a reflex contraction of the muscle and intrafusal fibres. As soon as the intrafusal fibres contract, tension in them is lowered and the discharge of sensory impulses from them is reduced. The annulo-spiral endings are, however, deformed when the diameter of the intrafusal fibres increases as they shorten. This deformation causes a further volley of impulses in the sensory neurone leaving the muscle spindle. By these means the nervous system is made aware both of stretching and active contraction of muscle. There is a further type of sense organ in the tendons (the Golgi tendon organ) which is sensitive to tension, whether produced by active contraction or passive stretch.

By these means the nervous system receives information about the changing pattern of stretch, tension or contraction, which is occurring in the various muscles of the body. This information is vital in the orientation of the body in space (see also p. 338) and in the modification of patterns of muscular activity during locomotion and in the maintenance of posture.

THE METABOLISM OF MUSCLE

The molecular basis of contraction

The basic structure of skeletal muscle has already been described in chapter IV. Here we shall consider the finer structure of muscle as shown by electron microscope studies and relate this to the mode of contraction and relaxation of muscle.

The sarcolemma. When seen through the light microscope the sarcolemma appears to be a structureless dense membrane about 0.1μ thick, but electron microscope studies show a more complex structure including the presence of tunnel like invaginations of the sarcolemma called caveolae. It is thought that these tunnels may permit substances to pass into the muscle fibre without having to move *through* the sarcolemma. Molecules may become bound to the surface of the sarcolemma and the sarcolemma may infold at these points to form caveolae. The caveolae may then become detached from the surface and pass inwards into the sarcoplasm of the fibre where the vesicle gives up the contained substances to other elements of the fibre e.g. by fusion of the vesicle with other internal membranes of the cell such as those of the endoplasmic reticulum or nuclear membrane. Experiments with large particles such as stained fat droplets show that these can cross the sarcolemma indicating that a type of vesiculation mechanism may well be involved, since fat droplets are too large to pass through the membrane by more usual methods.

Endoplasmic reticulum. The muscle cell has a complex endoplasmic reticulum consisting of a series of transverse networks passing across the fibre. The longitudinally aligned myofibrils of the fibre penetrate the meshwork of this system. It seems possible that this reticulum, extending inwards from the sarcolemma, is concerned with the spread of excitation from the sarcolemma inwards to the myofibrils. This is indicated by the fact that the threshold for electrical stimulation of the contractile mechanism is much lower at regions where the sarcolemma is in contact with the endoplasmic reticulum.

The myofibrils. The myofibrils are independent contractile units, less than 0.1μ thick. The structure of the myofibril as seen under the light microscope is shown in fig. 264.

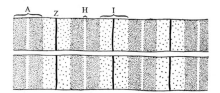

Fig. 264. *Diagram of two myofibrils.*

The 'A' or anisotropic band is birefringent in polarized light and stains darkly in conventionally stained preparations of muscle. The 'H' band is a narrow area of less density occupying the centre of the 'A' band, and is particularly apparent in preparations of stretched muscles. The 'I' or isotropic band is not birefringent and stains paler than the 'A' band. The 'I' band has a darkly staining 'Z' band passing through it. The significance of this repeating pattern along the length of the myofibril is seen on examination of the fibril under the electron microscope when the myofibril is seen to contain two series of overlapping protein filaments. In the 'A' band are thick filaments of the protein myosin. Thinner filaments of the protein actin occupy the 'I' bands and extend into adjacent 'A' bands between the myosin filaments. There is a gap in the middle of the 'A' band which is occupied by thick myosin filaments only. This is the reason for the paler area in the centre of the 'A' band seen under the light microscope. The 'Z' band is formed by a thickening of the thin actin filaments in this region and by a collection of amorphous material between the filaments. The thick myosin filaments bear a series of regularly placed lateral projections which touch adjacent thin actin filaments.

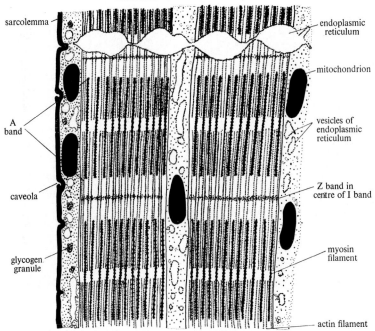

sarcolemma

endoplasmic reticulum

mitochondrion

A band

vesicles of endoplasmic reticulum

caveola

Z band in centre of I band

myosin filament

glycogen granule

actin filament

Fig. 265. *Diagram of longitudinal section of mammalian striated muscle made from several electron micrographs showing two myofibrils of a muscle fibre, the sarcolemma of the fibre and the sarcoplasm with its various components.*

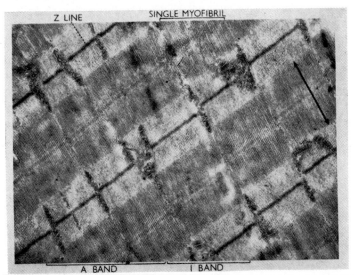

Z LINE

SINGLE MYOFIBRIL

A BAND

I BAND

Fig. 266. *L.P. electron-micrograph of an L.S. of striated muscle showing the banding pattern of the myofibrils.* The arrowed line shows the longitudinal axis of the myofibrils. (*By courtesy of Dr. S. Bradbury.*)

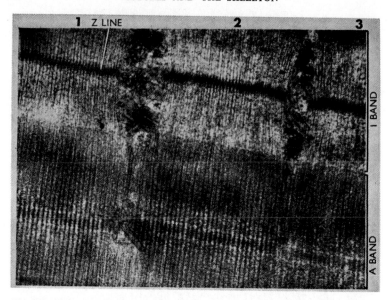

Fig. 267. *H.P. electron-micrograph of an L.S. of striated muscle showing part of A and I bands of three adjacent myofibrils.* Note the thick filaments with lateral projections extending throughout the A band. The thin filaments of the I band extend only partly into the A band to interdigitate with the thick filaments. The Z lines are clearly visible. The longitudinal axis of the myofibrils is in the long axis of the page. (*By courtesy of Dr. S. Bradbury.*)

Changes in the filaments during contraction and relaxation. During contraction or extension of the muscle fibre it seems that there is little change in length of these protein filaments actin and myosin. Instead there is a sliding of the actin filaments in between the myosin filaments of adjacent 'A' bands, so that the tips of the actin filaments approach the centre of the 'A' band. The 'H' band in the centre of the 'A' band thus shrinks during contraction of the myofibrils. If the fibre is stretched then the actin filaments move out of the adjacent 'A' bands. Thus the 'I' bands elongate and the ´H´ zone of the 'A' band widens. This is the so called sliding model of muscular contraction (fig. 267).

Muscle contraction requires a supply of energy and this is provided in the form of A.T.P. In resting muscle the energy output is low and the muscle can be easily stretched because the actin filaments are free to glide out of the myosin filaments. The first sign of the contractile process is a sudden resistance to being stretched. It is thought that this first sign of activation of the contractile mechanism is due to a splitting of some of the A.T.P. of muscle at the sites of the lateral projections on the myosin filaments. This results in a cross union between the actin and myosin filaments at these sites and the muscle develops resistance to being stretched. This development of cross linkages may exert some tension on the actin filaments and when the cross linkages are broken by the arrival of new supplies of A.T.P. at the sites of the cross union, the tension on the actin filaments manifests itself by a sliding of the actin filaments to the centre of the adjacent 'A' bands. Now new sites on the actin filaments are opposite the lateral projections on the myosin filaments. Further splitting of A.T.P. occurs and cross unions between the two series of filaments are remade, only to be broken again as new supplies of A.T.P. reach the region. Thus linkage between actin and myosin filaments is associated with the splitting of A.T.P. The protein myosin is probably itself the enzyme associated with the splitting of A.T.P.

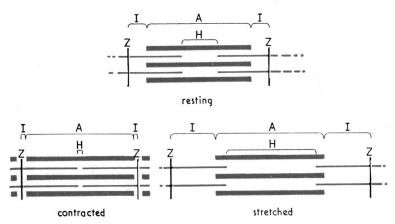

Fig. 268. *Diagram showing the changing relationships between thick and thin filaments during contraction and stretching of muscle.* Only one A band and half of two adjacent I bands is shown. Note that on contraction the thin filaments slide further into the A band, which results in a narrowing of both I and H bands. In stretching of a muscle the thin filaments slide out of the A band causing a widening of both I and H bands.

Relaxation occurs in the presence of free A.T.P. Resting muscle contains free A.T.P. and is extensible (i.e. it has no cross linkages). The A.T.P.-ase activity of myosin seems to be inhibited in the relaxed muscle by a substance present in the sarcoplasm of the fibre called 'relaxing factor'. The myosin A.T.P.-ase requires the presence of free calcium ions before splitting of A.T.P. occurs and the relaxing factor probably operates by binding calcium ions. Activation of the muscle in contraction is probably due to an influx of free calcium ions.

In the development of contraction of striated muscle a series of events occurs.

1. The development of the end plate potential due to the liberation of acetyl choline at the motor end plate. The acetyl choline

increases the permeability of the sarcolemma of the end plate to all ions, which now move along their diffusion gradients. This flow of ions 'short circuits' the resting transmembrane potential in this region. If this process is carried far enough the flow of current which is passing out through neighbouring regions of the sarcolemma initiates a propagated wave of depolarization of the sarcolemma. This involves an increase in the permeability of the sarcolemma specific to sodium ions, which now diffuse inwards along their diffusion gradient from the extracellular to intracellular fluid. This mechanism is identical to that which causes an action potential in a neurone.

2. These events in the sarcolemma are probably linked to the activation of the contractile mechanism by an influx of calcium ions into the muscle cell. This is called the 'coupling' process.

3. The influx of calcium ions activates the myosin-A.T.P.-ase action. Splitting of A.T.P. occurs and cross union of myosin and actin filaments develops, exerting tension upon the actin filaments. These links are broken as new supplies of A.T.P. reach sites and the sliding process begins. Sliding of the actin filaments (manifested as contraction) continues as long as splitting of A.T.P. persists, i.e. as long as myosin A.T.P.-ase activity continues. Relaxation occurs when the free calcium ions of the fibre become bound by the relaxing factor.

Smooth muscle. Smooth muscles of mammals show marked differences in structure and behaviour from striated muscle. The structural unit of smooth muscle is the single uninucleate cell contrasting with the multinucleate fibre of skeletal muscle. Although the cytoplasm of the smooth cell shows fine longitudinally aligned fibrils there is no apparent periodicity of structure of the fibrils, in contrast to skeletal muscle. Fundamentally however, the contractile mechanism of smooth muscle is the same as that of voluntary muscle. Both muscles contain, among others, the proteins actin and myosin. Both have similar energy providing mechanisms. The fact that the fibrils of smooth muscle show no periodicity of structure has led some to doubt the accuracy of the sliding model of contraction for striated muscle.

Spontaneous activity and membrane potential

Smooth muscles characteristically show slow contraction and relaxation and many mammalian smooth muscles show prolonged rhythmical activity. If one removes a piece of striated muscle from a mammal so that its nerves have no connection with the central nervous system then the muscle shows no activity at all unless it is stimulated, either indirectly by its nerves or directly by electrodes. However if one removes a piece of intestine or uterus, containing smooth muscle, and suspends it in a solution of physiological saline then the piece of organ shows spontaneous rhythmical activity independent of any stimulation.

Whether or not a muscle fibre shows spontaneous activity depends upon the transmembrane potential of the cell. This transmembrane potential is generated by active, energy consuming, processes in the cell (page 284). Activation of the muscle cell membrane to initiate a contraction depends upon lowering this transmembrane potential to a critical level.

The reaction of a muscle fibre to stimulation depends upon the membrane potential in excess of the critical level. Muscle fibres with a high membrane potential (e.g. of 90 milli volts) show no spontaneous activity. Activity reqr ires the depolarization of the membrane by means of lowering the end plate potential, until a critical level is reached, at which point the propagated wave of depolarization of the muscle membrane occurs. The muscle membrane of striated muscle cells thus shows an all or none phenomenon. If it is not depolarized to a critical level by the end plate potential, no propagated wave of membrane depolarization occurs and no contraction of the muscle develops. The first step in the contraction of mammalian striated muscle is a membrane event, a propagated wave of depolarization of the sarcolemma initiated by the end plate potential.

In some invertebrate muscles and in some amphibian striated muscles which contract slowly. the contraction of the muscle fibres is not initiated by a propagated wave of depolarization in the sarcolemma. The contraction is initiated directly by the end plate potential itself without any action of the sarcolemma. The end plate potential can be graded according to the amount of acetyl choline liberated at the end plate (which is in turn dependent upon the frequency of nerve stimulation to the muscle), and the muscle response is thereby graded by the size of the end plate potential. These muscles do not show the 'all or none' phenomenon. In striated muscles which obey the 'all or none law', contractions are graded by varying the number of fibres active, ratĥer than by altering the strength of contraction of individual fibres.

The membrane potential of many mammalian smooth muscles is low (e.g. 30–50 m.v.) and is close to the critical level at which a propagated wave of depolarization of the cell membrane occurs. Slow phasic shifts of the membrane potential occur spontaneously, probably as a result of changes in the rate of energy liberating processes inside the cell. These changes are manifested by changes in the rate of extrusion of sodium ions from the interior of the cell, thus leading to changes in the transmembrane potential. The phasic shifts in the transmembrane potential may reach the critical level at which point a propagated wave of depolarization of the membrane starts. In smooth muscle there is little electrical resistance between individual cells and therefore depolarization can easily spread from cell to cell and a wave of contraction will pass along the muscle bundles.

Since the value of the membrane potential shows shifts around the critical level, the activity of smooth muscles can be controlled in both directions. Reduction of the level of the membrane potential soon brings it to the critical level and contraction occurs. If the membrane potential is increased (i.e. hyperpolarized) this makes the muscle much less likely to depolarize spontaneously to the critical level and makes it less susceptible to stimulation i.e. inhibition occurs. Inhibition can thus be achieved in mammalian smooth muscles by substances which stimulate the energy providing mechanisms of the cell and thereby increase the activity of the sodium pump and thus increase the value of the transmembrane potential. Adrenaline for example causes inhibition of spontaneous activity of the small intestine by increasing the membrane potential of the smooth muscles cells.

The muscle cells of striated muscle already have a high and stable membrane potential. These cells can only be regulated in one direction —by depolarization towards the critical level of firing. Smooth muscles are innervated by fibres of the autonomic nervous system. The transmitter substances liberated from the terminals of the nerves, (noradrenaline and adrenaline from sympathetic fibres and acetyl choline from parasympathetic fibres), regulate the activity of the muscles by controlling the level of the membrane potential. Cells are also sensitive to other hormones which reach the muscle via the blood e.g. adrenaline and noradrenaline from the adrenal medulla, vasopressin and oxytocin from the posterior pituitary gland, and in the case of smooth muscle cells of the genital tract, androgenic and oestrogenic sex hormones from the gonads. Thus on the uterus for example adrenaline and the sex hormone progesterone may produce the same effect by stimulating the energy liberating mechanisms in the cell, resulting in an increased activity of the sodium pump and the development of an increased transmembrane potential, with a resulting fall in the spontaneous activity of the muscle cells. This action of progesterone is the basis for the 'progesterone defence mechanism' of pregnancy. Because of the elevated transmembrane potential, spontaneous activity is inhibited and the foetus is much less likely to be prematurely ejected from the uterus.

Stores of energy rich phosphate bonds. As in all cells the quantity of A.T.P. at any one time is very small in muscle tissue. In other tissues the metabolism is dependent upon a steady production of A.T.P., from the oxidation of carbohydrates and fats (see Chapter IX). But in muscle, large amounts of energy are needed at short notice and to supply this need, muscles contain stores of energy rich phosphate bonds in the form of creatine phosphate (phosphagen). This substance is present in striated, cardiac and smooth muscle of all vertebrate animals.

$$\underset{\underset{CH_3}{\overset{|}{}}{N.CH_2COOH}}{\overset{\overset{NH\sim P}{\diagup}}{HN{=}C}}\diagdown \qquad \text{creatine phosphate}$$

Invertebrate animals contain a similar store of energy rich phosphate bonds in the form of arginine phosphate.

During muscular contraction, creatine phosphate is broken down releasing energy rich phosphate bonds which are used to resynthesize A.T.P., and creatine accumulates in the tissues.

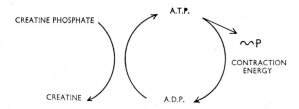

The creatine is gradually reconverted to creatine phosphate and this continues after the muscle has ceased contracting until the stores of phosphagen are built up again. Creatine is converted to creatine phosphate by the action of A.T.P. produced during the breakdown of the carbohydrate glycogen within the muscle.

The breakdown of glycogen and the formation of lactic acid

The glycogen of muscle forms a store of carbohydrate. The first step in the breakdown of glycogen consists in breaking off six-carbon fragments from the long molecular chain. The presence of phosphoric acid is necessary and the reaction, which is reversible, is catalysed by an enzyme phosphorylase. The product of the phosphorylation of glycogen is glucose-1-phosphate, which is then converted into glucose-6-phosphate by an enzyme phosphoglucomutase.

The reverse path is taken when glucose, brought to the muscle in the blood, is deposited in the muscle in the form of glycogen.

Glucose-6-phosphate can now enter the pathway of oxidation to yield pyruvic acid and an increase in energy rich phosphate bonds in the form of A.T.P.; the pyruvic acid is then incorporated into Krebs' citric acid cycle to yield further amounts of A.T.P. These pathways of oxidation have already been described in detail in Chapter IX.

It will be remembered that during the oxidation of glucose-6-phosphate to pyruvic acid, D.P.N. (diphosphopyridine nucleotide) becomes reduced by accepting hydrogen. D.P.N. passes its hydrogen on to a system of hydrogen carriers and ultimately to oxygen by way of the enzyme cytochrome oxidase, and so D.P.N. becomes once again available for the oxidation of more glucose. In the muscle at rest and during moderate exercise, adequate amounts of A.T.P. are produced by these means and adequate amounts of oxygen are reaching the muscle to accept the hydrogen from reduced D.P.N. through the chain of hydrogen carriers. But at the onset of exercise, time is needed for the adjustments to be made in the circulation, and the ventilation of the lungs, in order to provide the increased amount of oxygen required by the muscles. Even when the circulation and ventilation of the lungs has become adapted to supply the muscles with increased amounts of oxygen there may be bursts of severe muscular activity which outstrip the rate of supply of oxygen to the muscles.

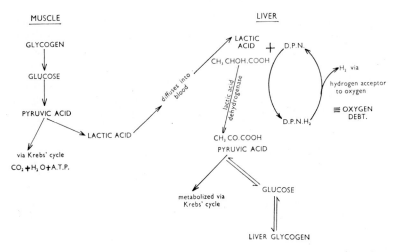

Fig. 269. *The fate of lactic acid produced during the anaerobic respiration of muscle.*

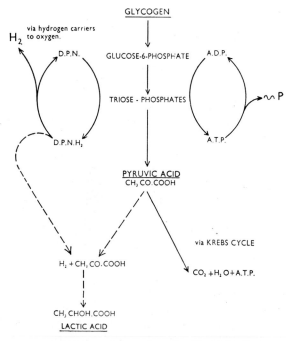

Fig. 270. *The metabolism of muscle. The broken lines indicate pathways used during temporary oxygen lack.*

Now there is only a limited amount of D.P.N. in the muscle tissues and once this has accepted hydrogen during the oxidation of glucose it can no longer participate in the oxidation of glucose, unless it is able to pass on its hydrogen to some hydrogen acceptor. During the initial phase of muscular activity and during severe exertion the supply of oxygen to the muscles is inadequate for the respiratory needs. In these circumstances reduced D.P.N. is able to pass on its hydrogen to pyruvic acid itself, so producing lactic acid (see fig. 270).

Lactic acid and oxygen debt. We have seen that in the initial phase of muscular exercise and during severe muscular exercise e.g. a sprint, the muscles obtain some of their energy by anaerobic means with the production of lactic acid. This lactic acid does not remain in the muscles but diffuses out into the tissue fluids and into the blood, and measurements of blood lactic acid shows that it increases progressively during severe muscular exertion. After the period of muscular activity

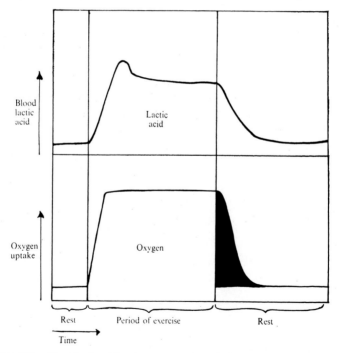

Fig. 271. *Blood lactic acid and oxygen intake, before, during and after severe exercise. The shaded area of the oxygen uptake graph represents the amount of oxygen which must be taken in after exercise has finished in order to metabolize the lactic acid produced during exercise. This amount of oxygen is called the oxygen debt.*

this lactic acid is, in the mammal, metabolized in the liver. Here the hydrogen which was accepted from reduced D.P.N. in the muscle is passed on to D.P.N. in the liver with the aid of an enzyme, lactic acid dehydrogenase, and the hydrogen is passed, through a system of hydrogen acceptors to oxygen. The pyruvic acid so produced is then metabolized, either passing into Krebs' cycle or being converted into glucose and perhaps glycogen (see fig. 270).

Thus it is apparent that during the period of rest after severe exercise increased amounts of oxygen are required in order to accept the hydrogen from the lactic acid produced during the severe exercise. The lactic acid is, as it were, acting as a temporary store of hydrogen until adequate amounts of oxygen are available. The increased amounts of oxygen needed during the period of rest, over and above the resting needs of the animal, is termed the oxygen debt. There is a physiological limit to the degree of oxygen debt that can be incurred and in man this amounts to about 16 litres of oxygen (see fig. 271).

The sprint versus the marathon. We can now summarize the metabolic changes that occur in muscles during exercise by comparing the sprint and the marathon. In the sprint large quantities of energy may be required over a short period of time. The available A.T.P. and phosphagen of the muscle is rapidly used up and the supply of oxygen to the muscle becomes inadequate to cope with the reduced D.P.N. which is being produced at a rapid rate from the breakdown of glucose to pyruvic acid. Reduced D.P.N. is temporarily relieved of its hydrogen by some of the pyruvic acid produced, which becomes converted into lactic acid. This lactic acid is oxidized later in the liver and thus an oxygen debt is incurred during the sprint which is paid off during the resting period.

Compared to the fast sprinter, the good marathon runner does not acquire an oxygen debt. The marathon runner is in oxygen equilibrium, oxygen uptake being equivalent to oxygen requirement. The energy requirement of the muscle is provided by the concurrent oxidation of carbohydrates. The increased oxygen required is made available by an opening up of the vascular bed of the muscle, by an increased output of the heart, an increased ventilation of the lungs, and by an increased abstraction of oxygen from the blood by the muscle tissues which is determined by the altered properties of haemoglobin in a medium of increased acidity, carbon dioxide tension and a rise in temperature (see p. 205). The efficiency of the marathon runner depends upon how well he can utilize all these mechanisms to ensure sufficient oxygen supply to the muscles, and this in some measure is determined by training. The sudden outbursts of severe muscular effort of the sprinter, or of the herbivorous animal being chased by a carnivore, occur before these adaptations of the cardiovascular and respiratory systems have time to develop, and the ability to incur an oxygen debt is of great adaptive significance. But the marathon runner could not possibly run the distances he does if he acquired any serious oxygen debt.

Muscle haemoglobin. In certain muscles of the mammal which are engaged in slow, repeated powerful movements (e.g. the hearts of large mammals, back muscles) there is a special adaptation which enables the muscle cells to obtain oxygen during a sustained contraction when the flow of blood through the contracting muscle is interrupted. This lies in the presence in these muscles of a haemoglobin (called myoglobin) with different properties from blood haemoglobin. Myoglobin has a greater affinity for oxygen than has blood haemoglobin, and it is saturated with oxygen at the oxygen pressures of venous blood, and becomes unloaded at lower partial pressures of oxygen.

The muscle fibres containing myoglobin are thick and red in colour contrasting with the pale, thin, rapidly contracting fibres which do not contain myoglobin. Although some mammals possess muscles consisting exclusively of one type of fibre, in man the muscles consist of a mixture of the two types.

THE SKELETAL SYSTEM

The general functions of the skeleton. Unlike invertebrate animals where the supporting system of the body is external (exo-skeleton) with inwardly directed projections for the attachment of muscles, the vertebrate animals possess an internal skeleton (endo-skeleton). This endo-skeleton serves several functions but basically it is a supporting system, supporting the weight of the animal, in its various positions, against the force of gravity. The action of the skeleton is reinforced by the action of the muscles attached to it in supporting the weight of the animal. The skeleton is not a rigid structure but is broken into units, jointed together, to permit movement. The exact form of the skeleton varies according to the needs of the particular species, depending on whether the animal is quadrupedal or bipedal, whether it walks and runs, swims or flies, on whether it is supported by water or is free living in air.

In addition to serving the function of support and movement the endoskeleton is also protective. The bony skull houses and protects the soft tissues of the brain, and the vital sense organs of sight and smell. The bony vertebral column contains the delicate spinal cord.

The bony cage of the thorax protects the heart and large blood vessels and the lungs.

Many of the bones of the body contain the red bone marrow in which the red blood cells and the white cells of the granulocyte series are being continually produced. In addition to these more obvious functions the bones contain vast quantities of calcium, which by the action of the hormone parathormone (see p. 641) form an easily mobilized store of calcium on which the body can draw in times of need.

The materials of the skeleton: cartilage and bone

Cartilage. Cartilage, like bone, is a connective tissue in which the cells secrete a matrix which forms a predominant part of the tissue. The amount of cartilage in the skeleton varies according to the species of vertebrate and the age of the animal. In some of the lower vertebrates e.g. Cyclostome fishes (lamprey etc.), the adult skeleton is composed entirely of cartilage, but this is regarded as a degenerative condition. In mammals cartilage plays a more important role in the embryo and young animal, and in the adult it becomes restricted to a few sites such as the free ends of long bones, the trachea and bronchi, external ear and nose, and parts of the ribs.

THE STRUCTURE OF CARTILAGE, HYALINE CARTILAGE. Hyaline cartilage may be regarded as having the basic structure of cartilage (fig. 272) which may be modified in certain regions of the body to serve particular functions. In the gross it is a semi-transparent, opalescent, elastic material. It is found in the body in the supporting rings and plates of the trachea and bronchi, at the joint surfaces of bones, and at the ventral ends of the ribs where the ribs join the sternum.

Cartilage is covered on its free surface, except where it abuts into the cavity of joints, by a layer of dense connective tissue, the perichondrium. From the perichondrium, fibroblasts differentiate into cartilage cells or chondroblasts and thus the cartilage grows from its surface. Cartilage is capable of a degree of internal expansion by division of cells within its substance. The newly formed chondroblasts undergo mitotic division several times in rapid succession to form groups of cells. These then begin to secrete the intercellular substance or matrix. Thus in mature cartilage groups of cells occur separated by areas of cartilaginous matrix. The chondroblasts lie in cavities within the matrix; in life they completely fill these cavities although in stained sections they may shrink away from the walls of the cavity or may even fall out, leaving an empty space in the matrix. Unlike the structure of bone there are no processes from the chondroblasts which extend into the matrix.

The interstitial matrix of hyaline cartilage appears structureless when seen in ordinary fixed and stained sections under the microscope. However, by special methods of silver impregnation it has been shown that there are many fine fibrils permeating the matrix, which may be organised into orientated bundles. The matrix stains with basic dyes

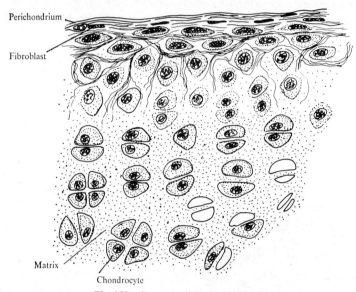

Fig. 272. *Structure of hyaline cartilage.*

because it contains a chondro-mucoid protein; this protein is a glyco-protein in which the carbohydrate component is a sulphonated poly-saccharide, chondroitin sulphate, the acidity of which explains the staining properties of the matrix. The matrix around the chondroblasts often stain more deeply with basic dyes, presumably because it has a higher concentration of the glyco-protein. The nutrition of the cartilage cells is dependent upon diffusion of substances through the matrix; unlike bone, cartilage has no blood vessels of its own, except in the perichondrium.

MODIFICATIONS OF THE BASIC PATTERN. The mechanical properties of hyaline cartilage are a reflection of its structure. The matrix itself forms an elastic and somewhat compressible material, but the fluid material of the chondroblasts enclosed in bundles of fibres confers a resistance to compression forces. Thus hyaline cartilage can sustain great weight in its situation at the end of the long bones and at the same time it permits the smooth movement of one bone upon another.

Greater flexibility and elasticity of the cartilage is conferred by the presence of elastic fibres in the matrix. Elastic cartilage which is yellower in colour and more opaque than hyaline cartilage is found in the pinna of the ear, in the eustachian tube and the epiglottis.

Greater tensile strength is present in fibro-cartilage in which there are dense bundles of collagen fibres in the matrix. Fibro-cartilage is intermediate in structure between hyaline cartilage and connective tissue. It is present as a resilient pad, the intervertebral disc between the bodies of the vertebrae, and at the symphysis pubis and associated with the connective tissues of capsules and ligaments of joints.

Bone. Bone is a hard connective tissue in which the matrix produced by the bone cells (osteoblasts) is heavily impregnated with salts of calcium to produce a rigid material. Its architecture, to be described

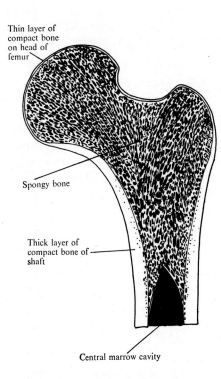

Thin layer of compact bone on head of femur

Spongy bone

Thick layer of compact bone of shaft

Central marrow cavity

Fig. 273. *Longitudinal section through the upper end of the femur showing the distribution of spongy and compact bone.*

subsequently, is adapted to withstand compression strains. In spite of its inert appearance and its low metabolic rate bone is a highly adaptable material and is able to adapt (mould) to meet changing mechanical requirements.

In gross appearance the bones of mammals may appear spongy or compact. In both types of bone there is the same fundamental structure. In spongy bone there is a meshwork of interconnecting bars of bone, the spaces between the meshes being filled with marrow. In this way rigidity of structure is attained with minimum weight. Compact bone, as its name implies, consists of a hard mass in which spaces cannot be seen macroscopically. This type of structure ensures maximal ability to withstand compression and is found, for example, in the shaft of long bones. Most bones contain an admixture of the two types of bone and in fig. 273 a long bone is illustrated to show the distribution of these two types of bone. The shaft of the long bone (diaphysis) consists of compact bone; the shaft is, however, not solid and contains a wide marrow cavity, another adaptation to reduce the weight of the bone. The epiphysis of the bone at the end of the shaft contains spongy bone, limited externally by a layer of compact bone. In the figure a mature bone is represented in which the cartilaginous plate, which in the growing mammal separated the epiphysis from the diaphysis, has been converted into bone.

HISTOLOGICAL STRUCTURE OF BONE. When a transverse section of compact bone is examined under the microscope a series of disc shaped structures is seen (fig. 274) each with a cavity in the centre. Each disc is the end view of a cylindrical structure called an haversian system. The canal in the centre is called the haversian canal and contains the blood vessels which supply the particular haversian system. When examined closely each disc is seen to consist of a series of concentric plates of bone or lamellae. Between the lamellae are small cavities or lacunae which house the osteoblasts, and radiating from the lacunae are many fine channels, the canaliculi, which interconnect the lacunae and promote the diffusion of tissue fluids around the osteoblasts.

The haversian systems are not simple cylinders of bony material, but they are irregularly branched and anastomosing structures which form a meshwork. Because they are directed mainly in the longitudinal axis of the bone a transverse section of bone shows them to be disc shaped structures. In longitudinal section the Haversian canals appear as long slits.

The disc shaped end views of the haversian systems do not compose the entire transverse section of the bone. Both on the outer surface of the bone and also lying around the marrow cavity there are lamellae which pass circumferentially, the basic or circumferential lamellae. Penetrating these circumferential lamellae, from the free surface of the bone or from the marrow cavity are canals containing blood vessels which connect with those in the Haversian canals; these vascular channels are called the canals of Volkmann.

The outer layer of bone is closely covered by a layer of dense connective tissue called the periosteum. A layer of connective tissue also

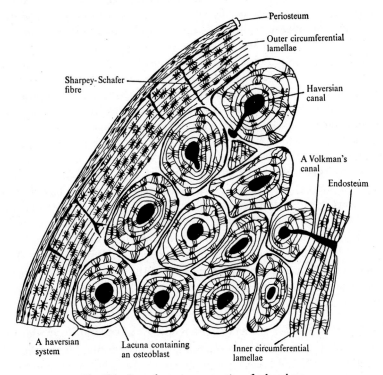

Fig. 274. *Part of a transverse section of a long bone.*

covers the bone facing the marrow cavity. Penetrating the bone from the periosteum are bundles of collagenous fibres called Sharpey's fibres. These ensure a close connection of the periosteum with the underlying bone, which is of vital importance in regions where tendons are inserted into the periosteum.

The bone matrix. The hard bone matrix secreted by the osteoblasts contains two components, an inorganic base amounting to about 65% of the weight of dry bone and an organic base amounting to about 35% of the weight of dry bone.

The inorganic component of bone consists of a lattice of submicroscopic crystals of hydroxy-apatite, $[Ca_3(PO_4)_2]_3.Ca(OH)_2$. In addition to calcium and phosphorus, bone also contains magnesium, sodium, carbonate and citrate. The organic component of bone is called ossein or bone collagen. The collagen fibres are manufactured and aligned by modified connective tissue cells called osteoblasts. When there is an adequate concentration of calcium and phosphate in the vicinity of the fibres these interact at specific sites on the collagen fibres and form small hydroxyapatite crystals. These small crystals slowly grow and in this phase the calcium and phosphate of the crystals is freely exchangeable with the tissue fluids; if the concentration of calcium and phosphorus in the tissue fluids falls then calcium and phosphate ions can pass from the crystals into the tissue fluids so maintaining the composition of the internal environment. As the crystals grow the bone becomes more compact and fluid is excluded. This solid bone does not release its ions readily to the internal environment, forming what is called non-exchangeable bone.

The inorganic and organic components of bone can be demonstrated relatively easily. The inorganic component of the matrix can be removed to a great extent by immersing a bone for some time in a dilute mineral acid, such as hydrochloric acid, when the calcium is gradually removed from the bone in the form of a soluble salt. When this happens the bone retains its external form but becomes pliable and rubber-like. The organic matrix can be removed by igniting the bone when the organic matrix is burnt away leaving behind the mineral components. The bone still retains its form but it becomes very brittle.

THE DEVELOPMENT OF BONE. Bone develops by a transformation of connective tissue. It may arise directly from connective tissue when it is called membrane bone e.g. the bones of the skull, or it may arise from a modification of pre-existing cartilage.

The development of bone from pre-existing cartilage (endochondral ossification). The development of bone from pre-existing cartilaginous structures has been considered to be a recapitulation, during the embryonic period (ontogeny), of what has occurred during the history of vertebrates (phylogeny). This concept arose because of the existence of present day primitive vertebrates in which the adult skeleton is composed entirely of cartilage (e.g. lamprey). It is now thought that this condition is a degenerate one and does not represent a stage in the development of the bony skeleton in phylogeny. The cartilaginous stage in the development of bones is a vital one, however, and reflects the need of the skeleton of the young mammal to serve two functions simultaneously, to function as a skeleton and to grow. The proper

functioning of the skeleton depends upon the articulation of one bone
with another and upon the attachment of muscles and tendons. Now
bone is a material which can grow only from its surface and if the
skeleton were to grow in this manner then there would be such a disturbance of the relations between jointed ends of bone and a disturbance
of the relations with tendons and muscles that growth would be incompatible with function. Cartilage however is capable of internal
expansion and can provide for growth of 'bones' without disturbing
their surface relations. The ways in which this occurs will be subsequently described.

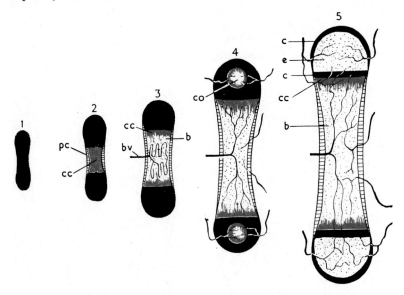

Fig. 275 (*for legend see opposite*).

We can now study the endochondral ossification of a long bone to
show how a bone is formed from a pre-existing cartilaginous model
and how the mature histological structure of bone arises. The general
shape of the cartilaginous model follows the future shape of the bone
(fig. 275 (1)). The first change in this structure is in the perichondrium
which surrounds the cartilage. It occurs first in the perichondrium midway along the shaft (the diaphysis) where a layer of membrane bone is
produced. There is a transformation of connective tissue cells of the
perichondrium into osteoblasts which secrete lamellae of osteoid
material which becomes calcified; this is the essential manner of membrane bone formation. This process results in the formation of a ring

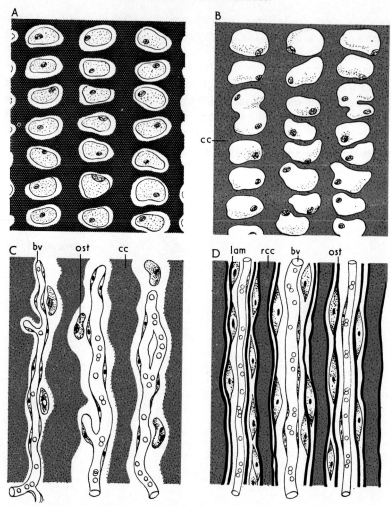

Fig. 275. *Diagrams illustrating the development of a long bone. 1. The cartilaginous model, showing the general shape of the long bone to be. A. High power view of a portion of cartilage from the shaft showing the longitudinal arrangement of cartilage cells. 2. The development of a periosteal cuff of bone* (p.c.) *in the diaphysis, together with the calcification of the underlying cartilage* (c.c.). *B. Shows the calcified cartilage with swollen vesicular cartilage cells. 3. The invasion of the calcified cartilage of the shaft by blood vessels* (b.v.) *and connective tissue cells from the periosteal cuff of bone. C. Shows how the invading blood vessels and connective tissue convert the columns of swollen cartilage cells into longitudinally aligned tubes separated by remains of cartilage. 4. Centres of ossification* (c.o.) *are being laid down at the ends of the bone—the epiphyseal centres* (e). *The connective tissue which has invaded the shaft has been converted into osteoblasts, producing bone. D. Shows these osteoblasts* (ost.) *laying down concentric lamellae* (l.) *of bone on the walls of the cavities produced by the breakdown of the cartilage cells—producing the primary haversian systems. Some remnants of calcified cartilage* (r.c.c.) *are shown between the haversian systems. 5. The entire cartilaginous model has been converted into bone except for an area of cartilage between the epiphysis* (e.) *and diaphysis. This is the site of growth in length. The cartilage is continually proliferating and as cartilage cells are pushed away from the active area they become calcified and invaded by blood vessels and osteoblasts from the diaphysis. Growth in length of the bone ceases when this epiphyseal plate of cartilage is converted into bone. Cartilage is shown black, calcified cartilage red.*

of bone in the diaphyseal region and it strengthens the shaft so that subsequent changes in the underlying cartilage do not unduly weaken the structure.

Whilst this process is going on there are changes in the underlying cartilage of the diaphysis. The cells of the cartilage swell and the matrix between them becomes calcified. This calcified cartilage is then invaded by connective tissue and blood vessels from the band of bone in the outer layers of the diaphysis (figs. 275 (3), 276). This tissue invades the cavities containing the swollen cartilage cells, and since the latter tend to be arranged in columns the opened up capsules of the cartilage cells eventually become hollow columns containing blood vessels and connective tissue, (the embryonic bone marrow) and surrounded by calcified cartilage. Connective tissue cells of the embryonic bone marrow transform into osteoblasts which are bone forming cells and therefore coat this calcified cartilage with lamellae of bone. This process extends up the shaft and at the same time there is an extension of the formation of membrane bone in the perichondrium (now called the periosteum).

Growth in length of the bone is occurring during this time by the activity of a layer of proliferating cartilage cells situated at the ends of

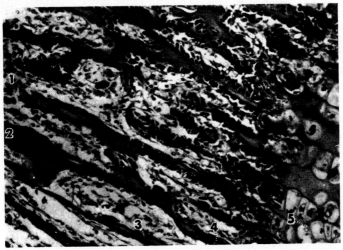

Fig. 276. *Photomicrograph of an L.S. of developing bone at the epiphysis showing invasion of columns of cartilage cells by blood vessels and osteoblasts.* The remnants of calcified cartilage are surrounded by osteoblasts. 1. primitive bone marrow. 2. spicule of new bone containing embedded osteocytes. 3. lamella of new bone laid down on remnant of calcified cartilage. 4. remnant of calcified cartilage covered by osteoblasts. 5. column of cartilage cells invaded at upper end by osteoblasts and blood vessels. (*By courtesy of Dr. S. Bradbury.*)

the shaft, between the shaft and the cartilaginous epiphysis. Later, centres of ossification develop in the epiphyses so that the entire cartilaginous model has become replaced by bone except for an area of proliferating cartilage between the epiphysis and the diaphysis, which is the site of growth in length and the site at which the growth hormone of the anterior pituitary gland exerts its effects. In the mature animal this zone of proliferating cartilage is transformed into bone and the possibility of growth in length ceases. The bones increase in width by the deposition of bone by the periosteum.

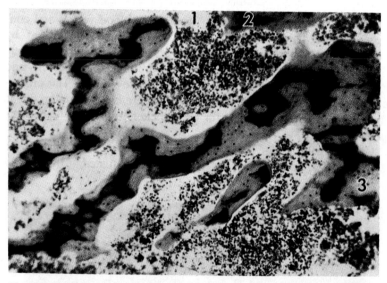

Fig. 277. *Photomicrograph of a section of developing endochondral bone.* The remnants of calcified cartilage are shown as dark regions in the centre of the spicules of new bone which also contain embedded osteoblasts. The spicules of bone are separated by primitive bone marrow. 1. bone marrow. 2. darkly stained remnant of calcified cartilage. 3. spicule of new bone containing osteocytes. (*By courtesy of Dr. S. Bradbury.*)

Reconstruction of bone. From the earliest period of endochondral ossification of the cartilaginous model, bone absorption is going on at the same time as bone formation. Bone absorption, both of matrix and inorganic elements, is thought to be carried out by certain cells of bone called osteoclasts. These are always present in large numbers in any area of bone resorption and disappear when this process ceases, perhaps becoming transformed into osteoblasts. In areas of bone resorption the osteoclasts may be seen to lie in small pits of bone which they have eroded, and these are called Howship's lacunae.

During the early development of the bone, primary haversian systems were formed by the laying down of bone on the walls of the eroded cartilage cells (fig. 275 (D)). Increasing accumulations of bone reduce the initially wide central cavity into a narrow channel, the primary haversian canal. These primary haversian systems are broken down by the action of osteoclasts working from within the central cavity, and this is followed by the reconstruction of new haversian systems by the laying down of bone on the inner surfaces of the new cavities produced. This reconstruction continues throughout life, but at a much slower rate in the adult animal. Reconstruction also involves the circumferential lamellae of bone, produced by the periosteum and the endosteum, and these are replaced by haversian systems as new circumferential lamellae are being formed.

The constant reconstruction of the architecture of the bone is the basis of the ability of bones to modify their internal architecture according to the mechanical stresses imposed on them. This is important in those species in which there is a change in the mode of progression or weight bearing during development. Following a fracture in which the ends of the bones are in poor anatomical alignment these remodelling processes may even bring about the reconstruction of a near normal bone, particularly in the young; this does not apply to cases in which there is a change in the length of the bone due to overlapping of the two pieces.

Joints

Joints occur where different bony or cartilaginous elements of the skeleton are in apposition. Two or more elements of the skeleton may be closely joined together with little possibility of movement between them e.g. the bones of the skull, and such a joint is termed a synarthrosis. The material which binds the elements together may be cartilage or fibrous tissue. In the other type of joint the skeletal elements are capable of movement on each other e.g. the hip joint, and such a joint is termed a diarthrosis.

Synarthrodial joints. As mentioned above a synarthrosis is a joint in which the skeletal elements are firmly attached to one another and in which movement is relatively limited. In the skull the separate bones are joined by synarthrodial joints in which there is a thin layer of fibrous tissue between the bones; this fibrous tissue is continuous with the overlying periosteum of the bones and the underlying dura mater covering the brain (fig. 278). Virtually no movement occurs between the bones of the skull and these joints are known as sutures. In some synarthrodial joints there is a much greater amount of connective tissue between the bony elements and some movement is possible e.g. the inferior tibio-fibular joint in man (fig. 279).

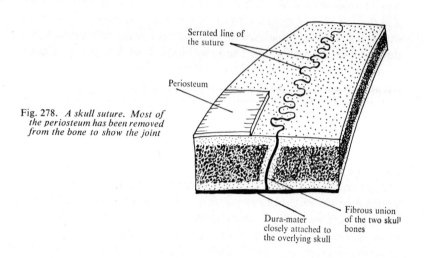

Serrated line of the suture

Periosteum

Fig. 278. *A skull suture. Most of the periosteum has been removed from the bone to show the joint*

Dura-mater closely attached to the overlying skull

Fibrous union of the two skull bones

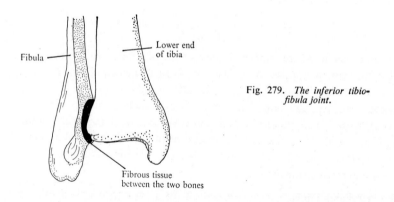

Fibula

Lower end of tibia

Fig. 279. *The inferior tibio-fibula joint.*

Fibrous tissue between the two bones

Diarthrodial joints. A diagram of a diarthrodial joint is shown in fig. 280. This is a highly specialized structure enabling the bones to move on one another. The exact form of the joint and the range of movement permitted at the joint varies from one part of the body to another and according to the species of animal. If in addition to movement the joint has to take part in weight bearing then other modifications in structure may appear.

The ends of the articulating bones are shaped to permit ease of movement between the two bones, the particular form varying from joint to joint. Thus in the hip joint of man the rounded head of the femur is sunk deep into a cup in the innominate bone; this arrangement permits ease of movement and at the same time the margins of the cup support the joint, which is weight bearing. At the shoulder joint stability is sacrificed in order to maintain mobility and the rounded head of the humerus fits into a shallow cup on the scapula; stability of the joint is maintained by muscles—but this joint in man is dislocated with relative ease.

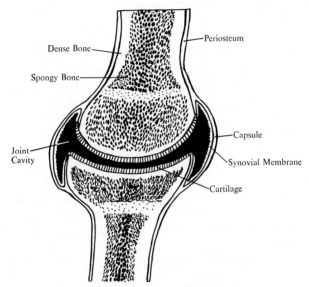

Fig. 280. *Diagram of a diarthrodial joint.*

The end of the bone which takes part in the joint consists of spongy bone surrounded by a layer of particularly compact bone. That part of the bone which comes into contact with the other member of the joint is covered by a layer of hyaline cartilage. This contains no nerves or blood vessels and is nourished by diffusion of substances from the thin layer of fluid in the joint cavity and from the synovial membrane.

The joint is strengthened by a strong connective tissue capsule which extends from the periosteum at one side of the joint, across the joint cavity to be attached to the periosteum at the other side. There are usually condensations of particularly dense connective tissue in the capsule and their disposition varies from joint to joint, relative to the stresses experienced by the particular joint. There may be projections of fibrocartilaginous material from the capsule into the cavity of the joint and these may extend right across the cavity of the joint as a pad of tissue between the two bones; this forms a resilient pad between the two bones. In the case of the human knee joint they form crescentic projections into the joint—the so-called 'cartilages' of the knee. The capsule of the joint is lined by a highly vascular velvety tissue, the synovial membrane, which is reflected onto the periosteum and reaches the edge of the hyaline cartilage. This produces a viscous colourless secretion called synovial fluid, which lubricates the joint surfaces and so reduces friction between them. It is present in small amounts only and there is less than 0·5 ml. in the human knee joint.

The metabolism of calcium

Functions of calcium. In addition to forming the important rigid component of bone, calcium salts are vital for the normal functioning of all cell membranes. The membranes of cells control the passage of substances into and out of the cell. Cells contain less sodium and more potassium ions than the extracellular fluid. These differences in ionic constitution of the cells and the fluid around them is determined by the properties of the cell membrane and by active mechanisms which extrude sodium from the interior of the cell and retain potassium. In excitable cells such as neurones and muscle cells, marked changes occur in the properties of the cell membrane during excitation. Acetyl choline, liberated at motor end plates on muscle fibres or at synapses impinging upon neurones, increases the permeability of the cell membrane allowing some or all the ions of the intra- and extra-cellular fluid to move along their diffusion gradients. Recovery after conduction in a neurone or recovery after contraction of a muscle fibre is associated with the re-establishment of the relatively impermeable properties of the cell membrane and the re-establishment of the ionic differences between the intra- and extra-cellular fluids. The calcium ion is important in maintaining the relative impermeability of the resting cell membrane. If neurones or muscle cells are bathed in solutions of low calcium concentration the cell membrane becomes increasingly permeable to ions so that differences in the ionic constitution of the intra- and extra-cellular fluids cannot be fully maintained. The potential difference across the cell membrane falls so that excitation of neurones and contraction and relaxation of muscle fibres is disturbed. In solutions of low calcium content, muscle cells may thus undergo spontaneous depolarization, without the need of activation by acetyl-choline, so that spontaneous twitching and spasms of the muscle occur. Because of the sustained permeability of the cell membrane, relaxation of the muscle fibre may not occur and sustained contractions of muscles may develop. This condition is called tetany.

In addition to this stabilizing influence of the calcium ion on the cell membrane, calcium ions are also concerned in another aspect of excitation of muscle cells. When a muscle cell is excited by acetyl choline there is not only a flux of sodium ions across the cell membrane but there is also an inward flux of calcium ions into the cell. This inward flux of calcium ions, by activating the A.T.P. splitting properties of the muscle protein myosin may be the actual triggering mechanism for contraction. Calcium ions are also important in the secretory activity of some types of cell e.g. adrenal medullary cells, secreting adrenaline and noradrenaline, and the cells in the hypothalamus which secrete anti-diuretic hormone. In solutions weak in calcium these cells do not give the normal response of outpouring of their hormones in response to nerve stimulation. Probably, as in muscle cells, there is an inward flux of calcium ions into these cells on nerve stimulation which triggers off the release of hormone.

Calcium ions also form an essential component of the mechanism which clots blood (p. 216).

The regulation of body calcium. A normal level of calcium ions and a proper ratio of calcium to phosphate (which is necessary for normal bone growth) in the circulating blood is maintained in two ways. First, an adequate dietary intake of these elements is essential. How the body deals with this dietary intake, that is whether the elements are excreted or retained by the body, and if retained, whether they are taken up into the circulating calcium and phosphate of the blood or are incorporated into bone, depends upon the activities of two hormones, parathormone and vitamin D. Vitamin D is here regarded as a hormone because it is produced in an organ, the skin, under the influence of sunlight, from which it passes into the blood stream to have its effects on distant organs. The body is dependent upon an external source of this hormone because of inadequate contact of sunlight with skin.

The parathyroid glands and calcium metabolism. In man the parathyroid glands consist of two pairs of yellowish-brown structures which are closely applied to the back of the thyroid gland (fig. 281). In development they arise from the third and fourth pharyngeal pouches. The secretion of these endocrine glands is a hormone called parathormone which is discharged into the blood vessels of the gland. Parathormone has now been purified and found to be a protein of rather low molecular weight (9500) consisting of a chain of 83 amino acids.

The parathyroid glands unlike many other endocrines is not under the control of the anterior pituitary gland. The secretory activity of the gland is controlled directly by the level of calcium ions in the blood stream, just as the islets of Langerhans of the pancreas are under the

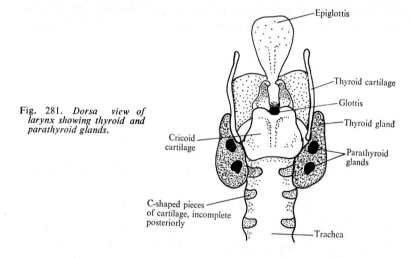

Fig. 281. *Dorsa view of larynx showing thyroid and parathyroid glands.*

direct control of the level of blood glucose. The parathyroid glands are stimulated to secrete parathormone when the concentration of calcium in the blood falls. The hormone acts at a variety of sites in the body, the net result of the actions at these various sites being to return the concentration of calcium ions in the blood to normal levels. The actions of the hormone include:

1. An increased absorption of calcium from the intestine.

2. An increased retention of calcium by the renal tubules.

3. An action on bone. When parathormone acts upon bone there is an appearance of bone destroying cells called osteoclasts. These cells actively destroy the old non-exchangeable bone and release the stores of calcium ions into the circulating blood.

4. There is a further effect on the kidney resulting in an excretion of increased quantities of phosphates. This offsets the rise in phosphate which would occur from a release of calcium and phosphate from bone.

The effect of persistently high rates of secretion of parathormone is seen in cases of tumours of the parathyroid gland. In this condition of hyperparathyroidism there is a progressive demineralisation of bone and cystic areas may appear in the bones containing large numbers of osteoclasts, and bones may fracture spontaneously at these sites of weakness. Because of a persistently high excretion of phosphate and calcium in the urine, calcium phosphate stones may appear in the kidneys which predispose to urinary infections.

In the opposite condition of hypoparathyroidism there is an inadequate production of parathormone and because the bones can no

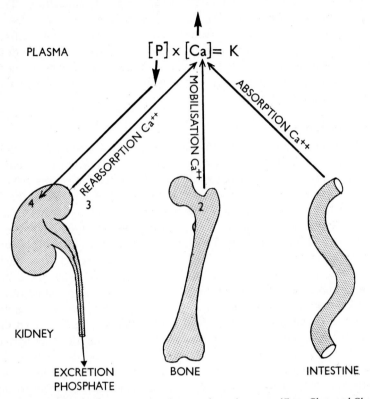

Fig. 282. *Diagram indicating the sites of action of parathormone.* (*From Clegg and Clegg, 'Hormones, Cells and Organisms', Heinemann Educational Books Ltd.*)

longer act as an easily mobilized source of calcium the plasma calcium values fall to abnormally low levels. This develops rapidly in the young animal where the growing bones are taking up calcium avidly. The nervous system becomes increasingly irritable with the falling blood calcium and muscular spasms may appear. If untreated, the condition is eventually fatal. In man the condition of hypoparathyroidism may appear after removal of the thyroid gland (done for example because of a goitre or tumour of the gland) since the parathyroid glands are so closely associated with the thyroid that they may all be removed accidentally.

The various effects of parathormoen have been difficult to elucidate and until recently the changes in bone were regarded as secondary to the low blood phosphate caused by increased renal excretion of phosphate. That the hormone does have a direct effect on bone is shown by the fact that osteoclasts appear and erosion of bone occurs when the hormone is added to bone isolated and growing in tissue culture. That the hormone has a direct renal effect has been demonstrated by injecting the hormone into one renal artery and collecting the urine from the two kidneys separately for analysis of phosphate. If this is done then it is found that the kidney receiving an injection of parathormone into the renal artery produces a urine with a higher phosphate content than the opposite kidney.

Thyrocalcitonin

It has often seemed remarkable that the fine regulation of the calcium concentration in blood can be achieved by means of a single hormone, parathormone. If the amount of calcium is artificially lowered by injecting a calcium binding agent into the blood of a mammal this is followed by a gradual return of calcium concentration to normal levels; this is due to the mobilization of parathormone secretion which frees calcium from the bones. But there is no temporary overshoot of calcium concentration above normal levels as one might expect if a single hormone were involved in the regulation of calcium. This and other observations gave rise to speculation that there must be some other factor acting as a *hypo*calcaemic agent to oppose the *hyper*calcaemic action of parathormone. In 1962 a worker called Copp described the presence of a fast-acting hypocalcaemic factor leaving the parathyroid glands when these were perfused with blood outside the body. The blood perfusing the glands was enriched with calcium artificially and when this blood had perfused the isolated glands and returned to the experimental animal it caused a rapid fall of the concentration of calcium in the blood of the animal (fig. 283).

We can visualize the regulation of the calcium content of blood as being the result of the action of two hormones having opposing effects. Parathormone can raise the calcium concentration of blood by mobilizing the ion from bone. Calcitonin reduces the calcium concentration of blood, probably by increasing the uptake of calcium from the blood by bones. Which of these two hormones is secreted into the blood depends on the calcium content of blood.

Because the parathyroid glands are small and do not have a discrete blood supply distinguishable from that to the thyroid gland Copp perfused thyroid tissue in which the parathyroid glands were embedded.

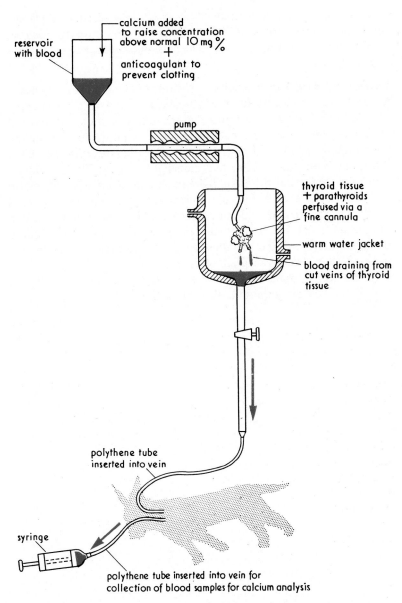

reservoir
with blood

calcium added
to raise concentration
above normal 10 mg %
+
anticoagulant to
prevent clotting

pump

thyroid tissue
+ parathyroids
perfused via a
fine cannula

warm water jacket

blood draining from
cut veins of thyroid
tissue

polythene tube
inserted into vein

syringe

polythene tube inserted into vein for
collection of blood samples for calcium analysis

Fig. 283. *Diagram representing the technique used by Copp when he demonstrated that the parathyroids in addition to producing parathormone also release a fast-acting hypocalcaemic factor into the circulation (calcitonin).* Parathyroid tissue is dissected from a dog and placed in a warm moist chamber as shown. This tissue is perfused with blood from the reservoir and the calcium content of the blood can be artificially increased or decreased. Blood leaving the isolated parathyroid glands is then passed into the vein of a dog and its effect on the calcium content of the circulating blood of the dog can be measured after taking a blood sample (sampling illustrated). When the calcium content of blood perfusing the isolated glands was raised above normal levels then the blood draining the glands caused a *fall* in the calcium content of the blood of the recipient dog.

He assumed that the hormone calcitonin arose in the parathyroid glands. later work has shown that calcitonin comes not from parathyroid glands but from the thyroid itself. The hormone has thus been renamed thyrocalcitonin.

Vitamin D and calcium metabolism. The role of vitamin D in the metabolism of calcium has already been discussed on p. 417. Summarizing, the role of the vitamin is to ensure that the total amount of calcium and phosphate ions in the internal environment is adequate to maintain bone formation. This it does by increasing the absorption of these substances from the intestine and reducing the losses by the kidneys. When the supplies of the vitamin are inadequate there is a progressive fall in the body content of both calcium and phosphate since there is a failure to absorb and retain these elements. The falling concentration of calcium in the blood stimulates the parathyroid gland to produce more parathormone and the blood levels of calcium are maintained at the expense of demineralization of bone.

In the normal animal parathormone and vitamin D act together in homeostasis on the important element calcium.

Index